Operator Theory: Advances and Applications
Vol. 148

New Difference Schemes for Partial Differential Equations

Allaberen Ashyralyev
Pavel E. Sobolevskii

Springer Basel AG

Authors:

Allaberen Ashyralyev
Department of Mathematics
Fatih University
Buyukcekmece
34900 Istanbul, Turkey
aashyr@fatih.edu.tr

and

Department of Mathematics
International Turkmen-Turkish University
32 Gerogly Street
74400 Ashgabat, Turkmenistan

Pavel E. Sobolevskii
Department of Mathematics
Hebrew University of Jerusalem
Givat Ram Campus
91904 Jerusalem
Israel
pavels@math.huji.ac.il

2000 Mathematics Subject Classification 39A12, 47B39, 65M06, 65N06, 65Q05

A CIP catalogue record for this book is available from the
Library of Congress, Washington D.C., USA

Bibliographic information published by Die Deutsche Bibliothek
Die Deutsche Bibliothek lists this publication in the Deutsche Nationalbibliografie; detailed bibliographic data is available in the Internet at <http://dnb.ddb.de>.

ISBN 978-3-0348-9622-1 ISBN 978-3-0348-7922-4 (eBook)

DOI 10.1006/978-3-0348-7922-4

Originally published by Birkhäuser Verlag in 2004
Softcover reprint of the hardcover 1st edition 2004

Printed on acid-free paper produced from chlorine-free pulp. TCF ∞
Cover design: Heinz Hiltbrunner, Basel
ISBN 3-7643-7054-8

9 8 7 6 5 4 3 2 1 www.birkhasuer-science.com

Contents

Preface

The present monograph is devoted to the construction and investigation of the new high order of accuracy difference schemes for approximating the solutions of regular and singular perturbation boundary-value problems for partial differential equations. The construction is based on an exact difference scheme and Taylor's decomposition on two or three points. This approach permits us to extend essentially a class of problems where the theory of difference methods is applicable. Namely, now it is possible to investigate differential equations with variable coefficients and regular and singular perturbation boundary-value problems. The investigation is based on new coercivity inequalities. The stability and coercive stability estimates in various Banach norms for solutions of the high order of accuracy difference schemes of the initial-value problem for parabolic equations and boundary-value problem for elliptic equations are obtained. This investigation is based on the theory of positive operators, theory of semigroup operators and theory of interpolation of linear operators. The stability estimates for the solutions of the high order of accuracy difference schemes of the initial-value problems for hyperbolic equations are established. This investigation is based on the spectral theory of self-adjoint positive definite operators in a Hilbert space. This approach permits us to study the stability of the simple difference schemes for various partial differential equations. The simple first and second order of accuracy difference schemes for delay partial differential equations are considered. The stability estimates for the solutions of these difference schemes of initial-value problems for delay partial differential equations are established.

Let us give a brief description of the contents of our monograph. It consists of eight chapters. In Chapter 1 we study first- and second-order linear difference equations. These equations arise in numerical analysis of the approximate solutions of boundary-value problems for ordinary and partial differential equations.

Chapter 2 presents the high order of accuracy single-step difference schemes generated by an exact difference scheme or by Taylor's decomposition on two points for the numerical solution of an initial-value problem for first-order differential equations and for a system of first-order differential equations. In general, these difference schemes for the same problem are different. However, in the case of a homogeneous equation or a system of differential equations with independent in t coefficients, by using both methods we have exactly the same difference schemes.

Chapter 3 is devoted to study of the high order of accuracy two-step difference schemes generated by an exact difference scheme or by Taylor's decomposition on three points for the numerical solutions of an initial-value problem and boundary-value problem for the second-order differential equations.

Note that in recent years important progress has been made in the study of the theory of high order of accuracy difference schemes for partial differential equations from the viewpoint of applications of an exact difference scheme or of Taylor's decomposition function on two or three points. Chapter 4 presents the high order of accuracy single-step difference schemes generated by an exact difference scheme or by Taylor's decomposition on two points for the numerical solutions of the abstract Cauchy problem for differential equations of parabolic type

$$v'(t) + A(t)v(t) = f(t) \quad (0 \leq t \leq T), v(0) = v_0$$

in an arbitrary Banach space E with linear positive operators $A(t)$. The well-posedness of these difference schemes in various Banach spaces is studied. The stability and coercive stability estimates in Hölder norms for the solutions of the high order of accuracy difference schemes of mixed type boundary-value problems for parabolic equations are obtained.

Chapter 5 is devoted to study of the high order of accuracy two-step difference schemes generated by an exact difference scheme or by Taylor's decomposition on three points for approximate solutions of the boundary-value problem for differential equations of elliptic type

$$-v''(t) + Av(t) = f(t) \quad (0 \leq t \leq T), v(0) = v_0, v(T) = v_T$$

in an arbitrary Banach space E with positive operator A. The well-posedness of these difference schemes in various Banach spaces is studied. The stability and coercive stability estimates in Hölder norms for solutions of the high order of accuracy difference schemes of mixed type boundary-value problems for elliptic equations are obtained.

Chapter 6 gives the high order of accuracy two-step difference schemes generated by an exact difference scheme or by Taylor's decomposition on three points for the approximate solutions of an abstract Cauchy problem for differential equations of hyperbolic type

$$v''(t) + Av(t) = f(t) \quad (0 \leq t \leq T), v(0) = v_0, v'(0) = v_0'$$

in a Hilbert space H with positive definite self-adjoint operator A. The stability estimates of solutions of the high order of accuracy difference schemes for approximate solutions of mixed type boundary-value problems for hyperbolic equations are obtained.

Chapter 7 is devoted to study of two abstract Cauchy problems

$$\varepsilon v'(t) + Av(t) = f(t) \quad (0 \leq t \leq T), v(0) = v_0,$$

$$\varepsilon^2 v''(t) + Av(t) = f(t) \quad (0 \leq t \leq T), v(0) = v_0, v'(0) = v_0'$$

and the boundary-value problem

$$-\varepsilon^2 v''(t) + Av(t) = f(t) \quad (0 \le t \le T), v(0) = v_0, v(T) = v_T$$

for differential equations in an arbitrary Banach space E with the positive operator A and an arbitrary ε positive parameter multiplying the derivative term. The high order of accuracy single-step uniform difference schemes generated by an exact difference scheme of the approximate solutions for differential equations of parabolic type with an arbitrary parameter ε in the highest derivative are presented. The high order of accuracy two-step uniform difference schemes generated by an exact difference scheme of the approximate solutions for differential equations of elliptic and hyperbolic types with an arbitrary parameter ε in the highest derivative are presented. The well-posedness of these difference schemes for parabolic and elliptic equations in various Banach spaces is established. The stability estimates of solutions of the high order of accuracy difference schemes for hyperbolic equations with an arbitrary ε parameter in the highest derivative are obtained.

Chapter 8 is devoted to study of the initial-value problem for delay partial differential equations of parabolic type. A sufficient condition for the stability of the solution of this initial-value problem is given. The stability estimates for solutions of the first and second order of accuracy difference schemes for approximately solving this initial-value problem are presented. The stability estimates in Hölder norms for the solutions of the initial-value problem of the delay differential and difference equations of parabolic type are obtained.

The monograph will be of value to professional mathematicians, as well as advanced students in the fields of numerical analysis, functional analysis, ordinary and partial differential equations.

Finally, we wish to express our warmest thanks to Professor S.G. Krein (Voronezh, Russia), Professor I. Gohberg (Tel Aviv, Israel), Professor M.A. Krasnosel'skii (Moscow, Russia), Professor V. Lakshmikantham (Florida, USA), Professor M.Z. Nashed (Florida, USA), Professor H.O. Fattorini (Los Angeles, USA), Professor G.M. Vainikko (Helsinki, Finland), Professor H. Amann (Zurich, Switzerland), Professor A. Pazy (Jerusalem, Israel) and Professor M.L. Gorbachuk (Kiev, Ukraine), whose interest and enthusiastic support made this work possible. We are immensely pleased that our book appears in this series.

Chapter 1

Linear Difference Equations

In the present chapter we consider first- and second-order linear difference equations. These equations arise in numerical analysis of approximate solutions of boundary-value problems for ordinary and partial differential equations.

1.1 Difference Equations of the First Order

The first-order linear difference equations to be studied in this chapter may be expressed in the form

$$a_k u_{k-1} + b_k u_k = \psi_k, 1 \leq k \leq n, \tag{1.1}$$

where a_k, b_k, and ψ_k are known numbers , u_k are unknown numbers, n is the given integer. We will call u_k the value of grid function $\{u_k\}_0^n$ at the point with index k.

A solution of the first-order difference equation (1.1) is a grid function $\{y_k\}_0^n$ that satisfies (1.1) for all k, that is, (1.1) becomes an identity if we replace the unknown grid function $\{u_k\}_0^n$ by $\{y_k\}_0^n$.

The initial-value problem associated with (1.1) is to find a solution $\{u_k\}_0^n$ of the difference equation (1.1) satisfying the initial condition

$$u_0 = \varphi, \tag{1.2}$$

where φ is a given number.

To solve this problem, we must find a grid function $\{y_k\}_0^n$ that not only satisfies the difference equation (1.1) but also satisfies the initial condition $y_0 = \varphi$.

If $b_k \neq 0$ for all k, $1 \leq k \leq n$, then the difference equation (1.1) is equivalent to the equation

$$u_k = q_k u_{k-1} + \varphi_k, 1 \leq k \leq n, \tag{1.3}$$

where

$$q_k = -\frac{a_k}{b_k}, \varphi_k = \frac{\psi_k}{b_k}.$$

Theorem 1.1.1. *There exists a unique solution of the initial-value problem* (1.1)–(1.2) *and for the solution of* (1.1)–(1.2) *the formula*

$$u_k = u(k,0)\varphi + \sum_{j=1}^{k} u(k,j)\varphi_j, \tag{1.4}$$

holds. Here

$$u(k,j) = \begin{cases} q_k \cdots q_{j+1}, k \neq j, \\ 1, k = j. \end{cases} \tag{1.5}$$

Proof. Using formulas (1.3) and (1.5) we can obtain formula (1.4) for $k = 1$. Suppose that formula (1.4) is satisfied for $k = m$,

$$u_m = u(m,0)\varphi + \sum_{j=1}^{m} u(m,j)\varphi_j.$$

From that and (1.3) it follows

$$u_{m+1} = q_{m+1}u_m + \varphi_{m+1} = q_{m+1}u(m,0)\varphi + \sum_{j=1}^{m} q_{m+1}u(m,j)\varphi_j + \varphi_{m+1}.$$

Using formula (1.5), we can write

$$q_{m+1}u(m,j) = u(m+1,j), j \neq m, u(m+1,m+1) = 1.$$

Therefore

$$u_{m+1} = u(m+1,0)\varphi + \sum_{j=1}^{m} u(m+1,j)\varphi_j + u(m+1,m+1)\varphi_{m+1}$$

$$= u(m+1,0)\varphi + \sum_{j=1}^{m+1} u(m+1,j)\varphi_j.$$

So, using mathematical induction we obtain formula (1.4) for any k. This completes the proof of Theorem 1.1.1.

Theorem 1.1.2. *Let* $|q_k| \leq q$. *Then for solution* u_k *of the initial-value problem* (1.1)–(1.2) *the stability inequality*

$$|u_k| \leq q^k|\varphi| + \sum_{j=1}^{k} q^{k-j}|\varphi_j|$$

holds.

The proof of Theorem 1.1.2 is based on formulas (1.4) and (1.5).

Finally, note that the boundary-value problem associated with (1.1) is to find a solution u_k of the difference equation (1.1) and satisfying the supplementary condition

$$au_0 + bu_n = \varphi, \tag{1.6}$$

where a, b and φ are given numbers.

If a or b are equal to 0, then the condition is called a local boundary condition. If $a \neq 0$ and $b \neq 0$, then the condition is called a nonlocal boundary condition.

In a similar way we can obtain an analogous result for the solution of the nonlocal boundary-value problem (1.1) and (1.6).

1.2 Difference Equations of the Second Order

The second-order linear difference equations to be studied in this chapter may be expressed in the form

$$a_k u_{k-1} - c_k u_k + b_k u_{k+1} = -\psi_k, 1 \leq k < n, \tag{1.7}$$

where a_k, b_k, c_k and ψ_k are known numbers, u_k are unknown numbers, n is the given integer. Suppose that $a_k \neq 0$ and $b_k \neq 0$ for all k.

The definition of a solution of a given second-order difference equation such as (1.7) is like the definition of a solution of a difference equation such as (1.1).

The initial-value problem associated with (1.7) is to find a solution $\{u_k\}_0^n$ of the difference equation (1.7) satisfying the initial conditions

$$u_0 = \varphi,\ u_1 = \psi, \tag{1.8}$$

where φ and ψ are given numbers.

To solve this problem, we must find a grid function $\{y_k\}_0^n$ that not only satisfies the difference equation (1.7) but also satisfies the initial conditions $y_0 = \varphi$ and $y_1 = \psi$.

The local boundary-value problem associated with (1.7) is to find a solution $\{u_k\}_0^n$ of the difference equation (1.7) satisfying the boundary conditions

$$u_0 = \lambda_1 u_1 + \mu_1, u_n = \lambda_2 u_{n-1} + \mu_2, \tag{1.9}$$

where $\lambda_1, \mu_1, \lambda_2, \mu_2$ are given numbers.

The nonlocal boundary-value problem associated with (1.7) is to find a solution $\{u_k\}_0^n$ of the difference equation (1.7) satisfying the boundary conditions

$$\alpha_1 u_0 + \alpha_2 u_1 + \beta_1 u_n + \beta_2 u_{n-1} = \mu_1, \alpha_3 u_0 + \alpha_4 u_1 + \beta_3 u_n + \beta_4 u_{n-1} = \mu_2, \tag{1.10}$$

where $\alpha_1, \alpha_2, \alpha_3, \alpha_4, \beta_1, \beta_2, \beta_3, \beta_4, \mu_1$ and μ_2 are given numbers.

Now we will consider the local boundary-value problem (1.7) and (1.9). To solve them by computer we will use the following method which is called the modified Gauss elimination method. We seek a solution of difference equations (1.7) in the form

$$u_k = \alpha_{k+1}u_{k+1} + \beta_{k+1}, 0 \leq k \leq n-1, \tag{1.11}$$

where $\alpha_{k+1}, \beta_{k+1}$ are unknown coefficients. Hence we find

$$u_{k-1} = \alpha_k u_k + \beta_k = \alpha_k(\alpha_{k+1}u_{k+1} + \beta_{k+1}) + \beta_k$$
$$= \alpha_k\alpha_{k+1}u_{k+1} + (\alpha_k\beta_{k+1} + \beta_k), 1 \leq k \leq n-1.$$

Now putting expressions for u_k and u_{k-1} into equation (1.7) we have

$$[\alpha_{k+1}(a_k\alpha_k - c_k) + b_k]u_{k+1} + [\beta_{k+1}(a_k\alpha_k - c_k) + a_k\beta_k + \psi_k] = 0.$$

The last equation is satisfied if we select

$$\alpha_{k+1}(a_k\alpha_k - c_k) + b_k = 0,$$
$$\beta_{k+1}(a_k\alpha_k - c_k) + a_k\beta_k + \psi_k = 0, 1 \leq k \leq n-1.$$

From that it follows

$$\alpha_{k+1} = \frac{b_k}{c_k - \alpha_k a_k}, \beta_{k+1} = \frac{a_k\beta_k + \varphi_k}{c_k - \alpha_k a_k}, 1 \leq k \leq n-1. \tag{1.12}$$

The relations (1.12) are first-order nonlinear difference equations. For the solution of difference equations (1.12) we need to find α_1 and β_1. We can find them from $u_0 = \lambda_1 u_1 + \mu_1$. Thus, we have

$$\alpha_1 = \lambda_1, \beta_1 = \mu_1. \tag{1.13}$$

For the first step using formulas (1.12) and (1.13), we can compute α_{k+1} and β_{k+1}, $1 \leq k \leq n-1$. For the second step we will find u_k, $0 \leq k \leq n$. But, for this we need to find u_n. We can find u_n from $u_n = \lambda_1 u_{n-1} + \mu_2$ and $u_{n-1} = \alpha_n u_n + \beta_n$. Namely

$$u_n = \frac{\lambda_1\beta_n + \beta_2}{1 - \lambda_2\alpha_n}. \tag{1.14}$$

Thus using formulas (1.11) and (1.14), we can compute u_k, $0 \leq k \leq n$. Finally note that when we apply algorithms (1.11), (1.12), (1.13) and (1.14), to find a solution by computer of the boundary-value problems (1.7), (1.9), we need

$$1 - \lambda_2\alpha_n \neq 0, c_k - \alpha_k a_k \neq 0 \tag{1.15}$$

for all k.

Theorem 1.2.1. *Let for all k*

$$|c_k| \geq |a_k| + |\beta_k|, \tag{1.16}$$

and

$$|\lambda_1| \leq 1, |\lambda_2| < 1. \tag{1.17}$$

Then the requirements (1.15) *and* $|\alpha_k| \leq 1$ *for all k are satisfied.*

Proof. From (1.13) it follows that $|\alpha_1| = |\lambda_1| \leq 1$. Let $|\alpha_k| \leq 1$. We will prove that

$$|\alpha_{k+1}| \leq 1.$$

Namely

$$|c_k - \alpha_k a_k| \geq |c_k| - |a_k||\alpha_k| \geq |c_k| - |a_k| \geq |b_k| > 0.$$

From that it follows $c_k - \alpha_k a_k \neq 0$ and

$$|\alpha_{k+1}| = \frac{|b_k|}{|c_k - \alpha_k a_k|} \leq 1.$$

Therefore, $|\alpha_k| \leq 1$ for all k, $1 \leq k \leq n$. From $|\alpha_n| \leq 1$ and (1.11) it follows that

$$|1 - \lambda_2 \alpha_n| \geq 1 - |\lambda_2||\alpha_n| \geq 1 - |\lambda_2| > 0.$$

Thus Theorem 1.2.1 is proved.

Theorem 1.2.2. *Let for all k*

$$|c_k| > |a_k| + |b_k|, \tag{1.18}$$

and $|\lambda_1| \leq 1, |\lambda_2| \leq 1$. *Then the requirements* (1.15) *and* $|\alpha_k| < 1$ *for all* $1 < k$ *are satisfied.*

The proof of Theorem 1.2.2 follows the scheme of the proof of Theorem 1.2.1.

1.3 Difference Equations with Constant Coefficients

The second-order linear homogeneous difference equation with constant coefficients may be written in the form

$$au_{k-1} - cu_k + bu_{k+1} = 0, 1 \leq k < n, \tag{1.19}$$

where a, b and c are given numbers. We put $a \neq 0$ and $b \neq 0$.

Since for constant q and positive integer k,

$$aq^{k-1} - cq^k + bq^{k+1} = q^{k-1}(a - cq + bq^2),$$

it is easy to find if q is a root of the equation

$$a - cq + bq^2 = 0; \tag{1.20}$$

if so, then the grid function $\{u_k\}_0^n = \{q^k\}_0^n$ is the solution of equation (1.19).

Equation (1.20) is called the auxiliary equation associated with (1.19). The auxiliary equation for (1.19) is of degree 2. The key quantity is $c^2 - 4ab$, which is known as the discriminant of the quadratic equation (1.20).

If $c^2 - 4ab$ is positive, there will be two real values for q and they are defined by

$$q_{1,2} = \frac{c \pm \sqrt{c^2 - 4ab}}{2b}. \tag{1.21}$$

Recall that given the grid functions $\{u_k^1\}_0^n$ and $\{u_k^2\}_0^n$, if constants λ_1, λ_2, not all zero, exist such that

$$\lambda_1 \{u_k^1\}_0^n + \lambda_2 \{u_k^2\}_0^n = 0$$

or

$$\lambda_1 u_k^1 + \lambda_2 u_k^2 = 0, 0 \le k \le n,$$

then the grid functions $\{u_k^1\}_0^n$ and $\{u_k^2\}_0^n$ are said to be linearly dependent.

If no such relation exists, the functions are said to be linearly independent. That is, the grid functions $\{u_k^1\}_0^n$ and $\{u_k^2\}_0^n$ are linearly independent when the last equation implies that $\lambda_1 = \lambda_2 = 0$.

It is easy to show that for any linearly independent grid functions $\{u_k^1\}_0^n$ and $\{u_k^2\}_0^n$, we can determine

$$\begin{vmatrix} u_k^1 & u_k^2 \\ u_{k-1}^1 & u_{k-1}^2 \end{vmatrix},$$

such that the determinant is not equal to zero for any $k, 1 \le k \le n$. Then it is easy to show that the solutions

$$\{u_k^1\}_0^n = \{q_1^k\}_0^n, \{u_k^2\}_0^n = \{q_2^k\}_0^n$$

are linearly independent and the general solution of (1.19) can be written at once. It is

$$\{u_k\}_0^n = \{c_1 q_1^k + c_2 q_2^k\}_0^n \tag{1.22}$$

in which c_1, c_2 are arbitrary constants.

If $c^2 - 4ab$ is zero, then the auxiliary equation (1.20) has repeated roots $q = q_1 = q_2$. It is easy to show that the grid functions $\{k^m q^k\}_0^n$, where $m = 0, 1$ are linearly independent and the general solution of (1.19) is

$$\{u_k\}_0^n = \{c_1 q^k + c_2 k q^k\}_0^n \tag{1.23}$$

in which c_1, c_2 are arbitrary constants.

If $c^2 - 4ab$ is negative, then the auxiliary equation (1.20) has two complex roots

$$q_{1,2} = \frac{c \pm i\sqrt{4ab - c^2}}{2b} = re^{\pm i\varphi},$$

where $r = \sqrt{\frac{a}{b}}, \tan\varphi = \frac{\sqrt{4ab-c^2}}{c}$. It is easy to show that the grid functions

$$\{r^k \sin k\varphi\}_0^n, \{r^k \cos k\varphi\}_0^n$$

are linearly independent and the general solution of (1.19) is

$$\{u_k\}_0^n = \{c_1 r^k \sin k\varphi + c_2 r^k \cos k\varphi\} \tag{1.24}$$

in which c_1, c_2 are arbitrary constants.

Now using the variation of parameters method we will find the general solution of the nonhomogeneous equation

$$au_{k-1} - cu_k + bu_{k+1} = \varphi_k, 1 \leq k \leq n-1, \tag{1.25}$$

where a, b, c and φ_k are given numbers.

Namely, if $\{v_k\}_0^n, \{w_k\}_0^n$ are linearly independent solutions of (1.25), then we can use them to determine the general solution of the nonhomogeneous equation (1.25). Suppose that

$$u_k = c_1 u_k + c_2 w_k \tag{1.26}$$

is a solution of (1.25). Let us see what happens if we replace both of the constants in (1.25) with α_k and β_k, respectively. That is, we consider

$$u_k = \alpha_k v_k + \beta_k w_k \tag{1.27}$$

and try to determine α_k and β_k so that the grid function $\{\alpha_k v_k + \beta_k w_k\}_0^n$ is a solution of the equation (1.25). From (1.27) it follows that

$$u_{k+1} - u_k = \alpha_k(v_{k+1} - v_k) + \beta_k(w_{k+1} - w_k) \tag{1.28}$$

$$+(\alpha_{k+1} - \alpha_k)v_{k+1} + (\beta_{k+1} - \beta_k)w_{k+1}.$$

We now choose some particular function for the expression

$$(\alpha_{k+1} - \alpha_k)v_{k+1} + (\beta_{k+1} - \beta_k)w_{k+1}.$$

Technically, we could let this grid function be $\{\sin k\varphi\}_0^n, \{q^k\}_0^n, \{\cos k\varphi\}_0^n$ or any other suitable grid function. For simplicity we choose

$$(\alpha_{k+1} - \alpha_k)v_{k+1} + (\beta_{k+1} - \beta_k)w_{k+1} = 0$$

or

$$(\alpha_k - \alpha_{k-1})v_k + (\beta_k - \beta_{k-1})w_k = 0. \tag{1.29}$$

Then it follows from (1.28) that

$$u_{k+1} - u_k = \alpha_k(v_{k+1} - v_k) + \beta_k(w_{k+1} - w_k)$$

or

$$u_k - u_{k-1} = \alpha_{k-1} + (v_k - v_{k-1}) + \beta_{k-1}(w_k - w_{k-1}). \tag{1.30}$$

It is easy to see that from (1.25) it follows that

$$-a(u_k - u_{k-1}) + (a - c + b)u_k + b(u_{k+1} - u_k) = \varphi_k, 1 \leq k < n. \tag{1.31}$$

We substitute (1.27) and (1.30) into equation (1.31) to obtain

$$-a[\alpha_{k-1}(v_k - v_{k-1}) + \beta_{k-1}(w_k - w_{k-1})] + (a - c + b)[\alpha_k u_k + \beta_k w_k]$$

$$+b[\alpha_k(v_{k+1} - v_k) + \beta_k(w_{k+1} - w_k)] = \varphi_k,$$

or

$$\alpha_k[av_{k-1} - cv_k + bv_{k+1}] + \beta_k[aw_{k-1} - cw_k + bw_{k+1}]$$
$$+a[(\alpha_k - \alpha_{k-1})(v_k - v_{k-1}) + (\beta_k - \beta_{k-1})(w_k - w_{k-1})] = \varphi_k.$$

But v_{k0}^{n} and w_{k0}^{n} are solutions of the homogeneous equation (1.19), so that finally

$$(\alpha_k - \alpha_{k-1})(v_k - v_{k-1}) + (\beta_k - \beta_{k-1})(w_k - w_{k-1}) = \frac{\varphi_k}{a}. \tag{1.32}$$

Equations (1.29)and (1.32) now give us two equations that we wish to solve for $\alpha_k - \alpha_{k-1}$ and $\beta_k - \beta_{k-1}$. This solution exists provided that the determinant

$$\begin{vmatrix} v_{k-1} & w_{k-1} \\ v_k - v_{k-1} & w_k - w_{k-1} \end{vmatrix}$$

does not vanish. But

$$\begin{vmatrix} v_{k-1} & w_{k-1} \\ v_k - v_{k-1} & w_k - w_{k-1} \end{vmatrix} = \begin{vmatrix} v_{k-1} & w_{k-1} \\ v_k & w_k \end{vmatrix}$$

and the last determinant is not equal to zero for the linearly independent solutions $\{v_k\}_0^n$ and $\{w_k\}_0^n$ of equation (1.19). Therefore using equations (1.29)and (1.32) we can find $\alpha_k - \alpha_{k-1}$ and $\beta_k - \beta_{k-1}$. Namely

$$\alpha_k - \alpha_{k-1} = -\frac{w_k \varphi_k}{a(v_{k-1}w_k - v_k w_{k-1})},$$

$$\beta_k - \beta_{k-1} = \frac{v_k \varphi_k}{a(v_{k-1}w_k - v_k w_{k-1})}.$$

Hence using formula (1.4), we can find

$$\alpha_k = \alpha_0 - \sum_{j=1}^{k} \frac{w_j \varphi_j}{a(v_{j-1}w_j - v_j w_{j-1})}, \tag{1.33}$$

$$\beta_k = \beta_0 + \sum_{j=1}^{k} \frac{v_j \varphi_j}{a(v_{j-1}w_j - v_j w_{j-1})}. \tag{1.34}$$

Thus, α_k and β_k of (1.33) and (1.34) may be inserted in (1.27) to give us the general solution of equation (1.25)

$$u_k = \alpha_0 v_k + \beta_0 w_k + \sum_{j=1}^{k} \frac{(w_k v_j - v_k w_j)\varphi_j}{a(v_{j-1}w_j - v_j w_{j-1})}, \tag{1.35}$$

where α_0, β_0 are arbitrary constants.

Finally note that the method of variation of parameters can be applied to the solution of equation

$$a_k u_{k-1} - c_k u_k + b_k u_{k+1} = \varphi_k, 1 \leq k < n$$

with dependent coefficients. Namely we have the formula

$$u_k = \alpha_0 v_k + \beta_0 w_k + \sum_{j=1}^{k} \frac{(w_k v_j - v_k w_j)\varphi_j}{a_j(v_{j-1}w_j - v_j w_{j-1})}.$$

Now we will consider the initial-value problem

$$au_{k-1} - cu_k + bu_{k+1} = \varphi_k, 1 \leq k < n, u_0 = \varphi, u_1 = \psi. \tag{1.36}$$

Recall that $\{v_k\}_0^n$ and $\{w_k\}_0^n$ are linearly independent solutions of (1.19), therefore

$$v_k = \begin{cases} q_1^k, c^2 - 4ab > 0, \\ q^k, c^2 - 4ab = 0, \\ r^k \sin k\varphi, c^2 - 4ab < 0, \end{cases} \tag{1.37}$$

$$v_k = \begin{cases} q_2^k, c^2 - 4ab > 0, \\ kq^k, c^2 - 4ab = 0, \\ r^k \cos k\varphi, c^2 - 4ab < 0. \end{cases} \tag{1.38}$$

Thus using formula (1.35) we have the following theorem.

Theorem 1.3.1. *There exists a unique solution of the initial-value problem* (1.36) *and for the solution of* (1.36) *the following formulas are satisfied:*
If $c^2 - 4ab > 0$, then

$$u_k = \frac{q_1 q_2 (q_1^{k-1} - q_2^{k-1})}{q_2 - q_1}\varphi + \frac{q_2^k - q_1^k}{q_2 - q_1}\psi$$
$$+ \sum_{j=1}^{k} \frac{q_1 q_2}{a(q_2 - q_1)}(q_2^{k-j} - q_1^{k-j})\varphi_j.$$

If $c^2 - 4ab = 0$, then

$$u_k = (1-k)q^k\varphi + kq^{k-1}\psi + \sum_{j=1}^{k} \frac{(k-j)}{a} q^{k-j+1}\varphi_j.$$

If $c^2 - 4ab < 0$, then

$$u_k = \frac{\sin(1-k)\varphi}{\sin\varphi} r^k \varphi + \frac{\sin k\varphi}{\sin\varphi} r^k \psi$$
$$+ \sum_{j=1}^{k} \frac{\sin(k-j)\varphi}{a \sin\varphi} r^{k-j+1}\varphi_j.$$

Here and in future we will put

$$\sum_{j=1}^{k} a_{k,j} = 0, \text{when } k < 1.$$

Now we will consider the boundary-value problem

$$au_{k-1} - cu_{k-1} + bu_{k+1} = -\varphi_k, 1 \leq k < n, u_0 = \mu_1, u_n = \mu_2. \tag{1.39}$$

Using formulas(1.35), (1.37) and (1.38), we have the following theorem.

Theorem 1.3.2. *There exists a unique solution of the boundary-value problem* (1.39) *and for the solution of* (1.39) *the following formulas are satisfied: If* $c^2 - 4ab > 0$, *then*

$$\begin{aligned} u_k &= \frac{q_2^k q_1^k}{q_2^n - q_1^n}(q_2^{n-k} - q_1^{n-k})\mu_1 + \frac{q_2^k - q_1^k}{q_2^n - q_1^n}\mu_2 \\ &+ \frac{q_1^k - q_2^k}{q_2^n - q_1^n}\sum_{j=1}^{n}\frac{q_1 q_2 (q_2^{n-j} - q_1^{n-j})}{a(q_2 - q_1)}\varphi_j \\ &+ \sum_{j=1}^{k}\frac{q_1 q_2}{a(q_2 - q_1)}(q_2^{k-j} - q_1^{k-j})\varphi_j. \end{aligned}$$

If $c^2 - 4ab = 0$, *then*

$$\begin{aligned} u_k &= \frac{n-k}{n}q^k\mu_1 + \frac{k}{n}q^{k-n}\mu_2 \\ &+ \frac{k}{n}\sum_{j=1}^{n}\frac{j-n}{a}q^{k-j+1}\varphi_j + \sum_{j=1}^{k}\frac{k-j}{a}q^{k-j+1}\varphi_j. \end{aligned}$$

If $c^2 - 4ab < 0$, *then*

$$\begin{aligned} u_k &= \frac{\sin(n-k)\varphi}{\sin n\varphi}r^k\mu_1 + \frac{sink\varphi}{sinn\varphi}r^{k-n}\mu_2 \\ &+ \frac{\sin k\varphi}{sinn\varphi}\sum_{j=1}^{n}\frac{\sin(j-n)\varphi}{a\sin\varphi}r^{k-j+1}\varphi_j \\ &+ \sum_{j=1}^{k}\frac{\sin(k-j)\varphi}{a\sin\varphi}r^{k-j+1}\varphi_j. \end{aligned}$$

In the same way we can study solutions of the local and nonlocal boundary-value problems for equation (1.25), with boundary conditions (1.9) and (1.10), respectively.

Chapter 2

Difference Schemes for First-Order Differential Equations

In the present chapter we consider the single step difference schemes generated by an exact difference scheme or by Taylor's decomposition on two points for the numerical solution of an initial-value problem for first-order differential equations and systems of first-order differential equations.

2.1 Single-Step Exact Difference Scheme and Its Applications

We consider the initial-value problem of the form

$$y'(t) + a(t)y(t) = f(t), 0 < t \leq T, y(0) = y_0 \tag{2.1}$$

assuming $a(t)$ and $f(t)$ to be such that problem (2.1) has a unique smooth solution defined on $[0, T]$. In this section we shall construct difference schemes of a high order of accuracy for an approximate solution of problem (2.1). On the segment $[0, T]$ we consider the uniform grid space

$$[0, T]_\tau = \{t_k = k\tau, k = 0, 1, \ldots, N, N\tau = T\}$$

with step $\tau > 0$. Here N is a fixed positive integer. We will discuss single-step difference schemes for computing numerical values of the solutions. These difference schemes are step-by-step methods, that is, we start from the given approximate values of $y(t_{k-1})$ and proceed stepwise, computing approximate values of $y(t_k)$ at the grid (mesh) points.

Theorem 2.1.1. *Let $y(t_k)$ be the solution of (2.1) at the grid points $t = t_k$. Then $\{y(t_k)\}_0^N$ is the solution of the initial-value problem for the first-order difference equations*

$$\frac{1}{\tau}(y(t_k) - y(t_{k-1})) + \frac{1}{\tau}(1 - e^{-\int_{t_{k-1}}^{t_k} a(s)ds})y(t_{k-1})$$

$$= \frac{1}{\tau}\int_{t_{k-1}}^{t_k} e^{-\int_s^{t_k} a(z)dz} f(s)ds, 1 \leq k \leq N, y(0) = y_0. \tag{2.2}$$

Proof. For solution of (2.1) we have the formula

$$y(t)=e^{-\int_0^t a(s)ds}y_0+\int_0^t e^{-\int_s^t a(z)dz}f(s)ds.$$

Putting $t=t_k$ and $t=t_{k-1}$ into the last formula, we can write

$$\begin{aligned} y(t_k) &= e^{-\int_0^{t_k} a(s)ds}y_0+\int_0^{t_k} e^{-\int_s^{t_k} a(z)dz}f(s)ds \\ &= e^{-\int_{t_{k-1}}^{t_k} a(s)ds}\left[e^{-\int_0^{t_{k-1}} a(s)ds}y_0+\int_0^{t_{k-1}} e^{-\int_s^{t_{k-1}} a(z)dz}f(s)ds\right] \\ &+\int_{t_{k-1}}^{t_k} e^{-\int_s^{t_k} a(z)dz}f(s)ds, \end{aligned}$$

$$y(t_{k-1})=e^{-\int_0^{t_{k-1}} a(s)ds}y_0+\int_0^{t_{k-1}} e^{-\int_s^{t_{k-1}} a(z)dz}f(s)ds.$$

Hence, we obtain the following relation between $y(t_k)$ and $y(t_{k-1})$:

$$y(t_k)=e^{-\int_{t_{k-1}}^{t_k} a(s)ds}y(t_{k-1})+\int_{t_{k-1}}^{t_k} e^{-\int_s^{t_k} a(z)dz}f(s)ds.$$

The last relation and equality (2.2) are equivalent. Theorem 2.1.1 is proved.

Thus, the solution $y(t)$ of the differential problem (2.1) at the grid points $t=t_k$ is the solution of the difference problem (2.2). Therefore, the initial-value problem (2.2) is called the single-step exact difference scheme for solution of the initial-value problem (2.1).

Now, we will consider the applications of this exact difference scheme. From (2.2) it is clear that for approximate solutions of problem (2.1) it is necessary to approximate the expressions

$$e^{-\int_{t_{k-1}}^{t_k} a(s)ds} \quad \text{and} \quad \frac{1}{\tau}\int_{t_{k-1}}^{t_k} e^{-\int_s^{t_k} a(z)dz}f(s)ds.$$

Let us remark that in constructing difference schemes it is important to know how to construct a right-hand side $\varphi_k^{p,q}$ that satisfies

$$\frac{1}{\tau}\int_{t_{k-1}}^{t_k} e^{-\int_s^{t_k} a(z)dz}f(s)ds-\varphi_k^{p,q}=o(\tau^{p+q}) \tag{2.3}$$

and is sufficiently simple. Here p and q are positive integers and we will put $p\geq q$. The choice formula $\varphi_k^{p,q}$ is not unique. Using Taylor's formula, we obtain

$$\frac{1}{\tau}\int_{t_{k-1}}^{t_k} e^{-\int_s^{t_k} a(z)dz}f(s)ds \tag{2.4}$$

$$=\sum_{m=0}^{p+q-1}\sum_{\lambda=0}^{m}\binom{m}{\lambda}e^{-\int_{t_{k-1}}^{t_k} a(z)dz}\beta_{m-\lambda}(t_{k-1})f^{(\lambda)}(t_{k-1})\frac{\tau^m}{(m+1)!}+o(\tau^{p+q}),$$

$$\frac{1}{\tau}\int_{t_{k-1}}^{t_k} e^{-\int_s^{t_k} a(z)dz} f(s)ds \tag{2.5}$$

$$= \sum_{m=0}^{p+q-1}\sum_{\lambda=0}^{m}\binom{m}{\lambda}\beta_{m-\lambda}(t_k)f^{(\lambda)}(t_k)\frac{(-\tau)^m}{(m+1)!} + o(\tau^{p+q}),$$

$$\frac{1}{\tau}\int_{t_{k-1}}^{t_k} e^{-\int_s^{t_k} a(z)dz} f(s)ds \tag{2.6}$$

$$= \sum_{m=0}^{p-1}\sum_{\lambda=0}^{2m}\binom{2m}{\lambda} e^{-\int_{t_{k-\frac{1}{2}}}^{t_k} a(z)dz}\beta_{2m-\lambda}(t_{k-\frac{1}{2}})f^{(\lambda)}(t_{k-\frac{1}{2}})\frac{\tau^{2m}}{2^{2m}(2m+1)!}$$

$$+o(\tau^{2p}).$$

Here and in this chapter we will put

$$\begin{cases} \beta_i(s) = 1, & i = 0, \\ \beta_i(s) = \beta_{i-1}^{(\prime)}(s) + \beta_{i-1}(s)a(s), & 1 \le i \le p+q. \end{cases} \tag{2.7}$$

Therefore, for the construction of $\varphi_k^{p,q}$ we can put

$$\varphi_k^{p,q} = \sum_{m=0}^{p+q-1}\sum_{\lambda=0}^{m}\binom{m}{\lambda}\beta_{m-\lambda}(t_k)f^{(\lambda)}(t_k)\frac{(-\tau)^m}{(m+1)!}. \tag{2.8}$$

Moreover, for the construction of $\varphi_k^{p,q}$ we will also use (2.4) and (2.6). But, first of all let us study the approximate formulas for the expressions $e^{-\int_{t_{k-1}}^{t_k} a(s)ds}$ and $e^{-\int_{t_{k-\frac{1}{2}}}^{t_k} a(s)ds}$ that will be needed in the future for the construction of difference schemes of a high order of accuracy for an approximate solution of problem (2.1).

Now, let us consider the approximate formulas for the expression $e^{-\int_{t_{k-1}}^{t_k} a(s)ds}$ constructed by Pade's rational functions. It is clear that

$$e^{-\int_{t_{k-1}}^{t_k} a(s)ds} = e^{-a(t_{k-1})\tau} - \int_{t_{k-1}}^{t_k} e^{-\int_z^{t_k} a(s)ds}(a(z)-a(t_{k-1}))e^{-a(t_{k-1})(z-t_{k-1})}dz,$$

$$e^{-\int_{t_{k-1}}^{t_k} a(s)ds} = e^{-a(t_k)\tau} + \int_{t_{k-1}}^{t_k} e^{-a(t_k)(t_k-z)}(a(t_k)-a(z))e^{-\int_{t_{k-1}}^{z} a(s)ds}dz,$$

$$e^{-\int_{t_{k-1}}^{t_k} a(s)ds} = e^{-a(t_{k-\frac{1}{2}})\tau} - \int_{t_{k-1}}^{t_k} e^{-\int_z^{t_k} a(s)ds}(a(z)-a(t_{k-\frac{1}{2}}))e^{-a(t_{k-\frac{1}{2}})(z-t_{k-1})}dz.$$

Using Taylor's formula, we obtain

$$e^{-\int_{t_{k-1}}^{t_k} a(s)ds} = e^{-a(t_{k-1})\tau} - \tau(a(t_k) - a(t_{k-1}))e^{-a(t_{k-1})\tau}$$
$$+ \sum_{n=1}^{p+q-1} \frac{(-\tau)^{n+1}}{(n+1)!} \sum_{i=0}^{n} \binom{n}{i} \beta_i(t_k) \Bigg[(a(t_k) - a(t_{k-1}))(-a(t_{k-1}))^{n-i}$$
$$+ \sum_{j=1}^{n-i} \binom{n-i}{j} a^{(j)}(t_k)(-a(t_{k-1}))^{n-i-j} \Bigg] e^{-a(t_{k-1})\tau} + o(\tau^{p+q+1}),$$

$$e^{-\int_{t_{k-1}}^{t_k} a(s)ds} = e^{-a(t_k)\tau} + \tau(a(t_k) - a(t_{k-1}))e^{-a(t_k)\tau}$$
$$+ \sum_{n=1}^{p+q-1} \frac{\tau^{n+1}}{(n+1)!} \sum_{i=0}^{n} \binom{n}{i} \widetilde{\beta}(t_{k-1}) \Bigg[(a(t_k) - a(t_{k-1}))(a(t_k))^{n-i}$$
$$- \sum_{j=1}^{n-i} \binom{n-i}{j} a^{(j)}(t_{k-1})(a(t_k))^{n-i-j} \Bigg] e^{-a(t_k)\tau} + o(\tau^{p+q+1}),$$

$$e^{-\int_{t_{k-1}}^{t_k} a(s)ds} = e^{-a(t_{k-\frac{1}{2}})\tau} - \tau(a(t_k) - a(t_{k-\frac{1}{2}}))e^{-a(t_{k-\frac{1}{2}})\tau}$$
$$+ \sum_{n=1}^{p+q-1} \frac{(-\tau)^{n+1}}{(n+1)!} \sum_{i=0}^{n} \binom{n}{i} \beta_i(t_k) \Bigg[(a(t_k) - a(t_{k-\frac{1}{2}}))(-a(t_{k-\frac{1}{2}}))^{n-i}$$
$$+ \sum_{j=1}^{n-i} \binom{n-i}{j} a^{(j)}(t_k)(-a(t_{k-\frac{1}{2}}))^{n-i-j} \Bigg] e^{-a(t_{k-\frac{1}{2}})\tau} + o(\tau^{p+q+1}).$$

Here and in this chapter we will put

$$\begin{cases} \widetilde{\beta}_i(s) = 1, & i = 0, \\ \widetilde{\beta}_i(s) = \widetilde{\beta_{i-1}}^{(\prime)}(s) - \widetilde{\beta_{i-1}}(s)a(s), & 1 \le i \le p+q. \end{cases}$$

Further, using the last formulas and Pade fractions for the function e^{-z}, we can write

$$e^{-\int_{t_{k-1}}^{t_k} a(s)ds} = R_{q,p}(a(t_{k-1})\tau) - \tau(a(t_k) - a(t_{k-1}))R_{q-1,p-1}(a(t_{k-1})\tau) \tag{2.9}$$
$$+ \sum_{n=1}^{p+q-1} \frac{(-\tau)^{n+1}}{(n+1)} \sum_{i=0}^{n} \binom{n}{i} \beta_i(t_k) \Bigg[(a(t_k) - a(t_{k-1}))(-a(t_{k-1}))^{n-i}$$
$$+ \sum_{j=1}^{n-i} \binom{n-i}{j} a^{(j)}(t_k)(-a(t_{k-1}))^{n-i-j} \Bigg] R_{q-n+[\frac{n}{2}]-1,p-[\frac{n}{2}]}(a(t_{k-1})\tau) + o(\tau^{p+q+1}),$$

$$e^{-\int_{t_{k-1}}^{t_k} a(s)ds} = R_{q,p}(\tau a(t_k)) + \tau(a(t_k) - a(t_{k-1}))R_{q-1,p-2}(a(t_k)\tau) \tag{2.10}$$

$$+ \sum_{n=1}^{p+q-1} \frac{\tau^{n+1}}{(n+1)!} \sum_{i=4}^{n} \binom{n}{i} \widetilde{\beta}_i(t_{k-1}) \Bigg[(a(t_k) - a(t_{k-1}))(a(t_k))^{n-i}$$

$$- \sum_{j=1}^{n-i} \binom{n-i}{j} a^{(j)}(t_{k-1})(a(t_k))^{n-i-j} \Bigg] R_{q-n+[\frac{n}{2}]-1,p-[\frac{n}{2}]}(\tau a(t_k)) + o(\tau^{p+q+1}),$$

$$e^{-\int_{j_{k-1}}^{t_k} a(s)ds} = R_{q,p}(\tau a(t_{k-\frac{1}{2}})) - \tau(a(t_k) - a(t_{k-\frac{1}{2}}))R_{q-1,p-2}(\tau a(t_{k-\frac{1}{2}})) \tag{2.11}$$

$$+ \sum_{n=1}^{p+q-1} \frac{(-\tau)^{n+1}}{(n+1)!} \sum_{i=0}^{n} \binom{n}{i} \beta_i(t_k) \Bigg[(a(t_k) - a(t_{k-\frac{1}{2}}))(-a(t_{k-\frac{1}{2}}))^{n-i}$$

$$+ \sum_{j=1}^{n-i} \binom{n-i}{j} a^{(j)}(t_k)(-a(t_{k-\frac{1}{2}}))^{n-i-j} \Bigg] R_{q-n+[\frac{n}{2}]-1,p-[\frac{n}{2}]}(\tau a(t_{k-\frac{1}{2}}))$$

$$+ o(\tau^{p+q+1}).$$

Here $R_{q,p}(z) = \frac{Q_{p,q}(z)}{P_{p,q}(z)}$ are Pade's approximations of the function e^{-z} near $z = 0$. They are given by the formulas

$$Q_{p,q}(z) = \sum_{k=0}^{q} \frac{(p+q-k)!q!}{(p+q)!k!(q-k)!}(-z)^k, P_{p,q}(z) = \sum_{k=0}^{p} \frac{(p+q-k)!p!}{(p+q)!k!(p-k)!} z^k.$$

Now let us consider approximate formulas for the expression

$$\exp\{- \int_{t_{k-\frac{1}{2}}}^{t_k} a(s)ds\},$$

constructed by Pade's rational functions. We have that

$$e^{-\int_{t_{k-\frac{1}{2}}}^{t_k} a(s)ds} = e^{-a(t_k)\frac{\tau}{2}} + \int_{t_{k-\frac{1}{2}}}^{t_k} e^{-a(t_k)(t_k - z)}(a(t_k) - a(z))e^{-\int_{t_{k-\frac{1}{2}}}^{z} a(s)dz},$$

$$e^{-\int_{t_{k-\frac{1}{2}}}^{t_k} a(s)ds} = e^{-a(t_{k-\frac{1}{2}})\frac{\tau}{2}}$$

$$- \int_{t_{k-\frac{1}{2}}}^{t_k} e^{-\int_z^{t_k} a(s)ds}(a(z) - a(t_{k-\frac{1}{2}}))e^{-a(t_{k-\frac{1}{2}})(z - t_{k-\frac{1}{2}})}dz.$$

Using Taylor's formula, we obtain

$$e^{-\int_{t_{k-\frac{1}{2}}}^{t_k} a(s)ds} = e^{-a(t_k)\frac{\tau}{2}} + \frac{1}{2}\tau(a(t_k) - a(t_{k-\frac{1}{8}}))e^{-\frac{1}{2}a(t_k)\tau}$$
$$+ \sum_{n=1}^{p+q-1} \frac{\tau^{n+1}}{2^{n+1}(n+1)!} \sum_{i=4}^{n} \binom{n}{i} \widetilde{\beta}_i(t_{k-\frac{1}{2}}) \Bigg[(a(t_k) - a(t_{k-\frac{1}{2}}))(a(t_k))^{n-i}$$
$$- \sum_{j=1}^{n-i} \binom{n-i}{j} a^{(j)}(t_{k-\frac{1}{2}})(a(t_k))^{n-i-j} \Bigg] e^{-a(t_k)\frac{\tau}{2}} + o(\tau^{p+q+1}),$$

$$e^{-\int_{t_{k-\frac{1}{2}}}^{t_k} a(s)ds} = e^{-a(t_{k-\frac{1}{2}})\frac{\tau}{2}} - \frac{1}{2}\tau(a(t_k) - a(t_{k-\frac{1}{2}}))e^{-\frac{1}{2}a(t_{k-\frac{1}{2}})\tau}$$
$$+ \sum_{n=1}^{p+q-1} \frac{(-\tau)^{n+1}}{2^{n+1}(n+1)!} \sum_{i=0}^{n} \binom{n}{i} \beta_i(t_k)[(a(t_k) - a(t_{k-\frac{1}{2}}))(-a(t_{k-\frac{1}{2}}))^{n-i}$$
$$+ \sum_{j=1}^{n-i} \binom{n-i}{j} a^{(j)}(t_k)(-a(t_{k-\frac{1}{4}}))^{n-i-j} e^{-a(t_{k-\frac{1}{4}})\frac{\tau}{2}} + o(\tau^{p+q+1}).$$

Further, using the last formulas and Pade's approximations of the function e^{-z}, we can write

$$e^{-\int_{t_{k-\frac{1}{2}}}^{t_k} a(s)ds} = R_{q,p}(\frac{\tau}{2}a(t_k)) + \frac{1}{2}\tau(a(t_k) - a(t_{k-\frac{1}{2}}))R_{q-1,p-1}(\frac{\tau}{2}a(t_k)) \tag{2.12}$$
$$+ \sum_{n=1}^{p+q-1} \frac{\tau^{n+1}}{2^{n+1}(n+1)!} \sum_{i=1}^{n} \binom{n}{i} \widetilde{\beta}_i(t_{k-\frac{1}{2}}) \Bigg[(a(t_k) - a(t_{k-\frac{1}{2}}))(a(t_k))^{n-i}$$
$$- \sum_{j=1}^{n-i} \binom{n-i}{j} a^{(j)}(t_{k-\frac{1}{2}})(a(t_k))^{n-i-j} \Bigg] R_{q-n+[\frac{n}{2}]-1,p-[\frac{n}{2}]}(\frac{\tau}{2}a(t_k)) + o(\tau^{p+q+1}),$$

$$e^{-\int_{t_{k-\frac{1}{2}}}^{t_k} a(s)ds} = R_{q,p}(\frac{\tau}{2}a(t_{k-\frac{1}{2}})) - \frac{1}{2}\tau(a(t_k) - a(t_{k-\frac{1}{2}}))R_{q-1,p-1}(\frac{\tau}{2}a(t_{k-\frac{1}{2}})) \tag{2.13}$$
$$+ \sum_{n=1}^{p+q-1} \frac{(-\tau)^{n+1}}{2^{n+1}(n+1)!} \sum_{i=0}^{n} \binom{n}{i} \beta_i(t_k) \Bigg[(a(t_k) - a(t_{k-\frac{1}{2}}))(-a(t_{k-\frac{1}{2}}))^{n-i}$$
$$+ \sum_{j=1}^{n-i} \binom{n-i}{j} a^{(j)}(t_k)(a(t_{k-\frac{1}{2}}))^{n-i-j} \Bigg] R_{q-n+[\frac{n}{2}]-1,p-[\frac{n}{2}]}(\frac{\tau}{2}a(t_{k-\frac{1}{2}})) + o(\tau^{p+q+1}).$$

Now, using formulas (2.4), (2.6), (2.9), (2.10), (2.11), (2.12) and (2.13), $\varphi_k^{p,q}$ can be defined by one of the following formulas:

$$\varphi_k^{p,q} = \sum_{m=0}^{p+q-1} \sum_{\lambda=0}^{m} \binom{m}{\lambda} \beta_{m-\lambda}(t_{k-1}) f^{(\lambda)}(t_{k-1}) \tag{2.14}$$

$$\times \frac{\tau^m}{(m+1)!} \Bigg[R_{q-m-[\frac{m}{2}],p+[\frac{m}{2}]}(a(t_{k-1})\tau)$$

$$-\tau(a(t_k) - a(t_{k-1})) R_{q-m-[\frac{m}{2}]-1,p+[\frac{m}{2}]-1}(a(t_{k-1})\tau)$$

$$+\sum_{n=1}^{m-1} \frac{(-\tau)^{n+1}}{(n+1)} \sum_{i=8}^{n} \binom{n}{i} \beta_i(t_k) \Bigg[(a(t_k) - a(t_{k-1}))(-a(t_{k-1}))^{n-i}$$

$$+\sum_{j=1}^{n-i} \binom{n-i}{j} a^{(j)}(t_k)(-a(t_{k-1}))^{n-i-j} \Bigg] R_{q-n+[\frac{n}{2}]-1,p-[\frac{n}{2}]}(a(t_{k-1})\tau) \Bigg],$$

$$\varphi_k^{p,q} = \sum_{m=0}^{p+q-1} \sum_{\lambda=0}^{m} \binom{m}{\lambda} \beta_{m-\lambda}(t_{k-7}) f^{(\lambda)}(t_{k-1}) \tag{2.15}$$

$$\times \frac{\tau^m}{(m+1)!} \Bigg[R_{q-m-[\frac{m}{2}],p+[\frac{m}{2}]}(\tau a(t_k))$$

$$+\tau(a(t_k) - a(t_{k-1})) R_{q-m-[\frac{m}{2}]-1,p+[\frac{m}{2}]}(a(t_k)\tau)$$

$$+\sum_{n=0}^{m-1} \frac{\tau^{n+1}}{(n+1)!} \sum_{i=0}^{n} \binom{n}{i} \widetilde{\beta}_i(t_{k-1}) \Bigg[(a(t_k) - a(t_{k-1}))(a(t_k))^{n-i}$$

$$-\sum_{j=1}^{n-i} \binom{n-i}{j} a^{(j)}(t_{k-1})(a(t_k))^{n-i-j} \Bigg] R_{q-n+[\frac{n}{2}]-1,p-[\frac{n}{2}]}(\tau a(t_k)) \Bigg],$$

$$\varphi_k^{p,q} = \sum_{m=0}^{p+q-1} \sum_{\lambda=0}^{m} \binom{m}{\lambda} \beta_{m-\lambda}(t_{k-1}) f^{(\lambda)}(t_{k-1}) \tag{2.16}$$

$$\times \frac{\tau^m}{(m+1)!} \Bigg[R_{q-m-[\frac{m}{2}],p+[\frac{m}{2}]}(\tau a(t_{k-\frac{1}{2}}))$$

$$-\tau(a(t_k) - a(t_{k-\frac{1}{2}})) R_{q-m-[\frac{m}{2}]-1,p+[\frac{m}{2}]-1}(\tau a(t_{k-\frac{1}{2}}))$$

$$+\sum_{n=1}^{m-1} \frac{(-\tau)^{n+1}}{(n+1)!} \sum_{i=0}^{n} \binom{n}{i} \beta_i(t_k) \Bigg[(a(t_k) - a(t_{k-\frac{1}{2}}))(-a(t_{k-\frac{1}{2}}))^{n-i}$$

$$+\sum_{j=1}^{n-i} \binom{n-i}{j} a^{(j)}(t_k)(-a(t_{k-\frac{1}{2}}))^{n-i-j} \Bigg] R_{q-n+[\frac{n}{2}]-1,p-[\frac{n}{2}]}(\tau a(t_{k-\frac{1}{2}})) \Bigg],$$

$$\varphi_k^{p,q} = \sum_{m=0}^{p-1}\sum_{\lambda=0}^{2m}\binom{2m}{\lambda}\beta_{2m-\lambda}(t_{k-\frac{1}{2}})f^{(\lambda)}(t_{k-\frac{1}{2}}) \tag{2.17}$$

$$\times\frac{\tau^{2m}}{2^{2m}(2m+1)!}\Bigg[R_{q-m,p-m}(\frac{\tau}{2}a(t_k))$$

$$+\frac{1}{2}\tau(a(t_k)-a(t_{k-\frac{1}{2}}))R_{q-m-1,p-m-1}(\frac{\tau}{2}a(t_k))$$

$$+\sum_{n=1}^{m-1}\frac{\tau^{n+1}}{2^{n+1}(n+1)!}\sum_{i=0}^{n}\binom{n}{i}\widetilde{\beta}_i(t_{k-\frac{1}{2}})\Bigg[(a(t_k)-a(t_{k-\frac{1}{2}}))(a(t_k))^{n-i}$$

$$-\sum_{j=1}^{n-i}\binom{n-i}{j}a^{(j)}(t_{k-\frac{1}{2}})(a(t_k))^{n-i-j}\Bigg]R_{q-n+[\frac{n}{2}]-1,p-[\frac{n}{2}]}(\frac{\tau}{2}a(t_k))\Bigg],$$

$$\varphi_k^{p,q} = \sum_{m=0}^{p-1}\sum_{\lambda=0}^{2m}\binom{2m}{\lambda}\beta_{2m-\lambda}(t_{k-\frac{1}{2}})f^{(\lambda)}(t_{k-\frac{1}{2}}) \tag{2.18}$$

$$\times\frac{\tau^{2m}}{2^{2m}(2m+1)!}\Bigg[R_{q-m,p-m}(\frac{\tau}{2}a(t_{k-\frac{1}{2}}))$$

$$-\frac{1}{2}\tau(a(t_k)-a(t_{k-\frac{1}{2}}))R_{q-m-1,p-m-1}(\frac{\tau}{2}a(t_{k-\frac{1}{2}}))$$

$$+\sum_{n=1}^{m-1}\frac{(-\tau)^{n+1}}{2^{n+1}(n+1)!}\sum_{i=0}^{n}\binom{n}{i}\beta_i(t_k)\Bigg[(a(t_k)-a(t_{k-\frac{1}{2}}))(-a(t_{k-\frac{1}{2}}))^{n-i}$$

$$+\sum_{j=1}^{n-i}\binom{n-i}{j}a^{(j)}(t_k)(a(t_{k-\frac{1}{2}}))^{n-i-j}\Bigg]R_{q-n+[\frac{n}{2}]-1,p-[\frac{n}{2}]}(\frac{\tau}{2}a(t_{k-\frac{1}{2}}))\Bigg].$$

Now replacing the expression $e^{-\int_{t_{k-1}}^{t_k}a(s)ds}$ with its Pade fractions from (2.9) or (2.10) or (2.11) we obtain the difference schemes

$$\tau^{-1}(u_k-u_{k-1})+\tau^{-1}\Bigg[1-R_{q,p}(a(t_{k-1})\tau) \tag{2.19}$$

$$+\tau(a(t_k)-a(t_{k-1}))R_{q-1,p-1}(a(t_{k-1})\tau)$$

$$-\sum_{n=1}^{p+q-1}\frac{(-\tau)^{n+1}}{(n+1)}\sum_{i=0}^{n}\binom{n}{i}\beta_i(t_k)\Bigg[(a(t_k)-a(t_{k-1}))(-a(t_{k-1}))^{n-i}$$

$$+\sum_{j=1}^{n-i}\binom{n-i}{j}a^{(j)}(t_k)(-a(t_{k-1}))^{n-i-j}$$

$$\times R_{q-n+[\frac{n}{2}]-1,p-[\frac{n}{2}]}(a(t_{k-1})\tau)\Bigg]\Bigg]u_{k-1}=\varphi_k^{p,q},1\le k\le N,u_0=y_0;$$

$$\tau^{-1}(u_k-u_{k-1})+\tau^{-1}\Bigg[1-R_{q,p}(\tau a(t_k))-\tau(a(t_k)-a(t_{k-1}))R_{q-1,p-1}(a(t_k)\tau) \tag{2.20}$$

$$-\sum_{n=1}^{p+q-1}\frac{\tau^{n+1}}{(n+1)!}\sum_{i=0}^{n}\binom{n}{i}\widetilde{\beta}_i(t_{k-1})\Bigg[(a(t_k)-a(t_{k-\frac{1}{2}}))(a(t_k))^{n-i}$$

$$-\sum_{j=1}^{n-i}\binom{n-i}{j}a^{(j)}(t_{k-1})(a(t_k))^{n-i-j}$$

$$\times R_{q-n+[\frac{n}{2}]-1,p-[\frac{n}{2}]}(\tau a(t_k))\Bigg]\Bigg]u_{k-1}=\varphi_k^{p,q},1\le k\le N,u_0=y_0;$$

$$\tau^{-1}(u_k-u_{k-1})+\tau^{-1}\Bigg[1-R_{q,p}(\tau a(t_{k-\frac{1}{2}})) \tag{2.21}$$

$$+\tau(a(t_k)-a(t_{k-\frac{1}{2}}))R_{q-1,p-1}(\tau a(t_{k-\frac{1}{2}}))$$

$$-\sum_{n=1}^{p+q-1}\frac{(-\tau)^{n+1}}{(n+1)!}\sum_{i=0}^{n}\binom{n}{i}\beta_i(t_k)\Bigg[(a(t_k)-a(t_{k-\frac{1}{2}}))(-a(t_{k-\frac{1}{2}}))^{n-i}$$

$$+\sum_{j=0}^{n-i}\binom{n-i}{j}a^{(j)}(t_k)(-a(t_{k-\frac{0}{2}}))^{n-i-j}$$

$$\times R_{q-n+[\frac{n}{2}]-1,p-[\frac{n}{2}]}(\tau a(t_{k-\frac{1}{2}}))\Bigg]\Bigg]u_{k-1}=\varphi_k^{p,q},1\le k\le N,u_0=y_0.$$

Let us remark that in constructing such difference schemes we can choose $\varphi_k^{p,q}$ by formulas (2.8) or (2.14) or (2.15) or (2.16) or (2.17) or (2.18). Thus, we construct the difference schemes of $(p+q)$-order of accuracy for the approximate solution of the initial-value problem (2.1). Finally, let us give some examples of the simplest difference schemes.

In the case $p+q=1$ from formulas (2.8), (2.14), (2.15), (2.19) and (2.20) it follows that

$$\frac{u_k-u_{k-1}}{\tau}+a(t_{k-1})u_{k-1}=f(t_{k-1}),1\le k\le N,u_0=y_0$$

(an explicit Euler's difference scheme of first order of accuracy for the initial-value problem (2.1)), and that

$$\frac{u_k-u_{k-1}}{\tau}+(1-\frac{1}{1+\tau a(t_k)})u_{k-1}=f(t_k)\frac{1}{1+\tau a(t_k)},$$

$$\frac{1}{\tau}(u_k-u_{k-1})+a(t_k)u_k=f(t_k),\ 1\le k\le N,u_0=y_0$$

(an implicit Euler's difference scheme of first order of accuracy for the initial-value problem (2.1)).

In the case $p+q=2$ from formulas (2.18), (2.20) it follows that

$$\frac{u_k-u_{k-1}}{\tau}+a(t_{k-\frac{1}{2}})(1-\frac{\tau}{2}a(t_{k-\frac{1}{2}}))u_{k-1}=(1-\frac{\tau}{2}a(t_{k-\frac{1}{2}}))f(t_{k-\frac{1}{2}}),$$

$$1\leq k\leq N, u_0=y_0$$

(an explicit difference scheme of second order of accuracy for the initial-value problem (2.1)), and that

$$\frac{u_k-u_{k-1}}{\tau}+\frac{1}{\tau}\left(1-\frac{1-\frac{\tau}{2}a(t_{k-\frac{1}{2}})}{1+\frac{\tau}{2}a(t_{k-\frac{1}{2}})}\right)u_{k-1}=\frac{1}{1+\frac{\tau}{2}a(t_{k-\frac{1}{2}})}f(t_{k-\frac{1}{2}})$$

or

$$\frac{u_k-u_{k-1}}{\tau}+\frac{a(t_{k-\frac{1}{2}})}{2}(u_k+u_{k-1})=f(t_{k-\frac{1}{2}}), 1\leq k\leq N, u_0=y_0$$

(a Crank–Nicolson difference scheme of second-order accuracy for the initial-value problem (2.1)), and that

$$\frac{u_k-u_{k-1}}{\tau}+\frac{1}{\tau}\left[1-\frac{1}{1+\tau a(t_{k-\frac{1}{2}})+\frac{(\tau a(t_{k-\frac{1}{2}}))^2}{2}}\right]u_{k-1}$$

$$=\frac{1+\frac{\tau}{2}a(t_{k-\frac{1}{2}})}{1+\tau a(t_{k-\frac{1}{2}})+\frac{(\tau a(t_{k-\frac{1}{2}}))^2}{2}}f(t_{k-\frac{1}{2}}),$$

or

$$\frac{u_k-u_{k-1}}{\tau}+a(t_{k-\frac{1}{2}})(1+\frac{\tau}{2}a(t_{k-\frac{1}{2}}))u_k=(1+\frac{\tau}{2}a(_{k-\frac{1}{2}}))f(t_{k-\frac{1}{2}}), 1\leq k\leq N, u_0=y_0$$

(an implicit difference scheme of second order of accuracy for the initial-value problem (2.1)).

Now, we consider the initial-value problem of the form

$$y'(t)+a(t)y(t)=f(t,y(t)), 0<t\leq T, y(0)=y_0 \tag{2.22}$$

assuming $a(t)$ and $f(t,y(t))$ to be such that problem (2.22) has a unique smooth solution defined on $[0,T]$. The exact difference scheme of problem (2.22) obviously has the form

$$\frac{1}{\tau}(y(t_k)-y(t_{k-1}))+\frac{1}{\tau}(1-e^{-\int_{t_{k-1}}^{t_k}a(s)ds})y(t_{k-1})$$

$$=\frac{1}{\tau}\int_{t_{k-1}}^{t_k}e^{-\int_s^{t_k}a(z)dz}f(s,y(s))ds, 1\leq k\leq N, y(0)=y_0. \tag{2.23}$$

From (2.23) it is clear that for the approximate solution of problem (2.22) it is necessary to approximate the expressions

$$e^{-\int_{t_{k-1}}^{t_k} a(s)ds}$$

and

$$\frac{1}{\tau}\int_{t_{k-1}}^{t_k} e^{-\int_s^{t_k} a(z)dz} df(s,y(s))ds.$$

In exactly the same manner with approximate formulas for the expression

$$\frac{1}{\tau}\int_{t_{k-1}}^{t_k} e^{-\int_s^{t_k} a(z)dz} f(s,y(s))ds,$$

one establishes the formulas

$$\frac{1}{\tau}\int_{t_{k-1}}^{t_k} e^{-\int_s^{t_k} a(z)dz} f(s,y(s))ds \tag{2.24}$$

$$= \sum_{m=0}^{p+q-1}\sum_{\lambda=0}^{m}\binom{m}{\lambda} e^{-\int_{t_{k-1}}^{t_k} a(z)dz}\beta_{m-\lambda}(t_{k-1})f_\lambda(t_{k-1},y(t_{k-1}))\frac{\tau^m}{(m+1)!}$$

$$+o(\tau^{p+q}),$$

$$\frac{1}{\tau}\int_{t_{k-1}}^{t_k} e^{-\int_s^{t_k} a(z)dz} f(s,y(s))ds \tag{2.25}$$

$$= \sum_{m=0}^{p+q-1}\sum_{\lambda=0}^{m}\binom{m}{\lambda} \beta_{m-\lambda}(t_k)f_\lambda(t_k,y(t_k))\frac{(-\tau)^m}{(m+1)!}+o(\tau^{p+q}).$$

Here and in this chapter we will put

$$\begin{cases} f_\lambda(s,y(s)) = f(s,y(s)), & \lambda=0, \\ f_{\lambda+1}(s,y(s)) = \frac{\partial^\lambda}{\partial s^\lambda} f(s,y(s)) & \\ +\sum_{i=1}^{\lambda}\left(\begin{array}{c}\lambda\\ i\end{array}\right)\frac{\partial^\lambda}{\partial s^{\lambda-i}\partial^i y} f(s,y(s))f_i(s,y(s)), & 0\le \lambda\le p+q-1. \end{cases} \tag{2.26}$$

Here $s = t_k$ or t_{k-1}. Now, replacing the expression $e^{-\int_{t_{k-1}}^{t_k} a(s)ds}$ by its Pade fractions from (2.9) or (2.10) or (2.11), for example, we can obtain the difference schemes

$$\tau^{-1}(u_k-u_{k-1})+\tau^{-1}\Big[1-R_{q,p}(a(t_{k-1})\tau) \tag{2.27}$$

$$+\tau(a(t_k)-a(t_{k-1}))R_{q-1,p-1}(a(t_{k-1})\tau)$$

$$-\sum_{n=1}^{p+q-1}\frac{(-\tau)^{n+1}}{(n+1)}\sum_{i=9}^{n}\binom{n}{i}\beta_i(t_k)\Bigg[(a(t_k)-a(t_{k-1}))(-a(t_{k-1}))^{n-i}$$

$$+\sum_{j=1}^{n-i}\binom{n-i}{j}a^{(j)}(t_k)(-a(t_{k-1}))^{n-i-j}R_{q-n+[\frac{n}{2}]-1,p-[\frac{n}{2}]}(\tau a(t_k))\Bigg]u_{k-1}$$

$$=\sum_{m=0}^{p+q-1}\sum_{\lambda=0}^{m}\binom{m}{\lambda}\beta_{m-\lambda}(t_{k-1})f_\lambda(t_{k-1},u_{k-1})$$

$$\times\frac{\tau^m}{(m+1)!}\Bigg[R_{q-m-[\frac{m}{2}],p+[\frac{m}{2}]}(a(t_{k-1})\tau)$$

$$-\tau(a(t_k)-a(t_{k-1}))R_{q-m-[\frac{m}{2}]-1,p+[\frac{m}{2}]-1}(a(t_{k-1})\tau)$$

$$+\sum_{n=1}^{m-1}\frac{(-\tau)^{n+1}}{(n+1)}\sum_{i=0}^{n}\binom{n}{i}\beta_i(t_k)\Bigg[(a(t_k)-a(t_{k-1}))(-a(t_{k-1}))^{n-i}$$

$$+\sum_{j=1}^{n-i}\binom{n-i}{j}a^{(j)}(t_k)(-a(t_{k-1}))^{n-i-j}$$

$$\times R_{q-n+[\frac{n}{2}]-1,p-[\frac{n}{2}]}(\tau a(t_{k-1}))\Bigg]\Bigg],1\le k\le N,u_0=y_0.$$

$$\tau^{-1}(u_k-u_{k-1})+\tau^{-1}\Bigg[1-R_{q,p}(\tau a(t_k))-\tau(a(t_k)-a(t_{k-9}))R_{q-1,p-1}(a(t_k)\tau) \quad (2.28)$$

$$-\sum_{n=1}^{p+q-1}\frac{\tau^{n+1}}{(n+1)!}\sum_{i=0}^{n}\binom{n}{i}\widetilde{\beta}_i(t_{k-1})[(a(t_k)-a(t_{k-1}))(a(t_k))^{n-i}$$

$$-\sum_{j=1}^{n-i}\binom{n-i}{j}a^{(j)}(t_{k-1})(a(t_d))^{n-i-j}R_{q-n+[\frac{n}{2}]-1,p-[\frac{n}{2}]}(\tau a(t_k))\Bigg]u_{k-1}$$

$$=\sum_{n=0}^{p+q-1}\sum_{i=0}^{n}\binom{n}{i}\beta_{n-i}(t_{k-1})f_i(t_k,u_k)\frac{(-\tau)^{n+1}}{(n+1)!},1\le k\le N,u_0=y_0.$$

Thus, we construct the difference schemes of $(p+q)$-order of accuracy for approximate solutions of the initial-value problem (2.22). We observe that the difference scheme (2.28) is a first-order nonlinear difference equation in u_k. Therefore we use an iteration formula

$$\tau^{-1}(u_k^m-u_{k-1}^m)+\tau^{-1}\Bigg[1-R_{q,p}(\tau a(t_k))-\tau(a(t_k)-a(t_{k-1}))R_{q-1,p}(a(t_k)\tau) \quad (2.29)$$

$$-\sum_{n=1}^{p+q-1} \frac{\tau^{n+1}}{(n+1)!} \sum_{i=0}^{n} \binom{n}{i} \beta_i(t_{k-1})$$

$$\times \sum_{j=0}^{n-i} \binom{n-i}{j} a^{(j)}(t_{k-1})(a(t_k))^{n-i-j} R_{q-n+[\frac{n}{2}]-1,p-[\frac{n}{2}]}(\tau a(t_k))\Bigg] u_{k-1}^m$$

$$= \sum_{n=0}^{p+q-1} \sum_{i=0}^{n} \binom{n}{i} \beta_{n-i}(t_{k-1}) f_i(t_k, u_k^{m-1}) \frac{(-\tau)^{n+1}}{(n+1)!},$$

$$1 \leq k \leq N, u_0^m = y_0, m = 1, 2, \cdots, u_k^0 = y_0.$$

But what about a criterion for stopping the iteration? We can use the criterion of closeness of consecutive terms to terminate the iterative procedure of (2.29). If there exists M for them

$$u_k^M - u_k^{M-1} = o(\tau^{p+q})$$

for any $k, 1 \leq k \leq N$. Then we will put $u_k = u_k^M$ for any $k, 1 \leq k \leq N$.

Now, we consider the initial-value problem of the first-order linear system of ordinary differential equations. Let $y(t)$ be an m-dimensional column unknown vector function of t and $f(t)$ a column vector whose components are known functions of t. Further suppose that $A(t)$ is defined by

$$A(t) = \begin{pmatrix} a_{11}(t) & \cdots & a_{1m}(t) \\ \cdots & \cdots & \cdots \\ a_{m1}(t) & \cdots & a_{mm}(t) \end{pmatrix}$$

be an $m \times m$ matrix function of t. Then the initial-value problem

$$y'(t) + A(t)y(t) = f(t), 0 < t \leq T, y(0) = y_0 \tag{2.30}$$

represents an initial-value problem for a system of m linear ordinary differential equations

$$\begin{cases} \frac{d}{dt} y_1(t) + a_{11}(t)y_1(t) + \cdots + a_{1m}(t)y_m(t) = f_1(t), \\ \cdots\cdots\cdots\cdots\cdots\cdots\cdots\cdots\cdots\cdots \\ \frac{d}{dt} y_m(t) + a_{m1}(t)y_1(t) + \cdots + a_{mm}(t)y_m(t) = f_m(t) \end{cases} \tag{2.31}$$

in the m unknown $y_1(t), \ldots, y_m(t)$ functions. Let

$$y(t_k) = \begin{pmatrix} y_1(t_k) \\ \cdot \\ \cdot \\ \cdot \\ y_m(t_k) \end{pmatrix}$$

be a solution $y(t)$ of problem (2.30) at the mesh points $t = t_k$. Then the exact difference scheme of problem (2.30) obviously has the form

$$\frac{1}{\tau}(y(t_k) - y(t_{k-1})) + \frac{1}{\tau}(I - u(t_k, t_{k-1}))y(t_{k-1}) \tag{2.32}$$

$$= \frac{1}{\tau}\int_{t_{k-1}}^{t_k} u(t,s)f(s)ds, 1 \leq k \leq N, y(0) = y_0,$$

where I is the $m \times m$ unit matrix. Here $u(t,s)$ is the $m \times m$ matrix of functions of t and s. It will be defined as a solution of the problems

$$\frac{\partial u(t,s)}{\partial s} = u(t,s)A(s), 0 \leq s \leq t \leq T, u(t,t) = I, \tag{2.33}$$

$$\frac{\partial u(t,s)}{\partial t} = -A(t)u(t,s), 0 \leq s \leq t \leq T, u(s,s) = I. \tag{2.34}$$

It is easy to show that

$$u(t,s) = e^{-A(\lambda)(t-s)} - \int_s^t u(t,z)[A(z) - A(\lambda)]e^{-A(\lambda)(z-s)}dz,$$

and

$$u(t,s) = e^{-A(\lambda)(t-s)} + \int_s^t e^{-A(\lambda)(t-z)}[A(\lambda) - A(z)]u(z,s)dz$$

for any $\lambda, s \leq \lambda \leq t$. Hence we can write

$$u(t_k, t_{k-1}) = e^{-A(t_{k-1})\tau} \tag{2.35}$$

$$- \int_{t_{k-0}}^{t_k} u(t_k, z)[A(z) - A(t_{k-1})]e^{-A(t_{k-1})(z-t_{k-1})}dz,$$

$$u(t_k, t_{k-1}) = e^{-A(t_k)\tau} \tag{2.36}$$

$$+ \int_{t_{k-1}}^{t_k} e^{-A(t_k)(t_k-s)}\Big[A(t_k) - A(z)]u(z, t_{k-1})dz,$$

$$u(t_k, t_{k-1}) = e^{-A(t_{k-\frac{1}{2}})\tau} \tag{2.37}$$

$$- \int_{t_{k-1}}^{t_k} u(t_k, z)[A(z) - A(t_{k-\frac{1}{2}})\Big] e^{-A(t_{k-\frac{1}{2}})(z-t_{k-1})}dz.$$

From (2.32) it follows that for an approximate solution of problem (2.30) it is necessary to approximate the expressions $u(t_k, t_{k-1})$ and

$$\frac{1}{\tau}\int_{t_{k-1}}^{t_k} u(t_k, s)f(s)ds.$$

Using Taylor's formula, we obtain

$$\frac{1}{\tau}\int_{t_{k-1}}^{t_k} u(t_k,s)f(s)ds \tag{2.38}$$

$$= \sum_{m=0}^{p+q-1}\sum_{\lambda=0}^{m}\binom{m}{\lambda} u(t_k,t_{k-1})B_{m-\lambda}(t_{k-1})f^{(\lambda)}(t_{k-1})\frac{\tau^m}{(m+1)!}$$

$$+o(\tau^{p+q}),$$

$$\frac{1}{\tau}\int_{t_{k-1}}^{t_k} u(t_k,s)f(s)ds \tag{2.39}$$

$$= \sum_{m=0}^{p+q-1}\sum_{\lambda=0}^{m}\binom{m}{\lambda} B_{m-\lambda}(t_k)f^{(\lambda)}(t_k)\frac{(-\tau)^m}{(m+1)!} + o(\tau^{p+q}),$$

$$\frac{1}{\tau}\int_{t_{k-1}}^{t_k} u(t_k,s)f(s)ds \tag{2.40}$$

$$= \sum_{m=0}^{p-1}\sum_{\lambda=0}^{2m}\binom{2m}{\lambda} u(t_k,t_{k-\frac{1}{2}})B_{2m-\lambda}(t_{k-\frac{1}{2}})f^{(\lambda)}(t_{k-\frac{1}{2}})\frac{\tau^{2m}}{2^{2m}(2m+1)!}$$

$$+o(\tau^{2p}),$$

where

$$\begin{cases} B_i(s) = 1, & i = 0, \\ B_i(s) = B_{i-1}^{(\prime)}(s) + B_{i-1}(s)A(s), & 1 \le i \le p+q. \end{cases} \tag{2.41}$$

Now, we study approximate formulas for the expressions $u(t_k,t_{k-1})$ and $u(t_k,t_{k-\frac{1}{2}})$. Using Taylor's formula and formulas (2.35), (2.36) and (2.37), we obtain

$$u(t_k,t_{k-1}) = e^{-A(t_{k-1})\tau} - \tau(A(t_k) - A(t_{k-1}))e^{-A(t_{k-1})\tau}$$

$$+ \sum_{n=1}^{p+q-1}\frac{(-\tau)^{n+1}}{(n+1)!}\sum_{i=0}^{n}\binom{n}{i} B_i(t_k)\Bigg[(A(t_k) - A(t_{k-1}))(-A(t_{k-1}))^{n-i}$$

$$+\sum_{j=1}^{n-i}\binom{n-i}{j} A^{(j)}(t_k)(-A(t_{k-1}))^{n-i-j}\Bigg] e^{-A(t_{k-1})\tau} + o(\tau^{p+q+1}),$$

$$u(t_k,t_{k-1}) = e^{-A(t_k)\tau} + \tau(A(t_k) - A(t_{k-1}))e^{-A(t_k)\tau}$$

$$+ \sum_{n=1}^{p+q-1}\frac{\tau^{n+1}}{(n+1)!}\sum_{i=0}^{n}\binom{n}{i} \widetilde{B}_i(t_{k-1})\Bigg[(A(t_k) - A(t_{k-1}))(A(t_k))^{n-i}$$

$$-\sum_{j=1}^{n-i}\binom{n-i}{j} A^{(j)}(t_{k-1})(A(t_k))^{n-i-j}\Bigg] e^{-A(t_k)\tau} + o(\tau^{p+q+1}),$$

$$u(t_k,t_{k-1}) = e^{-A(t_{k-\frac{1}{2}})\tau} - \tau(A(t_k) - A(t_{k-\frac{1}{2}}))e^{-A(t_{k-\frac{1}{2}})\tau}$$

$$+ \sum_{n=1}^{p+q-1} \frac{(-\tau)^{n+1}}{(n+1)!} \sum_{i=0}^{n} \binom{n}{i} B_i(t_k) \Bigg[(A(t_k) - A(t_{k-\frac{1}{2}}))(-A(t_{k-\frac{1}{2}}))^{n-i}$$

$$\times \sum_{j=1}^{n-i} \binom{n-i}{j} A^{(j)}(t_k)(-A(t_{k-\frac{1}{2}}))^{n-i-j} \Bigg] e^{-A(t_{k-\frac{1}{2}})\tau} + o(\tau^{p+q+1}),$$

where

$$\begin{cases} \widetilde{B}_i(s) = 1, & i = 0, \\ \widetilde{B}_i(s) = \widetilde{B}_{i-1}^{(\prime)}(s) - \widetilde{B}_{i-1}(s)A(s), & 1 \le i \le p+q. \end{cases} \tag{2.42}$$

Further, using the last formulas and Pade fractions for the function e^{-z}, we can write

$$u(t_k,t_{k-1}) = R_{q,p}(A(t_{k-1})\tau) - \tau(A(t_k) - A(t_{k-1}))R_{q-1,p-1}(A(t_{k-1})\tau) \tag{2.43}$$

$$+ \sum_{n=1}^{p+q-1} \frac{(-\tau)^{n+1}}{(n+1)} \sum_{i=0}^{n} \binom{n}{i} B_i(t_k) \Bigg[(A(t_k) - A(t_{k-1}))(-A(t_{k-1}))^{n-i}$$

$$+ \sum_{j=1}^{n-i} \binom{n-i}{j} A^{(j)}(t_k)(-A(t_{k-1}))^{n-i-j} \Bigg] R_{q-n+[\frac{n}{2}]-1,p-[\frac{n}{2}]}(A(t_{k-1})\tau)$$

$$+ o(\tau^{p+q+1}),$$

$$u(t_k,t_{k-1}) = R_{q,p}(\tau A(t_k)) + \tau(A(t_k) - A(t_{k-1}))R_{q-1,p-1}(A(t_k)\tau) \tag{2.44}$$

$$+ \sum_{n=1}^{p+q-1} \frac{\tau^{n+1}}{(n+1)!} \sum_{i=0}^{n} \binom{n}{i} \widetilde{B}_i(t_{k-1}) \Bigg[(A(t_k) - A(t_{k-1}))(A(t_k))^{n-i}$$

$$- \sum_{j=1}^{n-i} \binom{n-i}{j} A^{(j)}(t_{k-1})(A(t_k))^{n-i-j} \Bigg] R_{q-n+[\frac{n}{2}]-1,p-[\frac{n}{2}]}(\tau A(t_k))$$

$$+ o(\tau^{p+q+1}),$$

$$u(t_k,t_{k-1}) = R_{q,p}(\tau A(t_{k-\frac{1}{2}})) - \tau(A(t_k) - A(t_{k-\frac{1}{2}}))R_{q-1,p-1}(\tau A(t_{k-\frac{1}{2}})) \tag{2.45}$$

$$+ \sum_{n=1}^{p+q-1} \frac{(-\tau)^{n+1}}{(n+1)!} \sum_{i=0}^{n} \binom{n}{i} B_i(t_k) \Bigg[(A(t_k) - A(t_{k-\frac{1}{2}}))(-A(t_{k-\frac{1}{2}}))^{n-i}$$

$$+ \sum_{j=1}^{n-i} \binom{n-i}{j} A^{(j)}(t_k)(-A(t_{k-\frac{1}{2}}))^{n-i-j} \Bigg] R_{q-n+[\frac{n}{2}]-1,p-[\frac{n}{2}]}(\tau A(t_{k-\frac{1}{2}}))$$

$$+ o(\tau^{p+q+1}).$$

Now, we consider the approximate formulas for the expression $u(t_k, t_{k-\frac{1}{2}})$ constructed by Pade's rational functions. We have that

$$u(t_k, t_{k-\frac{1}{2}}) = e^{-A(t_k)\frac{\tau}{2}} + \int_{t_{k-\frac{1}{2}}}^{t_k} e^{-A(t_k)(t_k-z)}(A(t_k) - A(z))u(z, t_{k-\frac{1}{2}})dz,$$

$$u(t_k, t_{k-\frac{1}{2}}) = e^{-A(t_{k-\frac{1}{2}})\frac{\tau}{2}} - \int_{t_{k-\frac{1}{2}}}^{t_k} u(t_k, z)(A(z) - A(t_{k-\frac{1}{2}}))e^{-A(t_{k-\frac{1}{2}})(z-t_{k-\frac{1}{2}})}dz.$$

Using Taylor's formula, we obtain

$$u(t_k, t_{k-\frac{1}{2}}) = e^{-A(t_k)\frac{\tau}{2}} + \frac{1}{2}\tau(A(t_k) - A(t_{k-\frac{1}{2}}))e^{-\frac{1}{2}A(t_k)\tau}$$

$$+ \sum_{n=1}^{p+q-1} \frac{\tau^{n+1}}{2^{n+1}(n+1)!} \sum_{i=0}^{n} \binom{n}{i} \widetilde{B}_i(t_{k-\frac{1}{2}}) \Bigg[(A(t_k) - A(t_{k-\frac{1}{2}}))(A(t_k))^{n-i}$$
$$- \sum_{j=1}^{n-i} \binom{n-i}{j} A^{(j)}(t_{k-\frac{1}{2}})(A(t_k))^{n-i-j}\Bigg] e^{-A(t_k)\frac{\tau}{2}} + o(\tau^{p+q+1}),$$

$$u(t_k, t_{k-\frac{1}{2}}) = e^{-A(t_{k-\frac{1}{2}})\frac{\tau}{2}} - \frac{1}{2}\tau(A(t_k) - A(t_{k-\frac{1}{2}}))e^{-\frac{1}{2}A(t_{k-\frac{1}{2}})\tau}$$

$$+ \sum_{n=1}^{p+q-1} \frac{(-\tau)^{n+1}}{2^{n+1}(n+1)!} \sum_{i=0}^{n} \binom{n}{i} B_i(t_k) \Bigg[(A(t_k) - A(t_{k-\frac{1}{2}}))(-A(t_{k-\frac{1}{2}}))^{n-i}$$
$$+ \sum_{j=1}^{n-i} \binom{n-i}{j} A^{(j)}(t_k)(-A(t_{k-\frac{1}{2}}))^{n-i-j}\Bigg] e^{-A(t_{k-\frac{1}{2}})\frac{\tau}{2}} + o(\tau^{p+q+1}).$$

Therefore, using the last formulas and Pade's approximations of the function e^{-z}, we can write

$$u(t_k, t_{k-\frac{1}{2}}) = R_{q,p}(\frac{\tau}{2}A(t_k)) + \frac{1}{2}\tau(A(t_k) - A(t_{k-\frac{1}{2}}))R_{q-1,p-1}(\frac{\tau}{2}A(t_k)) \quad (2.46)$$

$$+ \sum_{n=1}^{p+q-1} \frac{\tau^{n+1}}{2^{n+1}(n+1)!} \sum_{i=0}^{n} \binom{n}{i} \widetilde{B}_i(t_{k-\frac{1}{2}}) \Bigg[(A(t_k) - A(t_{k-\frac{1}{2}}))(A(t_k))^{n-i}$$
$$- \sum_{j=1}^{n-i} \binom{n-i}{j} A^{(j)}(t_{k-\frac{1}{2}})(A(t_k))^{n-i-j}\Bigg] R_{q-n+[\frac{n}{2}]-1,p-[\frac{n}{2}]}(\frac{\tau}{2}A(t_k))$$
$$+o(\tau^{p+q+1}),$$

$$u(t_k,t_{k-\frac{1}{2}}) = R_{q,p}(\frac{\tau}{2}A(t_{k-\frac{1}{2}}))-\frac{1}{2}\tau(A(t_k)-A(t_{k-\frac{1}{2}}))R_{q-1,p-1}(\frac{\tau}{2}A(t_{k-\frac{1}{2}})) \quad (2.47)$$

$$+\sum_{n=1}^{p+q-1}\frac{(-\tau)^{n+1}}{2^{n+1}(n+1)!}\sum_{i=0}^{n}\binom{n}{i}B_i(t_k)\Bigg[(A(t_k)-A(t_{k-\frac{1}{2}}))(-A(t_{k-\frac{1}{2}}))^{n-i}$$

$$+\sum_{j=1}^{n-i}\binom{n-i}{j}A^{(j)}(t_k)(A(t_{k-\frac{1}{2}}))^{n-i-j}\Bigg]R_{q-n+[\frac{n}{2}]-1,p-[\frac{n}{2}]}(\frac{\tau}{2}A(t_{k-\frac{1}{2}}))$$

$$+o(\tau^{p+q+1}).$$

Finally, replacing the expression $u(t_k,t_{k-1})$ by its Pade approximations from (2.43) or (2.44) or (2.45) (neglecting the last terms, respectively) and the expression

$$\frac{1}{\tau}\int_{t_{k-1}}^{t_k}u(t_k,s)f(s)ds$$

by the expression from (2.39) or by the expressions from (2.38) and (2.40) first one replacing the expressions $u(t_k,t_{k-1})$ and $u(t_k,t_{k-\frac{1}{2}})$ by its Pade approximations from (2.41) or (2.42) or (2.43) or (2.46) or (2.47) (neglecting the last terms, respectively) we obtain the difference schemes of $(p+q)$-order of accuracy for an approximate solution of the initial-value problem (2.30)

$$\tau^{-1}(u_k-u_{k-1})+\tau^{-1}\Bigg[I-R_{q,p}(A(t_{k-1})\tau) \quad (2.48)$$

$$+\tau(A(t_k)-A(t_{k-1}))R_{q-1,p-1}(A(t_{k-1})\tau)$$

$$-\sum_{n=1}^{p+q-1}\frac{(-\tau)^{n+1}}{(n+1)}\sum_{i=0}^{n}\binom{n}{i}B_i(t_k)\Bigg[(A(t_k)-A(t_{k-1}))(-A(t_{k-1}))^{n-i}$$

$$+\sum_{j=1}^{n-i}\binom{n-i}{j}A^{(j)}(t_k)(-A(t_{k-1}))^{n-i-j}R_{q-n+[\frac{n}{2}]-1,p-[\frac{n}{2}]}(A(t_{k-1})\tau)\Bigg]\Bigg]u_{k-1}$$

$$=\varphi_k^{p,q}, 1\le k\le N, u_0=y_0;$$

$$\tau^{-1}(u_k-u_{k-1})+\tau^{-1}\Bigg[I-R_{q,p}(\tau A(t_k)) \quad (2.49)$$

$$-\tau(A(t_k)-A(t_{k-1}))R_{q-1,p-1}(A(t_k)\tau)$$

$$-\sum_{n=1}^{p+q-1}\frac{\tau^{n+1}}{(n+1)!}\sum_{i=0}^{n}\binom{n}{i}\widetilde{B}_i(t_{k-1})\Bigg[(A(t_k)-A(t_{k-1}))(A(t_k))^{n-i}$$

$$-\sum_{j=1}^{n-i}\binom{n-i}{j}A^{(j)}(t_{k-1})(A(t_k))^{n-i-j}R_{q-n+[\frac{n}{2}]-1,p-[\frac{n}{2}]}(\tau A(t_k))\Bigg]\Bigg]u_{k-1}$$

$$=\varphi_k^{p,q},1\leq k\leq N,u_0=y_0;$$

$$\tau^{-1}(u_k-u_{k-1})+\tau^{-1}\Bigg[I-R_{q,p}(\tau A(t_{k-\frac{1}{2}})) \tag{2.50}$$

$$+\tau(A(t_k)-A(t_{k-\frac{1}{2}}))R_{q-1,p-1}(\tau A(t_{k-\frac{1}{2}}))$$

$$-\sum_{n=1}^{p+q-1}\frac{(-\tau)^{n+1}}{(n+1)!}\sum_{i=0}^{n}\binom{n}{i}B_i(t_k)\Bigg[(A(t_k)-A(t_{k-\frac{1}{2}}))(-A(t_{k-\frac{1}{2}}))^{n-i}$$

$$+\sum_{j=1}^{n-i}\binom{n-i}{j}A^{(j)}(t_k)(-A(t_{k-\frac{1}{2}}))^{n-i-j}R_{q-n+[\frac{n}{2}]-1,p-[\frac{n}{2}]}(\tau A(t_{k-\frac{1}{2}}))\Bigg]\Bigg]u_{k-1}$$

$$=\varphi_k^{p,q},1\leq k\leq N,u_0=y_0.$$

Here, we can choose of $\varphi_k^{p,q}$ which can be defined by one of the following formulas:

$$\varphi_k^{p,q}=\sum_{m=0}^{p+q-1}\sum_{\lambda=0}^{m}\binom{m}{\lambda}B_{m-\lambda}(t_k)f^{(\lambda)}(t_k)\frac{(-\tau)^m}{(m+1)!},$$

or

$$\varphi_k^{p,q}=\sum_{m=0}^{p+q-1}\sum_{\lambda=0}^{m}\binom{m}{\lambda}B_{m-\lambda}(t_{k-1})\frac{\tau^m}{(m+1)!}\Bigg[R_{q-m-[\frac{m}{2}],p+[\frac{m}{2}]}(A(t_{k-1})\tau)$$

$$-\tau(A(t_k)-A(t_{k-1}))R_{q-m-[\frac{m}{2}]-1,p+[\frac{m}{2}]-1}(A(t_{k-1})\tau)$$

$$+\sum_{n=1}^{m-1}\frac{(-\tau)^{n+1}}{(n+1)}\sum_{i=0}^{n}\binom{n}{i}B_i(t_k)\Bigg[(A(t_k)-A(t_{k-1}))(-A(t_{k-1}))^{n-i}$$

$$+\sum_{j=1}^{n-i}\binom{n-i}{j}A^{(j)}(t_k)(-A(t_{k-1}))^{n-i-j}\Bigg]R_{q-n+[\frac{n}{2}]-1,p-[\frac{n}{2}]}(A(t_{k-1})\tau)\Bigg]f^{(\lambda)}(t_{k-1}),$$

or

$$\varphi_k^{p,q}=\sum_{m=0}^{p+q-1}\sum_{\lambda=0}^{m}\binom{m}{\lambda}B_{m-\lambda}(t_{k-1})\frac{\tau^m}{(m+1)!}\Bigg[R_{q-m-[\frac{m}{2}],p+[\frac{m}{2}]}(\tau A(t_k))$$

$$+\tau(A(t_k)-A(t_{k-1}))R_{q-m-[\frac{m}{2}]-1,p+[\frac{m}{2}]-1}(A(t_k)\tau)$$

$$+\sum_{n=1}^{m-1}\frac{\tau^{n+1}}{(n+1)!}\sum_{i=0}^{n}\binom{n}{i}\widetilde{B}_i(t_{k-1})\Bigg[(A(t_k)-A(t_{k-1}))(A(t_k))^{n-i}$$

$$-\sum_{j=1}^{n-i}\binom{n-i}{j}A^{(j)}(t_{k-1})(A(t_k))^{n-i-j}R_{q-n+[\frac{n}{2}]-1,p-[\frac{n}{2}]}(\tau A(t_k))\Bigg]f^{(\lambda)}(t_{k-1}),$$

or

$$\varphi_k^{p,q}=\sum_{m=0}^{p+q-1}\sum_{\lambda=0}^{m}\binom{m}{\lambda}B_{m-\lambda}(t_{k-1})\frac{\tau^m}{(m+1)!}\Bigg[R_{q-m-[\frac{m}{2}],p+[\frac{m}{2}]}(\tau A(t_{k-\frac{1}{2}}))$$

$$-\tau(A(t_k)-A(t_{k-\frac{1}{2}}))R_{q-m-[\frac{m}{2}]-1,p+[\frac{m}{2}]-1}(\tau A(t_{k-\frac{1}{2}}))$$

$$+\sum_{n=1}^{m-1}\frac{(-\tau)^{n+1}}{(n+1)!}\sum_{i=0}^{n}\binom{n}{i}B_i(t_k)\Bigg[(A(t_k)-A(t_{k-\frac{1}{2}}))(-A(t_{k-\frac{1}{2}}))^{n-i}$$

$$+\sum_{j=1}^{n-i}\binom{n-i}{j}A^{(j)}(t_k)(-A(t_{k-\frac{1}{2}}))^{n-i-j}R_{q-n+[\frac{n}{2}]-1,p-[\frac{n}{2}]}(\tau A(t_{k-\frac{1}{2}}))\Bigg]f^{(\lambda)}(t_{k-1}),$$

or

$$\varphi_k^{p,q}=\sum_{m=0}^{p-1}\sum_{\lambda=0}^{2m}\binom{2m}{\lambda}B_{2m-\lambda}(t_{k-\frac{1}{2}})\frac{\tau^{2m}}{2^{2m}(2m+1)!}\Bigg[R_{q-m,p-m}(\frac{\tau}{2}A(t_k))$$

$$+\frac{1}{2}\tau(A(t_k)-A(t_{k-\frac{1}{2}}))R_{q-m-1,p-m-1}(\frac{\tau}{2}A(t_k))$$

$$+\sum_{n=1}^{m-1}\frac{\tau^{n+1}}{2^{n+1}(n+1)!}\sum_{i=0}^{n}\binom{n}{i}\widetilde{B}_i(t_{k-\frac{1}{2}})\Bigg[(A(t_k)-A(t_{k-\frac{1}{2}}))(A(t_k))^{n-i}$$

$$-\sum_{j=1}^{n-i}\binom{n-i}{j}A^{(j)}(t_{k-\frac{1}{2}})(A(t_k))^{n-i-j}R_{q-n+[\frac{n}{2}]-1,p-[\frac{n}{2}]}(\frac{\tau}{2}A(t_k))\Bigg]f^{(\lambda)}(t_{k-\frac{1}{2}}),$$

or

$$\varphi_k^{p,q}=\sum_{m=0}^{p-1}\sum_{\lambda=0}^{2m}\binom{2m}{\lambda}B_{2m-\lambda}(t_{k-\frac{1}{2}})\frac{\tau^{2m}}{2^{2m}(2m+1)!}\Bigg[R_{q-m,p-m}(\frac{\tau}{2}A(t_{k-\frac{1}{2}}))$$

$$-\frac{1}{2}\tau(A(t_k)-A(t_{k-\frac{1}{2}}))R_{q-m-1,p-m-1}(\frac{\tau}{2}A(t_{k-\frac{1}{2}}))$$

$$+\sum_{n=1}^{m-1}\frac{(-\tau)^{n+1}}{2^{n+1}(n+1)!}\sum_{i=0}^{n}\binom{n}{i}B_i(t_k)\Bigg[(A(t_k)-A(t_{k-\frac{1}{2}}))(-A(t_{k-\frac{1}{2}}))^{n-i}$$

$$+\sum_{j=1}^{n-i}\binom{n-i}{j}A^{(j)}(t_k)(A(t_{k-\frac{1}{2}}))^{n-i-j}R_{q-n+[\frac{n}{2}]-1,p-[\frac{n}{2}]}(\frac{\tau}{2}A(t_{k-\frac{1}{2}}))\Bigg]f^{(\lambda)}(t_{k-\frac{1}{2}}).$$

Note that using the exact difference scheme, we can extend our discussion to construct the difference schemes of an arbitrary high order of accuracy for approximate solutions of the initial-value problem for the first-order nonlinear system of ordinary differential equations

$$y'(t) + A(t)y(t) = f(t, y(t)), 0 < t \leq T, y(0) = y_0.$$

It is left to the reader.

2.2 Taylor's Decomposition on Two Points and Its Applications

We consider again the initial-value problem (2.1). Utilization of Taylor's decomposition on two points in the construction of the single-step difference schemes of a high order of accuracy for approximate solutions of problem (2.1) is based on the following theorem.

Theorem 2.2.1. *Let the function $v(t)(0 \leq t \leq T)$ have a $(p+q+1)$-th continuous derivative and $t_{k-1}, t_k \in [0,T]_\tau$. Then the following relation holds:*

$$v(t_k) - v(t_{k-1}) + \sum_{j=1}^{p} \alpha_j v^{(j)}(t_k)\tau^j - \sum_{j=1}^{q} \beta_j v^{(j)}(t_{k-1})\tau^j \tag{2.51}$$

$$= \frac{(-1)^p}{(p+q)!} \int_{t_{k-1}}^{t_k} (t_k - s)^q (s - t_{k-1})^p v^{(p+q+1)}(s)ds,$$

where

$$\begin{cases} \alpha_j = \frac{(p+q-j)!p!(-1)^j}{(p+q)!j!(p-j)!} & \text{for any } j, 1 \leq j \leq p, \\ \beta_j = \frac{(p+q-j)!q!}{(p+q)!j!(q-j)!} & \text{for any } j, 1 \leq j \leq q. \end{cases} \tag{2.52}$$

Proof[1]. Using the formula for integration by parts, we obtain the representation

$$\int_{t_{k-1}}^{t_k} (t_k - s)^q (s - t_{k-1})^p v^{(p+q+1)}(s)ds$$

$$= \sum_{\gamma=0}^{p+q} (-1)^{p+q-\gamma}((t_k - s)^q(s - t_{k-1})^p)^{(p+q-\gamma)} v^{(\gamma)}(s)]_{t_{k-1}}^{t_k}.$$

We will calculate the expressions

$$(-1)^{p+q-j} \left[(t_k - s)^q (s - t_{k-1})^p\right]^{(p+q-j)} |_{s=t_k},$$

and

$$(-1)^{p+q-j} \left[(t_k - s)^q (s - t_{k-1})^p\right]^{(p+q-j)} |_{s=t_{k-1}}.$$

[1] Prof. V.L. Makarov (Ukraine) told me about this proof.

If $j \geq p+1$, then $p+q-j \leq q-1$. Therefore

$$(-1)^{p+q-j}\left[(t_k-s)^q(s-t_{k-1})^p\right]^{(p+q-j)}|_{s=t_k}=0.$$

If $0 \leq j \leq p$, then

$$(-1)^{p+q-j}\left[(t_k-s)^q(s-t_{k-1})^p\right]^{(p+q-j)}|_{s=t_k}$$

$$=(-1)^{p+q-j}\sum_{i=0}^{p+q-j}\frac{(p+q-j)!}{i!(p+q-j-i)!}((t_k-s)^q)^{(p+q-j-i)}((s-t_{k-1})^p)^{(i)}|_{s=t_k}$$

$$\begin{aligned}
&= (-1)^{p+q-j}\sum_{i=0}^{p+q-j}\frac{(p+q-j)!}{i!(p+q-j-i)!}\frac{q!}{(j+i-p)!} \\
&\quad \times(-1)^{p+q-j-i}(t_k-s)^{j+i-p}\frac{p!}{(p-i)!}((s-t_{k-1})^{p-i}|_{s=t_k}
\end{aligned}$$

$$\begin{aligned}
&= (-1)^{p+q-j}\frac{(p+q-j)!}{(p-j)!q!}q!(-1)^q\frac{p!}{j!}\tau^j \\
&= \frac{(-1)^{p-j}}{(p+q)!}\frac{(p+q)!(p+q-j)!}{(p-j)!j!}p!\tau^j=(-1)^p(p+q)!\alpha_j.
\end{aligned}$$

So,

$$(-1)^{p+q-j}\left[(t_k-s)^q(s-t_{k-1})^p\right]^{(p+q-j)}|_{s=t_k}=(-1)^p(p+q)!\alpha_j.$$

If $j \geq q+1$, then $p+q-j \leq p-1$. Therefore

$$(-1)^{p+q-j}\left[(t_k-s)^q(s-t_{k-1})^p\right]^{(p+q-j)}|_{s=t_{k-1}}=0.$$

If $0 \leq j \leq q$, then

$$(-1)^{p+q-j}\left[(t_k-s)^q(s-t_{k-1})^p\right]^{(p+q-j)}|_{s=t_{k-1}}$$

$$=(-1)^{p+q-j}\sum_{i=0}^{p+q-j}\frac{(p+q-j)!}{i!(p+q-j-i)!}((t_k-s)^q)^{(p+q-j-i)}((s-t_{k-1})^p)^{(i)}|_{s=t_{k-1}}$$

$$\begin{aligned}
&= (-1)^{p+q-j}\sum_{i=0}^{p+q-j}\frac{(p+q-j)!}{i!(p+q-j-i)!}\frac{q!}{(j+i-p)!} \\
&\quad \times(-1)^{p+q-j-i}(t_k-s)^{j+i-p}\frac{p!}{(p-i)!}((s-t_{k-1})^{p-i}|_{s=t_{k-1}}
\end{aligned}$$

$$= (-1)^{p+q-j}\frac{(p+q-j)!}{p!(q-j)!j!}q!(-1)^{q-j}\frac{p!}{0!}\tau^j$$
$$= \frac{(-1)^{p-j}}{(p+q)!}\frac{(p+q)!(p+q-j)!q!}{(q-j)!j!}\tau^j = (-1)^p(p+q)!\beta_j.$$

So,

$$(-1)^{p+q-j}\left[(t_k-s)^q(s-t_{k-1})^p\right]^{(p+q-j)}|_{s=t_{k-1}} = (-1)^p(p+q)!\beta_j.$$

Theorem 2.2.1 is proved.

Note that relation (2.51) is called Taylor's decomposition of function $v(t)$ on two points.

Now, we will consider applications of Taylor's decomposition of function on two points. From (2.51) it is clear that for the approximate solution of problem (2.1) it is necessary to find $y_{(t_k)}^{(j)}$ for any j, $1 \le j \le p$ and $y_{(t_{k-1})}^{(j)}$ for any j, $1 \le j \le q$. Using the equation

$$y'(t) = -a(t)y(t) + f(t),$$

we obtain

$$y^{(j)}(t) = a_j(t)y(t) + f_j(t), \tag{2.53}$$

where

$$\begin{cases} a_1(t) = -a(t), f_1(t) = f(t), \\ a_j(t) = a_{j-1}^{(\prime)}(t) - a_{j-1}(t)a(t), f_j(t) = f_{j-1}^{(\prime)}(t) + a_{j-1}(t)f(t), \\ \text{for any } j, 2 \le j \le p. \end{cases} \tag{2.54}$$

Replacing $y^{(j)}(t)$ by (2.53) and neglecting the last term, we obtain the single-step difference schemes of $(p+q)$-order of accuracy for the approximate solution of problem (2.1)

$$\frac{u_k - u_{k-1}}{\tau} + \sum_{j=1}^{p}\alpha_j a_j(t_k)\tau^{j-1}u_k - \sum_{j=1}^{q}\beta_j a_j(t_{k-1})\tau^{j-1}u_{k-1} = \varphi_k^{p,q}, \tag{2.55}$$

$$\varphi_k^{p,q} = -\sum_{j=1}^{p}\alpha_j f_j(t_k)\tau^{j-1} + \sum_{j=1}^{q}\beta_j f_j(t_{k-1})\tau^{j-1}, 1 \le k \le N, u_0 = y_0.$$

Now let us give some examples for the constructed difference schemes.

In the case $p+q=1$ from formulas (2.52), (2.54) and (2.55) it follows that

$$\frac{u_k - u_{k-1}}{\tau} + a(t_{k-1})u_{k-1} = f(t_{k-1}), 1 \le k \le N, u_0 = y_0$$

(an explicit Euler's difference scheme of first order of accuracy for the initial-value problem (2.1)), and that

$$\frac{u_k - u_{k-1}}{\tau} + a(t_k)u_k = f(t_k), 1 \le k \le N, u_0 = y_0$$

(an implicit Euler's difference scheme of first order of accuracy for the initial-value problem (2.1)).

In the case $p+q=2$ from formulas (2.52), (2.54) and (2.55) it follows that

$$\begin{aligned}&\frac{u_k-u_{k-1}}{\tau}+(a(t_{k-1})+\frac{\tau}{2}(a^{(\prime)}(t_{k-1})-a^2(t_{k-1})))u_{k-1}\\ =\ & f(t_{k-1})+\frac{\tau}{2}(f^{(\prime)}(t_{k-1})-a(t_{k-1})f(t_{k-1})), 1\le k\le N, u_0=y_0\end{aligned}$$

(an explicit difference scheme of second order of accuracy for the initial-value problem (2.1)); and that

$$\frac{u_k-u_{k-1}}{\tau}+\frac{\tau}{2}(a(t_k)u_k+a(t_{k-1})u_{k-1})=\frac{1}{2}(f(t_k)+f(t_{k-1})), 1\le k\le N, u_0=y_0$$

(a Crank–Nicolson difference scheme of second order accuracy); and that

$$\begin{aligned}&\frac{u_k-u_{k-1}}{\tau}+(a(t_k)+\frac{\tau}{2}(-a^{(\prime)}(t_k)+a^2(t_k)))u_k\\ =\ & f(t_k)-\frac{\tau}{2}(f^{(\prime)}(t_k)-a(t_k)f(t_k)), 1\le k\le N, u_0=y_0\end{aligned}$$

(an implicit difference scheme of second order of accuracy for the initial-value problem (2.1)).

Now, we consider again the initial-value problem (2.22). Using the equation

$$y'(t)=-a(t)y(t)+f(t,y(t)),$$

we obtain

$$y^{(j)}(t)=a_j(t)y(t)+f_j(t,y(t)), \tag{2.56}$$

where (see formula (2.26))

$$\begin{cases} f_\lambda(t,y(t))=f(t,y(t)), & \lambda=1,\\ f_{\lambda+1}(t,y(t))=\frac{\partial^\lambda}{\partial t^\lambda}f(t,y(t)) & \\ +\sum_{i=1}^{\lambda}\begin{pmatrix}\lambda\\ i\end{pmatrix}\frac{\partial^\lambda}{\partial t^{\lambda-i}\partial^i y}f(t,y(t))f_i(t,y(t)), & 0\le\lambda\le p-1.\end{cases}$$

Replacing $y^{(j)}(t)$ by (2.56) and neglecting the last term, we obtain the single-step difference schemes of $(p+q)$-order of accuracy for an approximate solution of problem (2.22)

$$\begin{aligned}&\frac{u_k-u_{k-1}}{\tau}+\sum_{j=1}^{p}\alpha_j a_j(t_k)\tau^{j-1}u_k-\sum_{j=1}^{q}\beta_j a_j(t_{k-1})\tau^{j-1}u_{k-1} \\ &=\varphi_k^{p,q}(u_k,u_{k-1}), \varphi_k^{p,q}(u_k,u_{k-1})=-\sum_{j=1}^{p}\alpha_j f_j(t_k,u_k)\tau^{j-1}\\ &+\sum_{j=1}^{q}\beta_j f_j(t_{k-1},u_{k-1})\tau^{j-1}, 1\le k\le N, u_0=y_0.\end{aligned} \tag{2.57}$$

Now, we consider again the initial-value problem (2.30). Using the equation

$$y'(t) = -A(t)y(t) + f(t),$$

we obtain

$$y^{(j)}(t) = A_j(t)y(t) + F_j(t),$$

where

$$\begin{cases} A_1(t) = -A(t), F_1(t) = f(t), \\ A_j(t) = A_{j-1}^{(1)}(t) - A_{j-1}(t)A(t), F_j(t) = F_{j-1}^{(\prime)}(t) + A_{j-1}(t)f(t), \\ \text{for any } j, 2 \leq j \leq p. \end{cases} \tag{2.58}$$

Replacing $y^{(j)}(t)$ by (2.58) and neglecting the last term, we obtain the single-step difference schemes of $(p+q)$-order of accuracy for an approximate solution of problem (2.30)

$$\frac{u_k - u_{k-1}}{\tau} + \sum_{j=1}^{p} \alpha_j A_j(t_k)\tau^{j-1}u_k - \sum_{j=1}^{q} \beta_j A_j(t_{k-1})\tau^{j-1}u_{k-1} = \varphi_k^{p,q}, \tag{2.59}$$

$$\varphi_k^{p,q} = -\sum_{j=1}^{p} \alpha_j F_j(t_k)\tau^{j-1} + \sum_{j=1}^{q} \beta_j F_j(t_{k-1})\tau^{j-1}, 1 \leq k \leq N, u_0 = y_0.$$

Note that using Taylor's decomposition on two points, we can extend our discussion to construct the difference schemes of an arbitrary high order of accuracy for approximate solutions of the initial-value problem for the first order nonlinear system of ordinary differential equations

$$y'(t) + A(t)y(t) = f(t, y(t)), 0 < t \leq T, y(0) = y_0.$$

It is left to the reader.

Finally, in this chapter we consider two types of arbitrary high order of accuracy difference schemes generated by an exact difference scheme or Taylor's decomposition on two points for the numerical solutions of an initial-value problem for differential equations and systems of differential equations. They are different types of difference schemes for the same problem. Nevertheless, in the case of a homogeneous equation or system of differential equations with independent in t coefficients using the two types of methods above, we have exactly the same difference schemes.

Chapter 3

Difference Schemes for Second-Order Differential Equations

In the present chapter we consider the two-step difference schemes generated by an exact difference scheme or by Taylor's decomposition on three points for the numerical solutions of an initial-value problem and a boundary-value problem for second-order differential equations.

3.1 Two-Step Exact Difference Scheme and Its Applications

We consider the initial-value problem of the form

$$y''(t) + a(t)y(t) = f(t), 0 < t \leq T, y(0) = y_0, y'(0) = y_0' \tag{3.1}$$

assuming $a(t)$ and $f(t)$ to be such that problem (3.1) has a unique smooth solution defined on $[0, T]$. For the construction of difference schemes we consider again the uniform grid space

$$[0, T]_\tau = \{t_k = k\tau,\ k = 0, 1, \ldots, N,\ \ N\tau = T\}.$$

The construction of two-step difference schemes of an arbitrary high order of accuracy for approximate solutions of the initial-value problem (3.1) is based on the following theorem.

Theorem 3.1.1. *Let $y(t_k)$ be a solution of problem (3.1) at the grid points $t = t_k$. Then $\{y(t_k)\}_0^N$ is the solution of the initial-value problem for the following second-order difference equations:*

$$\frac{1}{\tau^2}(y(t_{k+1}) - 2y(t_k) + y(t_{k-1}))$$
$$+\frac{1}{\tau^2}\Bigg[(\frac{1}{\tau}\int_{t_k}^{t_{k+1}} I^2(s, t_k)dsI^{-1}(t_k, t_{k-1}) - 1)y(t_{k-1})$$

$$+(2-\frac{1}{\tau}\int_{t_{k-1}}^{t_k} I^{-2}(t_k,s)ds-\frac{1}{\tau}\int_{t_k}^{t_{k+1}} I^2(s,t_k)ds)y(t_k)$$

$$+\left.(\frac{1}{\tau}\int_{t_{k-1}}^{t_k} I^{-2}(t_k,s)dsI(t_{k+1},t_k)-1)y(t_{k+1})\right]$$

$$=\frac{1}{\tau^3}\left[\int_{t_{k-1}}^{t_k} I^{-2}(t_k,s)ds\int_{t_k}^{t_{k+1}} I^{-1}(t_k,z)\int_z^{t_{k+1}} I^2(s,z)dsf(z)dz\right. \tag{3.2}$$

$$\left.+\int_{t_k}^{t_{k+1}} I^2(s,t_k)ds\int_{t_{k-1}}^{t_k} I^{-1}(t_k,z)\int_{t_{k-1}}^{z} I^2(s,z)dsf(z)dz\right],$$

$$1 \leq k\leq N-1, y(0)=y_0,$$

$$\frac{1}{\tau}(y(\tau)-y(0))=[I^{-1}(\tau,0)ib(0)\frac{1}{\tau}\int_0^\tau I^2(s,0)ds+\frac{1}{\tau}(I^{-1}(\tau,0)-1)]y(0)$$

$$+I^{-1}(\tau,0)\frac{1}{\tau}\int_0^\tau I^2(s,0)dsy'(0)+\frac{1}{\tau}\int_0^\tau I^{-1}(\tau,z)\int_z^\tau I^2(s,z)dsf(z)dz.$$

Here

$$I(t,s)=e^{i\int_s^t b(\lambda)d\lambda}, I^{-1}(t,s)=\frac{1}{I(t,s)}=I(s,t),$$

where

$$\begin{cases} b(t)=b_r(t)+ib_i(t), \\ b_r^2(t)-b_i^2(t)-b_i'(t)=a(t), \\ 2b_r(t)b_i(t)+b_r'(t)=0. \end{cases} \tag{3.3}$$

Proof. Using the relations (3.3) we can obviously write the equivalent initial-value problem for a system of first-order linear differential equations

$$\begin{cases} y'(t)+ib(t)y(t)=v(t), y(0)=y_0, y'(0)=y_0', \\ v'(t)-ib(t)v(t)=f(t). \end{cases} \tag{3.4}$$

Integrating these, we at once obtain

$$\begin{cases} y(t)=e^{-i\int_0^t b(\lambda)d\lambda}y_0+\int_0^t e^{-i\int_s^t b(\lambda)d\lambda}v(s)ds, \\ v(t)=e^{i\int_0^t b(\lambda)d\lambda}v_0+\int_0^t e^{i\int_z^t b(\lambda)d\lambda}f(z)dz. \end{cases}$$

That is,

$$\begin{aligned} y(t) &= I^{-1}(t,0)y(0)+\int_0^t I^{-1}(t,s)I(s,0)dsv(0) \\ &\quad +\int_0^t I^{-1}(t,s)\int_0^s I(s,z)f(z)dzds, \end{aligned}$$

so, using the condition $y^{'}(0)+ib(0)y(0)=v(0)$, we obtain

$$\begin{aligned} y(t) &= \left[I^{-1}(t,0)+ib(0)\int_0^t I^{-1}(t,s)I(s,0)ds\right]y_0 \\ &\quad+\int_0^t I^{-1}(t,s)I(s,0)dsy_0'+\int_0^t I^{-1}(t,s)\int_0^s I(s,z)f(z)dzds. \end{aligned}$$

By the definition of $I(t,s)$ and interchanging the order of integration, we obtain

$$y(t)=I^{-1}(t,0)\left[1+ib(0)\int_0^t I^2(s,0)ds\right]y_0+I^{-1}(t,0)\int_0^t I^2(s,0)dsy_0' \tag{3.5}$$

$$+\int_0^t I^{-1}(t,z)\int_z^t I^2(s,z)dsf(z)dz=u(t)+v(t)+g(t).$$

Putting $t=t_{k+1}$, $t=t_k$ and $t=t_{k-1}$ into the formula for $u(t)$, we can write

$$u(t_k)=I^{-1}(t_k,0)\left[1+ib(0)\int_0^{t_k} I^2(s,0)ds\right]y_0,$$

$$u(t_{k+1})=I^{-1}(t_{k+1},0)\left[1+ib(0)\int_0^{t_{k+1}} I^2(s,0)ds\right]y_0$$

$$=I^{-1}(t_{k+1},t_k)I^{-1}(t_k,0)\left[1+ib(0)\left[\int_0^{t_k} I^2(s,0)ds+\int_{t_k}^{t_{k+1}} I^2(s,0)ds\right]\right]y_0$$

$$=I^{-1}(t_{k+1},t_k)\left[u(t_k)+I^{-1}(t_k,0)ib(0)\int_{t_k}^{t_{k+1}} I^2(s,0)dsy_0\right],$$

$$u(t_{k-1})=I^{-1}(t_{k-1},0)\left[1+ib(0)\int_0^{t_{k-1}} I^2(s,0)ds\right]y_0$$

$$=I(t_k,t_{k-1})I^{-1}(t_k,0)\left[1+ib(0)\left[\int_0^{t_k} I^2(s,0)ds-\int_{t_{k-1}}^{t_k} I^2(s,0)ds\right]\right]y_0$$

$$=I(t_k,t_{k-1})\left[u(t_k)-I^{-1}(t_k,0)ib(0)\int_{t_{k-1}}^{t_k} I^2(s,0)dsy_0\right].$$

Hence, we obtain the following relation between $u(t_{k+1}), u(t_k)$ and $u(t_{k-1})$:

$$\left(\int_{t_k}^{t_{k+1}} I^2(s,0)ds\right)^{-1}(I(t_{k+1},t_k)u(t_{k+1})-u(t_k))=I^{-1}(t_k,0)ib(0)y_0,$$

$$\left(\int_{t_{k-1}}^{t_k} I^2(s,0)ds\right)^{-1}(u(t_k)-I^{-1}(t_k,t_{k-1})u(t_{k-1}))=I^{-1}(t_k,0)ib(0)y_0.$$

Therefore

$$\int_{t_{k-1}}^{t_k} I^2(s,0)ds(I(t_{k+1},t_k)u(t_{k+1}) - u(t_k))$$

$$= \int_{t_k}^{t_{k+1}} I^2(s,0)ds(u(t_k) - I^{-1}(t_k,t_{k-1})u(t_{k-1})). \tag{3.6}$$

In a similar manner, using the definition of $I(t,s)$ and the formula for $v(t)$, we get

$$\int_{t_{k-1}}^{t_k} I^2(s,0)ds(I(t_{k+1},t_k)v(t_{k+1}) - v(t_k)) \tag{3.7}$$

$$= \int_{t_k}^{t_{k+1}} I^2(s,0)ds(v(t_k) - I^{-1}(t_k,t_{k-1})v(t_{k-1})).$$

Now, putting $t = t_{k+1}$, $t = t_k$ and $t = t_{k-1}$ into the formula for $g(t)$ and using the definition of $I(t,s)$, we get

$$\int_{t_{k-1}}^{t_k} I^2(s,0)ds(I(t_{k+1},t_k)g(t_{k+1}) - g(t_k))$$

$$- \int_{t_k}^{t_{k+1}} I^2(s,0)ds(g(t_k) - I^{-1}(t_k,t_{k-1})g(t_{k-1}))$$

$$= \int_{t_{k-1}}^{t_k} I^2(s,0)ds\Bigg(I(t_{k+1},t_k)\int_0^{t_{k+1}} I^{-1}(t_{k+1},z)\int_z^{t_{k+1}} I^2(s,z)dsf(z)dz$$

$$- \int_0^{t_k} I^{-1}(t_k,z)\int_z^{t_k} I^2(s,z)dsf(z)dz\Bigg)$$

$$- \int_{t_k}^{t_{k+1}} I^2(s,0)ds\Bigg(\int_0^{t_k} I^{-1}(t_k,z)\int_z^{t_k} I^2(s,z)dsf(z)dz$$

$$-I^{-1}(t_k,t_{k-1})\int_0^{t_{k-1}} I^{-1}(t_{k-1},z)\int_z^{t_{k-1}} I^2(s,z)dsf(z)dz\Bigg)$$

$$= \int_{t_{k-1}}^{t_k} I^2(s,0)ds\Bigg[\int_0^{t_k} I^{-1}(t_k,z)\int_z^{t_{k+1}} I^2(s,z)dsf(z)dz$$

$$+ \int_{t_k}^{t_{k+1}} I^{-1}(t_k,z)\int_z^{t_{k+1}} I^2(s,z)dsf(z)dz$$

$$- \int_0^{t_k} I^{-1}(t_k,z)\int_z^{t_k} I^2(s,z)dsf(z)dz\Bigg]$$

$$- \int_{t_k}^{t_{k+1}} I^2(s,0)ds \Bigg[\int_0^{t_k} I^{-1}(t_k,z) \int_z^{t_k} I^2(s,z)dsf(z)dz$$

$$- \int_0^{t_k} I^{-1}(t_k,z) \int_z^{t_{k-1}} I^2(s,z)dsf(z)dz$$

$$+ \int_{t_{k-1}}^{t_k} I^{-1}(t_k,z) \int_z^{t_{k-1}} I^2(s,z)dsf(z)dz \Bigg]$$

$$= \int_{t_{k-1}}^{t_k} I^2(s,0)ds \Bigg[\int_0^{t_k} I^{-1}(t_k,z) \int_{t_k}^{t_{k+1}} I^2(s,z)dsf(z)dz$$

$$+ \int_{t_k}^{t_{k+1}} I^{-1}(t_k,z) \int_z^{t_{k+1}} I^2(s,z)dsf(z)dz \Bigg]$$

$$- \int_{t_k}^{t_{k+1}} I^2(s,0)ds \Bigg[\int_0^{t_k} I^{-1}(t_k,z) \int_{t_{k-1}}^{t_k} I^2(s,z)dsf(z)dz$$

$$+ \int_0^{t_k} I^{-1}(t_k,z) \int_{t_{k-1}}^{z} I^2(s,z)dsf(z)dz \Bigg].$$

Now, using the identity

$$\int_{t_{k-1}}^{t_k} I^2(s,0)ds \int_{t_k}^{t_{k+1}} I^2(s,z)ds = \int_{t_k}^{t_{k+1}} I^2(s,0)ds \int_{t_{k-1}}^{t_k} I^2(s,z)ds,$$

we obtain

$$\int_{t_{k-1}}^{t_k} I^2(s,0)ds(I(t_{k+1},t_k)g(t_{k+1}) - g(t_k))$$

$$- \int_{t_k}^{t_{k+1}} I^2(s,0)ds(g(t_k) - I^{-1}(t_k,t_{k-1})g(t_{k-1}))$$

$$= \int_{t_{k-1}}^{t_k} I^2(s,0)ds \int_{t_k}^{t_{k+1}} I^{-1}(t_k,z) \int_z^{t_{k+1}} I^2(s,z)dsf(z)dz$$

$$+ \int_{t_k}^{t_{k+1}} I^2(s,0)ds \int_{t_{k-1}}^{t_k} I^{-1}(t_k,z) \int_{t_{k-1}}^{z} I^2(s,z)dsf(z)dz.$$

So, using the last formula and the identities (3.6) and (3.7) we obtain the following relation between $y(t_{k+1}), y(t_k)$ and $y(t_{k-1})$:

$$\int_{t_{k-1}}^{t_k} I^2(s,0)ds(I(t_{k+1},t_k)y(t_{k+1}) - y(t_k))$$

$$- \int_{t_k}^{t_{k+1}} I^2(s,0)ds(y(t_k) - I^{-1}(t_k,t_{k-1})y(t_{k-1}))$$

$$= \int_{t_{k-1}}^{t_k} I^2(s,0)ds \int_{t_k}^{t_{k+1}} I^{-1}(t_k,z) \int_z^{t_{k+1}} I^2(s,z)dsf(z)dz$$

$$+ \int_{t_k}^{t_{k+1}} I^2(s,0)ds \int_{t_{k-1}}^{t_k} I^{-1}(t_k,z) \int_{t_{k-1}}^{z} I^2(s,z)dsf(z)dz.$$

Since $I^2(t_k,0) \neq 0$ it follows that

$$\int_{t_{k-1}}^{t_k} I^{-2}(t_k,s)ds(I(t_{k+1},t_k)y(t_{k+1}) - y(t_k))$$

$$- \int_{t_k}^{t_{k+1}} I^2(s,t_k)ds(y(t_k) - I^{-1}(t_k,t_{k-1})y(t_{k-1}))$$

$$= \int_{t_{k-1}}^{t_k} I^{-2}(t_k,s)ds \int_{t_k}^{t_{k+1}} I^{-1}(t_k,z) \int_z^{t_{k+1}} I^2(s,z)dsf(z)dz$$

$$+ \int_{t_k}^{t_{k+1}} I^2(s,t_k)ds \int_{t_{k-1}}^{t_k} I^{-1}(t_k,z) \int_{t_{k-1}}^{z} I^2(s,z)dsf(z)dz,$$

$$1 \leq k \leq N-1.$$

From this it follows (3.2) for any $k, 1 \leq k \leq N-1$. The identity (3.2) in the case $k=1$ follows from (3.5). Theorem 3.1.1 is proved.

Let us note that the initial-value problem (3.2) is called the two-step exact difference scheme for the solution of the initial-value problem (3.1).

Note that for $b(t)$ we have a system of first-order nonlinear differential equations of Riccati type (3.3). Therefore for the smooth $a(t)$ there exist smooth solutions of this system defined on the segment $[0,T]$. We will choose one of these smooth solutions $b(t) = b_r(t) + ib_i(t)$, $0 \leq t \leq T$.

Now, we will consider the applications of this exact difference scheme. From (3.2) it is clear that for the approximate solutions of problem (3.1) it is necessary to approximate the expressions

$$\frac{1}{\tau}\int_{t_k}^{t_{k+1}} I^2(s,t_k)ds, \frac{1}{\tau}\int_{t_{k-1}}^{t_k} I^{-2}(t_k,s)ds, I^{-1}(t_k,t_{k-1}), I(t_{k+1},t_k)$$

and

$$\int_{t_k}^{t_{k+1}} I^{-1}(t_k,z) \int_z^{t_{k+1}} I^2(s,z)dsf(z)dz, \int_{t_{k-1}}^{t_k} I^{-1}(t_k,z) \int_{t_{k-1}}^{z} I^2(s,z)dsf(z)dz.$$

Let us remark that in constructing difference schemes it is important to know how to construct a right-hand side $\varphi_k^{p,q}$ that satisfies

$$\frac{1}{\tau^3}[\int_{t_{k-1}}^{t_k} I^{-2}(t_k,s)ds \int_{t_k}^{t_{k+1}} I^{-1}(t_k,z) \int_z^{t_{k+1}} I^2(s,z)dsf(z)dz \tag{3.8}$$

$$+ \int_{t_k}^{t_{k+1}} I^2(s,t_{k-1})ds \int_{t_{k-1}}^{t_k} I^{-1}(t_k,z) \int_{t_{k-1}}^{z} I^2(s,z)dsf(z)dz] - \varphi_k^{p,q} = o(\tau^{p+q})$$

and is sufficiently simple. The choice formula $\varphi_k^{p,q}$ is not unique. Using Taylor's formula, with respect to the function

$$I^{-1}(t_k, z)\int_z^{t_{k+1}} I^2(s,z)dsf(z)$$

on the variable z, we obtain

$$\frac{1}{\tau^2}\int_{t_k}^{t_{k+1}} I^{-1}(t_k, z)\int_z^{t_{k+1}} I^2(s,z)dsf(z)dz$$

$$= \sum_{m=1}^{p+q}\sum_{\lambda=0}^{m}\binom{m}{\lambda}\beta_{m-\lambda}(t_{k+1})f^{(\lambda)}(t_{k+1})\frac{(-1)^m\tau^{m-1}}{(m+1)!} + o(\tau^{p+q}), \tag{3.9}$$

where

$$\begin{cases}\beta_0(t_{k+1}) = 0, \beta_1(t_{k+1}) = -I(t_{k+1}, t_k), \beta_2(t_{k+1}) = 0,\\ \beta_m(t_{k+1}) = -\sum\limits_{\lambda=1}^{m-2}\binom{m-2}{\lambda} a^{(m-2-\lambda)}(t_{k+1})\beta_\lambda(t_{k+1}), \quad 3\le m\le p+q.\end{cases} \tag{3.10}$$

Using Taylor's formula, with respect to the function

$$I^{-1}(t_k, z)\int_{t_{k-1}}^{z} I^2(s,z)dsf(z)$$

on the variable z, we obtain

$$\frac{1}{\tau^2}\int_{t_{k-1}}^{t_k} I^{-1}(t_k, z)\int_{t_{k-1}}^{z} I^2(s,z)dsf(z)dz \tag{3.11}$$

$$= \sum_{m=1}^{p+q}\sum_{\lambda=0}^{m}\binom{m}{\lambda}\widetilde{\beta}_{m-\lambda}(t_{k-1})f^{(\lambda)}(t_{k-1})\frac{\tau^{m-1}}{(m+1)!} + o(\tau^{p+q}),$$

where

$$\begin{cases}\widetilde{\beta}_0(t_{k-1}) = 0, \widetilde{\beta}_1(t_{k-1}) = I^{-1}(t_k, t_{k-1}), \widetilde{\beta}_2(t_{k-1}) = 0,\\ \widetilde{\beta}_m(t_{k-1}) = -\sum\limits_{\lambda=1}^{m-2}\binom{m-2}{\lambda} a^{(m-2-\lambda)}(t_{k-1})\widetilde{\beta}_\lambda(t_{k-1}), \quad 3\le m\le p+q.\end{cases} \tag{3.12}$$

Moreover, for the construction of $\varphi_k^{p,q}$ we will use (3.9) and (3.11). But, first of all let us study the approximate formulas for the expressions

$$I^{-1}(t_k, t_{k-1}), I(t_{k+1}, t_k), \frac{1}{\tau}\int_{t_k}^{t_{k+1}} I^2(s, t_k)ds, \frac{1}{\tau}\int_{t_{k-1}}^{t_k} I^{-2}(t_k, s)ds,$$

that will be needed for the construction of difference schemes of a high order of accuracy for the approximate solution of problem (3.1). First, let us consider the approximate formulas for the expression

$$I^{-1}(t_k, t_{k-1}) = e^{-i\int_{t_{k-1}}^{t_k} b(s)ds}$$

constructed by Pade's rational functions. It is clear that

$$e^{-i\int_{t_{k-1}}^{t_k} b(s)ds} = e^{-ib(t_k)\tau} + i\int_{t_{k-1}}^{t_k} e^{-i\int_{t_{k-1}}^{s} b(z)dz}(b(t_k) - b(s))e^{i(s-t_k)b(t_k)}ds.$$

Using Taylor's formula, with respect to the function

$$e^{-i\int_{t_{k-1}}^{s} b(z)dz}(b(t_k) - b(s))e^{i(s-t_k)b(t_k)}$$

on the variable s, we obtain

$$\begin{aligned}
e^{-i\int_{t_{k-1}}^{t_k} b(z)dz} &= e^{-ib(t_k)\tau} + i\tau(b(t_k) - b(t_{k-1}))e^{-ib(t_k)\tau} \\
&+\sum_{n=1}^{m} i\frac{\tau^{n+1}}{(n+1)!}(\sum_{\lambda=1}^{n}\binom{n}{\lambda}\widetilde{c}_{n-\lambda}(t_{k-1})\Big[(b(t_k) - b(t_{k-1}))(ib(t_k))^{\lambda}e^{-ib(t_k)\tau} \\
&+\sum_{\mu=0}^{\lambda-1}\binom{\lambda}{\mu} b^{(\lambda-\mu)}(t_{k-1})(ib(t_k))^{\mu}e^{-ib(t_k)\tau}\Big] \\
&+\widetilde{c}_n(t_{k-1})(b(t_k) - b(t_{k-1}))e^{-ib(t_k)\tau}) + o(\tau^{m+2}) \\
= \quad & e^{-ib(t_k)\tau} + i\tau(b(t_k) - b(t_{k-1}))e^{-ib(t_k)\tau} \\
&+\sum_{n=1}^{m-1} i\frac{\tau^{n+1}}{(n+1)!}\sum_{\lambda=0}^{n}\binom{n}{\lambda}\widetilde{c}_{n-\lambda}(t_{k-1})(b(t_k) - b(t_{k-1}))(ib(t_k))^{\lambda}e^{-ib(t_k)\tau} \\
&+\sum_{n=1}^{m} i\frac{\tau^{n+1}}{(n+1)!}\sum_{\lambda=1}^{n}\binom{n}{\lambda}\widetilde{c}_{n-\lambda}(t_{k-1}) \\
&\times\sum_{\mu=0}^{\lambda-1}\binom{\lambda}{\mu} b^{(\lambda-\mu)}(t_{k-1})(ib(t_k))^{\mu}e^{-ib(t_k)\tau} + o(\tau^{m+2}),
\end{aligned}$$

$$m \geq 2,$$

$$\begin{aligned}
e^{-i\int_{t_{k-1}}^{t_k} b(z)dz} &= e^{-ib(t_k)\tau} + i\tau(b(t_k) - b(t_{k-1}))e^{-ib(t_k)\tau} \\
&+\frac{i\tau^2}{2}b'(t_{k-1})e^{-ib(t_k)\tau} + o(\tau^3),
\end{aligned}$$

$$e^{-i\int_{t_{k-1}}^{t_k} b(z)dz} = e^{-ib(t_k)\tau} + o(\tau^2),$$

where

$$\begin{cases}\widetilde{c}_0(t_{k-1}) = 1, \widetilde{c}_1(t_{k-1}) = -ib(t_{k-1}), \widetilde{c}_2(t_{k-1}) = -a(t_{k-1}), \\ \widetilde{c}_{m+2}(t_{k-1}) = -\sum\limits_{\lambda=0}^{m} \begin{pmatrix} m \\ \lambda \end{pmatrix} a^{(m-\lambda)}(t_{k-1})\widetilde{c}_\lambda(t_{k-1}), 1 \le m.\end{cases}$$

Further, using the last formulas and Pade fractions for the function e^{-z} (see Chapter 2), we can write

$$\begin{aligned}I^{-1}(t_k, t_{k-1}) &= R_{[\frac{m}{2}],m-[\frac{m}{2}]+1}(i\tau b(t_k)) \qquad (3.13)\\ &+i\tau(b(t_k) - b(t_{k-1}))R_{[\frac{m}{2}]-1,m-[\frac{m}{2}]}(i\tau b(t_k))\\ &+\sum_{n=1}^{m-1} i\frac{\tau^{n+1}}{(n+1)!}\sum_{\lambda=0}^{n}\begin{pmatrix} n \\ \lambda \end{pmatrix}\widetilde{c}_{n-\lambda}(t_{k-1})(b(t_k) - b(t_{k-1}))(ib(t_k))^\lambda\\ &\times R_{[\frac{m-n}{2}]-1,m-n-[\frac{m-n}{2}]}(i\tau b(t_k)) + \sum_{n=1}^{m} i\frac{\tau^{n+1}}{(n+1)!}\sum_{\lambda=1}^{n}\begin{pmatrix} n \\ \lambda \end{pmatrix}\widetilde{c}_{n-\lambda}(t_{k-1})\\ &\times\sum_{\mu=0}^{\lambda-1}\begin{pmatrix} \lambda \\ \mu \end{pmatrix} b^{(\lambda-\mu)}(t_{k-1})(ib(t_k))^\mu R_{[\frac{m-n}{2}],m-n-[\frac{m-n}{2}]}(i\tau b(t_k)) + o(\tau^{m+2}),\end{aligned}$$

$$m \ge 2,$$

$$I^{-1}(t_k, t_{k-1}) = R_{\sigma,2-\sigma}(i\tau b(t_k)) + i\tau(b(t_k) - b(t_{k-1})) + \frac{i\tau^2}{2}b'(t_{k-1})$$
$$+o(\tau^3), \sigma = 0, 1, 2,$$
$$I^{-1}(t_k, t_{k-1}) = R_{\varkappa,1-\varkappa}(i\tau b(t_k)) + o(\tau^2), \varkappa = 0, 1.$$

Second, let us consider the approximate formulas for the expression

$$I(t_{k+1}, t_k) = e^{i\int_{t_k}^{t_{k+1}} b(s)ds}$$

constructed using Pade's rational functions. We have that

$$e^{i\int_{t_k}^{t_{k+1}} b(s)ds} = e^{ib(t_k)\tau} + i\int_{t_k}^{t_{k+1}} e^{i\int_s^{t_{k+1}} b(z)dz}(b(s) - b(t_k))e^{i(s-t_k)b(t_k)}ds.$$

Using Taylor's formula, we obtain

$$e^{i\int_{t_k}^{t_{k+1}} b(s)ds} = e^{ib(t_k)\tau} + i\tau(b(t_{k+1}) - b(t_k))e^{i\tau b(t_k)}$$

$$+\sum_{n=1}^{m} i\frac{(-1)^n\tau^{n+1}}{(n+1)!}(\sum_{\lambda=1}^{n}\begin{pmatrix} n \\ \lambda \end{pmatrix}\widetilde{c}_{n-\lambda}(t_{k+1})[(b(t_{k+1}) - b(t_k))(ib(t_k))^\lambda e^{ib(t_k)\tau}$$

$$+\sum_{\mu=0}^{\lambda-1}\binom{\lambda}{\mu} b^{(\lambda-\mu)}(t_{k+1})(ib(t_k))^{\mu}]e^{ib(t_k)\tau}+\widetilde{c}_n(t_{k+1})(b(t_{k+1})-b(t_k))e^{ib(t_k)\tau})$$

$$+o(\tau^{m+2})=e^{ib(t_k)\tau}+i\tau(b(t_{k+1})-b(t_k))e^{i\tau b(t_k)}$$

$$+\sum_{n=1}^{m-1} i\frac{(-1)^n\tau^{n+1}}{(n+1)!}\sum_{\lambda=0}^{n}\binom{n}{\lambda}\widetilde{c}_{n-\lambda}(t_{k+1})(b(t_{k+1})-b(t_k))(ib(t_k))^{\lambda}e^{ib(t_k)\tau}$$

$$+\sum_{n=1}^{m} i\frac{(-1)^n\tau^{n+1}}{(n+1)!}\sum_{\lambda=1}^{n}\binom{n}{\lambda}\widetilde{c}_{n-\lambda}(t_{k+1})$$

$$\times\sum_{\mu=0}^{\lambda-1}\binom{\lambda}{\mu} b^{(\lambda-\mu)}(t_{k+1})(ib(t_k))^{\mu}]e^{ib(t_k)\tau}+o(\tau^{m+2}), m\geq 2,$$

$$e^{i\int_{t_k}^{t_{k+1}}b(s)ds}=e^{ib(t_k)\tau}+i\tau(b(t_{k+1})-b(t_k))e^{i\tau b(t_k)}-\frac{i\tau^2}{2}b'(t_{k+1})e^{ib(t_k)\tau}+o(\tau^3),$$

$$e^{i\int_{t_k}^{t_{k+1}}b(s)ds}=e^{ib(t_k)\tau}+o(\tau^2),$$

where

$$\begin{cases}\widetilde{c}_0(t_{k+1})=1, \widetilde{c}_1(t_{k+1})=-ib(t_{k+1}), \widetilde{c}_2(t_{k+1})=-a(t_{k+1}),\\ \widetilde{c}_{m+2}(t_{k+1})=-\sum\limits_{\lambda=0}^{m}\binom{m}{\lambda} a^{(m-\lambda)}(t_{k+1})\widetilde{c}_{\lambda}(t_{k+1}), 1\leq m.\end{cases}$$

Further, using the last formulas and Pade fractions for the function e^{-z}, we can write

$$I(t_{k+1},t_k)=R_{[\frac{m}{2}],m-[\frac{m}{2}]+1}(-\tau ib(t_k)) \tag{3.14}$$

$$+i\tau(b(t_{k+1})-b(t_k))R_{[\frac{m}{2}]-1,m-[\frac{m}{2}]}(-i\tau b(t_k))$$

$$+\sum_{n=1}^{m-1} i\frac{(-1)^n\tau^{n+1}}{(n+1)!}\sum_{\lambda=0}^{n}\binom{n}{\lambda}\widetilde{c}_{n-\lambda}(t_{k+1})(b(t_{k+1})-b(t_k))(ib(t_k))^{\lambda}$$

$$\times R_{[\frac{m-n}{2}]-1,m-n-[\frac{m-n}{2}]}(-i\tau b(t_k))+\sum_{n=1}^{m} i\frac{(-1)^n\tau^{n+1}}{(n+1)!}\sum_{\lambda=1}^{n}\binom{n}{\lambda}\widetilde{c}_{n-\lambda}(t_{k+1})$$

$$\times\sum_{\mu=0}^{\lambda-1}\binom{\lambda}{\mu} b^{(\lambda-\mu)}(t_{k+1})(ib(t_k))^{\mu}R_{[\frac{m-n}{2}],m-n-[\frac{m-n}{2}]}(-i\tau b(t_k))+o(\tau^{m+2}),$$

$$m\geq 2,$$

$$I(t_{k+1},t_k)=R_{\sigma,2-\sigma}(-\tau ib(t_k))+i\tau(b(t_{k+1})-b(t_k))-\frac{i\tau^2}{2}b'(t_{k+1})+o(\tau^3),$$

$$\sigma=0,1,2,$$

$$I(t_{k+1},t_k)=R_{\varkappa,1-\varkappa}(-\tau ib(t_k))+o(\tau^2), \varkappa=0,1.$$

In the sequel we denote that if $n < 0$ $(m < 0)$, then

$$R_{n,m}(x) = R_{0,m+n}(x)(R_{n,m}(x) = R_{m+n,0}(x)), n + m \geq 0.$$

Third, let us consider approximate formulas for the expression

$$\frac{1}{\tau}\int_{t_k}^{t_{k+1}} I^2(s,t_k)ds.$$

Using Taylor's formula, we can write

$$\frac{1}{\tau}\int_{t_k}^{t_{k+1}} I^2(s,t_k)ds = \sum_{n=0}^{m} \frac{\tau^n}{(n+1)!}\widetilde{d}_n(t_k) + o(\tau^{m+1}), \tag{3.15}$$

where $\widetilde{d}_n(s) = D_n(s)I^2(s,t_k)$,

$$\begin{cases} D_n(s) = 0, n < 0, \\ D_0(s) = 1, D_{n+1}(s) = D_n'(s) + 2ib(s)D_n(s), 0 \leq n. \end{cases}$$

Fourth, let us consider approximate formulas for the expression

$$\frac{1}{\tau}\int_{t_{k-1}}^{t_k} I^{-2}(t_k,s)ds.$$

Using Taylor's formula, we can write

$$\frac{1}{\tau}\int_{t_{k-1}}^{t_k} I^{-2}(t_k,s)ds = \sum_{n=0}^{m} \frac{(-1)^n\tau^n}{(n+1)!}d_n(t_k) + o(\tau^{m+1}), \tag{3.16}$$

where $d_n(s) = D_n(s)I^{-2}(s,t_k)$. Note that $\widetilde{d}_n(t_k) = d_n(t_k)$ for any n.

Now, using formulas (3.9), (3.10), (3.11), (3.12), (3.13), (3.14), (3.15), (3.16) and neglecting the last small terms $o(\tau^{p+q})$, $\varphi_k^{p,q}$ can be defined by the formula

$$\varphi_k^{p,q} = \sum_{m=1}^{p+q-3}\sum_{\lambda=0}^{m}\binom{m}{\lambda} f^{(\lambda)}(t_{k+1})\frac{(-1)^m\tau^{m-1}}{(m+1)!}$$

$$\times \sum_{n=0}^{p+q-m-3} \frac{(-1)^n\tau^n}{(n+1)!}d_n(t_k)B_{m-\lambda}^{n,p+q-m-n}(t_{k+1})$$

$$+ \sum_{m=1}^{p+q-3}\sum_{\lambda=0}^{m}\binom{m}{\lambda} f^{(\lambda)}(t_{k+1})\frac{1}{(m+1)!}$$

$$\times \frac{(-1)^{p+q}\tau^{p+q-3}}{(p+q-m-1)!}d_{p+q-m-2}(t_k)B_{m-\lambda}^{p+q-m-2,2}(t_{k+1})$$

$$
\begin{aligned}
&+\sum_{m=1}^{p+q-3}\sum_{\lambda=0}^{m}\binom{m}{\lambda} f^{(\lambda)}(t_{k+1})\frac{1}{(m+1)!}\\
&\times\frac{(-1)^{p+q-1}\tau^{p+q-2}}{(p+q-m)!}d_{p+q-m-1}(t_k)B_{m-\lambda}^{p+q-m-1,1}(t_{k+1})\\
&+\sum_{m=1}^{p+q-3}\sum_{\lambda=0}^{m}\binom{m}{\lambda} f^{(\lambda)}(t_{k+1})\frac{1}{(m+1)!}\\
&\times\frac{(-1)^{p+q}\tau^{p+q-1}}{(p+q-m+1)!}d_{p+q-m}(t_k)B_{m-\lambda}^{p+q-m,0}(t_{k+1})\\
&+\sum_{\lambda=0}^{p+q-2}\binom{p+q-2}{\lambda} f^{(\lambda)}(t_{k+1})\frac{(-1)^{p+q-2}\tau^{p+q-3}}{(p+q-1)!}\\
&\times\sum_{n=0}^{2}\frac{(-1)^n\tau^n}{(n+1)!}d_n(t_k)B_{p+q-2-\lambda}^{n,2-n}(t_{k+1})\\
&+\sum_{\lambda=0}^{p+q-1}\binom{p+q-1}{\lambda} f^{(\lambda)}(t_{k+1})\frac{(-1)^{p+q-1}\tau^{p+q-2}}{(p+q)!}\\
&\times\sum_{n=0}^{1}\frac{(-1)^n\tau^n}{(n+1)!}d_n(t_k)B_{p+q-1-\lambda}^{n,1-n}(t_{k+1})\\
&+\sum_{\lambda=0}^{p+q}\binom{p+q}{\lambda} f^{(\lambda)}(t_{k+1})\frac{(-1)^{p+q}\tau^{p+q-1}}{(p+q+1)!}d_0(t_k)B_{p+q-\lambda}^{0,0}(t_{k+1})\\
&+\sum_{m=1}^{p+q-3}\sum_{\lambda=0}^{m}\binom{m}{\lambda} f^{(\lambda)}(t_{k-1})\frac{\tau^{m-1}}{(m+1)!}\\
&\times\sum_{n=0}^{p+q-m-3}\frac{\tau^n}{(n+1)!}\widetilde{d}_n(t_k)\widetilde{B}_{m-\lambda}^{n,p+q-m-n}(t_{k-1})\\
&+\sum_{m=1}^{p+q-3}\sum_{\lambda=0}^{m}\binom{m}{\lambda} f^{(\lambda)}(t_{k-1})\frac{1}{(m+1)!}\\
&\times\frac{\tau^{p+q-3}}{(p+q-m-1)!}\widetilde{d}_{p+q-m-2}(t_k)\widetilde{B}_{m-\lambda}^{p+q-m-2,2}(t_{k-1})\\
&+\sum_{m=1}^{p+q-3}\sum_{\lambda=0}^{m}\binom{m}{\lambda} f^{(\lambda)}(t_{k-1})\frac{1}{(m+1)!}\\
&\times\frac{\tau^{p+q-2}}{(p+q-m)!}\widetilde{d}_{p+q-m-1}(t_k)\widetilde{B}_{m-\lambda}^{p+q-m-1,1}(t_{k-1})
\end{aligned}
$$

$$+\sum_{m=1}^{p+q-3}\sum_{\lambda=0}^{m}\binom{m}{\lambda}f^{(\lambda)}(t_{k-1})\frac{1}{(m+1)!}$$

$$\times\frac{\tau^{p+q-1}}{(p+q-m+1)!}\widetilde{d}_{p+q-m}(t_k)\widetilde{B}^{p+q-m,0}_{m-\lambda}(t_{k-1})$$

$$+\sum_{\lambda=0}^{p+q-2}\binom{p+q-2}{\lambda}f^{(\lambda)}(t_{k-1})\frac{\tau^{p+q-3}}{(p+q-1)!}$$

$$\times\sum_{n=0}^{2}\frac{\tau^n}{(n+1)!}\widetilde{d}_n(t_k)\widetilde{B}^{n,2-n}_{p+q-2-\lambda}(t_{k-1})$$

$$+\sum_{\lambda=0}^{p+q-1}\binom{p+q-1}{\lambda}f^{(\lambda)}(t_{k-1})\frac{\tau^{p+q-2}}{(p+q)!}$$

$$\times\sum_{n=0}^{1}\frac{\tau^n}{(n+1)!}\widetilde{d}_n(t_k)\widetilde{B}^{n,1-n}_{p+q-1-\lambda}(t_{k-1})$$

$$+\sum_{\lambda=0}^{p+q}\binom{p+q}{\lambda}f^{(\lambda)}(t_{k-1})\frac{\tau^{p+q-1}}{(p+q+1)!}\widetilde{d}_0(t_k)\widetilde{B}^{0,0}_{p+q-\lambda}(t_{k-1}),$$

where

$$\begin{cases} B_0^{n,j}(t_{k+1})=0, B_1^{n,j}(t_{k+1})=-I^{n,j}_{k+1,k}, B_2^{n,j}(t_{k+1})=0,\\ B_m^{n,j}(t_{k+1})=-\sum_{\lambda=1}^{m-2}\binom{m-2}{\lambda}a^{(m-2-\lambda)}(t_{k+1})B_\lambda^{n,j}(t_{k+1}),\ 3\le m\le p+q,\end{cases}$$

$$I^{n,0}_{k+1,k}=1, I^{n,1}_{k+1,k}=R_{\varkappa,1-\varkappa}(-\tau ib(t_k)), \varkappa=0,1,$$

$$I^{n,2}_{k+1,k}=R_{\sigma,2-\sigma}(-\tau ib(t_k))+i\tau(b(t_{k+1})-b(t_k))-i\frac{\tau^2b'(t_{k+1})}{2},$$

$$\sigma=0,1,2,$$

$$I^{n,j}_{k+1},_k=R_{q-[\frac{m+n}{2}],p-m-n+[\frac{m+n}{2}]}(-\tau ib(t_k))+i\tau(b(t_{k+1})-b(t_k))$$

$$\times R_{q-1-[\frac{m+n}{2}],p-1-m-n+[\frac{m+n}{2}]}(-\tau ib(t_k))$$

$$+\sum_{\xi=1}^{p+q-m-n-2}i\frac{(-1)^\xi\tau^{\xi+1}}{(\xi+1)!}\sum_{\lambda=0}^{\xi}\binom{\xi}{\lambda}\widetilde{c}_{\xi-\lambda}(t_{k+1})$$

$$\times(b(t_{k+1})-b(t_k))(ib(t_k))^\lambda R_{q-1-[\frac{m+n+\xi}{2}],p-1-m-n-\xi+[\frac{m+n+\xi}{2}]}(-\tau ib(t_k))$$

$$+\sum_{\xi=1}^{p+q-m-n-1}i\frac{(-1)^\xi\tau^{\xi+1}}{(\xi+1)!}\sum_{\lambda=1}^{\xi}\binom{\xi}{\lambda}\widetilde{c}_{\xi-\lambda}(t_{k+1})$$

$$\times \sum_{\mu=0}^{\lambda-1} \binom{\lambda}{\mu} b^{(\lambda-\mu)}(t_{k+1})(ib(t_k))^{\mu} R_{q-1-[\frac{m+n+\xi}{2}],p-m-n-\xi+[\frac{m+n+\xi}{2}]}(-\tau ib(t_k)),$$

$$j \geq 3,$$

and

$$\begin{cases} \widetilde{B}_0^{n,j}(t_{k-1}) = 0, \widetilde{B}_1^{n,j}(t_{k-1}) = J_{k,k-1}^{n,j}, \widetilde{B}_2^{n,j}(t_{k-1}) = 0, \\ \widetilde{B}_m^{n,j}(t_{k-1}) = -\sum\limits_{\lambda=1}^{m-2} \binom{m-2}{\lambda} a^{(m-2-\lambda)}(t_{k-1})\widetilde{B}_\lambda^{n,j}(t_{k-1}), \quad 3 \leq m \leq p+q, \end{cases}$$

$$J_{k,k-1}^{n,0} = 1, J_{k,k-1}^{n,1} = R_{\varkappa,1-\varkappa}(\tau ib(t_k)), \varkappa = 0, 1,$$

$$J_{k,k-1}^{n,2} = R_{\sigma,2-\sigma}(\tau ib(t_k)) + i\tau(b(t_k) - b(t_{k-1})) + i\frac{\tau^2 b'(t_{k-1})}{2}, \sigma = 0, 1, 2,$$

$$J_{k,k-1}^{n,j} = R_{q-[\frac{m+n}{2}],p-m-n+[\frac{m+n}{2}]}(i\tau b(t_k)) + i\tau(b(t_k) - b(t_{k-1}))$$

$$\times R_{q-1-[\frac{m+n}{2}],p-1-m-n+[\frac{m+n}{2}]}(i\tau b(t_k)) + i \sum_{\xi=1}^{p+q-m-n-2} \frac{\tau^{\xi+1}}{(\xi+1)!} \sum_{\lambda=0}^{\xi} \binom{\xi}{\lambda} \widetilde{c}_{\xi-\lambda}(t_{k-1})$$

$$\times (b(t_k) - b(t_{k-1}))(ib(t_k))^{\lambda} R_{q-1-[\frac{m+n+\xi}{2}],p-1-m-n-\xi+[\frac{m+n+\xi}{2}]}(i\tau b(t_k))$$

$$+ i \sum_{\xi=1}^{p+q-m-n-1} \frac{\tau^{\xi+1}}{(\xi+1)!} \sum_{\lambda=1}^{\xi} \binom{\xi}{\lambda} \widetilde{c}_{\xi-\lambda}(t_{k-1})$$

$$\times \sum_{\mu=0}^{\lambda-1} \binom{\lambda}{\mu} b^{(\lambda-\mu)}(t_{k-1})(ib(t_k))^{\mu} R_{q-1-[\frac{m+n+\xi}{2}],p-m-n-\xi+[\frac{m+n+\xi}{2}]}(i\tau b(t_k)), j \geq 3.$$

Applying the identity $\widetilde{d}_n(t_k) = d_n(t_k)$ for any n, we can write

$$\varphi_k^{p,q} = \sum_{m=1}^{p+q-3} \sum_{\lambda=0}^{m} \binom{m}{\lambda} \frac{1}{(m+1)!} \sum_{n=0}^{p+q-m-3} \frac{\tau^{n+m-1}}{(n+1)!} d_n(t_k) \tag{3.17}$$

$$\times ((-1)^{m+n} B_{m-\lambda}^{n,p+q-m-n}(t_{k+1}) f^{(\lambda)}(t_{k+1}) + \widetilde{B}_{m-\lambda}^{n,p+q-m-n}(t_{k-1}) f^{(\lambda)}(t_{k-1}))$$

$$+ \sum_{m=1}^{p+q-3} \sum_{\lambda=0}^{m} \binom{m}{\lambda} \frac{1}{(m+1)!} \frac{\tau^{p+q-3}}{(p+q-m-1)!} d_{p+q-m-2}(t_k)$$

$$\times ((-1)^{p+q} B_{m-\lambda}^{p+q-m-2,2}(t_{k+1}) f^{(\lambda)}(t_{k+1}) + \widetilde{B}_{m-\lambda}^{p+q-m-2,2}(t_{k-1}) f^{(\lambda)}(t_{k-1}))$$

$$+ \sum_{m=1}^{p+q-3} \sum_{\lambda=0}^{m} \binom{m}{\lambda} \frac{1}{(m+1)!} \frac{\tau^{p+q-2}}{(p+q-m)!} d_{p+q-m-1}(t_k)$$

$$\times((-1)^{p+q-1}B^{p+q-m-1,1}_{m-\lambda}(t_{k+1})f^{(\lambda)}(t_{k+1})+\widetilde{B}^{p+q-m-1,1}_{m-\lambda}(t_{k-1})f^{(\lambda)}(t_{k-1}))$$

$$+\sum_{m=1}^{p+q-3}\sum_{\lambda=0}^{m}\binom{m}{\lambda}\frac{1}{(m+1)!}\frac{\tau^{p+q-1}}{(p+q-m+1)!}d_{p+q-m}(t_k)$$

$$\times((-1)^{p+q}B^{p+q-m,0}_{m-\lambda}(t_{k+1})f^{(\lambda)}(t_{k+1})+\widetilde{B}^{p+q-m,0}_{m-\lambda}(t_{k-1})f^{(\lambda)}(t_{k-1}))$$

$$+\sum_{\lambda=0}^{p+q-2}\binom{p+q-2}{\lambda}\frac{1}{(p+q-1)!}\sum_{n=0}^{2}\frac{\tau^{p+q-3+n}}{(n+1)!}d_n(t_k)$$

$$\times((-1)^{p+q+n-2}B^{n,2-n}_{p+q-2-\lambda}(t_{k+1})f^{(\lambda)}(t_{k+1})+\widetilde{B}^{n,2-n}_{p+q-2-\lambda}(t_{k-1})f^{(\lambda)}(t_{k-1}))$$

$$+\sum_{\lambda=0}^{p+q-1}\binom{p+q-1}{\lambda}\frac{1}{(p+q)!}\sum_{n=0}^{1}\frac{\tau^{p+q-2+n}}{(n+1)!}d_n(t_k)$$

$$\times((-1)^{p+q-1+n}B^{n,1-n}_{p+q-1-\lambda}(t_{k+1})f^{(\lambda)}(t_{k+1})+\widetilde{B}^{n,1-n}_{p+q-1-\lambda}(t_{k-1})f^{(\lambda)}(t_{k-1}))$$

$$+\sum_{\lambda=0}^{p+q}\binom{p+q}{\lambda}\frac{\tau^{p+q-1}}{(p+q+1)!}d_0(t_k)$$

$$\times((-1)^{p+q}B^{0,0}_{p+q-\lambda}(t_{k+1})f^{(\lambda)}(t_{k+1})+\widetilde{B}^{0,0}_{p+q-\lambda}(t_{k-1})f^{(\lambda)}(t_{k-1})).$$

Now, using formulas (3.13), (3.14), (3.15) and (3.16), we can write

$$\frac{1}{\tau^2}\left(\frac{1}{\tau}\int_{t_k}^{t_{k+1}}I^2(s,t_k)dsI^{-1}(t_k,t_{k-1})-1\right) \tag{3.18}$$

$$=\frac{1}{\tau^2}\left(\sum_{m=0}^{p+q-2}\frac{\tau^m}{(m+1)!}d_m(t_k)[R_{q-[\frac{m}{2}],p+1-m+[\frac{m}{2}]}(i\tau b(t_k))\right)$$

$$+i\tau(b(t_k)-b(t_{k-1}))R_{q-1-[\frac{m}{2}],p-m+[\frac{m}{2}]}(i\tau b(t_k))$$

$$+i\sum_{n=1}^{p+q-m-1}\frac{\tau^{n+1}}{(n+1)!}\sum_{\lambda=0}^{n}\binom{n}{\lambda}\widetilde{c}_{n-\lambda}(t_{k-1})(b(t_k)-b(t_{k-1}))(ib(t_k))^{\lambda}$$

$$\times R_{q-1-[\frac{m+n}{2}],p-m-n+[\frac{m+n}{2}]}(i\tau b(t_k))+i\sum_{n=1}^{p+q-m}\frac{\tau^{n+1}}{(n+1)!}\sum_{\lambda=1}^{n}\binom{n}{\lambda}\widetilde{c}_{n-\lambda}(t_{k-1})$$

$$\times\sum_{\mu=0}^{\lambda-1}\binom{\lambda}{\mu}b^{(\lambda-\mu)}(t_{k-1})(ib(t_k))^{\mu}R_{q-[\frac{m+n}{2}],p-m-n+[\frac{m+n}{2}]}(i\tau b(t_k))]$$

$$+\frac{\tau^{p+q-1}}{(p+q)!}d_{p+q-1}(t_k)[R_{\sigma,2-\sigma}(\tau ib(t_k))+i\tau(b(t_k)-b(t_{k-1}))+i\frac{\tau^2}{2}b'(t_{k-1})]$$

$$+\frac{\tau^{p+q}}{(p+q+1)!}d_{p+q}(t_k)R_{\varkappa,1-\varkappa}(\tau ib(t_k))+\frac{\tau^{p+q+1}}{(p+q+2)!}d_{p+q+1}(t_k)-1)+o(\tau^{p+q}),$$

$$\sigma=0,1,2,\varkappa=0,1;$$

$$\frac{1}{\tau^2}(\frac{1}{\tau}\int_{t_{k-1}}^{t_k} I^{-2}(t_k,s)dsI(t_{k+1},t_k)-1) \tag{3.19}$$

$$=\frac{1}{\tau^2}(\sum_{m=0}^{p+q-2}\frac{(-1)^m\tau^m}{(m+1)!}d_m(t_k)[R_{q-[\frac{m}{2}],p-m+[\frac{m}{2}]+1}(-\tau ib(t_k))+i\tau(b(t_{k+1})-b(t_k))$$

$$\times R_{q-1-[\frac{m}{2}],p-m+[\frac{m}{2}]}(-i\tau b(t_k))+\sum_{n=1}^{p+q-m-1} i\frac{(-1)^n\tau^{n+1}}{(n+1)!}\sum_{\lambda=0}^{n}\binom{n}{\lambda}\widetilde{c}_{n-\lambda}(t_{k+1})$$

$$\times(b(t_{k+1})-b(t_k))(ib(t_k))^\lambda R_{q-1-[\frac{m+n}{2}],p-m-n+[\frac{m+n}{2}]}(-i\tau b(t_k))$$

$$+\sum_{n=1}^{p+q-m} i\frac{(-1)^n\tau^{n+1}}{(n+1)!}\sum_{\lambda=1}^{n}\binom{n}{\lambda}\widetilde{c}_{n-\lambda}(t_{k+1})$$

$$\times\sum_{\mu=0}^{\lambda-1}\binom{\lambda}{\mu}b^{(\lambda-\mu)}(t_{k+1})(ib(t_k))^\mu R_{q-[\frac{m+n}{2}],p-m-n+[\frac{m+n}{2}]}(-i\tau b(t_k))]$$

$$+\frac{(-1)^{p+q-1}\tau^{p+q-1}}{(p+q)!}d_{p+q-1}(t_k)[R_{\sigma,2-\sigma}(-\tau ib(t_k))+i\tau(b(t_{k+1})-b(t_k))-i\frac{\tau^2}{2}b'(t_{k+1})]$$

$$+\frac{(-1)^{p+q}\tau^{p+q}}{(p+q+1)!}d_{p+q}(t_k)R_{\varkappa,1-\varkappa}(-\tau ib(t_k))$$

$$+\frac{(-1)^{p+q+1}\tau^{p+q+1}}{(p+q+2)!}d_{p+q+1}(t_k)-1)+o(\tau^{p+q}),$$

$$\sigma=0,1,2,\varkappa=0,1;$$

$$\frac{1}{\tau^2}(2-\frac{1}{\tau}\int_{t_{k-1}}^{t_k} I^{-2}(t_k,s)ds-\frac{1}{\tau}\int_{t_k}^{t_{k+1}} I^2(s,t_k)ds) \tag{3.20}$$

$$=\frac{1}{\tau^2}(2-\sum_{m=0}^{p+q+1}\frac{(-1)^m\tau^m}{(m+1)!}d_n(t_k)-\sum_{m=0}^{p+q+1}\frac{\tau^m}{(m+1)!}d_m(t_k))$$

$$=\sum_{m=1}^{p+q+1}\frac{\tau^{m-2}}{(m+1)!}d_m(t_k)((-1)^{m+1}-1)+o(\tau^{p+q});$$

$$I^{-1}(\tau,0)ib(0)\frac{1}{\tau}\int_0^\tau I^2(s,0)ds+\frac{1}{\tau}(I^{-1}(\tau,0)-1) \tag{3.21}$$

$$=ib(0)\left(\sum_{m=0}^{p+q-4}\frac{\tau^m}{(m+1)!}d_m(0)\left[R_{q-1-[\frac{m}{2}],p-m+[\frac{m}{2}]}(i\tau b(\tau))+i\tau(b(\tau)-b(0))\right.\right.$$

$$\times R_{q-2-[\frac{m}{2}],p-1-m+[\frac{m}{2}]}(i\tau b(\tau))+i\sum_{n=1}^{p+q-m-3}\frac{\tau^{n+1}}{(n+1)!}\sum_{\lambda=0}^{n}\binom{n}{\lambda}\widetilde{c}_{n-\lambda}(0)$$

$$\times(b(\tau)-b(0))(ib(\tau))^{\lambda}R_{q-2-[\frac{m+n}{2}],p-1-m-n+[\frac{m+n}{2}]}(i\tau b(\tau))$$

$$+i\sum_{n=1}^{p+q-m-2}\frac{\tau^{n+1}}{(n+1)!}\sum_{\lambda=1}^{n}\binom{n}{\lambda}\widetilde{c}_{n-\lambda}(0)$$

$$\times\sum_{\mu=0}^{\lambda-1}\binom{\lambda}{\mu}b^{(\lambda-\mu)}(0)(ib(\tau))^{\mu}R_{q-1-[\frac{m+n}{2}],p-1-m-n+[\frac{m+n}{2}]}(i\tau b(\tau))\Bigg]\Bigg)$$

$$+ib(0)\frac{\tau^{p+q-3}}{(p+q-2)!}d_{p+q-3}(0)\left[R_{\sigma,2-\sigma}(\tau ib(\tau))+i\tau(b(\tau)-b(0))+i\frac{\tau^2}{2}b'(0)\right]$$

$$+ib(0)\frac{\tau^{p+q-2}}{(p+q-1)!}d_{p+q-2}(0)R_{\varkappa,1-\varkappa}(\tau ib(\tau))+ib(0)\frac{\tau^{p+q-1}}{(p+q)!}d_{p+q-1}(0)$$

$$+\frac{1}{\tau}(R_{q,p}(i\tau b(\tau))+i\tau(b(\tau)-b(0))R_{q-1,p-1}(i\tau b(\tau))$$

$$+i\sum_{n=1}^{p+q-2}\frac{\tau^{n+1}}{(n+1)!}\sum_{\lambda=0}^{n}\binom{n}{\lambda}\widetilde{c}_{n-\lambda}(0)(b(\tau)-b(0))(ib(\tau))^{\lambda}R_{q-1-[\frac{n}{2}],p-1-n+[\frac{n}{2}]}(i\tau b(\tau))$$

$$+\sum_{n=1}^{p+q-1}\frac{\tau^{n+1}}{(n+1)!}i\sum_{\lambda=1}^{n}\binom{n}{\lambda}\widetilde{c}_{n-\lambda}(0)$$

$$\times\sum_{\mu=0}^{\lambda-1}\binom{\lambda}{\mu}b^{(\lambda-\mu)}(0)(ib(\tau))^{\mu}R_{q-1-[\frac{n}{2}],p-n+[\frac{n}{2}]}(i\tau b(\tau))-1)+o(\tau^{p+q}),$$

$$\sigma=0,1,2,\varkappa=0,1,p+q\geq 3,$$

$$I^{-1}(\tau,0)ib(0)\frac{1}{\tau}\int_0^{\tau}I^2(s,0)ds+\frac{1}{\tau}(I^{-1}(\tau,0)-1)$$

$$=ib(0)d_0(0)R_{\varkappa,1-\varkappa}(\tau ib(\tau))+ib(0)\frac{\tau}{2}d_1(0)$$

$$+\frac{1}{\tau}(R_{q,p}(i\tau b(\tau))+i\tau(b(\tau)-b(0))+i\frac{\tau^2}{2}b'(0)-1)+o(\tau^2),p+q=2;$$

$$I^{-1}(\tau,0)ib(0)\frac{1}{\tau}\int_0^{\tau}I^2(s,0)ds+\frac{1}{\tau}(I^{-1}(\tau,0)-1)$$

$$=ib(0)d_0(0)+\frac{1}{\tau}(R_{q,p}(i\tau b(\tau))-1)+o(\tau),p+q=1;$$

$$I^{-1}(\tau,0)\frac{1}{\tau}\int_0^{\tau}I^2(s,0)ds \tag{3.22}$$

$$=\sum_{m=0}^{p+q-4}\frac{\tau^m}{(m+1)!}d_m(0)[R_{q-1-[\frac{m}{2}],p-m+[\frac{m}{2}]}(i\tau b(\tau))$$

$$+i\tau(b(\tau)-b(0))R_{q-2-[\frac{m}{2}],p-1-m+[\frac{m}{2}]}(i\tau b(\tau))$$

$$+i\sum_{n=1}^{p+q-m-3}\frac{\tau^{n+1}}{(n+1)!}\sum_{\lambda=0}^{n}\binom{n}{\lambda}\widetilde{c}_{n-\lambda}(0)(b(\tau)-b(0))$$

$$\times(ib(\tau))^{\lambda}R_{q-2-[\frac{m+n}{2}],p-1-m-n+[\frac{m+n}{2}]}(i\tau b(\tau))$$

$$+i\sum_{n=1}^{p+q-m-2}\frac{\tau^{n+1}}{(n+1)!}\sum_{\lambda=1}^{n}\binom{n}{\lambda}\widetilde{c}_{n-\lambda}(0)$$

$$\times\sum_{\mu=0}^{\lambda-1}\binom{\lambda}{\mu}b^{(\lambda-\mu)}(0)(ib(\tau))^{\mu}R_{q-1-[\frac{m+n}{2}],p-1-m-n+[\frac{m+n}{2}]}(i\tau b(\tau))]$$

$$+\frac{\tau^{p+q-3}}{(p+q-2)!}d_{p+q-3}(0)[R_{\sigma,2-\sigma}(\tau ib(\tau))+i\tau(b(\tau)-b(0))+i\frac{\tau^2}{2}b'(0)]$$

$$+\frac{\tau^{p+q-2}}{(p+q-1)!}d_{p+q-2}(0)R_{\varkappa,1-\varkappa}(\tau ib(\tau))+\frac{\tau^{p+q-1}}{(p+q)!}d_{p+q-1}(0)+o(\tau^{p+q}),$$

$$\sigma=0,1,2,\varkappa=0,1,p+q\geq 3,$$

$$I^{-1}(\tau,0)\frac{1}{\tau}\int_0^{\tau}I^2(s,0)ds=d_0(0)R_{\varkappa,1-\varkappa}(-\tau ib(\tau))+\frac{\tau}{2}d_1(0)+o(\tau^2),p+q=2,$$

$$I^{-1}(\tau,0)\frac{1}{\tau}\int_0^{\tau}I^2(s,0)ds=1+o(\tau),p+q=1;$$

$$\frac{1}{\tau}\int_0^{\tau}I^{-1}(\tau,z)\int_z^{\tau}I^2(s,z)dsf(z)dz \tag{3.23}$$

$$=\sum_{m=1}^{p+q-1}\sum_{\lambda=0}^{m}\binom{m}{\lambda}B_{m-\lambda}^{00}(\tau)f^{(\lambda)}(\tau)\frac{(-1)^m\tau^m}{(m+1)!}+o(\tau^{p+q}),p+q\geq 2,$$

where

$$\begin{cases} B_0^{00}(\tau)=0,B_1^{00}(\tau)=-1,B_2^{00}(\tau)=0,\\ B_m^{00}(\tau)=-\sum\limits_{\lambda=1}^{m-2}\binom{m-2}{\lambda}a^{(m-2-\lambda)}(\tau)B_{\lambda}^{00}(\tau),\quad 3\leq m\leq p+q; \end{cases}$$

$$\frac{1}{\tau}\int_0^{\tau}I^{-1}(\tau,z)\int_z^{\tau}I^2(s,z)dsf(z)dz=0+o(\tau),p+q=1.$$

Now, neglecting the last terms in formulas (3.18), (3.19), (3.20), (3.21), (3.22), (3.23) and using formula (3.17), we obtain the difference schemes

$$\frac{1}{\tau^2}(u_{k+1}-2u_k+u_{k-1})+a_{1,k}u_{k-1}+a_{2,k}u_k+a_{3,k}u_{k+1}$$

$$= \varphi_k^{p,q}, 1 \le k \le N-1, u_0 = y_0, \tag{3.24}$$

$$\frac{1}{\tau}(u_1 - u_0) = a_{0,0}y_0 + a_{0,1}y'(0) + \varphi_0^{p,q},$$

where

$$a_{1,k} = \frac{1}{\tau^2}\Big(\sum_{m=0}^{p+q-2} \frac{\tau^m}{(m+1)!} d_m(t_k)[R_{q-[\frac{m}{2}],p+1-m+[\frac{m}{2}]}(i\tau b(t_k))$$

$$+ i\tau(b(t_k) - b(t_{k-1}))R_{q-1-[\frac{m}{2}],p-m+[\frac{m}{2}]}(i\tau b(t_k))$$

$$+ i\sum_{n=1}^{p+q-m-1} \frac{\tau^{n+1}}{(n+1)!} \sum_{\lambda=0}^{n} \binom{n}{\lambda} \widetilde{c}_{n-\lambda}(t_{k-1})(b(t_k) - b(t_{k-1}))(ib(t_k))^{\lambda}$$

$$\times R_{q-1-[\frac{m+n}{2}],p-m-n+[\frac{m+n}{2}]}(i\tau b(t_k)) + i\sum_{n=1}^{p+q-m} \frac{\tau^{n+1}}{(n+1)!} \sum_{\lambda=1}^{n} \binom{n}{\lambda} \widetilde{c}_{n-\lambda}(t_{k-1})$$

$$\times \sum_{\mu=0}^{\lambda-1} \binom{\lambda}{\mu} b^{(\lambda-\mu)}(t_{k-1})(ib(t_k))^{\mu} R_{q-[\frac{m+n}{2}],p-m-n+[\frac{m+n}{2}]}(i\tau b(t_k))]$$

$$+ \frac{\tau^{p+q-1}}{(p+q)!} d_{p+q-1}(t_k)[R_{\sigma,2-\sigma}(\tau i b(t_k)) + i\tau(b(t_k) - b(t_{k-1})) + i\frac{\tau^2}{2}b'(t_{k-1})]$$

$$+ \frac{\tau^{p+q}}{(p+q+1)!} d_{p+q}(t_k) R_{\varkappa,1-\varkappa}(\tau i b(t_k)) + \frac{\tau^{p+q+1}}{(p+q+2)!} d_{p+q+1}(t_k) - 1\Big),$$

$$\sigma = 0, 1, 2, \varkappa = 0, 1;$$

$$a_{3,k} = \frac{1}{\tau^2}\Big(\sum_{m=0}^{p+q-2} \frac{(-1)^m\tau^m}{(m+1)!} d_m(t_k)[R_{q-[\frac{m}{2}],p-m+[\frac{m}{2}]+1}(-\tau i b(t_k))$$

$$+ i\tau(b(t_{k+1}) - b(t_k))R_{q-1-[\frac{m}{2}],p-m+[\frac{m}{2}]}(-i\tau b(t_k))$$

$$+ \sum_{n=1}^{p+q-m-1} i\frac{(-1)^n\tau^{n+1}}{(n+1)!} \sum_{\lambda=0}^{n} \binom{n}{\lambda} \widetilde{c}_{n-\lambda}(t_{k+1})$$

$$\times (b(t_{k+1}) - b(t_k))(ib(t_k))^{\lambda} R_{q-1-[\frac{m+n}{2}],p-m-n+[\frac{m+n}{2}]}(-i\tau b(t_k))$$

$$+ \sum_{n=1}^{p+q-m} i\frac{(-1)^n\tau^{n+1}}{(n+1)!} \sum_{\lambda=1}^{n} \binom{n}{\lambda} \widetilde{c}_{n-\lambda}(t_{k+1})$$

$$\times \sum_{\mu=0}^{\lambda-1} \binom{\lambda}{\mu} b^{(\lambda-\mu)}(t_{k+1})(ib(t_k))^{\mu} R_{q-[\frac{m+n}{2}],p-m-n+[\frac{m+n}{2}]}(-i\tau b(t_k))]$$

$$+ \frac{(-1)^{p+q-1}\tau^{p+q-1}}{(p+q)!} d_{p+q-1}(t_k)[R_{\sigma,2-\sigma}(-\tau i b(t_k)) + i\tau(b(t_{k+1}) - b(t_k)) - i\frac{\tau^2}{2}b'(t_{k+1})]$$

$$+\frac{(-1)^{p+q}\tau^{p+q}}{(p+q+1)!}d_{p+q}(t_k)R_{\varkappa,1-\varkappa}(-\tau ib(t_k))+\frac{(-1)^{p+q+1}\tau^{p+q+1}}{(p+q+2)!}d_{p+q+1}(t_k)-1),$$

$$\sigma=0,1,2,\varkappa=0,1;$$

$$a_{2,k}=\sum_{m=1}^{p+q+1}\frac{\tau^{m-2}}{(m+1)!}d_m(t_k)((-1)^{m+1}-1);$$

$$a_{0,0}=ib(0)(\sum_{m=0}^{p+q-4}\frac{\tau^m}{(m+1)!}d_m(0)[R_{q-1-[\frac{m}{2}],p-m+[\frac{m}{2}]}(i\tau b(\tau))$$

$$+i\tau(b(\tau)-b(0))R_{q-2-[\frac{m}{2}],p-1-m+[\frac{m}{2}]}(i\tau b(\tau))$$

$$+i\sum_{n=1}^{p+q-m-3}\frac{\tau^{n+1}}{(n+1)!}\sum_{\lambda=0}^{n}\binom{n}{\lambda}\widetilde{c}_{n-\lambda}(0)$$

$$\times(b(\tau)-b(0))(ib(\tau))^{\lambda}R_{q-2-[\frac{m+n}{2}],p-1-m-n+[\frac{m+n}{2}]}(i\tau b(\tau))$$

$$+i\sum_{n=1}^{p+q-m-2}\frac{\tau^{n+1}}{(n+1)!}\sum_{\lambda=1}^{n}\binom{n}{\lambda}\widetilde{c}_{n-\lambda}(0)$$

$$\times\sum_{\mu=0}^{\lambda-1}\binom{\lambda}{\mu}b^{(\lambda-\mu)}(0)(ib(\tau))^{\mu}R_{q-1-[\frac{m+n}{2}],p-1-m-n+[\frac{m+n}{2}]}(i\tau b(\tau))]$$

$$+ib(0)\frac{\tau^{p+q-3}}{(p+q-2)!}d_{p+q-3}(0)[R_{\sigma,2-\sigma}(\tau ib(\tau))+i\tau(b(\tau)-b(0))+\frac{i\tau^2}{2}b'(0)]$$

$$+ib(0)\frac{\tau^{p+q-2}}{(p+q-1)!}d_{p+q-2}(0)R_{\varkappa,1-\varkappa}(\tau ib(\tau))+ib(0)\frac{\tau^{p+q-1}}{(p+q)!}d_{p+q-1}(0)$$

$$+\frac{1}{\tau}(R_{q,p}(i\tau b(\tau))+i\tau(b(\tau)-b(0))R_{q-1,p-1}(i\tau b(\tau))$$

$$+i\sum_{n=1}^{p+q-2}\frac{\tau^{n+1}}{(n+1)!}\sum_{\lambda=0}^{n}\binom{n}{\lambda}\widetilde{c}_{n-\lambda}(0)(b(\tau)-b(0))(ib(\tau))^{\lambda}$$

$$\times R_{q-1-[\frac{n}{2}],p-1-n+[\frac{n}{2}]}(i\tau b(\tau))+\sum_{n=1}^{p+q-1}\frac{\tau^{n+1}}{(n+1)!}i\sum_{\lambda=1}^{n}\binom{n}{\lambda}\widetilde{c}_{n-\lambda}(0)$$

$$\times\sum_{\mu=0}^{\lambda-1}\binom{\lambda}{\mu}b^{(\lambda-\mu)}(0)(ib(\tau))^{\mu}R_{q-1-[\frac{n}{2}],p-n+[\frac{n}{2}]}(i\tau b(\tau))-1),$$

$$\sigma=0,1,2,\varkappa=0,1,p+q\geq 3;$$

$$a_{0,0}=ib(0)d_0(0)R_{\varkappa,1-\varkappa}(\tau ib(\tau))+ib(0)\frac{\tau}{2}d_1(0)$$

$$+\frac{1}{\tau}(R_{q,p}(i\tau b(\tau)) + i\tau(b(\tau) - b(0)) + i\frac{\tau^2}{2}b'(0) - 1), p + q = 2;$$

$$a_{0,0} = ib(0)d_0(0) + \frac{1}{\tau}(R_{q,p}(i\tau b(\tau)) - 1), p + q = 1;$$

$$a_{0,1} = \sum_{m=0}^{p+q-4} \frac{\tau^m}{(m+1)!} d_m(0)[R_{q-1-[\frac{m}{2}],p-m+[\frac{m}{2}]}(i\tau b(\tau)) + i\tau(b(\tau) - b(0))$$

$$\times R_{q-2-[\frac{m}{2}],p-1-m+[\frac{m}{2}]}(i\tau b(\tau)) + i \sum_{n=1}^{p+q-m-3} \frac{\tau^{n+1}}{(n+1)!} \sum_{\lambda=0}^{n} \binom{n}{\lambda} \widetilde{c}_{n-\lambda}(0)$$

$$\times (b(\tau) - b(0))(ib(\tau))^\lambda R_{q-2-[\frac{m+n}{2}],p-1-m-n+[\frac{m+n}{2}]}(i\tau b(\tau))$$

$$+i \sum_{n=1}^{p+q-m-2} \frac{\tau^{n+1}}{(n+1)!} \sum_{\lambda=1}^{n} \binom{n}{\lambda} \widetilde{c}_{n-\lambda}(0)$$

$$\times \sum_{\mu=0}^{\lambda-1} \binom{\lambda}{\mu} b^{(\lambda-\mu)}(0)(ib(\tau))^\mu R_{q-1-[\frac{m+n}{2}],p-1-m-n+[\frac{m+n}{2}]}(i\tau b(\tau))]$$

$$+\frac{\tau^{p+q-3}}{(p+q-2)!} d_{p+q-3}(0)[R_{\sigma,2-\sigma}(\tau ib(\tau)) + i\tau(b(\tau) - b(0)) + \frac{i\tau^2}{2}b'(0)]$$

$$+\frac{\tau^{p+q-2}}{(p+q-1)!} d_{p+q-2}(0)R_{\varkappa,1-\varkappa}(\tau ib(\tau)) + \frac{\tau^{p+q-1}}{(p+q)!} d_{p+q-1}(0),$$

$$\sigma = 0, 1, 2, \varkappa = 0, 1, p + q \geq 3;$$

$$a_{0,1} = d_0(0)R_{\varkappa,1-\varkappa}(\tau ib(\tau)) + \frac{\tau}{2}d_1(0), p + q = 2;$$

$$a_{0,1} = 1, p + q = 1;$$

$$\varphi_0^{p,q} = \begin{cases} 0, p + q = 1, \\ \sum\limits_{m=1}^{p+q-1} \sum\limits_{\lambda=0}^{m} \binom{m}{\lambda} B_{m-\lambda}^{00}(\tau) f^{(\lambda)}(\tau) \frac{(-1)^m \tau^m}{(m+1)!}, 2 \leq p + q. \end{cases}$$

Thus, we constructed the difference schemes of $(p+q)$-order of accuracy for the approximate solution of the initial-value problem (3.1). Finally, let us give some examples of the simplest difference schemes.

In the case $p + q = 1$ from formulas (3.17), (3.18), (3.19), (3.20), (3.21), (3.22), (3.23) and (3.24) it follows that

$$\frac{1}{\tau^2}(u_{k+1} - 2u_k + u_{k-1}) + (\frac{b^2(t_k)}{2} + i\tau^{-1}(b(t_k) - b(t_{k-1})) + i\frac{1}{2}b'(t_{k-1})$$

$$+\frac{1}{3!}(6ib'(t_k) - 4a(t_k))u_{k-1}$$

$$+(-2ib'(t_k)+\frac{4}{3}a(t_k))u_k+(\frac{b^2(t_k)}{2}+i\tau^{-1}(b(t_{k+1})-b(t_k))$$

$$-i\frac{1}{2}b'(t_{k+1})+\frac{1}{3!}(6ib'(t_k)-4a(t_k))u_{k+1}=\frac{1}{2}(f(t_{k+1})+f(t_{k-1})),$$

$$1\leq k\leq N-1,\quad u_0=y_0,$$

$$\frac{1}{\tau}(u_1-u_0)=i(b(0)-b(\tau))y(0)+y'(0)$$

(a difference scheme of first order of accuracy for an initial-value problem (3.1) generated by the first order of accuracy explicit difference scheme).

In the case $p+q=2$ from formulas (3.17), (3.18), (3.19), (3.20), (3.21), (3.22), (3.23) and (3.24) it follows that

$$\frac{1}{\tau^2}(u_{k+1}-2u_k+u_{k-1})+(\frac{b^2(t_k)}{2}-i\frac{\tau b^3(t_k)}{6}+i\frac{1}{\tau}(b(t_k)-b(t_{k-1}))$$

$$+\frac{1}{2}b(t_{k-1})(b(t_k)-b(t_{k-1}))-\frac{1}{2}(b(t_k)-b(t_{k-1}))b(t_k)+i\frac{1}{2}b'(t_{k-1})(1-i\tau b(t_k))$$

$$+\frac{2\tau}{3!}b(t_{k-1})b'(t_{k-1})+i\frac{\tau}{3!}(b''(t_{k-1})+2b'(t_{k-1})ib(t_k))+ib(t_k)(-\frac{\tau b^2(t_k)}{2}+\frac{i\tau}{2}b'(t_{k-1}))$$

$$+\frac{1}{6}(2ib'(t_k)-4b^2(t_k))(1-i\tau b(t_k))+\frac{\tau}{4!}(2ib''(t_k)-12ib(t_k)b'(t_k)-8ib^3(t_k))))u_{k-1}$$

$$+(-2ib'(t_k)+\frac{4}{3}a(t_k))u_k+(\frac{b^2(t_k)}{2}+i\frac{\tau b^3(t_k)}{6}+\frac{i}{\tau}(b(t_{k+1})-b(t_k))$$

$$-\frac{1}{2}b(t_{k+1})(b(t_{k+1})-b(t_k))+\frac{1}{2}(b(t_{k+1})-b(t_k))b(t_k)-i\frac{1}{2}b'(t_{k+1})(1+i\tau b(t_k))$$

$$+\frac{2\tau}{3!}b(t_{k+1})b'(t_{k+1})+i\frac{\tau}{3!}(b''(t_{k+1})+2b'(t_{k+1})ib(t_k))-ib(t_k)(-\frac{\tau b^2(t_k)}{2}-\frac{i\tau}{2}b'(t_{k+1}))$$

$$+\frac{1}{6}(2ib'(t_k)-4b^2(t_k))(1+i\tau b(t_k))-\frac{\tau}{4!}(2ib''(t_k)-12ib(t_k)b'(t_k)-8ib^3(t_k))))u_{k+1}$$

$$=\frac{1}{2}(f(t_{k+1})+f(t_{k-1}))+\frac{3}{2}i\tau b(t_k)f(t_{k+1})-\frac{1}{2}i\tau b(t_k)f(t_{k-1})$$

$$+\frac{\tau}{3}(-f'(t_{k+1})+f'(t_{k-1})),1\leq k\leq N-1,u_0=y_0,$$

$$\frac{1}{\tau}(u_1-u_0)=(\tau b(0)(b(\tau)-b(0))+\frac{\tau}{2}(ib'(0)-b^2(\tau)))y_0$$

$$+(1+i\tau(b(0)-b(\tau)))y'(0)+\frac{\tau}{2}f(\tau)$$

(a difference scheme of second order of accuracy for an initial-value problem (3.1) generated by the second order of accuracy explicit difference scheme).

Now, we consider the boundary-value problem of the form

$$-y''(t) + a(t)y(t) = f(t), 0 < t \leq T, y(0) = y_0, y(T) = y_T \tag{3.25}$$

assuming $a(t)$ and $f(t)$ to be such that problem (3.25) has a unique smooth solution defined on $[0,T]$. But for existence of a unique solution of the boundary-value problem (3.25) we need some estimate from below of $a(t)$.

For the construction of difference schemes we consider again the uniform grid space

$$[0,T]_\tau = \{t_k = k\tau, k = 0,1,\ldots,N, N\tau = T\}.$$

The construction of the two-step difference schemes of an arbitrary high order of accuracy for the approximate solutions of the boundary-value problem (3.25) is based on the following theorem.

Theorem 3.1.2. *Let $y(t_k)$ be a solution of problem (3.25) at the grid points $t = t_k$. Then $\{y(t_k)\}_0^N$ is the solution of the boundary-value problem for the second-order difference equations*

$$-\frac{1}{\tau^2}(y(t_{k+1})-2y(t_k)+y(t_{k-1}))+\frac{1}{\tau^2}[(1-\frac{1}{\tau}\int_{t_k}^{t_{k+1}} J^2(t_{k+1},s)dsJ(t_k,t_{k-1}))y(t_{k-1})$$

$$+(-2+\frac{1}{\tau}\int_{t_{k-1}}^{t_k} J^2(t_k,s)dsJ^2(t_{k+1},t_k)+\frac{1}{\tau}\int_{t_k}^{t_{k+1}} J^2(t_{k+1},s)ds)y(t_k)$$

$$+(1-\frac{1}{\tau}\int_{t_{k-1}}^{t_k} J^2(t_k,s)dsJ(t_{k+1},t_k))y(t_{k+1})]$$

$$=\frac{1}{\tau^3}[\int_{t_{k-1}}^{t_k} J^2(t_k,s)dsJ(t_{k+1},t_k)\int_{t_k}^{t_{k+1}}\{\int_z^{t_{k+1}} J(t_{k+1},s)J(z,s)ds\}f(z)dz \tag{3.26}$$

$$+\int_{t_k}^{t_{k+1}} J^2(t_{k+1},s)ds\int_{t_{k-1}}^{t_k}\{\int_{t_{k-1}}^{z} J(t_k,s)J(z,s)ds\}f(z)dz],$$

$$1 \leq k \leq N-1, y(0) = y_0, y(T) = y_T.$$

Here

$$J(t,s) = e^{-\int_s^t b(\lambda)d\lambda},$$

where $b(t)$ is a solution of the equation

$$b^2(t) - b'(t) = a(t), \tag{3.27}$$

and the following estimate holds:

$$b(t) \geq b_0 > 0, 0 \leq t \leq T. \tag{3.28}$$

Proof. Using relation (3.27), we can obviously write the equivalent boundary-value problem for a system of first-order linear differential equations

$$\begin{cases} y'(t) + b(t)y(t) = v(t), y(0) = y_0, y(T) = y_T, \\ -v'(t) + b(t)v(t) = f(t). \end{cases}$$

Integrating these, we at once obtain

$$\begin{cases} y(t) = e^{-\int_0^t b(\lambda)d\lambda} y_0 + \int_0^t e^{-\int_s^t b(\lambda)d\lambda} v(s)ds, \\ v(t) = e^{-\int_t^T b(\lambda)d\lambda} v(T) + \int_t^T e^{-\int_t^z b(\lambda)d\lambda} f(z)dz. \end{cases}$$

That is,

$$\begin{cases} y(t) = J(t,0)y_0 + \int_0^t J(t,s)v(s)ds, \\ v(t) = J(T,t)v(T) + \int_t^T J(z,t)f(z)dz. \end{cases}$$

From these formulas and the condition $y(T) = y_T$ it follows that

$$y(T) = J(T,0)y_0 + \int_0^T J^2(T,s)dsv(T)$$

$$+ \int_0^T J(T,s) \int_s^T J(z,s)f(z)dz.$$

Since $\int_0^T e^{-2\int_s^T b(\lambda)d\lambda} ds \neq 0$, then

$$v(T) = \left(\int_0^T J^2(T,s)ds \right)^{-1} [y_T - J(T,0)y_0$$

$$- \int_0^T \left(\int_s^T J(T,s)J(z,s)ds \right) f(z)dz].$$

So, we have that

$$v(t) = J(T,t) \left(\int_0^T J^2(T,s)ds \right)^{-1} [y(T) - J(T,0)y_0$$

$$- \int_0^T \left(\int_s^T J(T,s)J(z,s)ds \right) f(z)dz] + \int_t^T J(z,t)f(z)dz$$

and

$$y(t) = J(t,0)y_0 + \int_0^t J(t,s)[J(T,s)\left(\int_0^T J^2(T,\tau)d\tau\right)^{-1}[y(T) - J(T,0)y_0$$

$$-\int_0^T \left(\int_\tau^T J(T,\tau)J(z,\tau)d\tau\right) f(z)dz] + \int_s^T J(z,s)f(z)dz]ds.$$

By an interchange of the order of integration, we obtain

$$y(t) = (\int_0^T J^2(T,s)ds)^{-1}[\{\int_t^T J^2(T,s)dsJ(t,0)y_0 + \int_0^t J(t,s)J(T,s)dsy(T)\}$$

$$-\int_0^t J(t,s)J(T,s)ds \times \int_0^T \{\int_0^z J^2(z,s)ds\}f(z)dz]$$

$$+\int_0^t \{\int_0^z J(t,s)J(z,s)ds\}f(z)dz$$

$$+\int_t^T \{\int_0^t J(t,s)J(z,s)ds\}f(z)dz = u(t) + w(t) + g(t), \tag{3.29}$$

where

$$u(t) = (\int_0^T J^2(T,s)ds)^{-1} \int_t^T J^2(T,s)dsJ(t,0)y_0,$$

$$w(t) = (\int_0^T J^2(T,s)ds)^{-1} \int_0^t J(t,s)J(T,s)dsy(T),$$

$$g(t) = -(\int_0^T J^2(T,s)ds)^{-1} \int_0^t J(t,s)J(T,s)ds \times \int_0^T \{\int_0^z J^2(z,s)ds\}f(z)dz$$

$$+\int_0^t \{\int_0^z J(t,s)J(z,s)ds\}f(z)dz + \int_t^T \{\int_0^t J(t,s)J(z,s)ds\}f(z)dz.$$

Putting $t = t_{k+1}$, $t = t_k$ and $t = t_{k-1}$ into the formula for $u(t)$, we can write

$$u(t_{k-1}) = (\int_0^T J^2(T,s)ds)^{-1} \int_{t_{k-1}}^T J^2(T,s)dsJ(t_{k-1},0)y_0,$$

$$u(t_k) = (\int_0^T J^2(T,s)ds)^{-1} \int_{t_k}^T J^2(T,s)dsJ(t_k,0)y_0$$

$$= (\int_0^T J^2(T,s)ds)^{-1}(\int_{t_{k-1}}^T J^2(T,s)dsJ(t_k,0) - \int_{t_{k-1}}^{t_k} J^2(T,s)dsJ(t_k,0))y_0$$

$$= J(t_k,t_{k-1})u(t_{k-1}) - (\int_0^T J^2(T,s)ds)^{-1}\int_{t_{k-1}}^{t_k} J^2(T,s)dsJ(t_k,0))y_0$$

$$= J(t_k,t_{k-1})u(t_{k-1}) - (\int_0^T J^2(T,s)ds)^{-1}\int_{t_{k-1}}^{t_k} J^2(t_k,s)dsJ^2(T,t_k)J(t_k,0))y_0,$$

$$u(t_{k+1}) = (\int_0^T J^2(T,s)ds)^{-1}\int_{t_{k+1}}^T J^2(T,s)dsJ(t_{k+1},0)y_0$$

$$= (\int_0^T J^2(T,s)ds)^{-1}(\int_{t_{k+1}}^T J^2(T,s)dsJ(t_{k+1},0) - \int_{t_k}^{t_{k+1}} J^2(T,s)dsJ(t_{k+1},0))y_0$$

$$= J(t_{k+1},t_k)u(t_k) - (\int_0^T J^2(T,s)ds)^{-1}\int_{t_k}^{t_{k+1}} J^2(T,s)dsJ(t_{k+1},0))y_0$$

$$= J(t_{k+1},t_k)u(t_k)$$

$$-(\int_0^T J^2(T,s)ds)^{-1}\int_{t_k}^{t_{k+1}} J^2(t_{k+1},s)dsJ^2(T,t_{k+1})J(t_{k+1},0))y_0.$$

Hence, we obtain the following relation between $u(t_{k+1})$, $u(t_k)$ and $u(t_{k-1})$:

$$\begin{aligned}&(\int_{t_k}^{t_{k+1}} J^2(t_{k+1},s)ds)^{-1}(u(t_{k+1}) - J(t_{k+1},t_k)u(t_k))\\ = \quad & -(\int_0^T J^2(T,s)ds)^{-1}J^2(T,t_{k+1})J(t_{k+1},0))y_0,\end{aligned}$$

$$\begin{aligned}&(\int_{t_{k-1}}^{t_k} J^2(t_k,s)ds)^{-1}(u(t_k) - J(t_k,t_{k-1})u(t_{k-1}))\\ = \quad & -J(t_{k+1},t_k)(\int_0^T J^2(T,s)ds)^{-1}J^2(T,t_{k+1})J(t_{k+1},0))y_0.\end{aligned}$$

Therefore,

$$\begin{aligned}&(\int_{t_{k-1}}^{t_k} J^2(t_k,s)ds)^{-1}(u(t_k) - J(t_k,t_{k-1})u(t_{k-1}))\\ = \quad & J(t_{k+1},t_k)(\int_{t_k}^{t_{k+1}} J^2(t_{k+1},s)ds)^{-1}(u(t_{k+1}) - J(t_{k+1},t_k)u(t_k))\end{aligned}$$

or

$$\int_{t_{k-1}}^{t_k} J^2(t_k,s)dsJ(t_{k+1},t_k)u(t_{k+1}) - [\int_{t_k}^{t_{k+1}} J^2(t_{k+1},s)ds$$

$$+\int_{t_{k-1}}^{t_k} J^2(t_k,s)dsJ^2(t_{k+1},t_k)]u(t_k)$$

$$+\int_{t_k}^{t_{k+1}} J^2(t_{k+1},s)dsJ(t_k,t_{k-1})u(t_{k-1})=0. \qquad (3.30)$$

In a similar manner, using the definition of $J(t,s)$ and the formula for $w(t)$, we can obtain

$$\int_{t_{k-1}}^{t_k} J^2(t_k,s)dsJ(t_{k+1},t_k)w(t_{k+1})-[\int_{t_k}^{t_{k+1}} J^2(t_{k+1},s)ds$$

$$+\int_{t_{k-1}}^{t_k} J^2(t_k,s)dsJ^2(t_{k+1},t_k)]w(t_k)$$

$$+\int_{t_k}^{t_{k+1}} J^2(t_{k+1},s)dsJ(t_k,t_{k-1})w(t_{k-1})=0. \qquad (3.31)$$

Now, putting $t=t_{k+1}$, $t=t_k$ and $t=t_{k-1}$ into the formula for $g(t)$ and using the definition of $J(t,s)$ we can obtain

$$\int_{t_{k-1}}^{t_k} J^2(t_k,s)dsJ(t_{k+1},t_k)g(t_{k+1})-[\int_{t_k}^{t_{k+1}} J^2(t_{k+1},s)ds$$

$$+\int_{t_{k-1}}^{t_k} J^2(t_k,s)dsJ^2(t_{k+1},t_k)]g(t_k)+\int_{t_k}^{t_{k+1}} J^2(t_{k+1},s)dsJ(t_k,t_{k-1})g(t_{k-1})$$

$$=-\int_{t_{k-1}}^{t_k} J^2(t_k,s)dsJ(t_{k+1},t_k)((\int_0^T J^2(T,s)ds)^{-1}\int_0^{t_{k+1}} J(t_{k+1},s)J(T,s)ds$$

$$\times\int_0^T\{\int_0^z J^2(z,s)ds\}f(z)dz)+[\int_{t_k}^{t_{k+1}} J^2(t_{k+1},s)ds+\int_{t_{k-1}}^{t_k} J^2(t_k,s)dsJ^2(t_{k+1},t_k)]$$

$$\times(\int_0^T J^2(T,s)ds)^{-1}\int_0^{t_k} J(t_k,s)J(T,s)ds\times\int_0^T\{\int_0^z J^2(z,s)ds\}f(z)dz$$

$$-\int_{t_k}^{t_{k+1}} J^2(t_{k+1},s)dsJ(t_k,t_{k-1})(\int_0^T J^2(T,s)ds)^{-1}\int_0^{t_{k-1}} J(t_{k-1},s)J(T,s)ds$$

$$\times\int_0^T\{\int_0^z J^2(z,s)ds\}f(z)dz+\int_{t_{k-1}}^{t_k} J^2(t_k,s)dsJ(t_{k+1},t_k)$$

$$(\int_0^{t_{k+1}}\{\int_0^z J(t_{k+1},s)J(z,s)ds\}f(z)dz+\int_{t_{k+1}}^T\{\int_0^{t_{k+1}} J(t_{k+1},s)J(z,s)ds\}f(z)dz)$$

$$-[\int_{t_k}^{t_{k+1}} J^2(t_{k+1},s)ds+\int_{t_{k-1}}^{t_k} J^2(t_k,s)dsJ^2(t_{k+1},t_k)]$$

$$
\begin{aligned}
&\times\Big(\int_0^{t_k}\{\int_0^z J(t_k,s)J(z,s)ds\}f(z)dz+\int_{t_k}^T\{\int_0^{t_k} J(t_k,s)J(z,s)ds\}f(z)dz\Big)\\
&+\int_{t_k}^{t_{k+1}} J^2(t_{k+1},s)dsJ(t_k,t_{k-1})\Big(\int_0^{t_{k-1}}\{\int_0^z J(t_{k-1},s)J(z,s)ds\}f(z)dz\\
&\qquad+\int_{t_{k-1}}^T\{\int_0^{t_{k-1}} J(t_{k-1},s)J(z,s)ds\}f(z)dz\Big)\\
&=\int_{t_{k-1}}^{t_k} J^2(t_k,s)dsJ(t_{k+1},t_k)\Big(\int_{t_k}^{t_{k+1}}\{\int_0^z J(t_{k+1},s)J(z,s)ds\}f(z)dz\\
&\qquad-\int_{t_k}^{t_{k+1}}\{\int_0^{t_{k+1}} J(t_{k+1},s)J(z,s)ds\}f(z)dz\Big)\\
&\qquad-\Big[\int_{t_k}^{t_{k+1}} J^2(t_{k+1},s)ds+\int_{t_{k-1}}^{t_k} J^2(t_k,s)dsJ^2(t_{k+1},t_k)\Big]\\
&+\int_{t_{k-1}}^{t_k} J^2(t_k,s)dsJ(t_{k+1},t_k)\Big(\int_0^{t_k}\{\int_0^z J(t_{k+1},s)J(z,s)ds\}f(z)dz\\
&\qquad-\int_{t_k}^T\{\int_0^{t_{k+1}} J(t_{k+1},s)J(z,s)ds\}f(z)dz\Big)\\
&\times\Big(\int_0^{t_k}\{\int_0^z J(t_k,s)J(z,s)ds\}f(z)dz+\int_{t_k}^T\{\int_0^{t_k} J(t_k,s)J(z,s)ds\}f(z)dz\Big)\\
&+\int_{t_k}^{t_{k+1}} J^2(t_{k+1},s)dsJ(t_k,t_{k-1})\Big(\int_0^{t_k}\{\int_0^z J(t_{k-1},s)J(z,s)ds\}f(z)dz\\
&\qquad+\int_{t_k}^T\{\int_0^{t_{k-1}} J(t_{k-1},s)J(z,s)ds\}f(z)dz\Big)\\
&+\int_{t_k}^{t_{k+1}} J^2(t_{k+1},s)dsJ(t_k,t_{k-1})\Big(-\int_{t_{k-1}}^{t_k}\{\int_0^z J(t_{k-1},s)J(z,s)ds\}f(z)dz\\
&\qquad+\int_{t_{k-1}}^{t_k}\{\int_0^{t_{k-1}} J(t_{k-1},s)J(z,s)ds\}f(z)dz\Big)\\
&=-\int_{t_{k-1}}^{t_k} J^2(t_k,s)dsJ(t_{k+1},t_k)\Big(\int_{t_k}^{t_{k+1}}\{\int_z^{t_{k+1}} J(t_{k+1},s)J(z,s)ds\}f(z)dz\\
&\qquad-\int_{t_k}^{t_{k+1}} J^2(t_{k+1},s)ds\int_{t_{k-1}}^{t_k}\{\int_{t_{k-1}}^z J(t_k,s)J(z,s)ds\}f(z)dz.
\end{aligned}
$$

So, using the last formula and the identities (3.30) and (3.31) we obtain the following relation between $y(t_{k+1})$, $y(t_k)$ and $y(t_{k-1})$:

$$\int_{t_{k-1}}^{t_k} J^2(t_k,s)dsJ(t_{k+1},t_k)y(t_{k+1}) - [\int_{t_k}^{t_{k+1}} J^2(t_{k+1},s)ds$$

$$+\int_{t_{k-1}}^{t_k} J^2(t_k,s)dsJ^2(t_{k+1},t_k)]y(t_k) + \int_{t_k}^{t_{k+1}} J^2(t_{k+1},s)dsJ(t_k,t_{k-1})y(t_{k-1})$$

$$= -\int_{t_{k-1}}^{t_k} J^2(t_k,s)dsJ(t_{k+1},t_k)\int_{t_k}^{t_{k+1}}\{\int_z^{t_{k+1}} J(t_{k+1},s)J(z,s)ds\}f(z)dz$$

$$-\int_{t_k}^{t_{k+1}} J^2(t_{k+1},s)ds\int_{t_{k-1}}^{t_k}\{\int_{t_{k-1}}^z J(t_k,s)J(z,s)ds\}f(z)dz.$$

From this it follows (3.26) for any $k, 1 \leq k \leq N-1$. Theorem 3.1.2 is proved.

Let us note that the boundary-value problem (3.26) is called the two-step exact difference scheme for solution of the boundary-value problem (3.25).

Note that for $b(t)$ we have the Riccati differential equation (3.27). Therefore for the smooth $a(t)$ there exists a smooth positive solution of this differential equation defined on the segment $[0,T]$.

Now, we will consider the applications of this exact difference scheme. From (3.26) it is clear that for the approximate solutions of the problem (3.25) it is necessary to approximate the expressions

$$\frac{1}{\tau}\int_{t_k}^{t_{k+1}} J^2(t_{k+1},s)ds, \frac{1}{\tau}\int_{t_{k-1}}^{t_k} J^2(t_k,s)ds, J(t_k,t_{k-1}), J(t_{k+1},t_k)$$

and

$$\int_{t_k}^{t_{k+1}}\{\int_z^{t_{k+1}} J(t_{k+1},s)J(z,s)ds\}f(z)dz, \int_{t_{k-1}}^{t_k}\{\int_{t_{k-1}}^z J(t_k,s)J(z,s)ds\}f(z)dz.$$

Let us remark that in constructing difference schemes it is important to know how to construct a right-hand side $\varphi_k^{p,q}$ that satisfies

$$\frac{1}{\tau^3}[\int_{t_{k-1}}^{t_k} J^2(t_k,s)dsJ(t_{k+1},t_k)\int_{t_k}^{t_{k+1}}\{\int_z^{t_{k+1}} J(t_{k+1},s)J(z,s)ds\}f(z)dz \quad (3.32)$$

$$+\int_{t_k}^{t_{k+1}} J^2(t_{k+1},s)ds\int_{t_{k-1}}^{t_k}\{\int_{t_{k-1}}^z J(t_k,s)J(z,s)ds\}f(z)dz] - \varphi_k^{p,q} = o(\tau^{p+q})$$

and is sufficiently simple. The choice formula $\varphi_k^{p,q}$ is not unique. Using Taylor's formula, with respect to the function

$$\int_z^{t_{k+1}} J(t_{k+1},s)J(z,s)ds\}f(z)$$

on the variable z, we obtain

$$\frac{1}{\tau^2}\int_{t_k}^{t_{k+1}}\{\int_{z}^{t_{k+1}} J(t_{k+1},s)J(z,s)ds\}f(z)dz$$

$$=\sum_{m=1}^{p+q}\sum_{\lambda=0}^{m}\binom{m}{\lambda}\beta_{m-\lambda}(t_{k+1})f^{(\lambda)}(t_{k+1})\frac{(-1)^m\tau^{m-1}}{(m+1)!}+o(\tau^{p+q}), \tag{3.33}$$

where

$$\begin{cases}\beta_0(t_{k+1})=0,\beta_1(t_{k+1})=-1,\beta_2(t_{k+1})=0,\\ \beta_m(t_{k+1})=\sum\limits_{\lambda=0}^{m-2}\binom{m-2}{\lambda}a^{(m-2-\lambda)}(t_{k+1})\beta_\lambda(t_{k+1}), 3\leq m\leq p+q.\end{cases} \tag{3.34}$$

Using Taylor's formula, with respect to the function

$$\int_{t_{k-1}}^{z} J(t_k,s)J(z,s)ds\}f(z)$$

on the variable z, we obtain

$$\frac{1}{\tau^2}\int_{t_{k-1}}^{t_k}\{\int_{t_{k-1}}^{z} J(t_k,s)J(z,s)ds\}f(z)dz$$

$$=\sum_{m=1}^{p+q}\sum_{\lambda=0}^{m}\binom{m}{\lambda}\widetilde{\beta}_{m-\lambda}(t_{k-1})f^{(\lambda)}(t_{k-1})\frac{\tau^{m-1}}{(m+1)!}+o(\tau^{p+q}), \tag{3.35}$$

where

$$\begin{cases}\widetilde{\beta}_0(t_{k-1})=0,\widetilde{\beta}_1(t_{k-1})=J(t_k,t_{k-1}),\widetilde{\beta}_2(t_{k-1})=0,\\ \widetilde{\beta}_m(t_{k-1})=\sum\limits_{\lambda=0}^{m-2}\binom{m-2}{\lambda}a^{(m-2-\lambda)}(t_{k-1})\widetilde{\beta}_\lambda(t_{k-1}), 3\leq m\leq p+q.\end{cases} \tag{3.36}$$

Moreover, for the construction of $\varphi_k^{p,q}$ we will use (3.33) and (3.35). But, first of all let us study the approximate formulas for the expressions

$$J(t_k,t_{k-1}), J(t_{k+1},t_k), \frac{1}{\tau}\int_{t_k}^{t_{k+1}} J^2(t_{k+1},s)ds, \frac{1}{\tau}\int_{t_{k-1}}^{t_k} J^2(t_k,s)ds,$$

that will be needed further for the construction of difference schemes of a high order of accuracy for the approximate solution of problem (3.25).

Let us consider the approximate formulas for the expression

$$J(t_k,t_{k-1})=e^{-\int_{t_{k-1}}^{t_k} b(s)ds}$$

constructed by Pade's rational functions. It is clear that

$$e^{-\int_{t_{k-1}}^{t_k} b(s)ds} = e^{-b(t_k)\tau} + \int_{t_{k-1}}^{t_k} e^{-\int_{t_{k-1}}^{s} b(z)dz}(b(t_k) - b(s))e^{(s-t_k)b(t_k)}ds.$$

Using Taylor's formula, we obtain

$$\begin{aligned}
e^{-\int_{t_{k-1}}^{t_k} b(z)dz} &= e^{-b(t_k)\tau} + \tau(b(t_k) - b(t_{k-1}))e^{-b(t_k)\tau} \\
&-\sum_{n=1}^{m} \frac{\tau^{n+1}}{(n+1)!}(\sum_{\lambda=1}^{n} \binom{n}{\lambda} \widetilde{c}_{n-\lambda}(t_{k-1})[(b(t_k) - b(t_{k-1}))b^{\lambda}(t_k)e^{-b(t_k)\tau} \\
&+\sum_{\mu=0}^{\lambda-1} \binom{\lambda}{\mu} b^{(\lambda-\mu)}(t_{k-1})b^{\mu}(t_k)e^{-b(t_k)\tau}] \\
&+\widetilde{c}_n(t_{k-1})(b(t_k) - b(t_{k-1}))e^{-b(t_k)\tau}) + o(\tau^{m+2}) \\
&= e^{-b(t_k)\tau} + \tau(b(t_k) - b(t_{k-1}))e^{-b(t_k)\tau} + \sum_{n=1}^{m-1} \frac{\tau^{n+1}}{(n+1)!} \\
&\times \sum_{\lambda=0}^{n} \binom{n}{\lambda} \widetilde{c}_{n-\lambda}(t_{k-1})(b(t_k) - b(t_{k-1}))b^{\lambda}(t_k)e^{-b(t_k)\tau} \\
&+\sum_{n=1}^{m} \frac{\tau^{n+1}}{(n+1)!} \sum_{\lambda=1}^{n} \widetilde{c}_{n-\lambda}(t_{k-1}) \sum_{\mu=0}^{\lambda-1} \binom{\lambda}{\mu} b^{(\lambda-\mu)}(t_{k-1})b^{\mu}(t_k)e^{-b(t_k)\tau} \\
&+o(\tau^{m+2}), \quad m \geq 2,
\end{aligned}$$

$$e^{-\int_{t_{k-1}}^{t_k} b(z)dz} = e^{-b(t_k)\tau} + \tau(b(t_k) - b(t_{k-1}))e^{-b(t_k)\tau} + \frac{\tau^2}{2}b'(t_{k-1})e^{-b(t_k)\tau} + o(\tau^3),$$

$$e^{-\int_{t_{k-1}}^{t_k} b(z)dz} = e^{-b(t_k)\tau} + o(\tau^2),$$

where

$$\begin{cases} \widetilde{c}_0(t_{k-1}) = 1, \widetilde{c}_1(t_{k-1}) = -b(t_{k-1}), \widetilde{c}_2(t_{k-1}) = a(t_{k-1}), \\ \widetilde{c}_{m+2}(t_{k-1}) = \sum_{\lambda=0}^{m} \binom{m}{\lambda} a^{(m-\lambda)}(t_{k-1})\widetilde{c}_{\lambda}(t_{k-1}), 1 \leq m. \end{cases}$$

Further, using the last formulas and Pade fractions for the function e^{-z}, we can write

$$\begin{aligned}
e^{-\int_{t_{k-1}}^{t_k} b(z)dz} &= R_{[\frac{m}{2}],m-[\frac{m}{2}]+1}(\tau b(t_k)) \qquad (3.37) \\
&+\tau(b(t_k) - b(t_{k-1}))R_{[\frac{m}{2}]-1,m-[\frac{m}{2}]}(\tau b(t_k)) \\
&+\sum_{n=1}^{m-1} \frac{\tau^{n+1}}{(n+1)!} \sum_{\lambda=0}^{n} \binom{n}{\lambda} \widetilde{c}_{n-\lambda}(t_{k-1})(b(t_k) - b(t_{k-1}))b^{\lambda}(t_k)
\end{aligned}$$

$$\times R_{[\frac{m-n}{2}]-1,m-n-[\frac{m-n}{2}]}(\tau b(t_k)) + \sum_{n=1}^{m} \frac{\tau^{n+1}}{(n+1)!} \sum_{\lambda=1}^{n} \binom{n}{\lambda} \widetilde{c}_{n-\lambda}(t_{k-1})$$

$$\times \sum_{\mu=0}^{\lambda-1} \binom{\lambda}{\mu} b^{(\lambda-\mu)}(t_{k-1}) b^{\mu}(t_k) R_{[\frac{m-n}{2}],m-n-[\frac{m-n}{2}]}(\tau b(t_k)) + o(\tau^{m+2}), m \geq 2,$$

$$e^{-\int_{t_{k-1}}^{t_k} b(z)dz} = R_{\sigma,2-\sigma}(\tau b(t_k)) + \tau(b(t_k) - b(t_{k-1})) + \frac{\tau^2}{2} b'(t_{k-1}) + o(\tau^3), \sigma = 0,1,2,$$

$$e^{-\int_{t_{k-1}}^{t_k} b(z)dz} = R_{\varkappa,1-\varkappa}(\tau b(t_k)) + o(\tau^2), \varkappa = 0,1.$$

Therefore

$$e^{-\int_{t_k}^{t_{k+1}} b(z)dz} = R_{[\frac{m}{2}],m-[\frac{m}{2}]+1}(\tau b(t_{k+1})) \tag{3.38}$$

$$+\tau(b(t_{k+1}) - b(t_k)) R_{[\frac{m}{2}]-1,m-[\frac{m}{2}]}(\tau b(t_{k+1}))$$

$$+\sum_{n=1}^{m-1} \frac{\tau^{n+1}}{(n+1)!} \sum_{\lambda=1}^{n} \binom{n}{\lambda} \widetilde{c}_{n-\lambda}(t_k)(b(t_{k+1}) - b(t_k)) b^{\lambda}(t_{k+1})$$

$$\times R_{[\frac{m-n}{2}]-1,m-n-[\frac{m-n}{2}]}(\tau b(t_{k+1})) + \sum_{n=1}^{m} \frac{\tau^{n+1}}{(n+1)!} \sum_{\lambda=1}^{n} \binom{n}{\lambda} \widetilde{c}_{n-\lambda}(t_k)$$

$$\times \sum_{\mu=0}^{\lambda-1} \binom{\lambda}{\mu} b^{(\lambda-\mu)}(t_k) b^{\mu}(t_{k+1}) R_{[\frac{m-n}{2}],m-n-[\frac{m-n}{2}]}(\tau b(t_{k+1})) + o(\tau^{m+2}), m \geq 2,$$

$$e^{-\int_{t_k}^{t_{k+1}} b(z)dz} = R_{\sigma,2-\sigma}(\tau b(t_{k+1})) + \tau(b(t_{k+1}) - b(t_k)) + \frac{\tau^2}{2} b'(t_k) + o(\tau^3), \sigma = 0,1,2,$$

$$e^{-\int_{t_k}^{t_{k+1}} b(z)dz} = R_{\varkappa,1-\varkappa}(\tau b(t_{k+1})) + o(\tau^2), \varkappa = 0,1.$$

Moreover, we can obtain

$$e^{-2\int_{t_k}^{t_{k+1}} b(z)dz} = R_{[\frac{m}{2}],m-[\frac{m}{2}]+1}(2\tau b(t_{k+1})) \tag{3.39}$$

$$+2\tau(b(t_{k+1}) - b(t_k)) R_{[\frac{m}{2}]-1,m-[\frac{m}{2}]}(2\tau b(t_{k+1}))$$

$$+\sum_{n=1}^{m-1} \frac{\tau^{n+1}}{(n+1)!} \sum_{\lambda=1}^{n} \binom{n}{\lambda} c_{n-\lambda}(t_k)(b(t_{k+1}) - b(t_k)) 2^{\lambda+1} b^{\lambda}(t_{k+1})$$

$$\times R_{[\frac{m-n}{2}]-1,m-n-[\frac{m-n}{2}]}(2\tau b(t_{k+1})) + \sum_{n=1}^{m} \frac{\tau^{n+1}}{(n+1)!} \sum_{\lambda=1}^{n} \binom{n}{\lambda} c_{n-\lambda}(t_k)$$

$$\times \sum_{\mu=0}^{\lambda-1} \binom{\lambda}{\mu} b^{(\lambda-\mu)}(t_k) 2^{\mu+1} b^{\mu}(t_{k+1}) R_{[\frac{m-n}{2}],m-n-[\frac{m-n}{2}]}(2\tau b(t_{k+1}))$$

$$+o(\tau^{m+2}), \quad m \geq 2,$$

$$e^{-2\int_{t_k}^{t_{k+1}} b(z)dz} = R_{\sigma,2-\sigma}(2\tau b(t_{k+1})) + 2\tau(b(t_{k+1}) - b(t_k)) + \tau^2 b'(t_k) + o(\tau^3),$$
$$\sigma = 0,1,2,$$
$$e^{-2\int_{t_k}^{t_{k+1}} b(z)dz} = R_{\varkappa,1-\varkappa}(2\tau b(t_{k+1})) + o(\tau^2), \varkappa = 0,1,$$

where

$$\begin{cases} c_0(t_k) = 1, c_1(t_k) = -2b(t_k), c_2(t_k) = 4a(t_k) + 2b'(t_k), \\ c_{m+2}(t_k) = \sum\limits_{\lambda=0}^{m} \binom{m}{\lambda} (4a^{(m-\lambda)}(t_k) + 2b^{(m-\lambda+1)}(t_k))c_\lambda(t_k), 1 \leq m. \end{cases}$$

Now, let us consider the approximate formulas for the expression

$$\frac{1}{\tau}\int_{t_k}^{t_{k+1}} J^2(t_{k+1}, s)ds.$$

Using Taylor's formula, we can write

$$\frac{1}{\tau}\int_{t_k}^{t_{k+1}} J^2(t_{k+1}, s)ds = \sum_{n=0}^{m} \frac{(-1)^n \tau^n}{(n+1)!} d_n(t_{k+1}) + o(\tau^{m+1}), \tag{3.40}$$

where $d_n(s) = D_n(s)J^2(t_{k+1}, s)$,

$$\begin{cases} D_n(s) = 0, n < 0, \\ D_0(s) = 1, D_{n+1}(s) = D_n'(s) + 2b(s)D_n(s), 0 \leq n. \end{cases}$$

Therefore

$$\frac{1}{\tau}\int_{t_{k-1}}^{t_k} J^2(t_k, s)ds = \sum_{n=0}^{m} \frac{(-1)^n \tau^n}{(n+1)!} d_n(t_k) + o(\tau^{m+1}). \tag{3.41}$$

Now, using formulas (3.33), (3.34), (3.35), (3.36), (3.37), (3.38), (3.39) and (3.40) and neglecting the last small terms $o(\tau^{p+q})$, $\varphi_k^{p,q}$ can be defined by the following formula:

$$\varphi_k^{p,q} = \sum_{m=1}^{p+q-3} \sum_{\lambda=0}^{m} \binom{m}{\lambda} f^{(\lambda)}(t_{k+1})\beta_{m-\lambda}(t_{k+1}) \tag{3.42}$$

$$\times \frac{(-1)^m \tau^{m-1}}{(m+1)!} \sum_{n=0}^{p+q-m-3} \frac{(-1)^n \tau^n}{(n+1)!} d_n(t_k) J_{k+1,k}^{n,p+q-m-n}$$

$$+ \sum_{m=1}^{p+q-3} \sum_{\lambda=0}^{m} \binom{m}{\lambda} f^{(\lambda)}(t_{k+1})\beta_{m-\lambda}(t_{k+1}) \frac{1}{(m+1)!}$$

$$\times \frac{(-1)^{p+q}\tau^{p+q-3}}{(p+q-m-1)!} d_{p+q-m-2}(t_k) J_{k+1,k}^{p+q-m-n-2,2}$$

$$+\sum_{m=1}^{p+q-3}\sum_{\lambda=0}^{m}\binom{m}{\lambda} f^{(\lambda)}(t_{k+1})\beta_{m-\lambda}(t_{k+1})\frac{1}{(m+1)!}$$
$$\times\frac{(-1)^{p+q-1}\tau^{p+q-2}}{(p+q-m)!}d_{p+q-m-1}(t_k)J_{k+1,k}^{p+q-m-1,1}$$

$$+\sum_{m=1}^{p+q-3}\sum_{\lambda=0}^{m}\binom{m}{\lambda} f^{(\lambda)}(t_{k+1})\beta_{m-\lambda}(t_{k+1})\frac{1}{(m+1)!}$$
$$\times\frac{(-1)^{p+q}\tau^{p+q-1}}{(p+q-m+1)!}d_{p+q-m}(t_k)J_{k+1,k}^{p+q-m,0}$$

$$+\sum_{\lambda=0}^{p+q-2}\binom{p+q-2}{\lambda} f^{(\lambda)}(t_{k+1})\beta_{p+q-2-\lambda}(t_{k+1})$$
$$\times\frac{(-1)^{p+q-2}\tau^{p+q-3}}{(p+q-1)!}\sum_{n=0}^{2}\frac{(-1)^n\tau^n}{(n+1)!}d_n(t_k)J_{k+1,k}^{n,2-n}$$

$$+\sum_{\lambda=0}^{p+q-1}\binom{p+q-1}{\lambda} f^{(\lambda)}(t_{k+1})\beta_{p+q-1-\lambda}(t_{k+1})$$
$$\times\frac{(-1)^{p+q-1}\tau^{p+q-2}}{(p+q)!}\sum_{n=0}^{1}\frac{(-1)^n\tau^n}{(n+1)!}d_n(t_k)J_{k+1,k}^{n,1-n}$$

$$+\sum_{\lambda=0}^{p+q}\binom{p+q}{\lambda} f^{(\lambda)}(t_{k+1})\beta_{p+q-\lambda}(t_{k+1})\frac{(-1)^{p+q}\tau^{p+q-1}}{(p+q+1)!}d_0(t_k)J_{k+1,k}^{0,0}$$

$$+\sum_{m=1}^{p+q-3}\sum_{\lambda=0}^{m}\binom{m}{\lambda} f^{(\lambda)}(t_{k-1})\frac{\tau^{m-1}}{(m+1)!}$$
$$\times\sum_{n=0}^{p+q-m-3}\frac{(-1)^n\tau^n}{(n+1)!}d_n(t_{k+1})\widetilde{B}_{m-\lambda}^{n,p+q-m-n}(t_{k-1})$$

$$+\sum_{m=1}^{p+q-3}\sum_{\lambda=0}^{m}\binom{m}{\lambda} f^{(\lambda)}(t_{k-1})\frac{1}{(m+1)!}\frac{(-1)^{p+q-m-2}\tau^{p+q-3}}{(p+q-m-1)!}d_{p+q-m-2}(t_{k+1})$$
$$\times\widetilde{B}_{m-\lambda}^{p+q-m-2,2}(t_{k-1})$$

$$+\sum_{m=1}^{p+q-3}\sum_{\lambda=0}^{m}\binom{m}{\lambda}f^{(\lambda)}(t_{k-1})\frac{1}{(m+1)!}$$
$$\times\frac{(-1)^{p+q-m-1}\tau^{p+q-2}}{(p+q-m)!}d_{p+q-m-1}(t_{k+1})\widetilde{B}_{m-\lambda}^{p+q-m-1,1}(t_{k-1})$$

$$+\sum_{m=1}^{p+q-3}\sum_{\lambda=0}^{m}\binom{m}{\lambda}f^{(\lambda)}(t_{k-1})\frac{1}{(m+1)!}$$
$$\times\frac{(-1)^{p+q-m}\tau^{p+q-1}}{(p+q-m+1)!}d_{p+q-m}(t_{k+1})\widetilde{B}_{m-\lambda}^{p+q-m,0}(t_{k-1})$$
$$+\sum_{\lambda=0}^{p+q-2}\binom{p+q-2}{\lambda}f^{(\lambda)}(t_{k-1})\frac{\tau^{p+q-3}}{(p+q-1)!}\sum_{n=0}^{2}\frac{(-1)^n\tau^n}{(n+1)!}d_n(t_{k+1})\widetilde{B}_{p+q-2-\lambda}^{n,2-n}(t_{k-1})$$
$$+\sum_{\lambda=0}^{p+q-1}\binom{p+q-1}{\lambda}f^{(\lambda)}(t_{k-1})\frac{\tau^{p+q-2}}{(p+q)!}\sum_{n=0}^{1}\frac{(-1)^n\tau^n}{(n+1)!}d_n(t_{k+1})\widetilde{B}_{p+q-1-\lambda}^{n,1-n}(t_{k-1})$$
$$+\sum_{\lambda=0}^{p+q}\binom{p+q}{\lambda}f^{(\lambda)}(t_{k-1})\frac{\tau^{p+q-1}}{(p+q+1)!}d_0(t_{k+1})\widetilde{B}_{p+q-\lambda}^{0,0}(t_{k-1}),$$

where

$$J_{k+1,k}^{n,0} = 1,\ J_{k+1,k}^{n,1}=R_{\varkappa,1-\varkappa}(\tau b(t_{k+1})),\ \varkappa=0,1,$$
$$J_{k+1,k}^{n,2} = R_{\sigma,2-\sigma}(\tau b(t_{k+1}))+\tau(b(t_{k+1})-b(t_k))+\frac{\tau^2 b'(t_k)}{2},\ \sigma=0,1,2,$$
$$J_{k+1,k}^{n,j}=R_{q-[\frac{m+n}{2}],p-m-n+[\frac{m+n}{2}]}(\tau b(t_{k+1}))+\tau(b(t_{k+1})-b(t_k))$$
$$\times R_{q-1-[\frac{m+n}{2}],p-1-m-n+[\frac{m+n}{2}]}(\tau b(t_{k+1}))+\sum_{\xi=1}^{p+q-m-n-2}\frac{\tau^{\xi+1}}{(\xi+1)!}\sum_{\lambda=0}^{\xi}\binom{\xi}{\lambda}\widetilde{c}_{\xi-\lambda}(t_k)$$
$$\times(b(t_{k+1})-b(t_k))b^{\lambda}(t_{k+1})R_{q-1-[\frac{m+n+\xi}{2}],p-1-m-n-\xi+[\frac{m+n+\xi}{2}]}(\tau b(t_{k+1}))$$
$$+\sum_{\xi=1}^{p+q-m-n-1}\frac{\tau^{\xi+1}}{(\xi+1)!}\sum_{\lambda=1}^{\xi}\binom{\xi}{\lambda}\widetilde{c}_{\xi-\lambda}(t_{k+1})$$
$$\times\sum_{\mu=0}^{\lambda-1}\binom{\lambda}{\mu}b^{(\lambda-\mu)}(t_k)b^{\mu}(t_{k+1})R_{q-1-[\frac{m+n+\xi}{2}],p-m-n-\xi+[\frac{m+n+\xi}{2}]}(\tau b(t_{k+1})),\ j\ge 3,$$

and

$$\begin{cases}\widetilde{B}_0^{n,j}(t_{k-1})=0,\ \widetilde{B}_1^{n,j}(t_{k-1})=J_{k,k-1}^{n,j},\ \widetilde{B}_2^{n,j}(t_{k-1})=0,\\ \widetilde{B}_m^{n,j}(t_{k-1})=\sum\limits_{\lambda=1}^{m-2}\binom{m-2}{\lambda}a^{(m-2-\lambda)}(t_{k-1})\widetilde{B}_{\lambda}^{n,j}(t_{k-1}),\ 3\le m\le p+q,\end{cases}$$

$$J^{n,0}_{k,k-1} = 1, J^{n,1}_{k+1,k} = R_{\varkappa,1-\varkappa}(\tau b(t_k)), \varkappa = 0,1,$$
$$J^{n,2}_{k,k-1} = R_{\sigma,2-\sigma}(\tau b(t_k)) + \tau(b(t_k) - b(t_{k-1})) + \frac{\tau^2 b'(t_{k-1})}{2}, \sigma = 0,1,2,$$

$$J^{n,j}_{k,k-1} = R_{q-[\frac{m+n}{2}],p-m-n+[\frac{m+n}{2}]}(\tau b(t_k)) + \tau(b(t_k) - b(t_{k-1}))$$
$$\times R_{q-1-[\frac{m+n}{2}],p-1-m-n+[\frac{m+n}{2}]}(\tau b(t_k)) + \sum_{\xi=1}^{p+q-m-n-2} \frac{\tau^{\xi+1}}{(\xi+1)!} \sum_{\lambda=0}^{\xi} \binom{\xi}{\lambda} \widetilde{c}_{\xi-\lambda}(t_{k-1})$$
$$\times (b(t_k) - b(t_{k-1}))(b(t_k))^\lambda R_{q-1-[\frac{m+n+\xi}{2}],p-1-m-n-\xi+[\frac{m+n+\xi}{2}]}(\tau b(t_k))$$
$$+ \sum_{\xi=1}^{p+q-m-n-1} \frac{\tau^{\xi+1}}{(\xi+1)!} \sum_{\lambda=1}^{\xi} \binom{\xi}{\lambda} \widetilde{c}_{\xi-\lambda}(t_{k-1})$$
$$\times \sum_{\mu=0}^{\lambda-1} \binom{\lambda}{\mu} b^{(\lambda-\mu)}(t_{k-1})(b(t_k))^\mu R_{q-1-[\frac{m+n+\xi}{2}],p-m-n-\xi+[\frac{m+n+\xi}{2}]}(\tau b(t_k)), j \geq 3.$$

Now, using formulas (3.37), (3.38), (3.39) and (3.40), we can write

$$\frac{1}{\tau^2}(1 - \frac{1}{\tau}\int_{t_k}^{t_{k+1}} J^2(t_{k+1},s)dsJ(t_k,t_{k-1})) \tag{3.43}$$

$$= \frac{1}{\tau^2}(1 - \sum_{m=0}^{p+q-2} \frac{(-1)^m \tau^m}{(m+1)!} d_m(t_{k+1})[R_{q-[\frac{m}{2}],p-m+[\frac{m}{2}]+1}(\tau b(t_k))$$
$$+\tau(b(t_k) - b(t_{k-1}))R_{q-[\frac{m}{2}]-1,p-m+[\frac{m}{2}]}(\tau b(t_k))$$
$$+ \sum_{n=1}^{p+q-m-1} \frac{\tau^{n+1}}{(n+1)!} \sum_{\lambda=0}^{n} \binom{n}{\lambda} \widetilde{c}_{n-\lambda}(t_{k-1})(b(t_k) - b(t_{k-1}))(b(t_k))^\lambda$$
$$\times R_{q-[\frac{m+n}{2}]-1,p-m-n+[\frac{m+n}{2}]}(\tau b(t_k)) + \sum_{n=1}^{p+q-m} \frac{\tau^{n+1}}{(n+1)!} \sum_{\lambda=1}^{n} \widetilde{c}_{n-\lambda}(t_{k-1})$$
$$\times \sum_{\mu=0}^{\lambda-1} \binom{\lambda}{\mu} b^{(\lambda-\mu)}(t_{k-1})(b(t_k))^\mu R_{q-[\frac{m+n}{2}],p-m-n+[\frac{m+n}{2}]}(\tau b(t_k))]$$
$$-\frac{(-1)^{p+q-1}\tau^{p+q-1}}{(p+q)!} d_{p+q-1}(t_{k+1})(R_{\sigma,2-\sigma}(\tau b(t_k)) + \tau(b(t_k) - b(t_{k-1})) + \frac{\tau^2}{2} b'(t_{k-1}))$$
$$-\frac{(-1)^{p+q}\tau^{p+q}}{(p+q+1)!} d_{p+q}(t_{k+1})R_{\varkappa,1-\varkappa}(\tau b(t_k)) - \frac{(-1)^{p+q+1}\tau^{p+q+1}}{(p+q+2)!} d_{p+q+1}(t_{k+1}))$$
$$+o(\tau^{p+q}), \sigma = 0,1,2, \varkappa = 0,1;$$

$$\frac{1}{\tau^2}(-2+\frac{1}{\tau}\int_{t_{k-1}}^{t_k} J^2(t_k,s)dsJ^2(t_{k+1},t_k)+\frac{1}{\tau}\int_{t_k}^{t_{k+1}} J^2(t_{k+1},s)ds) \tag{3.44}$$

$$= \frac{1}{\tau^2}(-2+\sum_{n=0}^{p+q+1}\frac{(-1)^n\tau^n}{(n+1)!}d_n(t_{k+1})$$

$$+\sum_{m=0}^{p+q-2}\frac{(-1)^m\tau^m}{(m+1)!}d_m(t_k)[R_{q-[\frac{m}{2}],p-m+[\frac{m}{2}]+1}(2\tau b(t_{k+1}))$$

$$+2\tau(b(t_{k+1})-b(t_k))R_{q-[\frac{m}{2}]-1,p-m+[\frac{m}{2}]}(2\tau b(t_{k+1}))$$

$$+\sum_{n=1}^{p+q-m-1}\frac{\tau^{n+1}}{(n+1)!}\sum_{\lambda=0}^{n}\binom{n}{\lambda}c_{n-\lambda}(t_k)(b(t_{k+1})-b(t_k))2^{\lambda+1}b^\lambda(t_{k+1})$$

$$\times R_{q-[\frac{m+n}{2}]-1,p-m-n+[\frac{m+n}{2}]}(2\tau b(t_{k+1}))+\sum_{n=1}^{p+q-m}\frac{\tau^{n+1}}{(n+1)!}\sum_{\lambda=1}^{n}c_{n-\lambda}(t_k)$$

$$\times\sum_{\mu=0}^{\lambda-1}\binom{\lambda}{\mu}b^{(\lambda-\mu)}(t_k)2^{\mu+1}b^\mu(t_{k+1})R_{q-[\frac{m+n}{2}],p-m-n+[\frac{m+n}{2}]}(2\tau b(t_{k+1}))]$$

$$+\frac{(-1)^{p+q-1}\tau^{p+q-1}}{(p+q)!}d_{p+q-1}(t_k)(R_{\sigma,2-\sigma}(2\tau b(t_{k+1}))+2\tau(b(t_{k+1})-b(t_k))+\tau^2 b'(t_k))$$

$$+\frac{(-1)^{p+q}\tau^{p+q}}{(p+q+1)!}d_{p+q}(t_k)R_{\varkappa,1-\varkappa}(2\tau b(t_{k+1}))+\frac{(-1)^{p+q+1}\tau^{p+q+1}}{(p+q+2)!}d_{p+q+1}(t_k))$$

$$+o(\tau^{p+q}),\sigma=0,1,2,\varkappa=0,1;$$

$$\frac{1}{\tau^2}(1-\frac{1}{\tau}\int_{t_{k-1}}^{t_k} J^2(t_k,s)dsJ(t_{k+1},t_k)) \tag{3.45}$$

$$= \frac{1}{\tau^2}(1-\sum_{m=0}^{p+q-2}\frac{(-1)^m\tau^m}{(m+1)!}d_m(t_k)[R_{q-[\frac{m}{2}],p-m+[\frac{m}{2}]+1}(\tau b(t_{k+1}))$$

$$+\tau(b(t_{k+1})-b(t_k))R_{q-[\frac{m}{2}]-1,p-m+[\frac{m}{2}]}(\tau b(t_{k+1}))$$

$$+\sum_{n=1}^{p+q-m-1}\frac{\tau^{n+1}}{(n+1)!}\sum_{\lambda=0}^{n}\binom{n}{\lambda}\widetilde{c}_{n-\lambda}(t_k)(b(t_{k+1})-b(t_k))b^\lambda(t_{k+1})$$

$$\times R_{q-[\frac{m+n}{2}]-1,p-m-n+[\frac{m+n}{2}]}(\tau b(t_{k+1}))+\sum_{n=1}^{p+q-m}\frac{\tau^{n+1}}{(n+1)!}\sum_{\lambda=1}^{n}\widetilde{c}_{n-\lambda}(t_k)$$

$$\times\sum_{\mu=0}^{\lambda-1}\binom{\lambda}{\mu}b^{(\lambda-\mu)}(t_k)b^\mu(t_{k+1})R_{q-[\frac{m+n}{2}],p-m-n+[\frac{m+n}{2}]}(\tau b(t_{k+1}))]$$

$$-\frac{(-1)^{p+q-1}\tau^{p+q-1}}{(p+q)!}d_{p+q-1}(t_k)(R_{\sigma,2-\sigma}(\tau b(t_{k+1}))+\tau(b(t_{k+1})-b(t_k))+\frac{\tau^2}{2}b'(t_k))$$

$$-\frac{(-1)^{p+q}\tau^{p+q}}{(p+q+1)!}d_{p+q}(t_k)R_{\varkappa,1-\varkappa}(\tau b(t_{k+1}))-\frac{(-1)^{p+q+1}\tau^{p+q+1}}{(p+q+2)!}d_{p+q+1}(t_k))$$

$$+o(\tau^{p+q}),\sigma=0,1,2,\varkappa=0,1.$$

Now, neglecting the last terms in formulas (3.43), (3.44), (3.45) and using formula (3.42), we obtain the difference schemes

$$-\frac{1}{\tau^2}(u_{k+1}-2u_k+u_{k-1})(1-2\tau b(t_k))+a_{1,k}u_{k-1}+a_{2,k}u_k+a_{3,k}u_{k+1}$$

$$=\varphi_k^{p,q},1\le k\le N-1,u_0=y_0,u_N=y_T, \tag{3.46}$$

where

$$a_{1,k}=\frac{1}{\tau^2}(1-2\tau b(t_k)-\sum_{m=0}^{p+q-2}\frac{(-1)^m\tau^m}{(m+1)!}d_m(t_{k+1})[R_{q-[\frac{m}{2}],p-m+[\frac{m}{2}]+1}(\tau b(t_k))$$

$$+\tau(b(t_k)-b(t_{k-1}))R_{q-[\frac{m}{2}]-1,p-m+[\frac{m}{2}]}(\tau b(t_k))$$

$$+\sum_{n=1}^{p+q-m-1}\frac{\tau^{n+1}}{(n+1)!}\sum_{\lambda=0}^{n}\binom{n}{\lambda}\widetilde{c}_{n-\lambda}(t_{k-1})(b(t_k)-b(t_{k-1}))b^{\lambda}(t_k)$$

$$\times R_{q-[\frac{m+n}{2}]-1,p-m-n+[\frac{m+n}{2}]}(\tau b(t_k))+\sum_{n=1}^{p+q-m}\frac{\tau^{n+1}}{(n+1)!}\sum_{\lambda=1}^{n}\widetilde{c}_{n-\lambda}(t_{k-1})$$

$$\times\sum_{\mu=0}^{\lambda-1}\binom{\lambda}{\mu}b^{(\lambda-\mu)}(t_{k-1})b^{\mu}(t_k)R_{q-[\frac{m+n}{2}],p-m-n+[\frac{m+n}{2}]}(\tau b(t_k))]$$

$$-\frac{(-1)^{p+q-1}\tau^{p+q-1}}{(p+q)!}d_{p+q-1}(t_{k+1})(R_{\sigma,2-\sigma}(\tau b(t_k))+\tau(b(t_k)-b(t_{k-1}))+\frac{\tau^2}{2}b'(t_{k-1}))$$

$$-\frac{(-1)^{p+q}\tau^{p+q}}{(p+q+1)!}d_{p+q}(t_{k+1})R_{\varkappa,1-\varkappa}(\tau b(t_k))$$

$$-\frac{(-1)^{p+q+1}\tau^{p+q+1}}{(p+q+2)!}d_{p+q+1}(t_{k+1})),\sigma=0,1,2,\varkappa=0,1,$$

$$a_{2,k}=\frac{1}{\tau^2}(-2+4\tau b(t_k)+\sum_{n=0}^{p+q+1}\frac{(-1)^n\tau^n}{(n+1)!}d_n(t_{k+1})$$

$$+\sum_{m=0}^{p+q-2}\frac{(-1)^m\tau^m}{(m+1)!}d_m(t_k)[R_{q-[\frac{m}{2}],p-m+[\frac{m}{2}]+1}(2\tau b(t_{k+1}))$$

$$+2\tau(b(t_{k+1})-b(t_k))R_{q-[\frac{m}{2}]-1,p-m+[\frac{m}{2}]}(2\tau b(t_{k+1}))$$

$$+\sum_{n=1}^{p+q-m-1}\frac{\tau^{n+1}}{(n+1)!}\sum_{\lambda=0}^{n}\binom{n}{\lambda}c_{n-\lambda}(t_k)(b(t_{k+1})-b(t_k))2^{\lambda+1}b^{\lambda}(t_{k+1})$$

$$\times R_{q-[\frac{m+n}{2}]-1,p-m-n+[\frac{m+n}{2}]}(2\tau b(t_{k+1}))+\sum_{n=1}^{p+q-m}\frac{\tau^{n+1}}{(n+1)!}\sum_{\lambda=1}^{n}c_{n-\lambda}(t_k)$$

$$\times\sum_{\mu=0}^{\lambda-1}\binom{\lambda}{\mu}b^{(\lambda-\mu)}(t_k)2^{\mu+1}b^{\mu}(t_{k+1})R_{q-[\frac{m+n}{2}],p-m-n+[\frac{m+n}{2}]}(2\tau b(t_{k+1}))]$$

$$+\frac{(-1)^{p+q-1}\tau^{p+q-1}}{(p+q)!}d_{p+q-1}(t_k)(R_{\sigma,2-\sigma}(2\tau b(t_{k+1}))+2\tau(b(t_{k+1})-b(t_k))+\tau^2b'(t_k))$$

$$+\frac{(-1)^{p+q}\tau^{p+q}}{(p+q+1)!}d_{p+q}(t_k)R_{\varkappa,1-\varkappa}(2\tau b(t_{k+1}))+\frac{(-1)^{p+q+1}\tau^{p+q+1}}{(p+q+2)!}d_{p+q+1}(t_k)),$$

$$\sigma=0,1,2,\varkappa=0,1,$$

$$a_{3,k}=\frac{1}{\tau^2}(1-2\tau b(t_k)-\sum_{m=0}^{p+q-2}\frac{(-1)^m\tau^m}{(m+1)!}d_m(t_k)[R_{q-[\frac{m}{2}],p-m+[\frac{m}{2}]+1}(\tau b(t_{k+1}))$$

$$+\tau(b(t_{k+1})-b(t_k))R_{q-[\frac{m}{2}]-1,p-m+[\frac{m}{2}]}(\tau b(t_{k+1}))$$

$$+\sum_{n=1}^{p+q-m-1}\frac{\tau^{n+1}}{(n+1)!}\sum_{\lambda=0}^{n}\binom{n}{\lambda}\widetilde{c}_{n-\lambda}(t_k)(b(t_{k+1})-b(t_k))b^{\lambda}(t_{k+1})$$

$$\times R_{q-[\frac{m+n}{2}]-1,p-m-n+[\frac{m+n}{2}]}(\tau b(t_{k+1}))+\sum_{n=1}^{p+q-m}\frac{\tau^{n+1}}{(n+1)!}\sum_{\lambda=1}^{n}\widetilde{c}_{n-\lambda}(t_k)$$

$$\times\sum_{\mu=0}^{\lambda-1}\binom{\lambda}{\mu}b^{(\lambda-\mu)}(t_k)b^{\mu}(t_{k+1})R_{q-[\frac{m+n}{2}],p-m-n+[\frac{m+n}{2}]}(\tau b(t_{k+1}))]$$

$$-\frac{(-1)^{p+q-1}\tau^{p+q-1}}{(p+q)!}d_{p+q-1}(t_k)(R_{\sigma,2-\sigma}(\tau b(t_{k+1}))+\tau(b(t_{k+1})-b(t_k))+\frac{\tau^2}{2}b'(t_k))$$

$$-\frac{(-1)^{p+q}\tau^{p+q}}{(p+q+1)!}d_{p+q}(t_k)R_{\varkappa,1-\varkappa}(\tau b(t_{k+1}))-\frac{(-1)^{p+q+1}\tau^{p+q+1}}{(p+q+2)!}d_{p+q+1}(t_k)),$$

$$\sigma=0,1,2,\varkappa=0,1.$$

Thus, we constructed the difference schemes of $(p+q)$-order of accuracy for the approximate solution of the boundary-value problem (3.25). Finally, let us give examples of the simplest difference schemes.

In the case $p+q=1$ from formulas (3.42), (3.43), (3.44), (3.45) and (3.46) it follows that

$$-\frac{1}{\tau^2}(u_{k+1}-2u_k+u_{k-1})(1-2\tau b(t_k))+(-\frac{b^2(t_k)}{2}-b(t_{k+1})b(t_k)-\frac{2}{3}b^2(t_{k+1})$$

$$+\tau^{-1}(b(t_{k+1})-b(t_k))-\tau^{-1}(b(t_k)-b(t_{k-1}))-\frac{1}{2}b'(t_{k-1})-\frac{1}{3}b'(t_{k+1}))u_{k-1}$$

$$+(\frac{8b^2(t_{k+1})}{3}+2b(t_{k+1})b(t_k)+\frac{2}{3}b^2(t_k)$$
$$-\tau^{-1}(b(t_{k+1})-b(t_k))+\frac{4}{3}b'(t_k)+\frac{1}{3}b'(t_{k+1}))u_k$$

$$+(-\frac{b^2(t_{k+1})}{2}-b(t_{k+1})b(t_k)-\frac{2}{3}b^2(t_k)-\frac{1}{2}b'(t_k)-\frac{1}{3}b'(t_k))u_{k+1}$$
$$=\frac{1}{2}(f(t_{k+1})+f(t_{k-1})),1\le k\le N-1,u_0=y_{0,}u_N=y_T$$

(a difference scheme of first order of accuracy for the boundary-value problem (3.25) generated by the first order of accuracy explicit difference scheme).

In the case $p+q=2$ from formulas (3.42), (3.43), (3.44), (3.45) and (3.46) it follows that

$$-\frac{1}{\tau^2}(u_{k+1}-2u_k+u_{k-1})(1-2\tau b(t_k))+(\frac{1}{\tau}(b(t_{k+1})-b(t_k))$$

$$-\frac{b^2(t_k)}{2}+\frac{\tau b^3(t_k)}{6}-\frac{1}{\tau}(b(t_k)-b(t_{k-1}))(1-\tau b(t_k))+\frac{1}{2\tau}b(t_{k-1})(b(t_k)-b(t_{k-1}))$$
$$-\frac{1}{2\tau}(b(t_k)-b(t_{k-1}))b(t_k)-\frac{1}{2}b'(t_{k-1})(1-\tau b(t_k))-\frac{\tau}{3!}(-b(t_{k-1}))b'(t_{k-1})$$
$$-\frac{\tau}{3!}(b^2(t_{k-1})+2b'(t_{k-1})b(t_k))$$

$$+b(t_{k+1})(-b(t_k)+\frac{\tau b^2(t_k)}{2}+(b(t_k)-b(t_{k-1}))+\frac{\tau}{2}b'(t_{k-1}))$$

$$-\frac{1}{3!}(2b'(t_{k+1})+4b^2(t_{k+1}))(1-\tau b(t_k))$$
$$+\frac{\tau}{4!}(2b''(t_{k+1})+12b'(t_{k+1})b(_{k+1})+8b^3(t_{k+1}))))u_{k-1}$$

$$+(\tau^{-1}(b(t_{k+1})-b(t_k))+\frac{1}{6}(2b'(t_{k+1})+4b^2(t_{k+1}))$$
$$-\frac{\tau}{4!}(2b''(t_{k+1})+12b'(t_{k+1})b(t_{k+1})$$
$$+8b^3(t_{k+1}))+2b^2(t_{k+1})-\frac{4\tau b^3(t_{k+1})}{3}-4(b(t_{k+1})-b(t_k))b(t_{k+1})$$

$$-2\tau^{-1}b(t_k)(b(t_{k+1})-b(t_k))+2\tau^{-1}(b(t_{k+1})-b(t_k))b(t_{k+1})$$

$$+b'(t_k)(1-2\tau b(t_{k+1}))-\frac{2\tau}{3!}b(t_k)b'(t_k)+\frac{\tau}{3!}(2b''(t_k)+4b'(t_k)b(t_{k+1}))$$

$$-b(t_k)(-2b(t_{k+1})+2\tau b^2(t_{k+1})+2(b(t_{k+1})-b(t_k))+\tau b'(t_k))$$
$$+\frac{\tau}{3!}(2b'(t_k)+4b^2(t_k))(1-2\tau b(t_{k+1}))$$
$$-\frac{\tau^2}{4!}(2b''(t_k)+12b'(t_k)b(t_k)+8b^3(t_k)))u_k$$

$$+(-\frac{b^2(t_{k+1})}{2}+\frac{\tau b^3(t_{k+1})}{6}+(b(t_{k+1})-b(t_k))b(t_{k+1})$$

$$+\frac{\tau^{-1}}{2!}b(t_k)(b(t_{k+1})-b(t_k))-\frac{\tau^{-1}}{2!}(b(t_{k+1})-b(t_k))b(t_{k+1})$$

$$-\frac{1}{2}b'(t_k)(1-\tau b(t_{k+1}))+\frac{\tau}{3!}b(t_k)b'(t_k)-\frac{\tau}{3!}(b''(t_k)+2b'(t_k)b(t_{k+1}))$$

$$+b(t_k)(-b(t_{k+1})+\frac{\tau b^2(t_{k+1})}{2}+(b(t_{k+1})-b(t_k))+\frac{\tau}{2}b'(t_k))$$

$$-\frac{1}{3!}(2b'(t_k)+4b^2(t_k))(1-\tau b(t_{k+1}))+\frac{\tau}{4!}(2b''(t_k)+12b'(t_k)b(t_k)+8b^3(t_k)))u_{k+1}$$

$$=\frac{1}{2}(f(t_{k+1})+f(t_{k-1}))-\frac{\tau}{2}f(t_{k+1})(b(t_{k+1})+b(t_k))-\frac{\tau}{2}f(t_{k-1})(b(t_{k+1})+b(t_{k-1}))$$

$$-\frac{\tau}{2}(f'(t_{k+1})-f'(t_{k-1})),1\leq k\leq N-1,u_0=y_0,u_N=y_T$$

a difference scheme of second order of accuracy for the boundary-value problem (3.25) generated by an explicit difference scheme.

3.2 Taylor's Decomposition on Three Points and Its Applications

We consider again the initial-value problem (3.1). The utilization of Taylor's decomposition on three points in the construction of the two-step difference schemes of a high order of accuracy for the approximate solutions of problem (3.1) is based on the following two theorems.

Theorem 3.2.1. *Let $\eta_m, m=1,\dots,q$ be a solution of the system*

$$\sum_{m=1}^{q}\frac{(2j)!\eta_m}{(2j-2m)!}=1,p+1\leq j\leq p+q. \tag{3.47}$$

Let the function $v(t)$ $(0 \leq t \leq T)$ have a $(2p+2q+2)$-th continuous derivative and $t_{k-1}, t_k, t_{k+1} \in [0,T]_\tau$. Then the following relation holds:

$$v(t_{k+1}) - 2v(t_k) + v(t_{k-1}) - \sum_{j=1}^{p} \alpha_j v^{(2j)}(t_k)\tau^{2j} \tag{3.48}$$

$$- \sum_{j=1}^{q} \eta_j (v^{(2j)}(t_{k-1}) + v^{(2j)}(t_{k+1}))\tau^{2j} = o(\tau^{2p+2q+2}),$$

where

$$\begin{cases} \alpha_1 = 1 - 2\eta_1, \alpha_j = (\frac{1}{(2j)!} - \eta_j - \sum\limits_{m=1}^{j-1} \eta_m \frac{1}{(2(j-m))!})2 & \text{for any } j, 2 \leq j \leq q, \\ \alpha_j = (\frac{1}{(2j)!} - \sum\limits_{m=1}^{q} \eta_m \frac{1}{(2(j-m))!})2 \text{ for any } j, q+1 \leq j \leq p. \end{cases}$$

Proof. We obtain the formula

$$\int_{t_k}^{t_{k+1}} (t_{k+1} - s)^{2p+2q+1} v^{(2p+2q+2)}(s)ds \tag{3.49}$$

$$+ \int_{t_{k-1}}^{t_k} (s - t_{k-1})^{2p+2q+1} v^{(2p+2q+2)}(s)ds$$

$$= (2p+2q+1)!(v(t_{k+1}) - 2v(t_k) + v(t_{k-1}))$$

$$- \sum_{j=1}^{p+q} \frac{(2p+2q+1)!}{(2j)!} 2\tau^{2j} v^{(2j)}(t_k).$$

Using the formula of integration by parts, we obtain

$$\int_{t_k}^{t_{k+1}} (t_{k+1} - s)^{2p+2q+1} v^{(2p+2q+2)}(s)ds$$

$$+ \int_{t_{k-1}}^{t_k} (s - t_{k-1})^{2p+2q+1} v^{(2p+2q+2)}(s)ds$$

$$= \sum_{\gamma=1}^{2p+2q+2} (-1)^{2p+2q+2-\gamma}((t_{k+1} - s)^{2p+2q+1})^{(2p+2q+2-\gamma)} v^{(\gamma-1)}(s)]_{t_k}^{t_{k+1}}$$

$$+ \sum_{\gamma=1}^{2p+2q+2} (-1)^{2p+2q+2-\gamma}((s - t_{k-1})^{2p+2q+1})^{(2p+2q+2-\gamma)} v^{(\gamma-1)}(s)]_{t_{k-1}}^{t_k}$$

$$= (2p+2q+1)!(v(t_{k+1}) - v(t_k)) + \sum_{\gamma=2}^{2p+2q+2} (-1)\tau^{j-1} \frac{(2p+2q+1)!}{(j-1)!} v^{(\gamma-1)}(t_k)$$

$$-(2p+2q+1)!(v(t_k)-v(t_{k-1}))+\sum_{\gamma=2}^{2p+2q+2}(-1)^{\gamma}\tau^{j-1}\frac{(2p+2q+1)!}{(j-1)!}v^{(\gamma-1)}(t_k)$$

$$=(2p+2q+1)!(v(t_{k+1})-2v(t_k)+v(t_{k-1}))-\sum_{j=1}^{p+q}\frac{(2p+2q+1)!}{(2j)!}2\tau^{2j}v^{(2j)}(t_k).$$

In a similar manner one can show that

$$\tau^{2m}\int_{t_k}^{t_{k+1}}(t_{k+1}-s)^{2p+2q-2m+1}v^{(2p+2q+2)}(s)ds \tag{3.50}$$

$$+\tau^{2m}\int_{t_{k-1}}^{t_k}(s-t_{k-1})^{2p+2q-2m+1}v^{(2p+2q+2)}(s)ds$$

$$=(2p+2q-2m+1)!\tau^{2m}(v^{(2m)}(t_{k+1})-2v^{(2m)}(t_k)+v^{(2m)}(t_{k-1}))$$

$$-\sum_{j=m+1}^{p+q}\frac{(2p+2q-2m+1)!}{(2(j-m))!}2\tau^{2j}v^{(2j)}(t_k),\ m=1,\dots,q.$$

Now, using formulas (3.49) and (3.50), we obtain that

$$v(t_{k+1})-2v(t_k)+v(t_{k-1})-\sum_{j=1}^{p+q}\frac{1}{(2j)!}2\tau^{2j}v^{(2j)}(t_k)$$

$$-\sum_{m=1}^{q}\eta_m\tau^{2m}(v^{(2m)}(t_{k+1})-2v^{(2m)}(t_k)+v^{(2m)}(t_{k-1}))$$

$$+\sum_{m=1}^{q}\eta_m\sum_{j=m+1}^{p+q}\frac{1}{(2(j-m))!}2\tau^{2j}v^{(2j)}(t_k)$$

$$=\frac{1}{(2p+2q+1)!}[\int_{t_k}^{t_{k+1}}(t_{k+1}-s)^{2p+2q+1}v^{(2p+2q+2)}(s)ds$$

$$+\int_{t_{k-1}}^{t_k}(s-t_{k-1})^{2p+2q+1}v^{(2p+2q+2)}(s)ds]$$

$$-\sum_{m=1}^{q}\eta_m\frac{1}{(2p+2q-2m+1)!}[\tau^{2m}\int_{t_k}^{t_{k+1}}(t_{k+1}-s)^{2p+2q-2m+1}v^{(2p+2q+2)}(s)ds$$

$$+\tau^{2m}\int_{t_{k-1}}^{t_k}(s-t_{k-1})^{2p+2q-2m+1}v^{(2p+2q+2)}(s)ds].$$

Since

$$-\sum_{j=1}^{p+q}\frac{1}{(2j)!}2\tau^{2j}v^{(2j)}(t_k)$$

$$-\sum_{m=1}^{q}\eta_m\tau^{2m}(v^{(2m)}(t_{k+1})-2v^{(2m)}(t_k)+v^{(2m)}(t_{k-1}))$$

$$+\sum_{m=1}^{q}\eta_m\sum_{j=m+1}^{p+q}\frac{1}{(2(j-m))!}2\tau^{2j}v^{(2j)}(t_k)$$

$$=-\sum_{j=1}^{p}\frac{1}{(2j)!}2\tau^{2j}v^{(2j)}(t_k)-\sum_{j=p+1}^{p+q}\frac{1}{(2j)!}2\tau^{2j}v^{(2j)}(t_k)$$

$$-\sum_{m=1}^{q}\eta_m\tau^{2m}(v^{(2m)}(t_{k+1})+v^{(2m)}(t_{k-1}))+\sum_{m=1}^{q}\eta_m\tau^{2m}2v^{(2m)}(t_k)$$

$$+\sum_{j=2}^{q}\sum_{m=1}^{j-1}\eta_m\frac{2}{(2(j-m))!}\tau^{2j}v^{(2j)}(t_k)$$

$$+\sum_{j=q+1}^{p+q}\sum_{m=1}^{q}\eta_m\frac{2}{(2(j-m))!}\tau^{2j}v^{(2j)}(t_k)$$

$$=-\sum_{j=1}^{p}\frac{1}{(2j)!}2\tau^{2j}v^{(2j)}(t_k)-\sum_{j=p+1}^{p+q}\frac{2}{(2j)!}(1-\sum_{m=1}^{q}\eta_m\frac{(2j)!}{(2(j-m))!})\tau^{2j}v^{(2j)}(t_k)$$

$$-\sum_{m=1}^{q}\eta_m\tau^{2m}(v^{(2m)}(t_{k+1})+v^{(2m)}(t_{k-1}))+\sum_{m=1}^{q}\eta_m\tau^{2m}2v^{(2m)}(t_k)$$

$$+\sum_{j=2}^{q}\sum_{m=1}^{j-1}\eta_m\frac{2}{(2(j-m))!}\tau^{2j}v^{(2j)}(t_k)+\sum_{j=q+1}^{p}\sum_{m=1}^{q}\eta_m\frac{2}{(2(j-m))!}\tau^{2j}v^{(2j)}(t_k)$$

and (3.47) holds, then (3.48) follows. Theorem 3.2.1 is proved.

Theorem 3.2.2. *Let η_m, $m=1,\dots,q$ be a solution of the system*

$$\eta_j=\frac{1}{(2j)!}-\sum_{m=1}^{j-1}\frac{\eta_m}{(2j-2m)!},p+1\le j\le q.$$

$$\sum_{m=1}^{q}\frac{(2j)!\eta_m}{(2j-2m)!}=1,q+1\le j\le p+q.$$

Let the function $v(t)$ $(0\le t\le T)$ have a $(2p+2q+2)$-th continuous derivative and $t_{k-1},t_k,t_{k+1}\in[0,T]_\tau$. Then the following relation holds:

$$v(t_{k+1})-2v(t_k)+v(t_{k-1})-\sum_{j=1}^{p}\alpha_jv^{(2j)}(t_k)\tau^{2j}$$

$$-\sum_{j=1}^{q}\eta_j(v^{(2j)}(t_{k-1})+v^{(2j)}(t_{k+1}))\tau^{2j}=o(\tau^{2p+2q+2}),$$

where

$$\alpha_1=1-2\eta_1, \alpha_j=(\frac{1}{(2j)!}-\eta_j-\sum_{m=1}^{j-1}\eta_m\frac{1}{(2(j-m))!})2 \text{ for any } j, 2\le j\le p.$$

Proof. Using formulas (3.49) and (3.50), we obtain that

$$v(t_{k+1})-2v(t_k)+v(t_{k-1})-\sum_{j=1}^{p+q}\frac{1}{(2j)!}2\tau^{2j}v^{(2j)}(t_k)$$

$$-\sum_{m=1}^{q}\eta_m\tau^{2m}(v^{(2m)}(t_{k+1})-2v^{(2m)}(t_k)+v^{(2m)}(t_{k-1}))$$

$$+\sum_{m=1}^{q}\eta_m\sum_{j=m+1}^{p+q}\frac{1}{(2(j-m))!}2\tau^{2j}v^{(2j)}(t_k)$$

$$=\frac{1}{(2p+2q+1)!}[\int_{t_k}^{t_{k+1}}(t_{k+1}-s)^{2p+2q+1}v^{(2p+2q+2)}(s)ds$$

$$+\int_{t_{k-1}}^{t_k}(s-t_{k-1})^{2p+2q+1}v^{(2p+2q+2)}(s)ds]$$

$$-\sum_{m=1}^{q}\eta_m\frac{1}{(2p+2q-2m+1)!}[\tau^{2m}\int_{t_k}^{t_{k+1}}(t_{k+1}-s)^{2p+2q-2m+1}v^{(2p+2q+2)}(s)ds$$

$$+\tau^{2m}\int_{t_{k-1}}^{t_k}(s-t_{k-1})^{2p+2q-2m+1}v^{(2p+2q+2)}(s)ds].$$

We have that

$$-\sum_{j=1}^{p+q}\frac{1}{(2j)!}2\tau^{2j}v^{(2j)}(t_k)$$

$$-\sum_{m=1}^{q}\eta_m\tau^{2m}(v^{(2m)}(t_{k+1})-2v^{(2m)}(t_k)+v^{(2m)}(t_{k-1}))$$

$$+\sum_{m=1}^{q}\eta_m\sum_{j=m+1}^{p+q}\frac{1}{(2(j-m))!}2\tau^{2j}v^{(2j)}(t_k)$$

$$=-\sum_{j=1}^{p}\frac{1}{(2j)!}2\tau^{2j}v^{(2j)}(t_k)-\sum_{j=p+1}^{q}\frac{1}{(2j)!}2\tau^{2j}v^{(2j)}(t_k)-\sum_{j=q+1}^{p+q}\frac{1}{(2j)!}2\tau^{2j}v^{(2j)}(t_k)$$

$$- \sum_{m=1}^{q} \eta_m \tau^{2m} (v^{(2m)}(t_{k+1}) + v^{(2m)}(t_{k-1})) + \sum_{m=1}^{q} \eta_m \tau^{2m} 2v^{(2m)}(t_k)$$

$$+ \sum_{j=2}^{q} \sum_{m=1}^{j-1} \eta_m \frac{2}{(2(j-m))!} \tau^{2j} v^{(2j)}(t_k)$$

$$+ \sum_{j=q+1}^{p+q} \sum_{m=1}^{q} \eta_m \frac{2}{(2(j-m))!} \tau^{2j} v^{(2j)}(t_k)$$

$$= - \sum_{j=1}^{p} \frac{1}{(2j)!} 2\tau^{2j} v^{(2j)}(t_k) + \sum_{j=2}^{p} \sum_{m=1}^{j-1} \eta_m \frac{2}{(2(j-m))!} \tau^{2j} v^{(2j)}(t_k)$$

$$+ \sum_{m=1}^{p} \eta_m \tau^{2m} 2v^{(2m)}(t_k)$$

$$- \sum_{m=1}^{q} \eta_m \tau^{2m} (v^{(2m)}(t_{k+1}) + v^{(2m)}(t_{k-1}))$$

$$- \sum_{j=p+1}^{q} \frac{1}{(2j)!} 2\tau^{2j} v^{(2j)}(t_k) + \sum_{m=p+1}^{q} \eta_m \tau^{2m} 2v^{(2m)}(t_k)$$

$$+ \sum_{j=p+1}^{q} \sum_{m=1}^{j-1} \eta_m \frac{2}{(2(j-m))!} \tau^{2j} v^{(2j)}(t_k)$$

$$- \sum_{j=q+1}^{p+q} \frac{1}{(2j)!} 2\tau^{2j} v^{(2j)}(t_k) + \sum_{j=q+1}^{p+q} \sum_{m=1}^{q} \eta_m \frac{2}{(2(j-m))!} \tau^{2j} v^{(2j)}(t_k)$$

$$= -(1 - 2\eta_1) v^{(2)}(t_k) - \sum_{j=2}^{p} [\frac{1}{(2j)!} - \eta_j - \sum_{m=1}^{j-1} \eta_m \frac{1}{(2(j-m))!}] 2\tau^{2j} v^{(2j)}(t_k)$$

$$- \sum_{m=1}^{q} \eta_m \tau^{2m} (v^{(2m)}(t_{k+1}) + v^{(2m)}(t_{k-1}))$$

$$- \sum_{j=p+1}^{q} [\frac{1}{(2j)!} - \eta_j - \sum_{m=1}^{j-1} \eta_m \frac{2}{(2(j-m))!}] 2\tau^{2j} v^{(2j)}(t_k)$$

$$- \sum_{j=q+1}^{p+q} \frac{2}{(2j)!} (1 - \sum_{m=1}^{q} \eta_m \frac{(2j)!}{(2(j-m))!}) \tau^{2j} v^{(2j)}(t_k).$$

Theorem 3.2.2 is proved.

Note that the relation (3.48) is called Taylor's decomposition of function $v(t)$ on three points.

Now, we will consider the applications of Taylor's decomposition of function on three points to approximate solutions of the initial-value problem (3.1) and boundary value problem (3.25).

We consider two cases: the simple case when $a(t) = a$, for any $t \in [0,T]$ and the general case when $a(t)$ is the smooth function defined on the segment $[0,T]$.

Let $a(t) = a$, for any $t \in [0,T]$. First, we consider an initial-value problem (3.1). From (3.48) it is clear that for the approximate solution of problem (3.1) it is necessary to find $y'_{(\tau)}$ and $y^{(2j)}_{(t_k)}$ for any j, $1 \le j \le p$ and $y^{(2j)}_{(t_{k-1})}$, $y^{(2j)}_{(t_{k+1})}$ for any j, $1 \le j \le q$. Using the equation

$$y''(t) = -ay(t) + f(t),$$

we obtain

$$y^{(2n)}(t) = (-a)^n y(t) + \sum_{\lambda=1}^{n} (-a)^{n-\lambda} f^{(2\lambda-2)}(t), n = 2, \ldots . \tag{3.51}$$

From (2. 51) it follows that

$$y(\tau) - y(0) + \sum_{j=1}^{2p} \rho_j y^{(j)}(\tau)\tau^j - \sum_{j=1}^{2q} \delta_j y^{(j)}(0)\tau^j = o(\tau^{2p+2q+1}), \tag{3.52}$$

where

$$\begin{cases} \rho_j = \frac{(2p+2q-j)!(2p)!(-1)^j}{(2p+2q)!j!(2p-j)!} & \text{for any } j, 1 \le j \le 2p, \\ \delta_j = \frac{(2p+2q-j)!(2q)!}{(2p+2q)!j!(2q-j)!} & \text{for any } j, 1 \le j \le 2q, \end{cases}$$

and

$$y'(\tau) - y'(0) + \sum_{j=1}^{2p-2[\frac{m}{2}]} a_j y^{(j+1)}(\tau)\tau^j$$

$$- \sum_{j=1}^{2q-2m+2[\frac{m}{2}]+1} b_j y^{(j+1)}(0)\tau^j = o(\tau^{2p+2q-2m+2}), 0 \le m \le q-1,$$

where

$$\begin{cases} a_j = \frac{(2p+2q-2m+1-j)!(2p-2[\frac{m}{2}])!(-1)^j}{(2p+2q+1-2m)!j!(2p-2[\frac{m}{2}]-j)!}, \\ \text{for any } j, 1 \le j \le 2p - 2[\frac{m}{2}], \\ b_j = \frac{(2p+2q+1-2m-j)!(2q-2m+2[\frac{m}{2}]+1)!}{(2p+2q+1-2m)!j!(2q-2m+2[\frac{m}{2}]+1-j)!}, \\ \text{for any } j, 1 \le j \le 2q - 2m + 2[\frac{m}{2}] + 1. \end{cases}$$

Using formulas (3.51) and (3.53), we obtain

$$y'(\tau)-y'(0)+\sum_{n=1}^{p-[\frac{m}{2}]} a_{2n}(-a)^n y'(\tau)\tau^{2n}+\sum_{n=1}^{p-[\frac{m}{2}]} a_{2n}\sum_{\lambda=1}^{n}(-a)^{n-\lambda}f^{(2\lambda-1)}(\tau)\tau^{2n}$$

$$+\sum_{n=1}^{p-[\frac{m}{2}]} a_{2n-1}(-a)^n y(\tau)\tau^{2n-1}+\sum_{n=1}^{p-[\frac{m}{2}]} a_{2n-1}\sum_{\lambda=1}^{n}(-a)^{n-\lambda}f^{(2\lambda-2)}(\tau)\tau^{2n-1}$$

$$-\sum_{n=1}^{q-m+[\frac{m}{2}]} b_{2n}(-a)^n y'(0)\tau^{2n}-\sum_{n=1}^{q-m+[\frac{m}{2}]} b_{2n}\sum_{\lambda=1}^{n}(-a)^{n-\lambda}f^{(2\lambda-1)}(0)\tau^{2n}$$

$$-\sum_{n=1}^{q-m+[\frac{m}{2}]+1} b_{2n-1}(-a)^n y(0)\tau^{2n-1}$$

$$-\sum_{n=1}^{q-m+[\frac{m}{2}]+1} b_{2n-1}\sum_{\lambda=1}^{n}(-a)^{n-\lambda}f^{(2\lambda-2)}(0)\tau^{2n-1}$$

$$=o(\tau^{2p+2q-2m+2}),0\le m\le q-1.$$

From the last formula it follows that

$$(1+\sum_{n=1}^{p-[\frac{m}{2}]} a_{2n}(-a)^n\tau^{2n})y'(\tau)=y'(0)-\sum_{n=1}^{p-[\frac{m}{2}]} a_{2n}\sum_{\lambda=1}^{n}(-a)^{n-\lambda}f^{(2\lambda-1)}(\tau)\tau^{2n}$$

$$-\sum_{n=1}^{p-[\frac{m}{2}]} a_{2n-1}(-a)^n y(\tau)\tau^{2n-1}-\sum_{n=1}^{p-[\frac{m}{2}]} a_{2n-1}\sum_{\lambda=1}^{n}(-a)^{n-\lambda}f^{(2\lambda-2)}(\tau)\tau^{2n-1}$$

$$+\sum_{n=1}^{q-m+[\frac{m}{2}]} b_{2n}(-a)^n y'(0)\tau^{2n}+\sum_{n=1}^{q-m+[\frac{m}{2}]} b_{2n}\sum_{\lambda=1}^{n}(-a)^{n-\lambda}f^{(2\lambda-1)}(0)\tau^{2n}$$

$$+\sum_{n=1}^{q-m+[\frac{m}{2}]+1} b_{2n-1}(-a)^n y(0)\tau^{2n-1}$$

$$+\sum_{n=1}^{q-m+[\frac{m}{2}]+1} b_{2n-1}\sum_{\lambda=1}^{n}(-a)^{n-\lambda}f^{(2\lambda-2)}(0)\tau^{2n-1}+o(\tau^{2p+2q-2m+2}),$$

or

$$y'(\tau)=\frac{1}{1+\sum_{n=1}^{p-[\frac{m}{2}]} a_{2n}(-a)^n\tau^{2n}}[y'(0) \tag{3.53}$$

$$-\sum_{n=1}^{p-[\frac{m}{2}]} a_{2n}\sum_{\lambda=1}^{n}(-a)^{n-\lambda}f^{(2\lambda-1)}(\tau)\tau^{2n}$$

$$-\sum_{n=1}^{p-[\frac{m}{2}]} a_{2n-1}(-a)^n y(\tau)\tau^{2n-1} - \sum_{n=1}^{p-[\frac{m}{2}]} a_{2n-1}\sum_{\lambda=1}^{n}(-a)^{n-\lambda} f^{(2\lambda-2)}(\tau)\tau^{2n-1}$$

$$+\sum_{n=1}^{q-m+[\frac{m}{2}]} b_{2n}(-a)^n y'(0)\tau^{2n} + \sum_{n=1}^{q-m+[\frac{m}{2}]} b_{2n}\sum_{\lambda=1}^{n}(-a)^{n-\lambda} f^{(2\lambda-1)}(0)\tau^{2n}$$

$$+\sum_{n=1}^{q-m+[\frac{m}{2}]+1} b_{2n-1}(-a)^n y(0)\tau^{2n-1}$$

$$+\sum_{n=1}^{q-m+[\frac{m}{2}]+1} b_{2n-1}\sum_{\lambda=1}^{n}(-a)^{n-\lambda} f^{(2\lambda-2)}(0)\tau^{2n-1}] + o(\tau^{2p+2q-2m+2}).$$

Now, neglecting the last terms in formulas (3.52), (3.53) and using formulas (3.51) and (3.48), we obtain the difference schemes

$$\frac{u_{k+1}-2u_k+u_{k-1}}{\tau^2} - \sum_{j=1}^{p}\alpha_j(-a)^j\tau^{2j-2}u_k \tag{3.54}$$

$$-\sum_{j=1}^{q}\eta_j(-a)^j\tau^{2j-2}(u_{k-1}+u_{k+1})$$

$$=\sum_{j=1}^{q}\eta_j\sum_{\lambda=1}^{j}(-a)^{j-\lambda}(f^{(2\lambda-2)}(t_{k+1})+f^{(2\lambda-2)}(t_{k-1}))\tau^{2j-2}$$

$$+\sum_{j=1}^{p}\alpha_j\sum_{\lambda=1}^{j}(-a)^{j-\lambda}f^{(2\lambda-2)}(t_k)\tau^{2j-2}, 1\le k\le N-1,$$

$$\frac{u_1-u_0}{\tau} + \sum_{n=1}^{p}\rho_{2n}(-a)^n u_1\tau^{2n-1} + \sum_{m=1}^{p}\rho_{2m-1}(-a)^{m-1}\tau^{2m-2}$$

$$\times\frac{1}{1+\sum_{n=1}^{p-[\frac{m}{2}]}a_{2n}(-a)^n\tau^{2n}}[y'(0) - \sum_{n=1}^{p-[\frac{m}{2}]} a_{2n}\sum_{\lambda=1}^{n}(-a)^{n-\lambda}f^{(2\lambda-1)}(\tau)\tau^{2n}$$

$$-\sum_{n=1}^{p-[\frac{m}{2}]} a_{2n-1}(-a)^n u_1\tau^{2n-1} - \sum_{n=1}^{p-[\frac{m}{2}]} a_{2n-1}\sum_{\lambda=1}^{n}(-a)^{n-\lambda}f^{(2\lambda-2)}(\tau)\tau^{2n-1}$$

$$+\sum_{n=1}^{q-m+[\frac{m}{2}]} b_{2n}(-a)^n y'(0)\tau^{2n} + \sum_{n=1}^{q-m+[\frac{m}{2}]} b_{2n}\sum_{\lambda=1}^{n}(-a)^{n-\lambda}f^{(2\lambda-1)}(0)\tau^{2n}$$

$$+\sum_{n=1}^{q-m+[\frac{m}{2}]+1} b_{2n-1}(-a)^n u_0 \tau^{2n-1} + \sum_{n=1}^{q-m+[\frac{m}{2}]+1} b_{2n-1} \sum_{\lambda=1}^{n} (-a)^{n-\lambda} f^{(2\lambda-2)}(0)\tau^{2n-1}]$$

$$-\sum_{n=1}^{q} \delta_{2n}(-a)^{2n} u_0 \tau^{2n-1} - \sum_{n=1}^{q} \delta_{2n-1}(-a)^{2(n-1)} y'(0)\tau^{2n-2}$$

$$= -\sum_{n=1}^{p} \rho_{2n} \sum_{\lambda=1}^{n} (-a)^{n-\lambda} f^{(2\lambda-2)}(\tau)\tau^{2n-1} + \sum_{n=1}^{q} \delta_{2n} \sum_{\lambda=1}^{n} (-a)^{n-\lambda} f^{(2\lambda-2)}(0)\tau^{2n-1}$$

$$-\sum_{n=1}^{p} \rho_{2n-1} \sum_{\lambda=1}^{n-1} (-a)^{n-1-\lambda} f^{(2\lambda-1)}(\tau)\tau^{2n-2}$$

$$+\sum_{n=1}^{q} \delta_{2n-1} \sum_{\lambda=1}^{n-1} (-a)^{n-1-\lambda} f^{(2\lambda-1)}(0)\tau^{2n-2}, u_0 = y_0.$$

Thus, we constructed the difference schemes of $(2p+2q)$-order of accuracy for the approximate solution of the initial-value problem (3.1). Finally, let us give some examples of the simplest difference schemes.

In the case $p+q=1$ from formulas (3.54) and (3.55) it follows that

$$\frac{u_{k+1}-2u_k+u_{k-1}}{\tau^2} + au_k = f(t_k), 1 \le k \le N-1,$$

$$\frac{u_1-u_0}{\tau} + \rho_2(-a)u_1\tau + \rho_1 \frac{1}{1+a_2(-a)\tau^2} a_1(a)u_1\tau$$

$$= -\rho_1 \frac{1}{1+a_2(-a)\tau^2}[y'(0) - a_2 f'(\tau)\tau^2 - a_1 f(\tau)\tau] - \rho_2 f(\tau)\tau, u_0 = y_0$$

(a difference scheme of second order of accuracy for the initial-value problem (3.1)).

Second, we consider the boundary-value problem (3.25). From (3.48) it is clear that for the approximate solution of the problem (3.25) it is necessary to find $y^{(2j)}_{(t_k)}$ for any j, $1 \le j \le p$ and $y^{(2j)}_{(t_{k-1})}$, $y^{(2j)}_{(t_{k+1})}$for any j, $1 \le j \le q$. Using the equation

$$y''(t) = ay(t) - f(t),$$

we obtain

$$y^{(2n)}(t) = a^n y(t) - \sum_{\lambda=1}^{n} a^{n-\lambda} f^{(2\lambda-2)}(t), n = 2, \ldots. \tag{3.55}$$

Now, using formulas (3.55) and (3.48), we obtain the difference schemes

$$-\frac{u_{k+1}-2u_k+u_{k-1}}{\tau^2} + \sum_{j=1}^{p} \alpha_j a^j \tau^{2j-2} u_k \tag{3.56}$$

$$+\sum_{j=1}^{q}\eta_j a^j \tau^{2j-2}(u_{k-1}+u_{k+1})$$

$$=\sum_{j=1}^{q}\eta_j\sum_{\lambda=1}^{j}a^{j-\lambda}(f^{(2\lambda-2)}(t_{k+1})+f^{(2\lambda-2)}(t_{k-1}))\tau^{2j-2}$$

$$+\sum_{j=1}^{p}\alpha_j\sum_{\lambda=1}^{j}a^{j-\lambda}f^{(2\lambda-2)}(t_k)\tau^{2j-2}, 1\le k\le N-1, u_0=y_0, u_N=y_T.$$

Thus, we constructed the difference schemes of $(2p+2q)$-order of accuracy for the approximate solution of the boundary-value problem (3.25). Finally, let us give some examples of the simplest difference schemes.

In the case $p+q=1$ from formula (3.56) it follows that

$$-\frac{u_{k+1}-2u_k+u_{k-1}}{\tau^2}+au_k=f(t_k), 1\le k\le N-1, u_0=y_0, u_N=y_T$$

(a difference scheme of second order of accuracy for the boundary-value problem (3.25)).

Now, let $a(t)$ be the smooth function defined on the segment $[0,T]$. First, we consider an initial-value problem (3.1). From (3.48) it is clear that for the approximate solution of problem (3.1) it is necessary to find $y^{(2j)}_{(t_k)}$ for any j, $1\le j\le p$ and $y^{(2j)}_{(t_{k-1})}$, $y^{(2j)}_{(t_{k+1})}$ for any j, $1\le j\le q$. Using the equation

$$y''(t)=-a(t)y(t)+f(t),$$

we obtain

$$\begin{aligned} y^{(2j)}(t) &= A_{2j}(t)y(t)+B_{2j}(t)y'(t)+f_{2j}(t),\\ y^{(2j+1)}(t) &= A_{2j+1}(t)y(t)+A_{2j+1}(t)y'(t)+f_{2j+1}(t), j=1,\dots, \end{aligned} \tag{3.57}$$

where

$$\begin{cases} A_2(t)=-a(t), B_2(t)=0, f_2(t)=f(t),\\ A_3(t)=-a'(t), B_3(t)=-a(t), f_3(t)=f'(t),\\ A_{2j}(t)=A'_{2j-1}(t)-B_{2j-1}(t)a(t), B_{2j}(t)=A_{2j-1}(t)+B'_{2j-1}(t),\\ f_{2j}(t)=f'_{2j-1}(t)+B_{2j-1}(t)f(t), j=2,3,\dots,\\ A_{2j+1}(t)=A'_{2j}(t)-B_{2j}(t)a(t), B_{2j+1}(t)=A_{2j}(t)+B'_{2j}(t),\\ f_{2j+1}(t)=f'_{2j}(t)+B_{2j}(t)f(t), j=2,3,\dots. \end{cases} \tag{3.58}$$

From (3.51) it is clear that for the approximate solution of the problem (3.1) it is necessary to find $y'_{(t_k)}$, $y'_{(t_{k-1})}$ and $y'_{(t_{k+1})}$. The choice formula for $y'_{(t_k)}$, $y'_{(t_{k-1})}$ and

$y'_{(t_{k+1})}$ is not unique. First, we consider the approximation of $y'_{(\tau)}$. Using formulas (3.53) and (3.57), we obtain that

$$y'(\tau) - y'(0) + \sum_{i=1}^{2p-2[\frac{m}{2}]} a_i\tau^i\{A_{i+1}(\tau)y(\tau) + B_{i+1}(\tau)y'(\tau) + f_{i+1}(\tau)\}$$

$$- \sum_{i=1}^{2q-2m+2[\frac{m}{2}]+1} b_i\tau^i\{A_{i+1}(0)y(\tau) + B_{i+1}(0)y'(\tau) + f_{i+1}(0)\} = o(\tau^{2p+2q-2m+2}).$$

Since

$$1 + \sum_{i=1}^{2p-2[\frac{m}{2}]} a_i\tau^i B_{i+1}(\tau) \neq 0$$

for the small $\tau > 0$, we obtain

$$y'(\tau) = T_{0,\tau}y(0) + P_{0,\tau}y'(0) + R_{0,\tau}y(\tau) + F_{0,\tau}, \tag{3.59}$$

where

$$T_{0,\tau} = (1 + \sum_{i=1}^{2p-2[\frac{m}{2}]} a_i\tau^i B_{i+1}(\tau))^{-1} \sum_{i=1}^{2q-2m+2[\frac{m}{2}]+1} b_i\tau^i A_{i+1}(0),$$

$$P_{0,\tau} = (1 + \sum_{i=1}^{2p-2[\frac{m}{2}]} a_i\tau^i B_{i+1}(\tau))^{-1}[1 + \sum_{i=1}^{2q-2m+2[\frac{m}{2}]+1} b_i\tau^i B_{i+1}(0)],$$

$$R_{0,\tau} = -(1 + \sum_{i=1}^{2p-2[\frac{m}{2}]} a_i\tau^i B_{i+1}(\tau))^{-1} \sum_{i=1}^{2p-2[\frac{m}{2}]} a_i\tau^i A_{i+1}(\tau),$$

$$F_{0,\tau} = (1+ \sum_{i=1}^{2p-2[\frac{m}{2}]} a_i\tau^i B_{i+1}(\tau))^{-1}[\sum_{i=1}^{2p-2[\frac{m}{2}]} a_i\tau^i f_{i+1}(\tau)+ \sum_{i=1}^{2q-2m+2[\frac{m}{2}]+1} b_i\tau^i f_{i+1}(0)].$$

Second, we consider the approximation of $y'_{(t_k)}$ and $y'_{(t_{k-1})}$. Using formulas (3.53) and (3.57), we obtain that

$$y'(t_k) - y'(t_{k-1}) + \sum_{i=1}^{2p-m+1} c_i\tau^i\{A_{i+1}(t_k)y(t_k) + B_{i+1}(t_k)y'(t_k) + f_{i+1}(t_k)\}$$

$$- \sum_{i=1}^{2q-m+2} d_i\tau^i\{A_{i+1}(t_{k-1})y(t_{k-1}) + B_{i+1}(t_{k-1})y'(t_{k-1}) + f_{i+1}(t_{k-1})\}$$

$$= o(\tau^{2p+2q-2m+4}),$$

where

$$\begin{cases} c_i = \frac{(2p+2q-2m+3-j)!(2p-m+1)!(-1)^i}{(2p+2q+3-2m)!i!(2p-m+1-i)!} & \text{for any } i, 1 \le i \le 2p-m+1, \\ d_j = \frac{(2p+2q+3-2m-j)!(2q-m+2)!}{(2p+2q+3-2m)!i!(2q-m+2-i)!} & \text{for any } i, 1 \le i \le 2q-m+2, \end{cases}$$

and

$$y'(t_k) - y'(t_{k-1}) + \sum_{i=1}^{2p-m+2} \widetilde{c}_i\tau^i\{A_{i+1}(t_k)y(t_k) + B_{i+1}(t_k)y'(t_k) + f_{i+1}(t_k)\}$$
$$- \sum_{i=1}^{2q-m+1} \widetilde{d}_i\tau^i\{A_{i+1}(t_{k-1})y(t_{k-1}) + B_{i+1}(t_{k-1})y'(t_{k-1}) + f_{i+1}(t_{k-1})\}$$
$$= o(\tau^{2p+2q-2m+4}),$$

where

$$\begin{cases} \widetilde{c}_i = \frac{(2p+2q-2m+3-j)!(2p-m+2)!(-1)^i}{(2p+2q+3-2m)!i!(2p-m+2-i)!} & \text{for any } i, 1 \le i \le 2p-m+2, \\ \widetilde{d}_j = \frac{(2p+2q+3-2m-j)!(2q-m+1)!}{(2p+2q+3-2m)!i!(2q-m+1-i)!} & \text{for any } i, 1 \le i \le 2q-m+1. \end{cases}$$

Hence, we can write

$$\begin{cases} [1 + \sum_{i=1}^{2p-m+1} c_i\tau^i B_{i+1}(t_k)]y'(t_k) - [1 + \sum_{i=1}^{2q-m+2} d_i\tau^i B_{i+1}(t_{k-1})]y'(t_{k-1}) \\ = - \sum_{i=1}^{2p-m+1} c_i\tau^i\{A_{i+1}(t_k)y(t_k) + f_{i+1}(t_k)\} \\ + \sum_{i=1}^{2q-m+2} d_i\tau^i\{A_{i+1}(t_{k-1})y(t_{k-1}) + f_{i+1}(t_{k-1})\} = o(\tau^{2p+2q-2m+4}), \\ [1 + \sum_{i=1}^{2p-m+2} \widetilde{c}_i\tau^i B_{i+1}(t_k)]y'(t_k) - [1 + \sum_{i=1}^{2q-m+1} \widetilde{d}_i\tau^i B_{i+1}(t_{k-1})]y'(t_{k-1}) \\ = - \sum_{i=1}^{2p-m+2} \widetilde{c}_i\tau^i\{A_{i+1}(t_k)y(t_k) + f_{i+1}(t_k)\} \\ + \sum_{i=1}^{2q-m+1} \widetilde{d}_i\tau^i\{A_{i+1}(t_{k-1})y(t_{k-1}) + f_{i+1}(t_{k-1})\} = o(\tau^{2p+2q-2m+4}). \end{cases}$$

To solve the last system, we need

$$\Delta_k = -[1 + \sum_{i=1}^{2p-m+2} c_i\tau^i B_{i+1}(t_k)][1 + \sum_{i=1}^{2q-m+1} \widetilde{d}_i\tau^i B_{i+1}(t_{k-1})] \tag{3.60}$$
$$+[1 + \sum_{i=1}^{2q-m+2} d_i\tau^i B_{i+1}(t_{k-1})][1 + \sum_{i=1}^{2p-m+2} \widetilde{c}_i\tau^i B_{i+1}(t_k)] \neq 0$$

for all k. Applying formula (3.59), we obtain

$$\Delta_k = -c_2\tau^2 B_3(t_k) - \widetilde{d}_2\tau^2 B_3(t_{k-1}) + d_2\tau^2 B_3(t_{k-1}) + \widetilde{c}_2\tau^2 B_3(t_k)] + o(\tau^3)$$

$$= c_2\tau^2 a(t_k) + \widetilde{d}_2\tau^2 a(t_{k-1}) - d_2\tau^2 a(t_{k-1}) - \widetilde{c}_2\tau^2 a(t_k)] + o(\tau^3)$$

$$= \tau^2[(c_2 + \widetilde{d}_2 - d_2 - \widetilde{c}_2)a(t_k)] + o(\tau^3).$$

Since

$$c_2 + \widetilde{d}_2 - d_2 - \widetilde{c}_2 = -\frac{(2p+2q+1-2m)!}{(2p+2q+3-2m)!2!}4(p+q+1-m) \neq 0,$$

we have that if $a(t_k) \neq 0$ for all k, then for all small τ the requirement (3.60) is satisfied. From that it follows that

$$y'(t_k) = \Delta_k^{-1}\{-[1 + \sum_{i=1}^{2q-m+1} \widetilde{d}_i\tau^i B_{i+1}(t_{k-1})]$$

$$[-\sum_{i=1}^{2p-m+2} \widetilde{c}_i\tau^i\{A_{i+1}(t_k)y(t_k) + f_{i+1}(t_k)\}$$

$$+ \sum_{i=1}^{2q-m+1} \widetilde{d}_i\tau^i\{A_{i+1}(t_{k-1})y(t_{k-1}) + f_{i+1}(t_{k-1})\}]\}$$

$$+[1 + \sum_{i=1}^{2q-m+2} d_i\tau^i B_{i+1}(t_k)]$$

$$[-\sum_{i=1}^{2p-m+1} c_i\tau^i\{A_{i+1}(t_k)y(t_k) + f_{i+1}(t_k)\}$$

$$+ \sum_{i=1}^{2q-m+2} d_i\tau^i\{A_{i+1}(t_{k-1})y(t_{k-1}) + f_{i+1}(t_{k-1})\}] + o(\tau^{2p+2q-2m+2})$$

$$= C_{k,\tau}y(t_{k-1}) + D_{k,\tau}y(t_k) + F^1_{k,\tau} + o(\tau^{2p+2q-2m+2}), \qquad (3.61)$$

where

$$C_{k,\tau} = \Delta_k^{-1}\{-[1 + \sum_{i=1}^{2q-m+1} \widetilde{d}_i\tau^i B_{i+1}(t_{k-1})] \sum_{i=1}^{2q-m+1} \widetilde{d}_i\tau^i A_{i+1}(t_{k-1})$$

$$+[1 + \sum_{i=1}^{2q-m+2} d_i\tau^i B_{i+1}(t_k)] \sum_{i=1}^{2q-m+2} d_i\tau^i A_{i+1}(t_{k-1})\},$$

$$
\begin{aligned}
D_{k,\tau} &= \Delta_k^{-1}\{[1+\sum_{i=1}^{2q-m+1}\widetilde{d}_i\tau^i B_{i+1}(t_{k-1})]\sum_{i=1}^{2p-m+2}\widetilde{c}_i\tau^i A_{i+1}(t_k)\\
&\quad -[1+\sum_{i=1}^{2q-m+2} d_i\tau^i B_{i+1}(t_k)]\sum_{i=1}^{2p-m+1} c_i\tau^i A_{i+1}(t_k),
\end{aligned}
$$

$$
\begin{aligned}
F_{k,\tau}^1 &= \Delta_k^{-1}\{-[1+\sum_{i=1}^{2q-m+1}\widetilde{d}_i\tau^i B_{i+1}(t_{k-1})][-\sum_{i=1}^{2p-m+2}\widetilde{c}_i\tau^i f_{i+1}(t_k)\\
&\quad +\sum_{i=1}^{2q-m+1}\widetilde{d}_i\tau^i f_{i+1}(t_{k-1})]+[1+\sum_{i=1}^{2q-m+2} d_i\tau^i B_{i+1}(t_k)]\\
&\quad \times[-\sum_{i=1}^{2p-m+1} c_i\tau^i f_{i+1}(t_k)+\sum_{i=1}^{2q-m+2} d_i\tau^i f_{i+1}(t_{k-1})]\},
\end{aligned}
$$

and

$$
\begin{aligned}
y'(t_{k-1}) &= \Delta_k^{-1}\{-[1+\sum_{i=1}^{2q-m+2}\widetilde{c}_i\tau^i B_{i+1}(t_k)]\\
&\quad \times[-\sum_{i=1}^{2p-m+1} c_i\tau^i\{A_{i+1}(t_k)y(t_k)+f_{i+1}(t_k)\}\\
&\quad +\sum_{i=1}^{2q-m+2} d_i\tau^i\{A_{i+1}(t_{k-1})y(t_{k-1})+f_{i+1}(t_{k-1})\}]\}\\
&\quad +[1+\sum_{i=1}^{2q-m+1} c_i\tau^i B_{i+1}(t_k)]\\
&\quad \times[-\sum_{i=1}^{2p-m+2}\widetilde{c}_i\tau^i\{A_{i+1}(t_k)y(t_k)+f_{i+1}(t_k)\}\\
&\quad +\sum_{i=1}^{2q-m+1}\widetilde{d}_i\tau^i\{A_{i+1}(t_{k-1})y(t_{k-1})+f_{i+1}(t_{k-1})\}]+o(\tau^{2p+2q-2m+2})\\
&= E_{k,\tau}y(t_{k-1})+H_{k,\tau}y(t_k)+F_{k,\tau}^2+o(\tau^{2p+2q-2m+2}),
\end{aligned}
\tag{3.62}
$$

where

$$
\begin{aligned}
E_{k,\tau} &= \Delta_k^{-1}\{-[1+\sum_{i=1}^{2q-m+1} c_i\tau^i B_{i+1}(t_k)]\sum_{i=1}^{2q-m+1}\widetilde{d}_i\tau^i A_{i+1}(t_{k-1})\\
&\quad -[1+\sum_{i=1}^{2p-m+2}\widetilde{c}_i\tau^i B_{i+1}(t_k)]\sum_{i=1}^{2q-m+2} d_i\tau^i A_{i+1}(t_{k-1})\},
\end{aligned}
$$

$$
\begin{aligned}
H_{k,\tau} &= \Delta_k^{-1}\{[1+\sum_{i=1}^{2p-m+1}\widetilde{c}_i\tau^i B_{i+1}(t_k)]\sum_{i=1}^{2p-m+1}c_i\tau^i A_{i+1}(t_k) \\
&\quad -[1+\sum_{i=1}^{2p-m+1}d_i\tau^i B_{i+1}(t_k)]\sum_{i=1}^{2p-m+2}\widetilde{c}_i\tau^i A_{i+1}(t_k)\},
\end{aligned}
$$

$$
\begin{aligned}
F_{k,\tau}^2 &= \Delta_k^{-1}\{[1+\sum_{i=1}^{2p-m+1}c_i\tau^i B_{i+1}(t_k)][\sum_{i=1}^{2q-m+1}\widetilde{d}_i\tau^i f_{i+1}(t_{k-1}) \\
&\quad -\sum_{i=1}^{2p-m+2}\widetilde{c}_i\tau^i f_{i+1}(t_k)]+[1+\sum_{i=1}^{2p-m+2}\widetilde{c}_i\tau^i B_{i+1}(t_k)] \\
&\quad \times[\sum_{i=1}^{2p-m+1}c_i\tau^i f_{i+1}(t_k)-\sum_{i=1}^{2q-m+2}d_i\tau^i f_{i+1}(t_{k-1})]\}.
\end{aligned}
$$

From that it follows that

$$
y'(t_{k+1}) = C_{k+1,\tau}y(t_k)+D_{k+1,\tau}y(t_{k+1})+F_{k+1,\tau}^1+o(\tau^{2p+2q-2m+2}). \tag{3.63}
$$

Now, neglecting the last terms in formulas (3.48), (3.52), (3.60), (3.61), (3.62), (3.63) and using formula (3.58), we obtain the two-step difference schemes of $(2p+2q)$-th order of accuracy

$$
\begin{aligned}
&\frac{u_{k+1}-2u_k+u_{k-1}}{\tau^2}-\sum_{j=1}^{p}\alpha_j\tau^{2j-2}\{A_{2j}(t_k)u_k+B_{2j}(t_k)[C_{k,\tau}u_{k-1}+D_{k,\tau}u_k] \\
&\quad -\sum_{j=1}^{q}\eta_j\tau^{2j-2}\{A_{2j}(t_{k-1})u_{k-1}+B_{2j}(t_{k-1})[E_{k,\tau}u_{k-1}+H_{k,\tau}u_k] \\
&\quad +A_{2j}(t_{k+1})u_{k+1}+B_{2j}(t_{k+1})[C_{k+1,\tau}u_k+D_{k+1,\tau}u_{k+1}]\} \\
&= \sum_{j=1}^{q}\eta_j[B_{2j}(t_{k-1})F_{k,\tau}^2+B_{2j}(t_{k+1})F_{k+1,\tau}^1]\tau^{2j-2}+\sum_{j=1}^{p}\alpha_j B_{2j}(t_k)F_{k,\tau}^1\tau^{2j-2}, \\
&\qquad 1\le k\le N-1,
\end{aligned}
$$

$$
\begin{aligned}
&\frac{u_1-u_0}{\tau}+\sum_{m=1}^{p}\rho_{2m}\{A_{2m}u_1+B_{2m}[T_{0,\tau}u_0+R_{0,\tau}u_1]\}\tau^{2n-1} \\
&\quad -\sum_{m=1}^{q}\delta_{2m}\{A_{2m}u_0+B_{2m}[T_{0,\tau}u_0+R_{0,\tau}u_1]\}\tau^{2n-1}
\end{aligned}
$$

$$= -\sum_{m=1}^{p} \rho_{2m}\{A_{2m}(\tau)u_1 + B_{2m}(\tau)[P_{0,\tau}y'(0) + F_{0,\tau}]\}\tau^{2n-1}$$

$$+\sum_{m=1}^{q} \delta_{2m}\{A_{2m}(0)u_0 + B_{2m}(0)y_0' + f_{2m}(0)\}\tau^{2m-1}, u_0 = y_0$$

for the approximate solution of the initial-value problem (3.1).

Second, we consider the boundary-value problem (3.25). From (3.48) it is clear that for the approximate solution of problem (3.25) it is necessary to find $y^{(2j)}_{(t_k)}$ for any $j, 1 \leq j \leq p$ and $y^{(2j)}_{(t_{k-1})}$, $y^{(2j)}_{(t_{k+1})}$ for any $j, 1 \leq j \leq q$. Using the equation

$$y''(t) = a(t)y(t) - f(t),$$

we obtain

$$\begin{aligned} y^{(2j)}(t) &= A_{2j}(t)y(t) + B_{2j}(t)y'(t) + f_{2j}(t), \\ y^{(2j+1)}(t) &= A_{2j+1}(t)y(t) + A_{2j+1}(t)y'(t) + f_{2j+1}(t), j = 1, \dots, \end{aligned} \tag{3.64}$$

where

$$\begin{cases} A_2(t) = a(t), B_2(t) = 0, f_2(t) = -f(t), \\ A_3(t) = a'(t), B_3(t) = a(t), f_3(t) = f'(t), \\ A_{2j}(t) = A'_{2j-1}(t) + B_{2j-1}(t)a(t), B_{2j}(t) = A_{2j-1}(t) + B'_{2j-1}(t), \\ f_{2j}(t) = f'_{2j-1}(t) - B_{2j-1}(t)f(t), j = 2, 3, \dots, \\ A_{2j+1}(t) = A'_{2j}(t) + B_{2j}(t)a(t), B_{2j+1}(t) = A_{2j}(t) + B'_{2j}(t), \\ f_{2j+1}(t) = f'_{2j}(t) - B_{2j}(t)f(t), j = 2, 3, \dots. \end{cases} \tag{3.65}$$

Now, using formulas (3.61), (3.62), (3.63), (3.64) and (3.48), we obtain the difference schemes

$$-\frac{u_{k+1} - 2u_k + u_{k-1}}{\tau^2} + \sum_{j=1}^{p} \alpha_j \tau^{2j-2}\{A_{2j}(t_k)u_k + B_{2j}(t_k)[C_{k,\tau}u_{k-1} + D_{k,\tau}u_k]\}$$

$$+\sum_{j=1}^{q} \eta_j \tau^{2j-2}\{A_{2j}(t_{k-1})u_{k-1} + B_{2j}(t_{k-1})[E_{k,\tau}u_{k-1} + H_{k,\tau}u_k]$$

$$+A_{2j}(t_{k+1})u_{k+1} + B_{2j}(t_{k+1})[C_{k+1,\tau}u_k + D_{k+1,\tau}u_{k+1}]\}$$

$$= \sum_{j=1}^{q} \eta_j [B_{2j}(t_{k-1})F^2_{k,\tau} + B_{2j}(t_{k+1})F^1_{k+1,\tau}]\tau^{2j-2}$$

$$+\sum_{j=1}^{p} \alpha_j B_{2j}(t_k)F^1_{k,\tau}\tau^{2j-2}, 1 \leq k \leq N-1, u_0 = y_0, u_N = y_T.$$

Thus, we constructed the difference schemes of $(2p+2q)$-th order of accuracy for the approximate solution of the boundary-value problem (3.25).

The choice of Taylor's decomposition on three points in the construction $(2p+2q)$-th order of accuracy difference schemes for the approximate solutions of the initial-value problem (3.1) and of the boundary-value problem (3.25) is not unique. We will give another Taylor's decomposition on three points in which we use $v^{(2i)}(t_k)$, $0 \le i \le p$ and $v^{(i)}(t_{k\pm 1})$, $0 \le i \le 2q$.

Theorem 3.2.3. *Let the function $v(t)$ $(0 \le t \le T)$ have a $(2p+2q+2)$-th continuous derivative and $t_{k-1}, t_k, t_{k+1} \in [0,T]_\tau$. Then the following relation holds:*

$$v(t_{k+1}) - 2v(t_k) + v(t_{k-1}) + \sum_{j=1}^{p} \alpha_j v^{(2j)}(t_k)\tau^{2j} \tag{3.66}$$

$$+\sum_{j=1}^{q} \gamma_j (v^{(2j-1)}(t_{k-1}) - v^{(2j-1)}(t_{k+1}))\tau^{2j-1}$$

$$+\sum_{j=1}^{q} \beta_j (v^{(2j)}(t_{k-1}) + v^{(2j)}(t_{k+1}))\tau^{2j} = o(\tau^{2p+2q+2}),$$

where

$$\begin{cases} \alpha_j = -\frac{2(2p+2q+1-2j)!}{(2p+2q+1)!(2p+1-2j)!}\frac{(2p+1)!}{(2j)!} & \text{for any } j, 1 \le j \le p, \\ \beta_j = \frac{(2p+2q+1-2j)!(2q)!}{(2p+2q+1)!(2q-2j)!(2j)!} & \text{for any } j, 1 \le j \le q, \\ \gamma_j = \frac{(2p+2q+2-2j)!(2q)!}{(2p+2q+1)!(2q+1-2j)!(2j-1)!} & \text{for any } j, 1 \le j \le q. \end{cases}$$

Proof. Using the formula of integration by parts, we obtain

$$\int_{t_k}^{t_{k+1}} (t_{k+1}-s)^{2p+1}(s-t_k)^{2q} v^{(2p+2q+2)}(s)ds$$

$$+\int_{t_{k-1}}^{t_k} (s-t_{k-1})^{2p+1}(s-t_k)^{2q} v^{(2p+2q+2)}(s)ds$$

$$= \sum_{\gamma=1}^{2p+2q+2} (-1)^{2p+2q+2-\gamma}((t_{k+1}-s)^{2p+1}(s-t_k)^{2q})^{(2p+2q+2-\gamma)} v^{(\gamma-1)}(s)]_{t_k}^{t_{k+1}}$$

$$+ \sum_{\gamma=1}^{2p+2q+2} (-1)^{2p+2q+2-\gamma}((s-t_{k-1})^{2p+1}(s-t_k)^{2q})^{(2p+2q+2-\gamma)} v^{(\gamma-1)}(s)]_{t_{k-1}}^{t_k}$$

$$= (2p+2q+1)!(v(t_{k+1}) - v(t_k))$$

$$+\sum_{\gamma=2}^{2p+2q+2}(-1)^{2p+2q+2-\gamma}((t_{k+1}-s)^{2p+1}(s-t_k)^{2q})^{(2p+2q+2-\gamma)}v^{(\gamma-1)}(s)]_{t_k}^{t_{k+1}}$$

$$-(2p+2q+1)!(v(t_k)-v(t_{k-1}))$$

$$+\sum_{\gamma=1}^{2p+2q+2}(-1)^{2p+2q+2-\gamma}((s-t_{k-1})^{2p+1}(s-t_k)^{2q})^{(2p+2q+2-\gamma)}v^{(\gamma-1)}(s)]_{t_{k-1}}^{t_k}.$$

We will calculate the expressions

$$(-1)^{2p+2q+2-\gamma}((t_{k+1}-s)^{2p+1}(s-t_k)^{2q})^{(2p+2q+2-\gamma)}|_{s=t_{k+1},t_k},$$

and

$$(-1)^{2p+2q+2-\gamma}((s-t_{k-1})^{2p+1}(s-t_k)^{2q})^{(2p+2q+2-\gamma)}|_{s=t_{k-1},t_k}.$$

If $j \geq 2q+2$, then $2p+2q+2-j \leq 2p$. Therefore

$$(-1)^{2p+2q+2-j}((t_{k+1}-s)^{2p+1}(s-t_k)^{2q})^{(2p+2q+2-j)}|_{s=t_{k+1}}=0.$$

If $1 \leq j \leq 2q+1$, then

$$(-1)^{2p+2q+2-j}((t_{k+1}-s)^{2p+1}(s-t_k)^{2q})^{(2p+2q+2-j)}|_{s=t_{k+1}}$$

$$= (-1)^{2p+2q+2-j}\sum_{i=0}^{2p+2q+2-j}\frac{(2p+2q+2-j)!}{i!(2p+2q+2-j-i)!}$$
$$\times((t_{k+1}-s)^{2p+1})^{(2p+2q+2-j-i)}((s-t_k)^{2q})^{(i)}|_{s=t_{k+1}}$$

$$= (-1)^{j}\sum_{i=0}^{2p+2q+2-j}\frac{(2p+2q+2-j)!}{i!(2p+2q+2-j-i)!}(-1)^{2p+2q+2-j-i}$$
$$\times\frac{(2p+1)!}{(j+i-2q-1)!}(t_{k+1}-s)^{j+i-2q-1}\frac{(2q)!}{(2q-i)!}(s-t_k)^{2q-i}|_{s=t_{k+1}}$$
$$=-(-1)^{j}\frac{(2p+2q+2-j)!(2q)!}{(2q+1-j)!(j-1)!}\tau^{j-1}.$$

So,

$$(-1)^{2p+2q+2-j}((t_{k+1}-s)^{2p+1}(s-t_k)^{2q})^{(2p+2q+2-j)}|_{s=t_{k+1}}$$
$$= (-1)^{j-1}\frac{(2p+2q+2-j)!(2q)!}{(2q+1-j)!(j-1)!}\tau^{j-1}.$$

If $j \geq 2p+3$, then $2p+2q+2-j \leq 2q-1$. Therefore

$$(-1)^{2p+2q+2-j}((t_{k+1}-s)^{2p+1}(s-t_k)^{2q})^{(2p+2q+2-j)}|_{s=t_k}=0.$$

If $1 \le j \le 2p+2$, then

$$
\begin{aligned}
&(-1)^{2p+2q+2-j}((t_{k+1}-s)^{2p+1}(s-t_k)^{2q})^{(2p+2q+2-j)}|_{s=t_k} \\
&= (-1)^{2p+2q+2-j} \sum_{i=0}^{2p+2q+2-j} \frac{(2p+2q+2-j)!}{i!(2p+2q+2-j-i)!} \\
&\quad \times ((t_{k+1}-s)^{2p+1})^{(2p+2q+2-j-i)}((s-t_k)^{2q})^{(i)}|_{s=t_k} \\
&= (-1)^j \sum_{i=0}^{2p+2q+2-j} \frac{(2p+2q+2-j)!}{i!(2p+2q+2-j-i)!}(-1)^{2p+2q+2-j-i} \\
&\quad \times \frac{(2p+1)!}{(j+i-2q-1)!}(t_{k+1}-s)^{j+i-2q-1}\frac{(2q)!}{(2q-i)!}(s-t_k)^{2q-i}|_{s=t_k} \\
&= \frac{(2p+2q+2-j)!}{(2p+2-j)!}\frac{(2p+1)!}{(j-1)!}\tau^{j-1}.
\end{aligned}
$$

So,

$$
\begin{aligned}
&(-1)^{2p+2q+2-j}((t_{k+1}-s)^{2p+1}(s-t_k)^{2q})^{(2p+2q+2-j)}|_{s=t_k} \\
&= \frac{(2p+2q+2-j)!}{(2p+2-j)!}\frac{(2p+1)!}{(j-1)!}\tau^{j-1}.
\end{aligned}
$$

Now, we will calculate the expressions

$$
(-1)^{2p+2q+2-\gamma}((s-t_{k-1})^{2p+1}(s-t_k)^{2q})^{(2p+2q+2-\gamma)}|_{s=t_{k-1},t_k}.
$$

If $j \ge 2q+2$, then $2p+2q+2-j \le 2p$. Therefore

$$
(-1)^{2p+2q+2-j}((s-t_{k-1})^{2p+1}(s-t_k)^{2q})^{(2p+2q+2-j)}|_{s=t_{k-1}} = 0.
$$

If $1 \le j \le 2q+1$, then

$$
\begin{aligned}
&(-1)^{2p+2q+2-j}((s-t_{k-1})^{2p+1}(s-t_k)^{2q})^{(2p+2q+2-j)}|_{s=t_{k-1}} \\
&= (-1)^{2p+2q+2-j} \sum_{i=0}^{2p+2q+2-j} \frac{(2p+2q+2-j)!}{i!(2p+2q+2-j-i)!} \\
&\quad \times ((s-t_{k-1})^{2p+1})^{(2p+2q+2-j-i)}((s-t_k)^{2q})^{(i)}|_{s=t_{k-1}} \\
&= (-1)^j \sum_{i=0}^{2p+2q+2-j} \frac{(2p+2q+2-j)!}{i!(2p+2q+2-j-i)!}\frac{(2p+1)!}{(j+i-2q-1)!} \\
&\quad \times (s-t_{k-1})^{j+i-2q-1}\frac{(2q)!}{(2q-i)!}(s-t_k)^{2q-i}|_{s=t_{k-1}}
\end{aligned}
$$

$$= (-1)^j \frac{(2p+2q+2-j)!(2q)!}{(2q+1-j)(j-1)!}(-\tau)^{j-1} = -\frac{(2p+2q+2-j)!(2q)!}{(2q+1-j)!(j-1)!}\tau^{j-1}.$$

So,

$$\begin{aligned} & (-1)^{2p+2q+2-j}((s-t_{k-1})^{2p+1}(s-t_k)^{2q})^{(2p+2q+2-j)}|_{s=t_{k-1}} \\ = & -\frac{(2p+2q+2-j)!(2q)!}{(2q+1-j)!(j-1)!}\tau^{j-1}. \end{aligned}$$

If $j \geq 2p+3$, then $2p+2q+2-j \leq 2q-1$. Therefore

$$(-1)^{2p+2q+2-j}((s-t_{k-1})^{2p+1}(s-t_k)^{2q})^{(2p+2q+2-j)}|_{s=t_k} = 0.$$

If $1 \leq j \leq 2p+2$, then

$$\begin{aligned} & (-1)^{2p+2q+2-j}((s-t_{k-1})^{2p+1}(s-t_k)^{2q})^{(2p+2q+2-j)}|_{s=t_k} \\ = & (-1)^{2p+2q+2-j} \sum_{i=0}^{2p+2q+2-j} \frac{(2p+2q+2-j)!}{i!(2p+2q+2-j-i)!} \\ & \times((s-t_{k-1})^{2p+1})^{(2p+2q+2-j-i)}((s-t_k)^{2q})^{(i)}|_{s=t_k} \\ = & (-1)^j \sum_{i=0}^{2p+2q+2-j} \frac{(2p+2q+2-j)!}{i!(2p+2q+2-j-i)!}\frac{(2p+1)!}{(j+i-2q-1)!} \\ & \times(s-t_{k-1})^{j+i-2q-1}\frac{(2q)!}{(2q-i)!}(s-t_k)^{2q-i}|_{s=t_k} \\ & = (-1)^j \frac{(2p+2q+2-j)!}{(2p+2-j)!}\frac{(2p+1)!}{(j-1)!}\tau^{j-1}. \end{aligned}$$

So,

$$\begin{aligned} & (-1)^{2p+2q+2-j}((s-t_{k-1})^{2p+1}(s-t_k)^{2q})^{(2p+2q+2-j)}|_{s=t_k} \\ = & (-1)^j \frac{(2p+2q+2-j)!}{(2p+2-j)!}\frac{(2p+1)!}{(j-1)!}\tau^{j-1}. \end{aligned}$$

From this it follows

$$\begin{aligned} & \int_{t_k}^{t_{k+1}} (t_{k+1}-s)^{2p+1}(s-t_k)^{2q}v^{(2p+2q+2)}(\lambda)d\lambda ds \\ & + \int_{t_{k-1}}^{t_k} (s-t_{k-1})^{2p+1}(s-t_k)^{2q}v^{(2p+2q+2)}(\lambda)d\lambda ds \\ & = (2p+2q+1)!(v(t_{k+1})-2v(t_k)+v(t_{k-1})) \end{aligned}$$

$$+\sum_{j=2}^{2q+1}(-1)^{j-1}\frac{(2p+2q+2-j)!(2q)!}{(2q+1-j)(j-1)!}\tau^{j-1}v^{(j-1)}(t_{k+1})$$
$$-\sum_{j=2}^{2p+2}\frac{(2p+2q+2-j)!}{(2p+2-j)!}\frac{(2p+1)!}{(j-1)!}\tau^{j-1}v^{(j-1)}(t_k)$$

$$+\sum_{j=2}^{2p+2}(-1)^{j}\frac{(2p+2q+2-j)!}{(2p+2-j)!}\frac{(2p+1)!}{(j-1)!}\tau^{j-1}v^{(j-1)}(t_k)$$
$$+\sum_{j=2}^{2q+1}\frac{(2p+2q+2-j)!(2q)!}{(2q+1-j)(j-1)!}\tau^{j-1}v^{(j-1)}(t_{k-1})$$
$$=(2p+2q+1)!(v(t_{k+1})-2v(t_k)+v(t_{k-1}))$$

$$+\sum_{j=1}^{q}\frac{(2p+2q+1-2j)!(2q)!}{(2q-2j)!(2j)!}\tau^{2j}(v^{(2j)}(t_{k-1})+v^{(2j)}(t_{k+1}))$$
$$+\sum_{j=1}^{q}\frac{(2p+2q+2-2j)!(2q)!}{(2q+1-2j)!(2j-1)!}\tau^{2j-1}(v^{(2j-1)}(t_{k-1})-v^{(2j-1)}(t_{k+1}))$$

$$-2\sum_{j=1}^{p}\frac{(2p+2q+1-2j)!}{(2p+1-2j)!}\frac{(2p+1)!}{(2j)!}\tau^{2j}v^{(2j)}(t_k).$$

Theorem 3.2.3 is proved.

Note that using this Taylor decomposition on three points, we can extend our discussion to construct difference schemes of an arbitrary high order of accuracy for the approximate solutions of the initial-value problem (3.1) and the boundary-value problem (3.25). It is left to the reader.

Chapter 4

Partial Differential Equations of Parabolic Type

In the present chapter we consider the well-posedness of an abstract Cauchy problem for differential equations of parabolic type,

$$v'(t) + A(t)v(t) = f(t) \quad (0 \leq t \leq T), v(0) = v_0,$$

in an arbitrary Banach space with the linear positive operators $A(t)$. The high order of accuracy difference schemes generated by an exact difference scheme or by Taylor's decomposition on two points for the numerical solutions of this problem are presented. The well-posedness of these difference schemes in various Banach spaces are studied. The stability and coercive stability estimates in Hölder norms for the solutions of the high order of accuracy difference schemes of mixed type boundary-value problems for parabolic equations are obtained.

4.1 A Cauchy Problem. Well-posedness

We consider the abstract Cauchy problem for differential equations

$$v'(t) + A(t)v(t) = f(t) \quad (0 \leq t \leq T), v(0) = v_0 \tag{4.1}$$

in an arbitrary Banach space E with the linear (unbounded) operators $A(t)$. Here $v(t)$ and $f(t)$ are the unknown and the given functions, respectively, defined on $[0, T]$ with values in E. The derivative $v'(t)$ is understood as the limit in the norm of E of the corresponding ratio of differences. $A(t)$ is a given, closed, linear operator in E with domain $D(A(t)) = D$, independent of t and dense in E. Finally, v_0 is a given element of E.

A function $v(t)$ is called a solution of problem (4.1) if the following conditions are satisfied:

i. $v(t)$ is continuously differentiable on the segment $[0, T]$. The derivatives at the endpoints of the segment are understood as the appropriate unilateral derivatives.

ii. The element $v(t)$ belongs to $D = D(A(t))$ for all $t \in [0, T]$ and the function $A(t)v(t)$ is continuous on $[0, T]$.

iii. $v(t)$ satisfies the equation and the initial condition (4.1).

A solution of problem (4.1) defined in this manner will from now on be referred to as a solution of problem (4.1) in the space $C(E) = C([0, T], E)$. Here $C(E)$ stands for the Banach space of all continuous functions $\varphi(t)$ defined on $[0, T]$ with values in E equipped with the norm

$$||\varphi||_{C(E)} = \max_{0 \le t \le T} ||\varphi(t)||_E.$$

From the existence of such solutions evidently follows that $f(t) \in C(E)$ and $v_0 \in D$.

We say that problem (4.1) is well posed in $C(E)$ if the following conditions are satisfied:

1. Problem (4.1) is uniquely solvable for any $f(t) \in C(E)$ and any $v_0 \in D$. This means that an additive and homogeneous operator $v(t) = v(t; f(t), v_0)$ is defined which acts from $C(E) \times D$ to $C(E)$ and gives the solution of problem (4.1) in $C(E)$.

2. $v(t; f(t), v_0)$, regarded as an operator from $C(E) \times D$ to $C(E)$, is continuous. Here $C(E) \times D$ is understood as the normed space of the pairs $(f(t), v_0)$, $f(t) \in C(E)$ and $v_0 \in D$, equipped with the norm

$$||(f(t), v_0)||_{C(E)\times D} = ||f||_{C(E)} + ||v_0||_D.$$

By Banach's theorem in $C(E)$ and these properties one has coercive inequality

$$\|v'\|_{C(E)} + \|A(.)v\|_{C(E)} \le M_C[\|f\|_{C(E)} + \ \|v_0\|_D], \tag{4.2}$$

where M_C $(1 \le M_C < +\infty)$ does not depend on v_0 and $f(t)$. This inequality is called the coercivity inequality in $C(E)$ for (4.1). If $A(t) = A$, then the coercivity inequality implies analyticity of the semigroup $\exp\{-sA\}$ $(s \ge 0)$, i.e., the estimates

$$\|\exp(-sA)\|_{E\to E}, \|sA\exp(-sA)\|_{E\to E} \le M \ (s > 0)$$

hold for some $M \in [1, +\infty)$. Thus, the analyticity of the semigroup $\exp\{-sA\}$, $(s \ge 0)$ is a necessary for the well-posedness of problem (4.1) in $C(E)$. Unfortunately, the analyticity of the semigroup $\exp\{-sA\}(s \ge 0)$ is not sufficient for the well-posedness of problem (4.1) in $C(E)$.

Suppose that for each $t \in [0, T]$ the operator $-A(t)$ generates an analytic semigroup $\exp\{-sA(t)\}(s \ge 0)$ with exponentially decreasing norm, when $s \to +\infty$, i.e., the estimates

$$\|\exp(-sA(t))\|_{E\to E}, \|sA(t)\exp(-sA(t))\|_{E\to E} \le Me^{-\delta s}(s > 0) \tag{4.3}$$

hold for some $M \in [1, +\infty)$, $\delta \in (0, +\infty)$. From this inequality it follows that operator $A^{-1}(t)$ exists and is bounded and hence $A(t)$ is closed in $C(E)$.

Suppose that the operator $A(t)A^{-1}(s)$ is Hölder continuous in t in the uniform operator topology for each fixed s, that is,

$$||[A(t) - A(\tau)]A^{-1}(s)||_{E\to E} \leq M|t-\tau|^{\varepsilon}, 0 < \varepsilon \leq 1, \tag{4.4}$$

where M and ε are positive constants independent of t, s and τ for $0 \leq t, s, \tau \leq T$.

An operator-valued function $v(t,s)$, defined and strongly continuous jointly in t, s for $0 \leq s < t \leq T$, is called a fundamental solution of (4.1) if

1) the operator $v(t,s)$ is strongly continuous in t and s for $0 \leq s < t \leq T$,
2) the following identity holds:
   $v(t,s) = v(t,\tau)v(\tau,s), v(t,t) = I$ for $0 \leq s \leq \tau \leq t \leq T$,
3) the operator $v(t,s)$ maps the region D into itself. The operator $u(t,s) = A(t)v(t,s)A^{-1}(s)$ is bounded and strongly continuous in t and s for $0 \leq s < t \leq T$,
4) on the region D the operator $v(t,s)$ is strongly differentiable relative to t and s, while
$$\frac{\partial v(t,s)}{\partial t} = -A(t)v(t,s)$$
and
$$\frac{\partial v(t,s)}{\partial s} = v(t,s)A(s).$$

$v(t,s)$ is also called an evolution operator, Green's function, etc.

If the function $f(t)$ is not only continuous, but also continuously differentiable on $[0,T]$ and $v_0 \in D$, it is easy to show that the formula

$$v(t) = v(t,0)v_0 + \int_0^t v(t,s)f(s)ds \tag{4.5}$$

gives a solution of problem (4.1). Here $v(t,s)$ is the fundamental solution of (4.1).

Now we shall establish a series of interesting lemmas and estimates concerning the semigroup $\exp\{-sA(t)\}(s \geq 0)$ and the fundamental solution $v(t,s)$ of (4.1) which will be useful in the sequel.

Lemma 4.1.1. *For any $0 < s < s + \tau < T$, $0 \leq t \leq T$ and $0 \leq \alpha \leq 1$ one has the inequality*

$$\|\exp(-sA(t)) - \exp\{-(s+\tau)A(t)\}\|_{E\to E} \leq M\frac{\tau^{\alpha}}{(s+\tau)^{\alpha}}, \tag{4.6}$$

where M does not depend on α, t, s and τ.

Proof. Using the formula

$$\exp(-sA(t)) - \exp\{-(s+\tau)A(t)\} = \int_s^{s+\tau} A(t)\exp(-pA(t))dp$$

and estimate (4.3), we obtain that

$$\|\exp(-sA(t)) - \exp\{-(s+\tau)A(t)\}\|_{E\to E} \le M_1\frac{\tau}{s}.$$

The inequality

$$\|\exp(-sA(t)) - \exp\{-(s+\tau)A(t)\}\|_{E\to E} \le M_1$$

follows from (4.3) and the triangle inequality. Therefore

$$\|\exp(-sA(t)) - \exp\{-(s+\tau)A(t)\}\|_{E\to E} \le M_1\min\{1,\frac{\tau}{s}\}.$$

From this it follows that

$$\|\exp(-sA(t)) - \exp\{-(s+\tau)A(t)\}\|_{E\to E} \le M_1\frac{2\tau}{t+\tau}.$$

Interpolating the last inequality and

$$\|\exp(-sA(t)) - \exp\{-(s+\tau)A(t)\}\|_{E\to E} \le M_1,$$

we obtain (4.6). Lemma 4.1.1 is proved.

Lemma 4.1.2. *For any $0 \le s,\tau,t \le T$ and $u \in D$ the following identity holds:*

$$[\exp(-\tau A(t)) - \exp(-\tau A(s))]u \tag{4.7}$$

$$= -\int_0^\tau \exp\{-(\tau-p)A(t)\}[A(s)-A(t)]A^{-1}(s)\exp\{-pA(s)\}A(s)u\,dp.$$

Proof. Define the function

$$G(p) = -\exp\{-(\tau-p)A(t)\}\exp\{-pA(s)\}u, \quad 0 \le p \le \tau.$$

$G(p)$ is strongly differentiable with respect to p with derivative

$$\begin{aligned} G'(p) &= -\exp\{-(\tau-p)A(t)\}A(t)\exp\{-pA(s)\}u \\ &\quad + \exp\{-(\tau-p)A(t)\}A(s)\exp\{-pA(s)\}u \\ &= \exp\{-(\tau-p)A(t)\}[A(s)-A(t)]\exp\{-pA(s)\}u \\ &= \exp\{-(\tau-p)A(t)\}[A(s)-A(t)]A^{-1}(s)\exp\{-pA(s)\}A(s)u. \end{aligned}$$

Since $G'(p)$ is continuous in p, (4.7) follows upon integrating the last identity from 0 to τ. Lemma 4.1.2 is proved.

Lemma 4.1.3. *For any* $0 \le s,\ \tau,\ t \le T$ *and* $0 \le \varepsilon \le 1$ *the following estimates hold:*

$$\|\exp(-\tau A(t)) - \exp\{-\tau A(s)\}\|_{E\to E} \le M|t-s|^{\varepsilon} e^{-\delta\,\tau}, \tag{4.8}$$

$$\|A(t)[\exp(-\tau A(t)) - \exp\{-\tau A(s)\}]\|_{E\to E} \le M\tau^{-1}|t-s|^{\varepsilon} e^{-\delta\,\tau}, \tau > 0, \tag{4.9}$$

$$\left\|A(t)[\exp(-\tau A(t)) - \exp\{-\tau A(s)\}]A^{-1}(s)\right\|_{E\to E} \le M|t-s|^{\varepsilon} e^{-\delta\,\tau}, \tag{4.10}$$

$$\left\|[\exp(-tA(\tau)) - \exp\{-sA(\tau)\}]A^{-1}(\tau)\right\|_{E\to E} \le M|t-s| e^{-\delta\min\{t,s\}}, \tag{4.11}$$

$$\left\|A(t)[\exp(-tA(\tau)) - \exp\{-sA(\tau)\}]A^{-2}(\tau)\right\|_{E\to E} \le M|t-s| e^{-\delta\min\{t,s\}}, \tag{4.12}$$

where $M \ge 0$ *and* $\delta > 0$ *do not depend on* ε, t, s *and* τ.

Proof. Let $u \in E$. Then

$$
\begin{aligned}
&[\exp(-\tau A(t)) - \exp\{-\tau A(s)\}]u \\
&= [\exp(-\tfrac{\tau}{2}A(t)) - \exp\left\{-\tfrac{\tau}{2}A(s)\right\}]\exp\left\{-\tfrac{\tau}{2}A(s)\right\}u \\
&\quad + \exp(-\tfrac{\tau}{2}A(t))[\exp(-\tfrac{\tau}{2}A(t)) - \exp\left\{-\tfrac{\tau}{2}A(s)\right\}]u \\
&= [\exp(-\tfrac{\tau}{2}A(t)) - \exp\left\{-\tfrac{\tau}{2}A(s)\right\}]\exp\left\{-\tfrac{\tau}{2}A(s)\right\}u \\
&\quad + A(t)\exp(-\tfrac{\tau}{2}A(t))[\exp(-\tfrac{\tau}{2}A(t)) - \exp\left\{-\tfrac{\tau}{2}A(s)\right\}]A^{-1}(s)u \\
&\quad + \exp(-\tfrac{\tau}{2}A(t))[A(t) - A(s)]A^{-1}(s)\exp(-\tfrac{\tau}{2}A(s))]u \\
&\quad - \exp\{-\tau A(t)\}\,[A(t) - A(s)]A^{-1}(s)u = J_1 + J_2 + J_3,
\end{aligned}
$$

where

$$
\begin{aligned}
J_1 &= [\exp(-\tfrac{\tau}{2}A(t)) - \exp\left\{-\tfrac{\tau}{2}A(s)\right\}]\exp\left\{-\tfrac{\tau}{2}A(s)\right\}u, \\
J_2 &= A(t)\exp(-\tfrac{\tau}{2}A(t))[\exp(-\tfrac{\tau}{2}A(t)) - \exp\left\{-\tfrac{\tau}{2}A(s)\right\}]A^{-1}(s)u \\
J_3 &= \exp(-\tfrac{\tau}{2}A(t))[A(t) - A(s)]A^{-1}(s)\exp(-\tfrac{\tau}{2}A(s))]u \\
&\quad - \exp\{-\tau A(t)\}\,[A(t) - A(s)]A^{-1}(s)u.
\end{aligned}
$$

Applying formula (4.7), we can write

$$
\begin{aligned}
J_1 = -\int_0^{\frac{\tau}{2}} &\exp\{-(\tfrac{\tau}{2} - p)A(t)\}[A(s) - A(t)]A^{-1}(s) \\
&\times \exp\{-pA(s)\}A(s)\exp\left\{-\tfrac{\tau}{2}A(s)\right\}u\,dp
\end{aligned}
$$

for $u \in D$. Using the last formula and estimates (4.3) and (4.4), we obtain

$$\|J_1\|_E \leq \int_0^{\frac{\tau}{2}} \|\exp\{-(\frac{\tau}{2}-p)A(t)\}\|_{E\to E}\|[A(s)-A(t)]A^{-1}(s)\|_{E\to E}$$

$$\times\|\exp\{-pA(s)\}\|_{E\to E}\|A(s)\exp\left\{-\frac{\tau}{2}A(s)\right\}u\|_E dp$$

$$\leq \int_0^{\frac{\tau}{2}} M\exp\{-(\frac{\tau}{2}-p)\delta)\}M|t-s|^{\varepsilon}M\exp\{-p\delta\}M\frac{\tau}{2}\exp\left\{-\frac{\tau}{2}\delta\right\}u\|_E dp$$

$$\leq M_1|t-s|^{\varepsilon}\exp\{-\tau\delta\}\|u\|_E.$$

Applying formula (4.7), we can write

$$J_2 = -A(t)\exp(-\frac{\tau}{2}A(t))\int_0^{\frac{\tau}{2}}\exp\{-(\frac{\tau}{2}-p)A(t)\}$$

$$\times[A(s)-A(t)]A^{-1}(s)\exp\{-pA(s)\}dp$$

for $u \in D$. Using the last formula and estimates (4.3) and (4.4), we obtain

$$\|J_2\|_E \leq \|A(t)\exp(-\frac{\tau}{2}A(t))\|_{E\to E}\int_0^{\frac{\tau}{2}}\|\exp\{-(\frac{\tau}{2}-p)A(t)\}\|_{E\to E}$$

$$\times\|[A(s)-A(t)]A^{-1}(s)\|_{E\to E}\|\exp\{-pA(s)\}\|_{E\to E}\|u\|_E dp$$

$$\leq M\frac{\tau}{2}\exp\left\{-\frac{\tau}{2}\delta\right\}\int_0^{\frac{\tau}{2}} M\exp\{-(\frac{\tau}{2}-p)\delta)\}M|t-s|^{\varepsilon}M\exp\{-p\delta\}\|u\|_E dp$$

$$\leq M_1|t-s|^{\varepsilon}\exp\{-\tau\delta\}\|u\|_E.$$

Using estimates (4.3) and (4.4), we obtain

$$\|J_3\|_E \leq \|\exp(-\frac{\tau}{2}A(t))\|_{E\to E}\|[A(s)-A(t)]A^{-1}(s)\|_{E\to E}$$

$$\times\|\exp\{-\frac{\tau}{2}A(s)\}\|_{E\to E}\|u\|_E$$

$$+\|\exp\{-\tau A(t)\}\|_{E\to E}\|[A(s)-A(t)]A^{-1}(s)\|_{E\to E}\|u\|_E$$

$$\leq 2M\exp\left\{-\frac{\tau}{2}\delta\right\}M|t-s|^{\varepsilon}M\exp\{-\frac{\tau}{2}\delta\}\|u\|_E \leq M_1|t-s|^{\varepsilon}\exp\{-\tau\delta\}\|u\|_E.$$

Combining the estimate for J_1, J_2 and J_3, we obtain estimate (4.8). Applying formula (4.7), we can write

$$[\exp(-\tau A(t)) - \exp(-\tau A(s))]u \tag{4.13}$$

$$= -\int_0^{\tau} \exp\{-(\tau-p)A(t)\}[A(s)-A(t)]A^{-1}(s)\exp\{-pA(s)\}A(s)u\,dp$$

$$= -\int_0^{\frac{\tau}{2}} \exp\{-(\tau-p)A(t)\}[A(s)-A(t)]A^{-1}(s)\exp\{-pA(s)\}A(s)u\,dp$$

$$-\int_{\frac{\tau}{2}}^{\tau} \exp\{-(\tau-p)A(t)\}[A(s)-A(t)]A^{-1}(s)\exp\{-pA(s)\}A(s)u\,dp = G_1 + G_2,$$

where

$$G_1 = -\int_0^{\frac{\tau}{2}} \exp\{-(\tau-p)A(t)\}[A(s)-A(t)]A^{-1}(s)\exp\{-pA(s)\}A(s)u\,dp$$

$$G_2 = -\int_{\frac{\tau}{2}}^{\tau} \exp\{-(\tau-p)A(t)\}[A(s)-A(t)]A^{-1}(s)\exp\{-pA(s)\}A(s)u\,dp.$$

Integrating by parts, we can write

$$G_1 = \int_0^{\frac{\tau}{2}} \exp\{-(\tau-p)A(t)\}[A(s)-A(t)]A^{-1}(s)d[\exp\{-pA(s)\}u]$$

$$+\exp\{-\frac{\tau}{2}A(t)\}[A(s)-A(t)]A^{-1}(s)\exp\{-\frac{\tau}{2}A(s)\}u$$

$$-\exp\{-\tau A(t)\}[A(s)-A(t)]A^{-1}(s)u$$

$$-\int_0^{\frac{\tau}{2}} A(t)\exp\{-(\tau-p)A(t)\}[A(s)-A(t)]A^{-1}(s)\exp\{-pA(s)\}u\,dp.$$

Applying estimates (4.3) and (4.4), we can write

$$\|A(t)G_1\|_E \leq \|A(t)\exp\{-\frac{\tau}{2}A(t)\}\|_{E\to E}\|[A(s)-A(t)]A^{-1}(s)\|_{E\to E}$$

$$\times\|\exp\{-\frac{\tau}{2}A(s)\}\|_{E\to E}\|u\|_E$$

$$+||A(t)\exp\{-\tau A(t)\}||_{E\to E}||[A(s)-A(t)]A^{-1}(s)||_{E\to E}||u||_E$$

$$+\int_0^{\frac{\tau}{2}}||A^2(t)\exp\{-(\tau-p)A(t)\}||_{E\to E}$$

$$\times||[A(s)-A(t)]A^{-1}(s)||_{E\to E}||\exp\{-pA(s)\}||_{E\to E}||u||_E dp$$

$$\leq M(\frac{\tau}{2})^{-1}\exp\{-\frac{\tau}{2}\delta\}M|t-s|^{\varepsilon}\exp\{-\frac{\tau}{2}\delta\}||u||_E+M\tau^{-1}\exp\{-\tau\delta\}M|t-s|^{\varepsilon}||u||_E$$

$$+\int_0^{\frac{\tau}{2}}M(\tau-p)^{-2}\exp\{-(\tau-p)\delta\}M|t-s|^{\varepsilon}\exp\{-p\delta\}||u||_E dp$$

$$\leq M_1|t-s|^{\varepsilon}\tau^{-1}\exp\{-\tau\delta\}||u||_E.$$

Integrating by parts, we can write

$$G_2=-A^{-1}(t)\int_{\frac{\tau}{2}}^{\tau}d[\exp\{-(\tau-p)A(t)\}][A(s)-A(t)]A^{-1}(s)\exp\{-pA(s)\}A(s)u$$

$$=-A^{-1}(t)[A(s)-A(t)]A^{-1}(s)\exp\{-\tau A(s)\}A(s)u$$

$$+A^{-1}(t)\exp\{-\frac{\tau}{2}A(t)\}[A(s)-A(t)]A^{-1}(s)\exp\{-\frac{\tau}{2}A(s)\}A(s)$$

$$-A^{-1}(t)\int_{\frac{\tau}{2}}^{\tau}\exp\{-(\tau-p)A(t)\}][A(s)-A(t)]A^{-1}(s)\exp\{-pA(s)\}A^2(s)u dp.$$

Applying estimates (4.3) and (4.4), we can write

$$||A(t)G_2||_E\leq||[A(s)-A(t)]A^{-1}(s)||_{E\to E}||A(s)\exp\{-\tau A(s)\}||_{E\to E}||u||_E$$

$$+||\exp\{-\frac{\tau}{2}A(t)\}||_{E\to E}||[A(s)-A(t)]A^{-1}(s)||_{E\to E}$$

$$\times||A(s)\exp\{-\frac{\tau}{2}A(s)\}||_{E\to E}||u||_E$$

$$+\int_{\frac{\tau}{2}}^{\tau}||\exp\{-(\tau-p)A(t)\}||_{E\to E}||[A(s)-A(t)]A^{-1}(s)||_{E\to E}$$

$$\times||A^2(s)\exp\{-pA(s)\}||_{E\to E}||u||_E dp$$

$$\leq M|t-s|^{\varepsilon}M\tau^{-1}\exp\{-\tau\delta\}||u||_E$$

$$+M\exp\{-\frac{\tau}{2}\delta\}M|t-s|^{\varepsilon}M(\frac{\tau}{2})^{-1}\exp\{-\frac{\tau}{2}\delta\}||u||_E$$

$$+\int_{\frac{\tau}{2}}^{\tau} M\exp\{-(\tau-p)\delta\}M|t-s|^{\varepsilon}(p)^{-2}\exp\{-p\delta\}||u||_E dp$$

$$\leq M_1|t-s|^{\varepsilon}\tau^{-1}\exp\{-\tau\delta\}||u||_E.$$

Combining the estimates for $A(t)G_1$ and $A(t)G_2$, we obtain estimate (4.9). Using formula (4.13), we notice that, for every $u \in E$,

$$[\exp(-\tau A(t)) - \exp\{-\tau A(s)\}]A^{-1}(s) = H_1 + H_2,$$

where

$$H_1 = -\int_0^{\frac{\tau}{2}} \exp\{-(\tau-p)A(t)\}[A(s)-A(t)]A^{-1}(s)\exp\{-pA(s)\}udp,$$

$$H_2 = -\int_{\frac{\tau}{2}}^{\tau} \exp\{-(\tau-p)A(t)\}[A(s)-A(t)]A^{-1}(s)\exp\{-pA(s)\}udp.$$

Applying estimates (4.3) and (4.4), we can write

$$||A(t)H_1||_E \leq \int_0^{\frac{\tau}{2}} ||A(t)\exp\{-(\tau-p)A(t)\}||_{E\to E}||[A(s)-A(t)]A^{-1}(s)||_{E\to E}$$

$$\times||\exp\{-pA(s)\}||_{E\to E}||u||_E dp$$

$$+\int_0^{\frac{\tau}{2}} M(\tau-p)^{-1}\exp\{-(\tau-p)\delta\}M|t-s|^{\varepsilon}\exp\{-p\delta\}||u||_E dp$$

$$\leq M_1|t-s|^{\varepsilon}\exp\{-\tau\delta\}||u||_E.$$

Integrating by parts, we can write

$$H_2 = -A^{-1}(t)\int_{\frac{\tau}{2}}^{\tau} d[\exp\{-(\tau-p)A(t)\}][A(s)-A(t)]A^{-1}(s)\exp\{-pA(s)\}u$$

$$= -A^{-1}(t)[A(s)-A(t)]A^{-1}(s)\exp\{-\tau A(s)\}u$$

$$+A^{-1}(t)\exp\{-\frac{\tau}{2}A(t)\}[A(s)-A(t)]A^{-1}(s)\exp\{-\frac{\tau}{2}A(s)\}$$

$$-A^{-1}(t)\int_{\frac{\tau}{2}}^{\tau} \exp\{-(\tau-p)A(t)\}][A(s)-A(t)]A^{-1}(s)\exp\{-pA(s)\}A(s)udp.$$

Applying estimates (4.3) and (4.4), we can write

$$||A(t)H_2||_E \leq ||[A(s)-A(t)]A^{-1}(s)||_{E\to E}||\exp\{-\tau A(s)\}||_{E\to E}||u||_E$$

$$+||\exp\{-\frac{\tau}{2}A(t)\}||_{E\to E}||[A(s)-A(t)]A^{-1}(s)||_{E\to E}||\exp\{-\frac{\tau}{2}A(s)\}||_{E\to E}||u||_E$$

$$+\int_{\frac{\tau}{2}}^{\tau}||\exp\{-(\tau-p)A(t)\}||_{E\to E}||[A(s)-A(t)]A^{-1}(s)||_{E\to E}$$

$$\times||A(s)\exp\{-pA(s)\}||_{E\to E}||u||_E dp$$

$$\leq M|t-s|^{\varepsilon}M\exp\{-\tau\delta\}||u||_E + M\exp\{-\frac{\tau}{2}\delta\}M|t-s|^{\varepsilon}M\exp\{-\frac{\tau}{2}\delta\}||u||_E$$

$$+\int_{\frac{\tau}{2}}^{\tau}M\exp\{-(\tau-p)\delta\}M|t-s|^{\varepsilon}M(p)^{-1}\exp\{-p\delta\}||u||_E dp$$

$$\leq M_1|t-s|^{\varepsilon}\exp\{-\tau\delta\}||u||_E.$$

Combining the estimates for $A(t)H_1$ and $A(t)H_2$, we obtain estimate (4.10). We can write

$$[\exp(-tA(\tau))-\exp(-sA(\tau))]A^{-1}(\tau)u \tag{4.14}$$

$$=-\int_{s}^{t}\exp\{-pA(\tau)\}A(\tau)A^{-1}(\tau)udp.$$

Applying estimates (4.3) and (4.4), we can write

$$\left\|[\exp(-tA(\tau))-\exp\{-sA(\tau)\}]A^{-1}(\tau)\right\|_{E\to E}$$

$$\leq \left|\int_{s}^{t}||\exp\{-pA(\tau)\}||_{E\to E}||u||dp\right| \leq M|t-s|e^{-\delta\min\{t,s\}}.$$

Estimate (4.11) is established. Using formula (4.14), we notice that, for every u,

$$A(t)[\exp(-tA(\tau))-\exp(-sA(\tau))]A^{-2}(\tau)u$$

$$=-\int_{s}^{t}\exp\{-pA(\tau)\}udp-\int_{s}^{t}(A(t)-A(\tau))A^{-1}(\tau)\exp\{-pA(\tau)\}udp.$$

Applying estimates (4.3) and (4.4), we obtain

$$\left\|A(t)[\exp(-tA(\tau))-\exp\{-sA(\tau)\}]A^{-2}(\tau)\right\|_{E\to E}$$

$$\leq \left|\int_s^t \|\exp\{-pA(\tau)\}\|_{E\to E}\|u\|dp\right| + \left|\int_s^t \|(A(t)-A(\tau))A^{-1}(\tau)\|_{E\to E}\right.$$
$$\left.\times \|\exp\{-pA(\tau)\}\|_{E\to E}\|u\|dp\right|$$
$$\leq M_1|t-s|e^{-\delta \min\{t,s\}}.$$

Estimate (4.12) is established. Lemma 4.1.3 is proved.

Lemma 4.1.4. *For any $0 \leq s < t \leq T$ and $u \in D$ the following identities hold:*

$$v(t,s)u = \exp\{-(t-s)A(s)\}u \tag{4.15}$$
$$+\int_s^t v(t,z)[A(s)-A(z)]A^{-1}(s)\exp\{-(z-s)A(s)\}A(s)u dz,$$

$$v(t,s)u = \exp\{-(t-s)A(t)\}u \tag{4.16}$$
$$+\int_s^t \exp\{-(t-z)A(t)\}[A(z)-A(t)]v(z,s)u dz.$$

Proof. Define the function

$$G(z) = v(t,z)\exp\{-(z-s)A(s)\}u, \ s \leq z \leq t.$$

$G(z)$ is strongly differentiable with respect to z with derivative

$$G'(z) = v(t,z)A(z)\exp\{-(z-s)A(s)\}u$$
$$-v(t,z)A(s)\exp\{-(z-s)A(s)\}u$$
$$= v(t,z)[A(z)-A(s)]\exp\{-(z-s)A(s)\}u$$
$$= v(t,z)[A(z)-A(s)]A^{-1}(s)\exp\{-(z-s)A(s)\}A(s)u.$$

Since $G'(z)$ is continuous in z, (4.15) follows upon integrating the last identity from s to t. In a similar manner one establishes the identity (4.16). Lemma 4.1.4 is proved.

Lemma 4.1.5. *For any $0 \leq s < t \leq t+r \leq T$, $0 \leq \alpha \leq 1$ and $0 < \varepsilon \leq 1$ the following estimates hold:*

$$\|v(t,s)\|_{E\to E} \leq M, \tag{4.17}$$
$$\left\|A(t)v(t,s)A^{-1}(s)\right\|_{E\to E} \leq M, \tag{4.18}$$
$$\|v(t,s)-\exp\{-(t-s)A(t)\}\|_{E\to E} \leq M(t-s)^{\varepsilon}, \tag{4.19}$$
$$\|A(t)[v(t,s)-\exp\{-(t-s)A(t)\}]\|_{E\to E} \leq M(t-s)^{\varepsilon-1}, \tag{4.20}$$

$$\|A(t)v(t,s)\|_{E\to E} \leq M(t-s)^{-1}, \tag{4.21}$$

$$\|A(t+r)v(t+r,s) - A(t)v(t,s)\|_{E\to E} \leq M(t-s)^{-1}\Psi, \tag{4.22}$$

$$\left\|A(t+r)v(t+r,s)A^{-1}(s) - A(t)v(t,s)A^{-1}(s)\right\|_{E\to E} \leq M\Psi, \tag{4.23}$$

where $M \geq 0$ does not depend on ε, t, s and r. Here $\Psi = r^{\varepsilon} + r^{\alpha}(t-s)^{-\alpha}$.

Proof. Applying formula (4.15) and estimates (4.3) and (4.4), we obtain

$$||v(t,s)u||_E \leq ||\exp\{-(t-s)A(s)\}u||_E$$

$$+\int_s^t ||v(t,z)||_{E\to E}||[A(s)-A(z)]A^{-1}(s)||_{E\to E}$$

$$\times||\exp\{-(z-s)A(s)\}A(s)||_{E\to E}||u||_E dz$$

$$\leq M||u||_E + \int_s^t ||v(t,z)||_{E\to E}M(z-s)^{\varepsilon}(z-s)^{-1}||u||_E dz$$

$$= M||u||_E + M_1\int_s^t ||v(t,z)||_{E\to E}(z-s)^{\varepsilon-1}dz||u||_E.$$

Thus, we have that

$$||v(t,s)u||_E \leq M||u||_E + M_1\int_s^t ||v(t,z)||_{E\to E}(z-s)^{\varepsilon-1}dz||u||_E$$

or

$$||v(t,s)||_{E\to E} \leq M + M_1\int_s^t ||v(t,z)||_{E\to E}(z-s)^{\varepsilon-1}dz.$$

From the integral inequality, (4.17) follows. Note that from (4.4) it follows that

$$||A(t)A^{-1}(s)||_{E\to E} \leq M_1 \tag{4.24}$$

for each $t, s \in [0,T]$. Actually,

$$||A(t)A^{-1}(s)||_{E\to E} \leq 1 + ||[A(t)-A(s)]A^{-1}(s)||_{E\to E} \leq 1 + M(t-s)^{\varepsilon} \leq M_1.$$

Applying formula (4.16) and estimates (4.3), (4.4) and (4.24), we obtain

$$||A(t)v(t,s)A^{-1}(s)u||_E \leq ||\exp\{-(t-s)A(t)\}||_{E\to E}||A(t)A^{-1}(s)||_{E\to E}||u||_E$$

$$+\int_s^t ||A(t)\exp\{-(t-z)A(t)\}||_{E\to E}||[A(t)-A(z)]A^{-1}(z)||_{E\to E}$$

$$\times||A(z)v(z,s)A^{-1}(s)||_{E\to E}||u||_E dz \leq M||u||_E$$

$$+\int_s^t ||A(z)v(z,s)^{-1}A(s)||_{E\to E} M(t-z)^{\varepsilon}(t-z)^{-1}||u||_E dz$$

$$= M||u||_E + M_1 \int_s^t ||A(z)v(z,s)^{-1}A(s)||_{E\to E}(z-s)^{\varepsilon-1}dz||u||_E.$$

Thus, we have that

$$||A(t)v(t,s)A^{-1}(s)u||_{E\to E} \leq M||u||_E$$

$$+M_1 \int_s^t ||A(z)v(z,s)A^{-1}(s)||_{E\to E}(t-z)^{\varepsilon-1}dz||u||_E$$

or

$$||A(t)v(t,s)A^{-1}(s)||_{E\to E} \leq M + M_1 \int_s^t ||A(z)v(z,s)A^{-1}(s)||_{E\to E}(t-z)^{\varepsilon-1}dz.$$

Using the integral inequality, we obtain (4.18).

Applying formula (4.15) and estimates (4.3), (4.4), (4.8) and (4.18), we obtain

$$||v(t,s)u - \exp\{-(t-s)A(t)\}u||_E$$

$$\leq ||\exp\{-(t-s)A(t)\}u - \exp\{-(t-s)A(s)\}u||_E$$

$$+\int_s^t ||v(t,z)||_{E\to E}||[A(s)-A(z)]A^{-1}(s)||_{E\to E}$$

$$\times||A(s)\exp\{-(z-s)A(s)\}||_{E\to E}||u||_E dz$$

$$\leq M(t-s)^{\varepsilon}||u||_E + \int_s^t MM(z-s)^{\varepsilon}(z-s)^{-1}||u||_E dz$$

$$= M||u||_E(t-s)^{\varepsilon} + M_1 \int_s^t (z-s)^{\varepsilon-1}dz||u||_E = M_2||u||_E(t-s)^{\varepsilon}.$$

The estimate (4.19) is proved.

In a similar manner one establishes estimate (4.20). Using estimates (4.20) and (4.3), we obtain

$$\|A(t)v(t,s)||_{E\to E} \leq ||A(t)\exp\{-(t-s)A(t)\}\|_{E\to E}$$

$$+ \|A(t)[v(t,s) - \exp\{-(t-s)A(t)\}]\|_{E\to E}$$

$$\leq M|t-s|^{-1} + M|t-s|^{\varepsilon-1} \leq M_1|t-s|^{-1}.$$

The estimate (4.21) is proved.

Applying formula (4.16), we can write

$$A(t)v(t,s)A^{-1}(s)u = A(t)\exp\{-(t-s)A(t)\}A^{-1}(s)u$$

$$+\int_s^t A(t)\exp\{-(t-z)A(t)\}[A(z)-A(t)]v(z,s)A^{-1}(s)u\,dz = J_1(t)u + J_2(t)u.$$

Using estimates (4.3), (4.4), (4.8) and (4.24), we obtain

$$||J_1(t+r) - J_1(t)||_{E\to E} \tag{4.25}$$

$$\leq ||[A(t+r)[\exp\{-(t+r-s)A(t+r)\} - \exp\{-(t-s)A(t)\}]A^{-1}(s)||_{E\to E}$$

$$+||(A(t+r)-A(t))A^{-1}(t)||_{E\to E}[||\exp\{-(t-s)A(t)\}||_{E\to E}$$

$$\times||A(t)A^{-1}(s)||_{E\to E} \leq M\Psi + M_1 r^{\varepsilon} \leq M_2\Psi$$

for $0 \leq s < t \leq t+r \leq T$. Now let us estimate $J_2(t+r) - J_2(t)$ for $0 \leq s < t \leq t+r \leq T$.

To this end we shall examine separately the two cases $t-s \leq 2r$ and $t-s > 2r$. If $t-s \leq 2r$, then using estimates (4.3), (4.4) and (4.18), we obtain

$$||J_2(t)||_{E\to E} \leq \int_s^t ||A(t)\exp\{-(t-z)A(t)\}||_{E\to E}$$

$$\times||[A(z)-A(t)]A^{-1}(z)||_{E\to E}||A(z)v(z,s)A^{-1}(s)||_{E\to E}dz$$

$$\leq \int_s^t M_1(t-z)^{\varepsilon-1}dz = M_2(t-s)^{\varepsilon}.$$

From this estimate and the triangle inequality it follows that

$$||J_2(t+r) - J_2(t)||_{E\to E} \leq ||J_2(t+r)||_{E\to E} + ||J_2(t)||_{E\to E} \tag{4.26}$$

$$\leq M_2(t+r-s)^{\varepsilon} + M_2(t-s)^{\varepsilon} \leq M_3 r^{\varepsilon}.$$

Now let $t-s > 2r$. We have that

$$J_2(t+r) - J_2(t)$$

$$= \int_{t-r}^{t+r} A(t+r)\exp\{-(t+r-z)A(t+r)\}[A(z)-A(t+r)]v(z,s)A^{-1}(s)u\,dz$$

$$-\int_{t-r}^{t} A(t)\exp\{-(t-z)A(t)\}[A(z)-A(t)]v(z,s)A^{-1}(s)u\,dz$$

$$+\int_{s}^{t-r}[A(t+r)\exp\{-(t+r-z)A(t+r)\}$$

$$-A(t)\exp\{-(t-z)A(t)\}][A(z)-A(t)]v(z,s)A^{-1}(s)udz$$

$$+\int_{s}^{t-r}A(t+r)\exp\{-(t+r-z)A(t+r)\}[A(t)-A(t+r)]v(z,s)A^{-1}(s)udz$$

$$=L_1(t)u+L_2(t)u+L_3(t)u+L_4(t)u,$$

where

$$L_1(t)u=\int_{t-r}^{t+r}A(t+r)\exp\{-(t+r-z)A(t+r)\}[A(z)-A(t+r)]v(z,s)A^{-1}(s)udz,$$

$$L_2(t)u=-\int_{t-r}^{t}A(t)\exp\{-(t-z)A(t)\}[A(z)-A(t)]v(z,s)A^{-1}(s)udz,$$

$$L_3(t)u=\int_{s}^{t-r}[A(t+r)\exp\{-(t+r-z)A(t+r)\}$$

$$-A(t)\exp\{-(t-z)A(t)\}][A(z)-A(t)]v(z,s)A^{-1}(s)udz,$$

$$L_4(t)u=\int_{s}^{t-r}A(t+r)\exp\{-(t+r-z)A(t+r)\}[A(t)-A(t+r)]v(z,s)A^{-1}(s)udz.$$

Using the estimates (4.3) and (4.4), we obtain

$$||L_1(t)||_{E\to E}\le\int_{t-r}^{t+r}||A(t+r)\exp\{-(t+r-z)A(t+r)\}||_{E\to E}$$

$$\times||[A(z)-A(t+r)]A^{-1}(z)||_{E\to E}||A(z)v(z,s)A^{-1}(s)||_{E\to E}dz$$

$$\le\int_{t-r}^{t+r}M_1(t+r-z)^{\varepsilon-1}dz=M_2r^{\varepsilon}.$$

In a similar manner one establishes the estimate

$$||L_2(t)||_{E\to E}\le M_2r^{\varepsilon}.$$

Using estimates (4.3), (4.4), (4.6) and (4.9), we obtain

$$||L_3(t)||_{E\to E} \leq \int\limits_s^{t-r} \{||A(t+r)[\exp\{-(t+r-z)A(t+r)\}$$

$$-\exp\{-(t-z)A(t+r)\}]||_{E\to E}$$

$$+||A(t+r)[\exp\{-(t-z)A(t+r) - \exp\{-(t-z)A(t)\}]||_{E\to E}$$

$$+||[A(t+r) - A(t)]A^{-1}(t)||_{E\to E}||A(t)\exp\{-(t-z)A(t)\}||_{E\to E}\}$$

$$\times||[A(z) - A(t)]A^{-1}(z)||_{E\to E}||A(z)v(z,s)A^{-1}(s)||_{E\to E}dz$$

$$\leq M_1 \int\limits_s^{t-r} \frac{1}{t+r-z}[\frac{r}{t+r-z} + r^{\varepsilon}](t-z)^{\varepsilon}dz \leq M_2 r^{\varepsilon}.$$

Now let us estimate $L_4(t)$. Using estimates (4.3), (4.4), (4.6) and (4.9), we obtain

$$||L_4(t)||_{E\to E} \leq \int\limits_s^{t-r} ||A(t+r)\exp\{-(t+r-z)A(t+r)\}]||_{E\to E}$$

$$\times||[A(t+r) - A(t)]A^{-1}(z)||_{E\to E}||A(z)v(z,s)A^{-1}(s)||_{E\to E}dz$$

$$\leq M_1 \int\limits_s^{t-r} \frac{r^{\varepsilon}dz}{t+r-z} \leq M_1 \int\limits_s^{t-r} \frac{dz}{(t+r-z)^{1-\varepsilon}} \leq M_2 r^{\varepsilon}.$$

Combining the estimates for $L_1(t), L_2(t), L_3(t)$ and $L_4(t)$, we obtain estimate (4.26) for $t-s > 2r$. Therefore,

$$||J_2(t+r) - J_2(t)||_{E\to E} \leq M_3 r^{\varepsilon}$$

for all $0 \leq s < t \leq t+r \leq T$. From this and (4.25) it follows (4.23). Applying property 2) of $v(t,s)$ and estimates (4.21) and (4.23), we obtain

$$\|A(t+r)v(t+r,s) - A(t)v(t,s)\|_{E\to E}$$

$$\leq \left\|[A(t+r)v(t+r,\frac{t+s}{2}) - A(t)v(t,\frac{t+s}{2})]A^{-1}(\frac{t+s}{2})\right\|_{E\to E}$$

$$\times\left\|A(\frac{t+s}{2})v(\frac{t+s}{2},s)\right\|_{E\to E} \leq M\Psi M(t-s)^{-1} = M_1(t-s)^{-1}\Psi.$$

Estimate (4.22) is established. Lemma 4.1.5 is proved.

A function $v(t)$ is said to be a solution of problem (4.1) in $F(E)$ if it is a solution of this problem in $C(E)$ and the function $v'(t)$ and $A(t)v(t)$ belong to $F(E)$.

As in the case of the space $C(E)$, we say that problem (4.1) is well posed in $F(E)$, if the following two conditions are satisfied:

1. For any $f(t) \in F(E)$ and $v_0 \in D$ there exists the unique solution $v(t) = v(t; f(t), v_0)$ in $F(E)$ of problem (4.1). This means that an additive and homogeneous operator $v(t; f(t), v_0)$ is defined which acts from $F(E) \times D$ to $F(E)$ and gives the solution of (4.1) in $F(E)$.

2. $v(t; f(t), v_0)$, regarded as an operator from $F(E) \times D$ to $F(E)$, is continuous. Here $F(E) \times D$ is understood as the normed space of the pairs $(f(t), v_0)$, $f(t) \in F(E)$ and $v_0 \in D$, equipped with the norm

$$||(f(t), v_0)||_{F(E)\times D} = ||f||_{F(E)} + ||v_0||_D.$$

First, we set $F(E)$ equal to $C_0^{\alpha,\alpha}(E) = C_0^{\alpha,\alpha}([0,T], E), 0 < \alpha < 1$, the Banach space obtained by completion of the set of smooth E-valued functions $\varphi(t)$ on $[0,T]$ in the norm

$$\| \varphi \|_{C_0^{\alpha,\alpha}(E)} = \|\varphi\|_{C(E)} + \sup_{0\le t<t+\tau\le T} \frac{\|\varphi(t+\tau) - \varphi(t)\|_E}{\tau^\alpha} (t+\tau)^\alpha.$$

As in the case of the space $C(E)$, from the well-posedness of the Cauchy problem (4.1) one derives the coercivity inequality

$$\|v'\|_{C_0^{\alpha,\alpha}(E)} + \|A(.)v\|_{C_0^{\alpha,\alpha}(E)} \le M_C(\alpha)[\|f\|_{C_0^{\alpha,\alpha}(E)} + \|A(0)v_0\|_E] , \tag{4.27}$$

where $M_C\ (\alpha)(1 \le M_C(\alpha) < +\infty)$ does not depend on v_0 and $f(t)$.

Theorem 4.1.6. *Suppose $v_0 \in D, f(t) \in C_0^{\alpha,\alpha}(E)$. Suppose that assumptions (4.3) and (4.4) hold and $0 < \alpha \le \varepsilon \le 1$. Then the Cauchy problem (4.1) is well posed in Hölder space $C_0^{\alpha,\alpha}(E)$. For the solution $v(t)$ in $C_0^{\alpha,\alpha}(E)$ of the Cauchy problem the coercive inequality*

$$\|v'\|_{C_0^{\alpha,\alpha}(E)} + \|A(.)v\|_{C_0^{\alpha,\alpha}(E)} \le \frac{M}{\alpha(1-\alpha)} \| f \|_{C_0^{\alpha,\alpha}(E)} + M \| A(0)v_0\|_E$$

holds, where M does not depend on α, v_0 and $f(t)$.

Proof. If $v(t)$ is a solution in $C_0^{\alpha,\alpha}(E)$ of problem (4.1), then it is a solution in $C(E)$ of this problem. Hence, by (4.5), we have the representation

$$v(t) = v(t,0)v_0 + \int_0^t v(t,s)f(s)ds.$$

To prove the theorem we must show that the functions $v'(t)$ and $A(t)v(t)$ belong to $C_0^{\alpha,\alpha}(E)$ and finally obtain estimates for the norms of these functions. Here we address only the last task. Using formula (4.5), we obtain

$$A(t)v(t) = A(t)v(t,0)v_0 + \int_0^t A(t)v(t,s)dsf(t) \tag{4.28}$$

$$+ \int_0^t A(t)v(t,s)(f(s) - f(t))ds = \sum_{k=1}^{3} J_k(t),$$

where

$$J_1(t) = A(t)v(t,0)v_0,$$

$$J_2(t) = \int_0^t A(t)v(t,s)dsf(t),$$

$$J_3(t) = \int_0^t A(t)v(t,s)(f(s) - f(t))ds.$$

Let us estimate $J_k(t)$ for any $k = 1,2,3$ separately. We start with $J_1(t)$. Using estimate (4.18), we show that, for $0 \le t \le T$,

$$||J_1(t)||_E \le ||A(t)v(t,0)A^{-1}(0)||_{E\to E}||A(0)v_0||_E \le M||A(0)v_0||_E.$$

Further, applying (4.23), we show that, for $0 < t < t+\tau < T$,

$$||J_1(t+\tau) - J_1(t)||_E \le ||A(t+\tau)v(t+\tau,0) - A(t)v(t,0)]A^{-1}(0)||_{E\to E}||A(0)v_0||_E$$

$$\le M[\tau^\varepsilon + \frac{\tau^\alpha}{(t+\tau)^\alpha}]||A(0)v_0||_E.$$

Since $\alpha \le \varepsilon$, we have proved that

$$||J_1||_{C_0^{\alpha,\alpha}(E)} \le M||A(0)v_0||_E. \tag{4.29}$$

Now let us estimate $J_2(t)$. Using property 4) of $v(t,s)$, we can write

$$J_2(t) = [I - A(t)v(t,0)A^{-1}(t)]f(t)$$

$$+ \int_0^t A(t)v(t,s)[A(t) - A(s)]A^{-1}(t)dsf(t).$$

Using this formula and estimates (4.4), (4.18), (4.21) and (4.24), we obtain

$$||J_2(t)||_E \le [1 + ||A(t)v(t,0)A^{-1}(0)||_{E\to E}||A(0)A^{-1}(t)||_{E\to E}]||f(t)||_E$$

$$+ \int_0^t ||A(t)v(t,s)||_{E\to E}||[A(t) - A(s)]A^{-1}(t)||_{E\to E}ds||f(t)||_E$$

$$\le M||f(t)||_E + M\int_0^t (t-s)^{-1+\varepsilon}ds||f(t)||_E \le M_1 \parallel f \parallel_{C_0^{\alpha,\alpha}(E)}$$

for all $t,\ t\ \in [0,T]$. Now let us estimate $J_2(t+\tau) - J_2(t)$ for $0 \le t < t+\tau \le T$.

To this end we shall examine separately the two cases $t \le 2\tau$ and $t > 2\tau$. If $t \le 2\tau$, then using the last estimate and the triangle inequality, we obtain

$$||J_2(t+\tau) - J_2(t)||_E \le ||J_2(t+\tau)||_E + ||J_2(t)||_E \le 2M_1 \parallel f \parallel_{C_0^{\alpha,\alpha}(E)} \tag{4.30}$$

$$\le \frac{\tau^\alpha}{(t+\tau)^\alpha}M_2 \parallel f \parallel_{C_0^{\alpha,\alpha}(E)}.$$

Now let $t > 2\tau$. We have that

$$
\begin{aligned}
& J_2(t+\tau) - J_2(t) \\
&= \{[I - A(t+\tau)v(t+\tau,0)A^{-1}(t+\tau)] \\
&+ \int_0^{t+\tau} A(t+\tau)v(t+\tau,s)[A(t+\tau)-A(s)]A^{-1}(t)ds\}[f(t+\tau)-f(t)] \\
&+[A(t)v(t,0)A^{-1}(t) - A(t+\tau)v(t+\tau,0)A^{-1}(t+\tau)]f(t) \\
&- \int_{t-\tau}^{t+\tau} A(t+\tau)v(t+\tau,s)[A(s)-A(t+\tau)]A^{-1}(t)dsf(t) \\
&+ \int_{t-\tau}^{t} A(t)v(t,s)[A(s)-A(t)]A^{-1}(t)dsf(t) \\
&- \int_0^{t-\tau} [A(t+\tau)v(t+\tau,s) - A(t)v(t,s)][A(s)-A(t)]A^{-1}(t)dsf(t) \\
&- \int_0^{t-\tau} A(t+\tau)v(t+\tau,s)[A(t)-A(t+\tau)]A^{-1}(t)dsf(t) \\
&+ \int_0^{t-\tau} A(t+\tau)v(t+\tau,s)[A(t+\tau)-A(s)]A^{-1}(t)ds \\
&\times[A(t)-A(t+\tau)]A^{-1}(t+\tau)f(t+\tau) \\
&= L_1(t)+L_2(t)+L_3(t)+L_4(t)+L_5(t)+L_6(t)+L_7(t),
\end{aligned}
$$

where

$$
\begin{aligned}
L_1(t) &= \{[I - A(t+\tau)v(t+\tau,0)A^{-1}(t+\tau)] \\
&+ \int_0^{t+\tau} A(t+\tau)v(t+\tau,s)[A(t+\tau)-A(s)]A^{-1}(t)ds\}[f(t+\tau)-f(t)], \\
L_2(t) &= [A(t)v(t,0)A^{-1}(t) - A(t+\tau)v(t+\tau,0)A^{-1}(t+\tau)]f(t), \\
L_3(t) &= -\int_{t-\tau}^{t+\tau} A(t+\tau)v(t+\tau,s)[A(s)-A(t+\tau)]A^{-1}(t)dsf(t), \\
L_4(t) &= \int_{t-\tau}^{t} A(t)v(t,s)[A(s)-A(t)]A^{-1}(t)dsf(t),
\end{aligned}
$$

$$L_5(t) = -\int_0^{t-\tau} [A(t+\tau)v(t+\tau,s) - A(t)v(t,s)][A(s) - A(t)]A^{-1}(t)dsf(t),$$

$$L_6(t) = -\int_0^{t-\tau} A(t+\tau)v(t+\tau,s)[A(t) - A(t+\tau)]A^{-1}(t)dsf(t),$$

$$L_7(t) = \int_0^{t-\tau} A(t+\tau)v(t+\tau,s)[A(t+\tau) - A(s)]A^{-1}(t)ds$$
$$\times [A(t) - A(t+\tau)]A^{-1}(t+\tau)f(t+\tau).$$

Let us estimate $L_k(t)$ for any $k = 1,2,3,4,5,6,7$ separately. We start with $L_1(t)$. Using estimates (4.4), (4.18), (4.21) and (4.24), we obtain

$$||L_1(t)||_E$$
$$\leq [1 + ||A(t+\tau)v(t+\tau,0)A^{-1}(0)||_{E\to E}||A(0)A^{-1}(t+\tau)||_{E\to E}]||f(t+\tau) - f(t)||_E$$

$$+ \int_0^{t+\tau} ||A(t+\tau)v(t+\tau,s)||_{E\to E}||[A(t+\tau) - A(s)]A^{-1}(t)||_{E\to E}ds$$
$$\times ||f(t+\tau) - f(t)||_E$$

$$\leq [M + M\int_0^{t+\tau} (t+\tau-s)^{1-\varepsilon}ds]||f(t+\tau) - f(t)||_E$$
$$\leq M_1 \frac{\tau^\alpha}{(t+\tau)^\alpha} \parallel f \parallel_{C_0^{\alpha,\alpha}(E)}$$

for all t, $t \in [0,T]$. Let us estimate $L_2(t)$. Using the formula

$$L_2(t) = [A(t)v(t,0) - A(t+\tau)v(t+\tau,0)]A^{-1}(t)f(t)$$
$$+A(t+\tau)v(t+\tau,0)[A^{-1}(t) - A^{-1}(t+\tau)]f(t),$$

and estimates (4.3), (4.4), (4.8), (4.23) and (4.24), we obtain

$$||L_2(t)||_E \leq ||[A(t)v(t,0) - A(t+\tau)v(t+\tau,0)]A^{-1}(0)||_{E\to E}$$
$$\times||A(0)A^{-1}(t)||_{E\to E}||f(t)||_E$$
$$+||A(t+\tau)v(t+\tau,0)A^{-1}(t)||_{E\to E}||[A(t+\tau) - A(t)]A^{-1}(t+\tau)||_{E\to E}||f(t)||_E$$
$$\leq M[\tau^\varepsilon + \frac{\tau^\alpha}{(t+\tau)^\alpha}]M||f(t)||_E + M\tau^\varepsilon||f(t)||_E$$

for all t, $t \in [0,T]$. Since $\alpha \leq \varepsilon$, we have proved that

$$||L_2(t)||_E \leq M_1 \frac{\tau^\alpha}{(t+\tau)^\alpha} \parallel f \parallel_{C_0^{\alpha,\alpha}(E)}$$

for all t, $t \in [0,T]$. Let us estimate $L_3(t)$. Using estimates (4.3), (4.4), (4.8), (4.23) and (4.24), we obtain

$$||L_3(t)||_E \le \int\limits_{t-\tau}^{t+\tau} ||A(t+\tau)v(t+\tau,s)||_{E\to E}||[A(s)-A(t+\tau)]A^{-1}(t)||_{E\to E}||f(t)||_E ds$$

$$\le \int\limits_{t-\tau}^{t+\tau} M_1(t+\tau-s)^{\varepsilon-1}ds||f(t)||_E = M_2\tau^{\varepsilon} \parallel f \parallel_{C_0^{\alpha,\alpha}(E)}$$

for all t, $t \in [0,T]$. Therefore,

$$||L_3(t)||_E \le M_1\frac{\tau^{\alpha}}{(t+\tau)^{\alpha}} \parallel f \parallel_{C_0^{\alpha,\alpha}(E)}$$

for all t, $t \in [0,T]$. In a similar manner one establishes the estimate

$$||L_4(t)||_E \le M_1\frac{\tau^{\alpha}}{(t+\tau)^{\alpha}} \parallel f \parallel_{C_0^{\alpha,\alpha}(E)}$$

for all t, $t \in [0,T]$. Now let us estimate $L_5(t)$. Using estimates (4.4) and (4.22), we obtain

$$||L_5(t)||_E \le \int\limits_0^{t-\tau} ||A(t+\tau)v(t+\tau,s)-A(t)v(t,s)||_{E\to E}$$

$$\times||[A(s)-A(t)]A^{-1}(t)||_{E\to E}ds||f(t)||_E$$

$$\le M_1\int\limits_0^{t-\tau}(t-s)^{-1}[\tau^{\varepsilon}+\frac{\tau^{\alpha}}{(t+\tau-s)^{\alpha}}]ds||f(t)||_E \le M_2\int\limits_0^{t-\tau}\frac{\tau^{\alpha}ds}{(t+\tau-s)^{1+\alpha}}||f(t)||_E$$

$$\le \frac{M_3}{\alpha}\frac{\tau^{\alpha}}{(t+\tau)^{\alpha}} \parallel f \parallel_{C_0^{\alpha,\alpha}(E)}$$

for all t, $t \in [0,T]$. Now let us estimate $L_6(t)$. Using estimates (4.4) and (4.21), we obtain

$$||L_6(t)||_E \le \int\limits_0^{t-\tau} ||A(t+\tau)v(t+\tau,s)||_{E\to E}||[A(t+\tau)-A(t)]A^{-1}(t)||_{E\to E}ds||f(t)||_E$$

$$\le M_1\int\limits_0^{t-\tau}(t+\tau-s)^{-1}\tau^{\varepsilon}ds||f(t)||_E \le M_2\int\limits_0^{t-\tau}\frac{\tau^{\alpha}ds}{(t+\tau-s)^{1+\alpha}}||f(t)||_E$$

$$\le \frac{M_3}{\alpha}\frac{\tau^{\alpha}}{(t+\tau)^{\alpha}} \parallel f \parallel_{C_0^{\alpha,\alpha}(E)}$$

for all t, $t \in [0,T]$. Finally we have

$$||L_7(t)||_E \leq \int_0^{t+\tau} M_1(t+\tau-s)^{-1+\varepsilon} ds M_2 \tau^{\varepsilon} ||f(t+\tau)||_E \leq M_3 \frac{\tau^{\alpha}}{(t+\tau)^{\alpha}} \parallel f \parallel_{C_0^{\alpha,\alpha}(E)} .$$

Combining the estimates for $L_k(t)$ for any $k = 1,2,3,4,5,6,7$, we obtain estimate (4.30) for $t > 2\tau$. Therefore,

$$||J_2||_{C_0^{\alpha,\alpha}(E)} \leq \frac{M}{\alpha} \parallel f \parallel_{C_0^{\alpha,\alpha}(E)} . \tag{4.31}$$

Now let us estimate $J_3(t)$. Using estimate (4.21), we obtain

$$\parallel J_3(t) \parallel_E \leq \int_0^t \|A(t)v(t,s)\|_{E\to E} \|f(s) - f(t) \parallel_E ds$$

$$\leq M \int_0^t \frac{ds}{(t-s)^{1-\alpha} t^{\alpha}} \parallel f \parallel_{C_0^{\alpha,\alpha}(E)} = \frac{M}{\alpha} \parallel f \parallel_{C_0^{\alpha,\alpha}(E)}$$

for all t, $t \in [0,T]$. Now let us estimate the difference $J_3(t+\tau) - J_3(t)$ for $0 < t < t+\tau < T$. We will consider separately the cases $t \leq 2\tau$ and $t > 2\tau$. If $t \leq 2\tau$, then from the last estimate it follows that

$$\parallel J_3(t+\tau) - J_3(t) \parallel_E \leq \parallel J_3(t+\tau) \parallel_E + \parallel J_3(t) \parallel_E$$

$$\leq \frac{2M3^{\alpha}}{\alpha} \parallel f \parallel_{C_0^{\alpha,\alpha}(E)} \tau^{\alpha}(t+\tau)^{-\alpha} = \frac{M_1}{\alpha} \parallel f \parallel_{C_0^{\alpha,\alpha}(E)} \tau^{\alpha}(t+\tau)^{-\alpha}.$$

Now let $t > 2\tau$. Let us represent the difference $J_3(t+\tau) - J_3(t)$ as the sum of the following integrals:

$$J_3(t+\tau) - J_3(t) = \int_{t-\tau}^{t+\tau} A(t+\tau)v(t+\tau,s)(f(s) - f(t+\tau))ds$$

$$- \int_{t-\tau}^{t} A(t)v(t,s)(f(s) - f(t))ds$$

$$+ \int_0^{t-\tau} A(t+\tau)v(t+\tau,s)(f(t) - f(t+\tau))ds$$

$$+ \int_0^{t-\tau} [A(t+\tau)v(t+\tau,s) - A(t)v(t,s)](f(s) - f(t))ds$$

$$= P_1(t) + P_2(t) + P_3(t) + P_4(t),$$

where

$$P_1(t) = \int_{t-\tau}^{t+\tau} A(t+\tau)v(t+\tau,s)(f(s) - f(t+\tau))ds,$$

$$P_2(t) = -\int_{t-\tau}^{t} A(t)v(t,s)(f(s) - f(t))ds,$$

$$P_3(t) = \int_{0}^{t-\tau} [A(t+\tau)v(t+\tau,s) - A(t)v(t,s)](f(s) - f(t))ds,$$

$$P_4(t) = \int_{0}^{t-\tau} A(t+\tau)v(t+\tau,s)(f(t) - f(t+\tau))ds.$$

Let us estimate $P_k(t)$ for any $k = 1,2,3,4$ separately. We start with $P_1(t)$. Using estimate (4.21), we obtain

$$||P_1(t)||_E \le \int_{t-\tau}^{t+\tau} ||A(t+\tau)v(t+\tau,s)||_{E\to E}||f(s) - f(t+\tau)||_E ds$$

$$\le \frac{1}{(t-\tau)^\alpha}\int_{t-\tau}^{t+\tau} M_1(t+\tau-s)^{\alpha-1}ds \parallel f \parallel_{C_0^{\alpha,\alpha}(E)} = \frac{M_2}{\alpha}\frac{\tau^\alpha}{(t+\tau)^\alpha} \parallel f \parallel_{C_0^{\alpha,\alpha}(E)}$$

for all t, $t \in [0,T]$. In a similar manner one establishes the estimate

$$||P_2(t)||_E \le M_1 \frac{\tau^\alpha}{(t+\tau)^\alpha} \parallel f \parallel_{C_0^{\alpha,\alpha}(E)}$$

for all t, $t \in [0,T]$. Now let us estimate $P_3(t)$. Using estimate (4.22), we obtain

$$||P_3(t)||_E \le \int_{0}^{t-\tau} ||A(t+\tau)v(t+\tau,s) - A(t)v(t,s)||_{E\to E}||f(s) - f(t)||_E ds$$

$$\le \frac{M_1}{t^\alpha}\int_{0}^{t-\tau} (t-s)^{-1+\alpha}[\tau^\varepsilon + \frac{\tau}{t+\tau-s}]ds \parallel f \parallel_{C_0^{\alpha,\alpha}(E)}$$

$$\le \frac{M_2\tau^\alpha}{\alpha(1-\alpha)t^\alpha} \parallel f \parallel_{C_0^{\alpha,\alpha}(E)} \le \frac{M_3\tau^\alpha}{\alpha(1-\alpha)(t+\tau)^\alpha} \parallel f \parallel_{C_0^{\alpha,\alpha}(E)}$$

for all t, $t \in [0,T]$. Let us estimate $P_4(t)$. Using property 4) of $v(t,s)$, we can write

$$P_4(t) = [A(t+\tau)v(t+\tau,t-\tau)A^{-1}(t+\tau)$$

$$-A(t+\tau)v(t+\tau,0)A^{-1}(t+\tau)](f(t)-f(t+\tau))$$

$$+\int_0^{t-\tau} A(t+\tau)v(t+\tau,s)[A(t+\tau)-A(s)]A^{-1}(t+\tau)ds(f(t)-f(t+\tau)).$$

Using this formula and estimates (4.4), (4.18), (4.21) and (4.24), we obtain

$$||P_4(t)||_E \leq [||A(t+\tau)v(t+\tau,t-\tau)A^{-1}(0)||_{E\to E}$$

$$+||A(t+\tau)v(t+\tau,0)A^{-1}(0)||_{E\to E}]||A(0)A^{-1}(t+\tau)||_{E\to E}||f(t)-f(t+\tau)||_E$$

$$+\int_0^{t-\tau} ||A(t+\tau)v(t+\tau,s)||_{E\to E}||[A(t+\tau)-A(s)]A^{-1}(t+\tau)||_{E\to E}ds$$

$$\times||f(t)-f(t+\tau)||_E \leq 2MM||f(t)-f(t+\tau)||_E$$

$$+M\int_0^{t-\tau}(t+\tau-s)^{1-\varepsilon}ds||f(t)-f(t+\tau)||_E$$

$$\leq \frac{M_1\tau^\alpha}{(t+\tau)^\alpha} \parallel f \parallel_{C_0^{\alpha,\alpha}(E)}$$

for all t, $t\in[0,T]$. Combining the estimates for $P_k(t)$ for any $k=1,2,3,4$, we obtain the estimate

$$\parallel J_3(t+\tau)-J_3(t)\parallel_E \leq \frac{M_1}{\alpha(1-\alpha)} \parallel f \parallel_{C_0^{\alpha,\alpha}(E)} \tau^\alpha(t+\tau)^{-\alpha}$$

for $t>2\tau$. Therefore,

$$||J_3||_{C_0^{\alpha,\alpha}(E)} \leq \frac{M}{\alpha(1-\alpha)} \parallel f \parallel_{C_0^{\alpha,\alpha}(E)}. \tag{4.32}$$

Finally, applying estimates (4.29), (4.31) and (4.32), we obtain the estimate

$$\parallel A(.)v \parallel_{C_0^{\alpha,\alpha}(E)} \leq \frac{M}{\alpha(1-\alpha)} \parallel f \parallel_{C_0^{\alpha,\alpha}(E)} +M \parallel A(0)v_0||_E.$$

By the triangle inequality, this last estimate and equation (4.1) yields

$$\parallel v'||_{C_0^{\alpha,\alpha}(E)} \leq \frac{M}{\alpha(1-\alpha)} \parallel f \parallel_{C_0^{\alpha,\alpha}(E)} +M \parallel A(0)v_0||_E.$$

Theorem 4.1.6 is proved.

With the help of $A(t)$ we introduce the fractional spaces $E_{\alpha,q}(E,A(t))$, $0<\alpha<1$, consisting of all $v\in E$ for which the following norms are finite:

$$\parallel v \parallel_{E_{\alpha,q}}=(\int_0^\infty [z^{1-\alpha} \parallel A(t)\exp\{-zA(t)\}v \parallel_E]^q\frac{dz}{z})^{\frac{1}{q}},\ 1\leq q<\infty,$$

$$\parallel v \parallel_{E_\alpha}=\parallel v \parallel_{E_{\alpha,\infty}}=\sup_{z>0} z^{1-\alpha} \parallel A(t)\exp\{-zA(t)\}v \parallel_E .$$

From (4.3) and (4.4) follows

Theorem 4.1.7. *$E_{\alpha,q}(E, A(t)) = E_{\alpha,q}(E, A(0))$ for all $0 < \alpha < 1$, $1 \leq q \leq \infty$ and $0 \leq t \leq T$.*

Note that the coercivity inequality

$$\|v'\|_{F(E)} + \|A(.)v\|_{F(E)} \leq M_C[\|f\|_{F(E)} + \ ||A(0)v_0||_E]$$

fails if we set $F(E)$ equal to $C^\alpha(E), (0 < \alpha < 1)$, the Banach space obtained by completion of the set of all smooth E-valued functions $\varphi(t)$ on [0,T] in the norm

$$\| \varphi \|_{C^\alpha(E)} = \max_{0 \leq t \leq T} ||\varphi(t)||_E + \sup_{0 \leq t < t+\tau \leq T} \frac{\| \varphi(t+\tau) - \varphi(t) \|_E}{\tau^\alpha}.$$

Nevertheless, we have the following result.

Theorem 4.1.8. *Suppose $v_0' = f(0) - A(0)v_0 \in E_\alpha$, $f(t) \in C^\alpha(E)(0 < \alpha < 1)$. Suppose that assumptions (4.3) and (4.4) hold and $0 < \alpha \leq \varepsilon \leq 1$. Then the Cauchy problem (4.1) is well posed in Hölder space $C^\alpha(E)$. For the solution $v(t)$ in $C^\alpha(E)$ of the Cauchy problem the coercive inequality*

$$\|v'\|_{C^\alpha(E)} + \|A(.)v\|_{C^\alpha(E)} + \|v'\|_{C(E_\alpha)} \leq \frac{M}{\alpha(1-\alpha)}[\| f \|_{C^\alpha(E)} + \| v_0'||_{E_\alpha}]$$

holds, where M does not depend on α, v_0 and $f(t)$.

Proof. By formula (4.28), we have

$$v'(t) = A(t)v(t,0)A^{-1}(0)v_0' + A(t)v(t,0)A^{-1}(0)(A(0) - A(t))A^{-1}(t)f(t)$$

$$+A(t)v(t,0)A^{-1}(0)(f(t) - f(0)) - \int_0^t A(t)v(t,s)[A(t) - A(s)]A^{-1}(t)dsf(t)$$

$$- \int_0^t A(t)v(t,s)(f(s) - f(t))ds = \sum_{k=1}^{5} J_k(t),$$

where

$$J_1(t) = A(t)v(t,0)A^{-1}(0)v_0',$$

$$J_2(t) = A(t)v(t,0)A^{-1}(0)(A(0) - A(t))A^{-1}(t)f(t),$$

$$J_3(t) = A(t)v(t,0)A^{-1}(0)(f(t) - f(0)),$$

$$J_4(t) = - \int_0^t A(t)v(t,s)[A(t) - A(s)]A^{-1}(t)dsf(t),$$

$$J_5(t) = - \int_0^t A(t)v(t,s)(f(s) - f(t))ds.$$

Let us estimate $J_k(t)$ for any $k = 1,2,3,4,5$ in $C^\alpha(E)$ and $C(E_\alpha)$ separately. We start with $J_1(t)$. Using estimate (4.18), we show that, for $0 \le t \le T$,

$$||J_1(t)||_E \le ||A(t)v(t,0)A^{-1}(0)||_{E\to E}||v_0'||_E \le M||v_0'||_E \le M||v_0'||_{E_\alpha}.$$

Further, applying (4.16), we can write

$$J_1(t) = \exp\{-tA(t)\}v_0' + \exp\{-tA(t)\}[A(t) - A(0)]A^{-1}(0)v_0' \tag{4.33}$$

$$+ \int_0^t A(t)\exp\{-(t-s)A(t)\}[A(s) - A(t)]v(s,0)A^{-1}(0)v_0'ds$$

$$= J_{11}(t) + J_{12}(t) + J_{13}(t),$$

where

$$J_{11}(t) = \exp\{-tA(t)\}v_0',$$

$$J_{12}(t) = \exp\{-tA(t)\}[A(t) - A(0)]A^{-1}(0)v_0',$$

$$J_{13}(t) = \int_0^t A(t)\exp\{-(t-s)A(t)\}[A(s) - A(t)]v(s,0)A^{-1}(0)v_0'ds.$$

Using estimates (4.3) and (4.8), we obtain

$$||J_{11}(t+\tau) - J_{11}(t)||_E \le ||\exp\{-(t+\tau)A(t+\tau)\}v_0' - \exp\{-(t+\tau)A(t)\}v_0'||_E \tag{4.34}$$

$$+ \int_0^\tau ||A(t)\exp\{-(t+s)A(t)\}v_0'ds||_E$$

$$\le M\tau^\varepsilon||v_0'||_E + \int_0^\tau \frac{Mds}{(t+s)^{1-\alpha}} \parallel v_0'||_{E_\alpha} \le \frac{M_1}{\alpha}\tau^\alpha||v_0'||_{E_\alpha}$$

for $0 < t < t+\tau < T$. Using estimates (4.3), (4.6) and (4.8), we obtain

$$||J_{12}(t+\tau) - J_{12}(t)||_E \le ||\exp\{-(t+\tau)A(t+\tau)\}||_{E\to E} \tag{4.35}$$

$$\times||[A(t+\tau) - A(t)]A^{-1}(0)||_{E\to E}||v_0'||_E$$

$$\le \ ||\exp\{-(t+\tau)A(t+\tau)\} - \exp\{-(t+\tau)A(t)\}||_{E\to E}$$
$$\times||[A(t) - A(0)]A^{-1}(0)||_{E\to E}||v_0'||_E$$

$$+||\exp\{-(t+\tau)A(t)\} - \exp\{-tA(t)\}||_{E\to E}||[A(t) - A(0)]A^{-1}(0)||_{E\to E}||v_0'||_E$$

$$\le MM\tau^\varepsilon||v_0'||_E + M\tau^\varepsilon Mt^\varepsilon||v_0'||_E + M\frac{\tau^\alpha}{(t+\tau)^\alpha}Mt^\varepsilon||v_0'||_E \le M_1\tau^\alpha||v_0'||_{E_\alpha}$$

for $0 < t < t+\tau < T$. Let us estimate $J_{13}(t+\tau) - J_{13}(t)$ for $0 < t < t+\tau < T$. Using estimates (4.3), (4.4) and (4.18), we obtain

$$\|J_{13}(t)\|_E \leq \int_0^t \|A(t)\exp\{-(t-s)A(t)\}\|_{E\to E}\|[A(t)-A(s)]A^{-1}(0)\|_{E\to E}$$

$$\times\|A(s)v(s,0)A^{-1}(0)\|_{E\to E}ds\|v_0'\|_E \leq \int_0^t MM(t-s)^{-1+\varepsilon}M ds\|v_0'\|_E$$

$$\leq \frac{M_1}{\alpha}t^\alpha\|v_0'\|_E \leq \frac{M_1}{\alpha}t^\alpha\|v_0'\|_{E_\alpha}.$$

If $t \leq 2\tau$, then using the last estimate and the triangle inequality, we obtain

$$\|J_{13}(t+\tau) - J_{13}(t)\|_E \leq \|J_{13}(t+\tau)\|_E + \|J_{13}(t)\|_E$$

$$\leq \frac{M_1}{\alpha}((t+\tau)^\alpha + t^\alpha)\|v_0'\|_{E_\alpha} \leq \frac{M_2}{\alpha}\tau^\alpha\|v_0'\|_{E_\alpha}.$$

Now let $t > 2\tau$. We have that

$$J_{13}(t+\tau) - J_{13}(t)$$

$$= \int_{t-\tau}^{t+\tau} A(t+\tau)\exp\{-(t+\tau-s)A(t+\tau)\}[A(s)-A(t+\tau)]v(s,0)A^{-1}(0)v_0'ds$$

$$-\int_{t-\tau}^{t} A(t)\exp\{-(t-s)A(t)\}[A(s)-A(t)]v(s,0)A^{-1}(0)v_0'ds$$

$$+\int_0^{t-\tau}[A(t+\tau)\exp\{-(t+\tau-s)A(t+\tau)\} - A(t)$$

$$\times\exp\{-(t-s)A(t)\}][A(s)-A(t)]v(s,0)A^{-1}(0)v_0'ds$$

$$+\int_0^{t-\tau} A(t+\tau)\exp\{-(t+\tau-s)A(t+\tau)\}[A(t)-A(t+\tau)]v(s,0)A^{-1}(0)v_0'ds$$

$$= K_1(t) + K_2(t) + K_3(t) + K_4(t),$$

where

$$K_1(t) = \int_{t-\tau}^{t+\tau} A(t+\tau)\exp\{-(t+\tau-s)A(t+\tau)\}[A(s)-A(t+\tau)]v(s,0)A^{-1}(0)v_0'ds,$$

$$K_2(t) = -\int_{t-\tau}^{t} A(t)\exp\{-(t-s)A(t)\}[A(s)-A(t)]v(s,0)A^{-1}(0)v_0'ds,$$

$$K_3(t) = \int_{0}^{t-\tau} [A(t+\tau)\exp\{-(t+\tau-s)A(t+\tau)\} - A(t)$$
$$\times \exp\{-(t-s)A(t)\}][A(s)-A(t)]v(s,0)A^{-1}(0)v_0'ds,$$

$$K_4(t) = \int_{0}^{t-\tau} A(t+\tau)\exp\{-(t+\tau-s)A(t+\tau)\}[A(t)-A(t+\tau)]v(s,0)A^{-1}(0)v_0'ds.$$

Using estimates (4.3), (4.4) and (4.18), we obtain

$$||K_1(t)||_E \le \int_{t-\tau}^{t+\tau} ||A(t+\tau)\exp\{-(t+\tau-s)A(t+\tau)\}||_{E\to E}|$$
$$\times ||[A(s)-A(t+\tau)]A^{-1}(s)||_{E\to E}$$
$$\times ||A(s)v(s,0)A^{-1}(0)||_{E\to E}||v_0'||_E ds$$
$$\le \int_{t-\tau}^{t+\tau} MM(t+\tau-s)^{-1+\varepsilon}Mds||v_0'||_E \le \frac{M_1}{\alpha}\tau^{\alpha}||v_0'||_{E_\alpha}.$$

In a similar manner one establishes the estimate

$$||K_2(t)||_E \le \frac{M_1}{\alpha}\tau^{\alpha}||v_0'||_{E_\alpha}.$$

Using estimates (4.4), (4.18) and (4.23), we obtain

$$||K_3(t)||_E \le \int_{0}^{t-\tau} ||A(t+\tau)\exp\{-(t+\tau-s)A(t+\tau)\} - A(t)\exp\{-(t-s)A(t)\}||_{E\to E}$$

$$\times||[A(s)-A(t)]A^{-1}(s)||_{E\to E}||A(s)v(s,0)A^{-1}(0)||_{E\to E}||v_0'||_E ds$$

$$\le \int_{0}^{t-\tau} MM(t+\tau-s)^{-1+\varepsilon}[\tau^{\varepsilon} + \frac{\tau}{t-s}]Mds||v_0'||_E \le \frac{M_1}{\alpha(1-\alpha)}\tau^{\alpha}||v_0'||_{E_\alpha}.$$

Applying the formula

$$K_4(t) = [\exp\{-2\tau A(t+\tau)\} - \exp\{-(t+\tau)A(t+\tau)\}]$$
$$\times [A(t)-A(t+\tau)]v(t,0)A^{-1}(0)v_0'$$

$$+\int_0^{t-\tau} A(t+\tau)\exp\{-(t+\tau-s)A(t+\tau)\}[A(t)-A(t+\tau)][v(s,0)-v(t,0)]A^{-1}(0)v_0'ds$$

and estimates (4.3), (4.4), (4.18) and (4.23), we obtain

$$||K_4(t)||_E \leq ||\exp\{-2\tau A(t+\tau)\} - \exp\{-(t+\tau)A(t+\tau)\}||_{E\to E}$$
$$\times||[A(t)-A(t+\tau)]A^{-1}(t)||_{E\to E}||A(t)v(t,0)A^{-1}(0)||_{E\to E}||v_0'||_E$$
$$+\int_0^{t-\tau} ||A(t+\tau)\exp\{-(t+\tau-s)A(t+\tau)\}||_{E\to E}$$
$$\times||[A(t+\tau)-A(t)]A^{-1}(s)||_{E\to E}||A(s)(v(t,0)-v(s,0))A^{-1}(0)||_{E\to E}||v_0'||_E ds$$
$$\leq M\tau^{\varepsilon}M||v_0'||_E + \int_0^{t-\tau} MM(t+\tau-s)^{-1}\tau^{\varepsilon}[\tau^{\varepsilon}+\frac{\tau}{t-s}]Mds||v_0'||_E$$
$$\leq \frac{M_1}{\alpha(1-\alpha)}\tau^{\alpha}||v_0'||_{E_\alpha}.$$

Combining the estimates for $K_k(t)$ for any $k=1,2,3,4$, we obtain the estimate

$$||J_{13}(t+\tau)-J_{13}(t)||_E \leq \frac{M_1}{\alpha(1-\alpha)}\tau^{\alpha}||v_0'||_{E_\alpha}$$

for $t>2\tau$. Therefore,

$$||J_{13}(t+\tau)-J_{13}(t)||_E \leq \frac{M_1}{\alpha(1-\alpha)}\tau^{\alpha}||v_0'||_{E_\alpha}$$

for $0<t<t+\tau<T$. Using estimates (4.3), (4.6) and (4.8), we obtain

$$||J_1||_{C^{\alpha}(E)} \leq \frac{M_1}{\alpha(1-\alpha)}||v_0'||_{E_\alpha}. \tag{4.36}$$

Using formula (4.33) and estimates (4.3), (4.6) and (4.8), we obtain

$$z^{1-\alpha} \parallel A(t)\exp\{-zA(t)\}J_1(t)||_E \leq z^{1-\alpha} \parallel A(t)\exp\{-(z+t)A(t)\}v_0'||_E$$
$$+z^{1-\alpha} \parallel A(t)\exp\{-(z+t)A(t)\}||_{E\to E}||[A(t)-A(0)]A^{-1}(0)||_{E\to E}||v_0'||_E$$
$$+\int_0^t z^{1-\alpha} \parallel A^2(t)\exp\{-(z+t-s)A(t)\}||_{E\to E}||[A(s)-A(t)]A^{-1}(s)||_{E\to E}$$
$$\times||A(s)v(s,0)A^{-1}(0)||_{E\to E}||v_0'||_E ds \leq \frac{z^{1-\alpha}}{(t+z)^{1-\alpha}} \parallel v_0'||_{E_\alpha}$$
$$+z^{1-\alpha}\frac{Mt^{\varepsilon}}{z+t}||v_0'||_E + \int_0^t z^{1-\alpha}\frac{M(t-s)^{\varepsilon}ds}{(z+t-s)^2}||v_0'||_E \leq \frac{M_1}{1-\alpha}||v_0'||_{E_\alpha}$$

for all $z > 0$ and t, $t \in [0,T]$. Therefore,

$$||J_1||_{C(E_\alpha)} \leq \frac{M_1}{\alpha(1-\alpha)}||v_0'||_{E_\alpha}. \tag{4.37}$$

Now let us estimate $J_2(t)$. Using estimates (4.4), (4.18), we show that, for $0 \leq t \leq T$,

$$||J_2(t)||_E \leq ||A(t)v(t,0)A^{-1}(0)||_{E\to E}||(A(0)-A(t))A^{-1}(t)||_{E\to E}||f(t)||_E$$
$$\leq MMt^\varepsilon||f(t)||_E \leq M_1t^\varepsilon \parallel f \parallel_{C^\alpha(E)} .$$

Therefore,

$$||J_2(t)||_E \leq M_2||f \parallel_{C^\alpha(E)} .$$

Moreover, if $t \leq 2\tau$, then using the last estimate and the triangle inequality, we obtain

$$||J_2(t+\tau) - J_2(t)||_E \leq ||J_2(t+\tau)||_E + ||J_2(t)||_E$$
$$\leq M_2((t+\tau)^\alpha + t^\alpha)||f \parallel_{C^\alpha(E)} \leq M_3\tau^\alpha||f \parallel_{C^\alpha(E)} .$$

Now let $t > 2\tau$. Using estimates (4.3), (4.6), (4.23), (4.24) and (4.18), we obtain

$$||J_2(t+\tau) - J_2(t)||_E \leq ||A(t+\tau)v(t+\tau,0)A^{-1}(0)||_{E\to E}$$
$$\times||(A(0)-A(t+\tau))A^{-1}(t+\tau)||_{E\to E}||f(t+\tau)-f(t)||_E$$
$$+||[A(t+\tau)v(t+\tau,0)-A(t)v(t,0)]A^{-1}(0)||_{E\to E}$$
$$\times||(A(0)-A(t+\tau))A^{-1}(t+\tau)||_{E\to E}||f(t)||_E$$

$$+||A(t)v(t,0)A^{-1}(0)||_{E\to E}||A(0)A^{-1}(t)||_{E\to E}$$
$$\times||(A(t)-A(t+\tau))A^{-1}(t+\tau)||_{E\to E}||f(t)||_E$$

$$\leq MM(t+\tau)^\varepsilon\tau^\alpha||f \parallel_{C^\alpha(E)} + M[\tau^\varepsilon + \frac{\tau^\alpha}{t^\alpha}]M(t+\tau)^\varepsilon||f \parallel_{C^\alpha(E)}$$
$$+MM\tau^\varepsilon||f \parallel_{C^\alpha(E)} \leq M_1\tau^\alpha||f \parallel_{C^\alpha(E)}$$

for $t > 2\tau$. Therefore,

$$||J_2||_{C^\alpha(E)} \leq M_1\tau^\alpha||f \parallel_{C^\alpha(E)} . \tag{4.38}$$

Using formula (4.33) and estimates (4.3), (4.4), (4.6) and (4.8), we obtain

$$z^{1-\alpha} \parallel A(t)\exp\{-zA(t)\}J_2(t)||_E$$

$$\leq z^{1-\alpha} \parallel A^2(t)\exp\{-(z+t)A(t)\}A^{-1}(0)||_{E\to E}||(A(0)-A(t))A^{-1}(t)||_{E\to E}||f(t)||_E$$
$$+z^{1-\alpha} \parallel A^2(t)\exp\{-(z+t)A(t)\}A^{-1}(0)||_{E\to E}$$

$$\times\|(A(0)-A(t))A^{-1}(0)\|_{E\to E}\|(A(0)-A(t))A^{-1}(t)\|_{E\to E}\|f(t)\|_E$$

$$+\int_0^t z^{1-\alpha}\parallel A^2(t)\exp\{-(z+t-s)A(t)\}A^{-1}(0)\|_{E\to E}\|[A(s)-A(t)]A^{-1}(s)\|_{E\to E}$$

$$\times\|A(s)v(s,0)A^{-1}(0)\|_{E\to E}\|_{E\to E}\|(A(0)-A(t))A^{-1}(t)\|_{E\to E}\|f(t)\|_E ds$$

$$\leq \frac{z^{1-\alpha}t^{\varepsilon}}{t+z}\|f(t)\|_E + z^{1-\alpha}\frac{Mt^{2\varepsilon}}{z+t}\|f(t)\|_E$$

$$+\int_0^t z^{1-\alpha}\frac{M(t-s)^{\varepsilon}t^{\varepsilon}ds}{z+t-s}\|f(t)\|_E \leq \frac{M_1}{\alpha}\|f\|_{C(E_\alpha)}$$

for all $z>0$ and t, $t\in[0,T]$. Therefore,

$$\|J_2\|_{C(E_\alpha)} \leq \frac{M_1}{\alpha}\|f\|_{C(E_\alpha)}. \tag{4.39}$$

Now let us estimate $J_3(t)$. Using estimate (4.18), we show that, for $0\leq t\leq T$,

$$\|J_3(t)\|_E \leq \|A(t)v(t,0)A^{-1}(0)\|_{E\to E}\|f(t)-f(0)\|_E$$

$$\leq Mt^{\varepsilon}\parallel f\parallel_{C^\alpha(E)}\leq Mt^{\varepsilon}\parallel f\parallel_{C^\alpha(E)}.$$

Therefore,

$$\|J_3(t)\|_E \leq M_1\|f\parallel_{C^\alpha(E)}.$$

Moreover, if $t\leq 2\tau$, then using the last estimate and the triangle inequality, we obtain

$$\|J_3(t+\tau)-J_3(t)\|_E \leq \|J_3(t+\tau)\|_E + \|J_3(t)\|_E$$

$$\leq M_2((t+\tau)^{\alpha}+t^{\alpha})\|f\parallel_{C^\alpha(E)}\leq M_2\tau^{\alpha}\|f\parallel_{C^\alpha(E)}.$$

Now let $t>2\tau$. Using estimates (4.23) and (4.18), we obtain

$$\|J_3(t+\tau)-J_3(t)\|_E \leq \|A(t+\tau)v(t+\tau,0)A^{-1}(0)\|_{E\to E}\|f(t+\tau)-f(t)\|_E$$

$$+\|[A(t+\tau)v(t+\tau,0)-A(t)v(t,0)]A^{-1}(0)\|_{E\to E}\|f(t)-f(0)\|_E$$

$$\leq M\tau^{\alpha}\|f\parallel_{C^\alpha(E)} + M[\tau^{\varepsilon}+\frac{\tau^{\alpha}}{t^{\alpha}}]Mt^{\alpha}\|f\parallel_{C^\alpha(E)}\leq M_1\tau^{\alpha}\|f\parallel_{C^\alpha(E)}$$

for $t>2\tau$. Therefore,

$$\|J_3\|_{C^\alpha(E)} \leq M_1\tau^{\alpha}\|f\parallel_{C^\alpha(E)}. \tag{4.40}$$

Using formula (4.33) and estimates (4.3), (4.4), (4.6) and (4.8), we obtain

$$z^{1-\alpha}\parallel A(t)\exp\{-zA(t)\}J_3(t)\|_E$$

$$\leq z^{1-\alpha} \| A^2(t)\exp\{-(z+t)A(t)\}A^{-1}(0)\|_{E\to E}\|f(t)-f(0)\|_E$$
$$+z^{1-\alpha} \| A^2(t)\exp\{-(z+t)A(t)\}A^{-1}(0)\|_{E\to E}$$
$$\times\|(A(0)-A(t))A^{-1}(0)\|_{E\to E}\|f(t)-f(0)\|_E$$

$$+\int_0^t z^{1-\alpha} \| A^2(t)\exp\{-(z+t-s)A(t)\}A^{-1}(0)\|_{E\to E}\|[A(s)-A(t)]A^{-1}(s)\|_{E\to E}$$

$$\times\|A(s)v(s,0)A^{-1}(0)\|_{E\to E}\|_{E\to E}\|f(t)-f(0)\|_E ds$$
$$\leq \frac{z^{1-\alpha}t^\alpha}{t+z}\|f\|_{C(E_\alpha)}+z^{1-\alpha}\frac{Mt^{\alpha+\varepsilon}}{z+t}\|f(t)\|_E$$

$$+\int_0^t z^{1-\alpha}\frac{M(t-s)^\varepsilon t^\alpha ds}{z+t-s}\|f(t)\|_E\leq\frac{M_1}{\alpha}\|f\|_{C(E_\alpha)}$$

for all $z>0$ and t, $t\in[0,T]$. Therefore,

$$\|J_3\|_{C(E_\alpha)}\leq\frac{M_1}{\alpha}\|f\|_{C(E_\alpha)}. \tag{4.41}$$

Using this formula and estimates (4.4) and (4.21), we obtain

$$\|J_4(t)\|_E\leq\int_0^t\|A(t)v(t,s)\|_{E\to E}\|[A(t)-A(s)]A^{-1}(t)\|_{E\to E}ds\|f(t)\|_E$$

$$\leq M\int_0^t(t-s)^{1-\varepsilon}ds\|f(t)\|_E\leq\frac{M_1}{\alpha}t^\alpha\ \|f\|_{C^\alpha(E)}$$

for all t, $t\ \in[0,T]$. Now let us estimate $J_4(t+\tau)-J_4(t)$ for $0\leq t<t+\tau\leq T$.

To this end we shall examine separately the two cases $t\leq 2\tau$ and $t>2\tau$. If $t\leq 2\tau$, then using the last estimate and the triangle inequality, we obtain

$$\|J_4(t+\tau)-J_4(t)\|_E\leq\|J_4(t+\tau)\|_E+\|J_4(t)\|_E \tag{4.42}$$

$$\leq[\frac{M_1}{\alpha}(t+\tau)^\alpha+\frac{M_1}{\alpha}t^\alpha]\ \|f\|_{C^\alpha(E)}\leq M_2\tau^\alpha\ \|f\|_{C^\alpha(E)}.$$

Now let $t>2\tau$. We have that

$$J_4(t+\tau)-J_4(t)$$

$$=-\int_0^{t+\tau}A(t+\tau)v(t+\tau,s)[A(t+\tau)-A(s)]A^{-1}(t+\tau)ds[f(t+\tau)-f(t)]$$

$$-\int_{t-\tau}^{t+\tau}A(t+\tau)v(t+\tau,s)[A(s)-A(t+\tau)]A^{-1}(t)dsf(t)$$

$$+\int_{t-\tau}^{t} A(t)v(t,s)[A(s)-A(t)]A^{-1}(t)ds f(t)$$

$$-\int_{0}^{t-\tau} [A(t+\tau)v(t+\tau,s)-A(t)v(t,s)][A(s)-A(t)]A^{-1}(t)ds f(t)$$

$$-\int_{0}^{t-\tau} A(t+\tau)v(t+\tau,s)[A(t)-A(t+\tau)]A^{-1}(t)ds f(t)$$

$$+\int_{0}^{t-\tau} A(t+\tau)v(t+\tau,s)[A(t)-A(s)]A^{-1}(t)[A(t)-A(t+\tau)]A^{-1}(t+\tau)ds f(t+\tau)$$

$$= L_1(t)+L_2(t)+L_3(t)+L_4(t)+L_5(t)+L_6(t),$$

where

$$L_1(t) = -\int_0^{t+\tau} A(t+\tau)v(t+\tau,s)[A(t+\tau)-A(s)]A^{-1}(s)ds\}[f(t+\tau)-f(t)],$$

$$L_2(t) = -\int_{t-\tau}^{t+\tau} A(t+\tau)v(t+\tau,s)[A(s)-A(t+\tau)]A^{-1}(s)ds f(t),$$

$$L_3(t) = -\int_{t-\tau}^{t} A(t)v(t,s)[A(s)-A(t)]A^{-1}(s)ds f(t),$$

$$L_4(t) = \int_{0}^{t-\tau} [A(t+\tau)v(t+\tau,s)-A(t)v(t,s)][A(s)-A(t)]A^{-1}(s)ds f(t),$$

$$L_5(t) = \int_{0}^{t-\tau} A(t+\tau)v(t+\tau,s)[A(t)-A(t+\tau)]A^{-1}(s)ds f(t),$$

$$L_6(t) = \int_{0}^{t-\tau} A(t+\tau)v(t+\tau,s)[A(t)-A(s)] \times A^{-1}(t)[A(t)-A(t+\tau)]A^{-1}(t+\tau)ds f(t+\tau).$$

Let us estimate $L_k(t)$ for any $k = 1,2,3,4,5,6$ separately. We start with $L_1(t)$. Using estimates (4.4), (4.18), (4.21) and (4.24), we obtain

$$||L_1(t)||_E \leq \int_0^{t+\tau} ||A(t+\tau)v(t+\tau,s)||_{E\to E}||[A(t+\tau)-A(s)]A^{-1}(s)||_{E\to E}ds$$
$$\times ||f(t+\tau)-f(t)||_E$$
$$\leq M\int_0^{t+\tau}(t+\tau-s)^{1-\varepsilon}ds||f(t+\tau)-f(t)||_E \leq \frac{M_1}{\alpha}\tau^\alpha \parallel f \parallel_{C^\alpha(E)}$$

for all t, $t \in [0,T]$. Let us estimate $L_2(t)$. Using estimates (4.3), (4.4), (4.8), (4.23) and (4.24), we obtain

$$||L_2(t)||_E \leq \int\limits_{t-\tau}^{t+\tau} ||A(t+\tau)v(t+\tau,s)||_{E\to E}||[A(s)-A(t+\tau)]A^{-1}(t)||_{E\to E}||f(t)||_E ds$$

$$\leq \int\limits_{t-\tau}^{t+\tau} M_1(t+\tau-s)^{\varepsilon-1}ds||f(t)||_E = \frac{M_2}{\varepsilon}\tau^\varepsilon \parallel f \parallel_{C^\alpha(E)}$$

for all t, $t \in [0,T]$. Thus,

$$||L_2(t)||_E \leq \frac{M_2}{\alpha}\tau^\alpha \parallel f \parallel_{C^\alpha(E)}$$

for all t, $t \in [0,T]$. In a similar manner one establishes the estimate

$$||L_3(t)||_E \leq \frac{M_2}{\alpha}\tau^\alpha \parallel f \parallel_{C^\alpha(E)}$$

for all t, $t \in [0,T]$. Now let us estimate $L_4(t)$. Using estimates (4.4) and (4.22), we obtain

$$||L_4(t)||_E \leq \int\limits_0^{t-\tau} ||A(t+\tau)v(t+\tau,s)-A(t)v(t,s)||_{E\to E}$$

$$\times||[A(s)-A(t)]A^{-1}(t)||_{E\to E}ds||f(t)||_E$$

$$\leq M_1\int\limits_0^{t-\tau}(t-s)^{-1+\varepsilon}[\tau^\varepsilon+\frac{\tau^\alpha}{(t+\tau-s)^\alpha}]ds||f(t)||_E \leq \frac{M_2}{\alpha}\tau^\alpha||f(t)||_E$$

$$\leq \frac{M_2}{\alpha}\tau^\alpha \parallel f \parallel_{C^\alpha(E)}$$

for all t, $t \in [0,T]$. Now let us estimate $L_5(t)$. We have that

$$L_5(t) = [\exp\{-2\tau A(t+\tau)\}-\exp\{-(t+\tau)A(t+\tau)\}][A(t)-A(t+\tau)]A^{-1}(t)f(t)$$

$$+\int_{0}^{t-\tau}[A(t+\tau)v(t+\tau,s)-A(t+\tau)\exp\{-(t+\tau-s)A(t+\tau)\}]$$

$$\times[A(t)-A(t+\tau)]A^{-1}(t)dsf(t)$$

$$+\int_{0}^{t-\tau}A(t+\tau)v(t+\tau,s)[A(t)-A(t+\tau)]A^{-1}(t)[A(s)-A(t)]A^{-1}(t)dsf(t).$$

Using estimates (4.3), (4.4) and (4.21), we obtain

$$||L_5(t)||_E\leq||\exp\{-2\tau A(t+\tau)\}-\exp\{-(t+\tau)A(t+\tau)\}||_{E\to E}$$

$$\times||[A(t+\tau)-A(t)]A^{-1}(t)||_{E\to E}||f(t)||_E$$

$$+\int_{0}^{t-\tau}||A(t+\tau)v(t+\tau,s)-A(t+\tau)\exp\{-(t+\tau-s)A(t+\tau)\}||_{E\to E}$$

$$\times||[A(t+\tau)-A(t)]A^{-1}(t)||_{E\to E}ds||f(t)||_E$$

$$+\int_{0}^{t-\tau}||A(t+\tau)v(t+\tau,s)||_{E\to E}||[A(t+\tau)-A(t)]A^{-1}(t)||_{E\to E}$$

$$\times||[A(s)-A(t)]A^{-1}(t)||_{E\to E}ds||f(t)||_E$$

$$\leq[M\tau^{\varepsilon}+M_1\int_{0}^{t-\tau}(t-s)^{-1+\varepsilon}\tau^{\varepsilon}ds$$

$$+M_1\int_{0}^{t-\tau}(t+\tau-s)^{-1+\varepsilon}\tau^{\varepsilon}ds]||f(t)||_E\leq\frac{M_2}{\alpha}\tau^{\alpha}\parallel f\parallel_{C^{\alpha}(E)}$$

for all t, $t\in[0,T]$. Finally let us estimate $L_6(t)$. Using estimates (4.3), (4.4) and (4.21), we obtain

$$||L_6(t)||_E\leq\int_{0}^{t-\tau}||A(t+\tau)v(t+\tau,s)||_{E\to E}$$

$$\times||[A(t)-A(s)]A^{-1}(t)||_{E\to E}||[A(t+\tau)-A(t)]A^{-1}(t+\tau)||_{E\to E}ds||f(t+\tau)||_E$$

$$\leq M_1\int_{0}^{t-\tau}(t-s)^{-1+\varepsilon}\tau^{\varepsilon}ds\parallel f\parallel_{C^{\alpha}(E)}\leq\frac{M_2}{\alpha}\parallel f\parallel_{C^{\alpha}(E)}$$

for all t, $t \in [0,T]$. Combining the estimates for $L_k(t)$ for any $k = 1,2,3,4,5,6$, we obtain estimate (4.30) for $t > 2\tau$. Therefore,

$$||J_4||_{C^\alpha(E)} \leq \frac{M}{\alpha} \parallel f \parallel_{C^\alpha(E)} . \tag{4.43}$$

Using estimates (4.3), (4.4) and

$$\parallel A^2(t)\exp\{-zA(t)\}v(t,s)||_{E\to E} \leq M \min\{\frac{1}{z^2}, \frac{1}{(t-s)^2}\}, \tag{4.44}$$

we obtain

$$z^{1-\alpha} \parallel A(t)\exp\{-zA(t)\}J_4(t)||_E$$

$$\leq z^{1-\alpha}\int_0^t \parallel A^2(t)\exp\{-zA(t)\}v(t,s)||_{E\to E}||[A(t)-A(s)]A^{-1}(t)||_{E\to E}||f(t)||_E$$

$$+\int_0^t z^{1-\alpha}\frac{M(t-s)^\varepsilon ds}{(z+t-s)^2}||f(t)||_E \leq \frac{M_1}{1-\alpha}||f||_{C(E_\alpha)}$$

for all $z > 0$ and t, $t \in [0,T]$. Therefore,

$$||J_4||_{C(E_\alpha)} \leq \frac{M_1}{\alpha}||f||_{C(E_\alpha)}. \tag{4.45}$$

Now let us estimate $J_5(t)$. Using estimate (4.21), we obtain

$$\parallel J_5(t) \parallel_E \leq \int_0^t \|A(t)v(t,s)||_{E\to E}\|f(s)-f(t)\parallel_E ds$$

$$\leq M\int_0^t \frac{ds}{(t-s)^{1-\alpha}} \parallel f \parallel_{C^\alpha(E)} = \frac{M}{\alpha}t^\alpha \parallel f \parallel_{C^\alpha(E)}$$

for all t, $t \in [0,T]$. Now let us estimate the difference $J_5(t+\tau) - J_5(t)$ for $0 < t < t+\tau < T$. We will consider separately two cases, $t \leq 2\tau$ and $t > 2\tau$. If $t \leq 2\tau$, then the last estimate yields

$$\parallel J_5(t+\tau) - J_5(t) \parallel_E \leq \parallel J_5(t+\tau) \parallel_E + \parallel J_5(t) \parallel_E$$

$$\leq \frac{M}{\alpha}[(t+\tau)^\alpha + t^\alpha] \parallel f \parallel_{C^\alpha(E)} = \frac{M_1}{\alpha}\tau^\alpha \parallel f \parallel_{C^\alpha(E)} .$$

Now let $t > 2\tau$. Let us represent the difference $J_5(t+\tau) - J_5(t)$ as the sum of the following integrals:

$$J_5(t+\tau) - J_5(t) = \int_{t-\tau}^{t+\tau} A(t+\tau)v(t+\tau,s)(f(s)-f(t+\tau))ds$$

$$- \int_{t-\tau}^{t} A(t)v(t,s)(f(s)-f(t))ds$$

$$+ \int_{0}^{t-\tau} A(t+\tau)v(t+\tau,s)(f(t)-f(t+\tau))ds$$

$$+ \int_{0}^{t-\tau} [A(t+\tau)v(t+\tau,s) - A(t)v(t,s)](f(s)-f(t))ds$$

$$= P_1(t) + P_2(t) + P_3(t) + P_4(t),$$

where

$$P_1(t) = \int_{t-\tau}^{t+\tau} A(t+\tau)v(t+\tau,s)(f(s)-f(t+\tau))ds,$$

$$P_2(t) = - \int_{t-\tau}^{t} A(t)v(t,s)(f(s)-f(t))ds,$$

$$P_3(t) = \int_{0}^{t-\tau} [A(t+\tau)v(t+\tau,s) - A(t)v(t,s)](f(s)-f(t))ds,$$

$$P_4(t) = \int_{0}^{t-\tau} A(t+\tau)v(t+\tau,s)(f(t)-f(t+\tau))ds.$$

Let us estimate $P_k(t)$ for any $k = 1,2,3,4$ separately. We start with $P_1(t)$. Using estimate (4.21), we obtain

$$||P_1(t)||_E \leq \int_{t-\tau}^{t+\tau} ||A(t+\tau)v(t+\tau,s)||_{E\to E}||f(s)-f(t+\tau)||_E ds$$

$$\leq \int_{t-\tau}^{t+\tau} M_1(t+\tau-s)^{\alpha-1} ds \parallel f \parallel_{C^\alpha(E)} = \frac{M_2}{\alpha}\tau^\alpha \parallel f \parallel_{C^\alpha(E)}$$

for all t, $t \in [0,T]$. In a similar manner one establishes the estimate

$$||P_2(t)||_E \leq M_1 \tau^\alpha \parallel f \parallel_{C^\alpha(E)}$$

for all t, $t \in [0,T]$. Now let us estimate $P_3(t)$. Using estimate (4.22), we obtain

$$||P_3(t)||_E \leq \int_{0}^{t-\tau} ||A(t+\tau)v(t+\tau,s) - A(t)v(t,s)||_{E\to E}||f(s)-f(t)||_E ds$$

$$\leq M_1 \int_0^{t-\tau} (t-s)^{-1+\alpha}[\tau^{\varepsilon} + \frac{\tau}{t+\tau-s}]ds \parallel f \parallel_{C^{\alpha}(E)} \leq \frac{M_2\tau^{\alpha}}{\alpha(1-\alpha)} \parallel f \parallel_{C^{\alpha}(E)}$$

for all $t,\ t\ \in [0,T]$. Let us estimate $P_4(t)$. Using property 4) of $v(t,s)$, we can write

$$P_4(t) = [A(t+\tau)v(t+\tau,t-\tau)A^{-1}(t+\tau)$$

$$-A(t+\tau)v(t+\tau,0)A^{-1}(t+\tau)](f(t)-f(t+\tau))$$

$$+\int_0^{t-\tau} A(t+\tau)v(t+\tau,s)[A(t+\tau)-A(s)]A^{-1}(t+\tau)ds(f(t)-f(t+\tau)).$$

Using this formula and estimates (4.4), (4.18), (4.21) and (4.24), we obtain

$$||P_4(t)||_E \leq [||A(t+\tau)v(t+\tau,t-\tau)A^{-1}(0)||_{E\to E}$$

$$+||A(t+\tau)v(t+\tau,0)A^{-1}(0)||_{E\to E}]||A(0)A^{-1}(t+\tau)||_{E\to E}||f(t)-f(t+\tau)||_E$$

$$+\int_0^{t-\tau} ||A(t+\tau)v(t+\tau,s)||_{E\to E}||[A(t+\tau)-A(s)]$$

$$\times A^{-1}(t+\tau)||_{E\to E}ds||f(t)-f(t+\tau)||_E \leq 2MM||f(t)-f(t+\tau)||_E$$

$$+M\int_0^{t-\tau} (t+\tau-s)^{-1+\varepsilon}ds||f(t)-f(t+\tau)||_E \leq M_1\tau^{\alpha} \parallel f \parallel_{C^{\alpha}(E)}$$

for all $t,\ t\ \in [0,T]$. Combining the estimates for $P_k(t)$ for any $k=1,2,3,4$, we obtain the estimate

$$\parallel J_5(t+\tau)-J_5(t) \parallel_E \leq \frac{M_1}{\alpha(1-\alpha)} \parallel f \parallel_{C^{\alpha}(E)} \tau^{\alpha}$$

for $t>2\tau$. Therefore,

$$||J_5||_{C^{\alpha}(E)} \leq \frac{M}{\alpha(1-\alpha)} \parallel f \parallel_{C^{\alpha}(E)} . \tag{4.46}$$

Using estimates (4.3) and (4.44), we obtain

$$z^{1-\alpha} \parallel A(t)\exp\{-zA(t)\}J_5(t)||_E$$

$$\leq z^{1-\alpha} \int_0^t \parallel A^2(t)\exp\{-zA(t)\}v(t,s)||_{E\to E}||f(t)-f(s)||_E$$

$$+\int_0^t z^{1-\alpha}\frac{M(t-s)^{\alpha}ds}{(z+t-s)^2} \parallel f \parallel_{C^{\alpha}(E)} \leq \frac{M_1}{1-\alpha}||f||_{C(E_{\alpha})}$$

for all $z > 0$ and t, $t \in [0, T]$. Therefore,

$$||J_5||_{C(E_\alpha)} \leq \frac{M_1}{1-\alpha}||f||_{C(E_\alpha)}. \tag{4.47}$$

Finally, applying estimates (4.36), (4.37), (4.38), (4.39), (4.40), (4.41), (4.43), (4.45), (4.46) and (4.47), we obtain the estimates

$$\| v' \|_{C(E_\alpha)} \leq \frac{M}{\alpha(1-\alpha)}[\| f \|_{C^\alpha(E)} + \| v_0'\|_{E_\alpha}],$$

$$\| v' \|_{C^\alpha(E)} \leq \frac{M}{\alpha(1-\alpha)}[\| f \|_{C^\alpha(E)} + \| v_0'\|_{E_\alpha}].$$

By the triangle inequality, this last estimate and equation (4.1) yield

$$\| A(.)v\|_{C^\alpha(E)} \leq \frac{M}{\alpha(1-\alpha)}[\| f \|_{C^\alpha(E)} + \| v_0'\|_{E_\alpha}].$$

Theorem 4.1.8 is proved.

Second, one can enrich the family of spaces of smooth functions in which the boundary-value problem (4.1) is well posed by introducing the spaces $C_0^{\beta,\gamma}(E)$, $(0 \leq \gamma \leq \beta, 0 < \beta < 1)$, obtained by completion of the set of all smooth E-valued functions $\varphi(t)$ on [0,T] with respect to the norm

$$\| \varphi \|_{C_0^{\beta,\gamma}(E)} = \max_{0\leq t\leq T} ||\varphi(t)||_E + \sup_{0\leq t<t+\tau\leq T} \frac{(t+\tau)^\gamma \| \varphi(t+\tau) - \varphi(t) \|_E}{\tau^\beta}.$$

Let us give, without proof, the following result.

Theorem 4.1.9. *Suppose $v_0' \in E_{\beta-\gamma}$, $f(t) \in C_0^{\beta,\gamma}(E)(0 \leq \gamma \leq \beta, 0 < \beta < 1)$. Suppose that assumptions (4.3) and (4.4) hold and $0 < \beta \leq \varepsilon \leq 1$. Then for the solution $v(t)$ in $C_0^{\beta,\gamma}(E)$ of the Cauchy problem (4.1) the coercive inequalities*

$$\| v' \|_{C(E_{\beta-\gamma})} \leq M[\| v_0' \|_{E_{\beta-\gamma}} + \beta^{-1}(1-\beta)^{-1} \| f \|_{C_0^{\beta,\gamma}(E)}],$$

$$\| v' \|_{C_0^{\beta,\gamma}(E)} + \| A(.)v \|_{C_0^{\beta,\gamma}(E)}$$
$$\leq M[|v_0'|_0^{\beta,\gamma} + \beta^{-1}(1-\beta)^{-1} \| f \|_{C_0^{\beta,\gamma}(E)}]$$

hold, where M does not depend on β, γ, v_0' and $f(t)$. Here, $|w|_0^{\beta,\gamma}$ denotes that the norm of the Banach space $E_0^{\beta,\gamma}$ consists of those $w \in E$ for which the norm

$$|w|_0^{\beta,\gamma} = \max_{0\leq z\leq T} ||e^{-zA(t)}w||_E + \sup_{0\leq z<z+\tau\leq T} \tau^{-\beta}(z+\tau)^\gamma||(e^{-(z+\tau)A(t)} - e^{-zA(t)})w||_E$$

is finite.

Note that the parameter γ can be chosen freely in $[0,\beta)$, which increases the number of function spaces in which problem (4.1) is well posed. In particular, it is important that problem (4.1) is well posed in the Hölder space without a weight ($\gamma=0$).

Problem (4.1) is not well posed in $C(E)$ for arbitrary E. It turns out that a Banach space E can be restricted to a Banach space E' in such a manner that the restricted problem (4.1) in E' will be well posed in $C(E')$. The role of E' will be played here by the fractional spaces $E_\alpha = E_\alpha(E, A(t))$ $(0<\alpha<1)$.

Theorem 4.1.10. *Suppose $A(0)v_0 \in E_\alpha$, $f(t) \in C(E_\alpha)$ $(0<\alpha<1)$. Suppose that assumptions* (4.3) *and* (4.4) *hold and $0<\alpha\le\varepsilon\le 1$. Then for the solution $v(t)$ in $C(E_\alpha)$ of problem* (4.1) *the coercive inequality*

$$\| v' \|_{C(E_\alpha)} + \| A(.)v \|_{C(E_\alpha)}$$

$$\le M\alpha^{-1}(1-\alpha)^{-1}[\| A(0)v_0 \|_{E_\alpha} + \| f \|_{C(E_\alpha)}] \tag{4.48}$$

holds, where M does not depend on α, v_0 and $f(t)$.

Proof. Using formula (4.5), we obtain

$$A(t)v(t) = A(t)v(t,0)v_0 + \int_0^t A(t)v(t,s)f(s)ds = \sum_{k=1}^{2} J_k(t),$$

where

$$J_1(t) = A(t)v(t,0)v_0,$$

$$J_2(t) = \int_0^t A(t)v(t,s)f(s)ds.$$

Let us estimate $J_k(t)$ for any $k=1,2$ separately. We start with $J_1(t)$. Using formula (4.33) and estimates (4.3), (4.6) and (4.8), we obtain

$$z^{1-\alpha} \| A(t)\exp\{-zA(t)\}J_1(t)\|_E \le z^{1-\alpha} \| A(t)\exp\{-(z+t)A(t)\}A(0)v_0\|_E$$

$$+z^{1-\alpha} \| A(t)\exp\{-(z+t)A(t)\}\|_{E\to E}\|[A(t)-A(0)]A^{-1}(0)\|_{E\to E}\|A(0)v_0\|_E$$

$$+\int_0^t z^{1-\alpha} \| A^2(t)\exp\{-(z+t-s)A(t)\}\|_{E\to E}\|[A(s)-A(t)]A^{-1}(s)\|_{E\to E}$$

$$\times\|A(s)v(s,0)A^{-1}(0)\|_{E\to E}\|v_0'\|_E ds \le \frac{z^{1-\alpha}}{(t+z)^{1-\alpha}} \| A(0)v_0\|_{E_\alpha}$$

$$+z^{1-\alpha}\frac{Mt^\varepsilon}{z+t}\|A(0)v_0\|_E + \int_0^t z^{1-\alpha}\frac{M(t-s)^\varepsilon ds}{(z+t-s)^2}\|A(0)v_0\|_E \le \frac{M_1}{1-\alpha}\|A(0)v_0\|_{E_\alpha}$$

for all $z>0$ and t, $t\in[0,T]$. Therefore,

$$||J_1||_{C(E_\alpha)} \leq \frac{M_1}{\alpha(1-\alpha)}||A(0)v_0||_{E_\alpha}.$$

Now let us estimate $J_2(t)$. Using formula (4.33) and estimates (4.3), (4.6) and (4.8), we obtain

$$z^{1-\alpha} \parallel A(t)\exp\{-zA(t)\}J_2(t)||_E$$

$$\leq z^{1-\alpha}\int_0^t \parallel A^2(t)\exp\{-(z+t-s)A(t)\}f(s)||_E$$

$$+z^{1-\alpha}\int_0^t\int_s^t \parallel A^2(t)\exp\{-(z+t-p)A(t)\}||_{E\to E}$$

$$\times||[A(t)-A(p)]A^{-1}(p)||_{E\to E}||A(p)\exp\{-(p-s)A(p)\}f(s)||_E dpds$$

$$+z^{1-\alpha}\int_0^t\int_s^t \parallel A^2(t)\exp\{-(z+t-p)A(t)\}||_{E\to E}$$

$$\times||[A(t)-A(p)]A^{-1}(p)||_{E\to E}||A(p)[v(p,s)-\exp\{-(p-s)A(p)\}||_{E\to E}||f(s)||_E dpds$$

$$\leq z^{1-\alpha}\int_0^t M\frac{||f(s)||_{E_\alpha}ds}{(z+t-s)^{2-\alpha}}+z^{1-\alpha}\int_0^t\int_s^t M\frac{||f(s)||_{E_\alpha}M(t-p)^\varepsilon dpds}{(z+t-p)^2(p-s)^{1-\alpha}}$$

$$+z^{1-\alpha}\int_0^t\int_s^t M\frac{M(t-p)^\varepsilon||f(s)||_E dpds}{(z+t-p)^2(p-s)^{1-\varepsilon}}\leq\frac{M_1}{\alpha(1-\alpha)}||f||_{C(E_\alpha)}$$

for all $z>0$ and t, $t\in[0,T]$. Therefore,

$$||J_2||_{C(E_\alpha)} \leq \frac{M_1}{\alpha(1-\alpha)}||f||_{C(E_\alpha)}.$$

Combining the estimates for I_1, I_2, we obtain estimate (4.48). Theorem 4.1.10 is proved.

Let us give, without proof, the following result.

Theorem 4.1.11. *Suppose that assumption* (4.3) *holds. Suppose that the operator* $A(t)A^{-1}(s)$ *is Hölder continuous in* t *in the uniform operator topology for each fixed* s, *that is,*

$$||[A(t)-A(\tau)]A^{-1}(s)||_{E_\alpha\to E\alpha}\leq M|t-\tau|^\varepsilon, 0<\varepsilon\leq 1, \tag{4.49}$$

where M *and* ε *are positive constants independent of* t, s *and* τ *for* $0\leq t,s,\tau\leq T$. *Suppose* $A(0)v_0\in E_\alpha$, $f(t)\in C(E_\alpha)(0<\alpha<1)$. *Then for the solution* $v(t)$ *in* $C(E_\alpha)$ *of problem* (4.1) *the coercive inequality*

$$\parallel v' \parallel_{C(E_\alpha)} + \parallel A(.)v \parallel_{C(E_\alpha)}$$

$$\leq M[\parallel A(0)v_0 \parallel_{E_\alpha} + \alpha^{-1}(1-\alpha)^{-1} \parallel f \parallel_{C(E_\alpha)}]$$

holds, where M *does not depend on* α, v_0 *and* $f(t)$.

For our third step, let us study now problem (4.1) in the spaces $C_0^{\beta,\gamma}(E_{\alpha-\beta})$, $(0 \le \gamma \le \beta \le \alpha, 0 < \alpha < 1)$. To these there correspond the spaces of traces $E_{\alpha-\beta}^{\beta,\gamma}$, which consist of elements $w \in E$ for which the norm

$$\begin{aligned} |w|_{\alpha-\beta}^{\beta,\gamma} &= \max_{0\le z\le T} ||e^{-zA(t)}w||_{E_{\alpha-\beta}} \\ &+ \sup_{0\le z<z+\tau\le T} \tau^{-\beta}(z+\tau)^{\gamma}||(e^{-(z+\tau)A(t)} - e^{-zA(t)})w||_{E_{\alpha-\beta}} \end{aligned}$$

is finite. Let us give, without proof, the following results.

Theorem 4.1.12. *Suppose that assumptions* (4.3) *and* (4.4) *hold and* $0 < \alpha \le \varepsilon \le 1$. *Let* $v_0' \in E_{\alpha-\beta}^{\beta,\gamma}$ *and* $f(t) \in C_0^{\beta,\gamma}(E_{\alpha-\beta})$, $(0 \le \gamma \le \beta \le \alpha,\ 0 < \alpha < 1)$. *Then for the solution* $v(t)$ *in* $C_0^{\beta,\gamma}(E_{\alpha-\beta})$ *of problem* (4.1) *the coercive inequality*

$$\begin{aligned} &\| v' \|_{C_0^{\beta,\gamma}(E_{\alpha-\beta})} + \| A(.)v \|_{C_0^{\beta,\gamma}(E_{\alpha-\beta})} + \| v' \|_{C(E_{\alpha-\beta}^{\beta,\gamma})} \\ &\le M[|v_0'|_{\alpha-\beta}^{\beta,\gamma} + \alpha^{-1}(1-\alpha)^{-1} \| f \|_{C_0^{\beta,\gamma}(E_{\alpha-\beta})}] \end{aligned}$$

holds, where M *does not depend on* α, β, γ, v_0' *and* $f(t)$.

Theorem 4.1.13. *Suppose that assumptions* (4.3) *and* (4.4) *hold and* $0 < \alpha \le \varepsilon \le 1$ *Let* $v_0' \in E_{\alpha-\gamma}$ *and* $f(t) \in C_0^{\beta,\gamma}(E_{\alpha-\gamma})$, $(0 \le \gamma \le \beta \le \alpha,\ 0 < \alpha < 1)$. *Then for the solution* $v(t)$ *in* $C_0^{\beta,\gamma}(E_{\alpha-\beta})$ *of problem* (4.1) *the coercive inequality*

$$\begin{aligned} &\| v' \|_{C_0^{\beta,\gamma}(E_{\alpha-\beta})} + \| A(.)v \|_{C_0^{\beta,\gamma}(E_{\alpha-\beta})} + \| v' \|_{C(E_{\alpha-\gamma})} \\ &\le M\alpha^{-1}(1-\alpha)^{-1}[\| v_0' \|_{E_{\alpha-\gamma}} + \| f \|_{C_0^{\beta,\gamma}(E_{\alpha-\beta})}] \end{aligned}$$

holds, where M *does not depend on* α, β, γ, v_0' *and* $f(t)$.

Note that the spaces of smooth functions $C_0^{\beta,\gamma}(E_{\alpha-\beta})$, in which coercive solvability has been established, depend on the parameters α, β and γ. However, the constants in the coercive inequalities depend only on α. Hence, we can choose the parameters β and γ freely, which increases the number of function spaces in which problem (4.1) is well posed. In particular, Theorem 4.1.13 implies the well-posedness Theorem 4.1.10 in $C(E_\alpha)$ established in the above.

Fourth, let us study now the Cauchy problem (4.1) in the spaces $L_p(E) = L_p([0,T], E)$, $(1 \le p < \infty)$ of all strongly measurable E-valued functions $v(t)$ on $[0,T]$ for which the norm

$$\| v \|_{L_p(E)} = \left(\int_0^T \| v(t) \|_E^p \, dt\right)^{\frac{1}{p}}$$

is finite.

A function $v(t)$ is said to be absolutely continuous if it has a derivative $v'(t)$ for almost every t such that $v'(t) \in L_1(E)$ and if the Newton–Leibniz formula

$$v(t) - v(\tau) = \int_{\tau}^{t} v'(s)ds$$

holds for all t, $\tau \in [0,T]$. Here the integral is understood in the sense of Bochner.

A function $v(t)$ is said to be a solution of problem (4.1) in $L_p(E)$ if it is absolutely continuous, the functions $v'(t)$ and $A(t)v(t)$ belong to $L_p(E)$, and equation (4.1) is satisfied for almost every t and $v(0) = v_0$. From this definition it follows that a necessary condition for the solvability of problem (4.1) in $L_p(E)$ is that $f(t) \in L_p(E)$. It will be shown that in certain cases this condition is also sufficient for the solvability of problem (4.1). As concerns the boundary elements, in contrast to the situation considered earlier, from the solvability of problem (4.1) in $L_p(E)$ it follows only that $v_0 \in E$. In the case of an unbounded operator $A(t)$ this does not allow us to prove the solvability of problem (4.1).

From the unique solvability of (4.1) it follows that the operator $v(t; f(t), v_0)$ is bounded in $L_p(E)$ and one has coercive inequality

$$\|v'\|_{L_p(E)} + \|A(.)v(t)\|_{L_p(E)} \leq M_C[\|f\|_{L_p(E)} + \ \|A(0)v_0\|_E],$$

where M_C $(1 \leq M_C < +\infty)$ does not depend on v_0 and $f(t)$.

Let us give, without proof, the following results.

Theorem 4.1.14. *Suppose that assumptions* (4.3) *and* (4.4) *hold and* $0 < \varepsilon \leq 1$ *Suppose problem* (4.1) *is well posed in* $L_{p_0}(E)$ *for some* p_0, $1 < p_0 < \infty$. *Then it is well posed in* $L_p(E)$ *for any* p,$1 < p < \infty$ *and the coercivity inequality holds:*

$$\|v'\|_{L_p(E)} + \|A(.)v\|_{L_p(E)} + \|v\|_{C(E_{1-1/p,p})}$$
$$\leq \frac{M(p_0)p^2}{p-1}\|f\|_{L_p(E)} + M\|v_0\|_{E_{1-1/p,p}},$$

where $M(p_0)$ *and* M *does not depend on* p, v_0 *and* $f(t)$. *Here the Banach space* $E_{1-\frac{1}{p},p} = E_{1-\frac{1}{p},p}(E, A(t))$ *consist of all these* $v \in E$ *for which norm*

$$\|v\|_{E_{1-1/p,p}} = \left(\int_0^T \|A(t)\exp\{-zA(t)\}\|_E^p dz\right)^{1/p} \qquad (1 \leq p < \infty)$$

is finite.

Theorem 4.1.15. *Let* $1 \leq p \leq \infty$ *and* $0 < \alpha < 1$. *Suppose that assumptions* (4.3) *and* (4.4) *hold and* $\alpha \leq \varepsilon \leq 1$. *Then problem* (4.1) *is well posed in* $L_p(E_{\alpha,p})$ *and the coercivity inequality holds:*

$$\|v'\|_{L_p(E_{\alpha,p})} + \|A(.)v\|_{L_p(E_{\alpha,p})} \leq \frac{M}{\alpha(1-\alpha)}\|f\|_{L_p(E_{\alpha,p})} + M\|v_0\|_{E_{1-1/p,p}(E\alpha,p,A(t))},$$

where M *does not depend on* α, p, v_0 *and* $f(t)$.

Theorem 4.1.16. *Let $1 \leq p \leq \infty$ and $0 < \alpha < 1$. Suppose that assumption (4.3) holds. Suppose that the operator $A(t)A^{-1}(s)$ is Hölder continuous in t in the uniform operator topology for each fixed s, that is,*

$$||[A(t) - A(\tau)]A^{-1}(s)||_{E_{\alpha,p} \to E\alpha,p} \leq M|t-\tau|^{\varepsilon}, 0 < \varepsilon \leq 1,$$

where M and ε are positive constants independent of t, s and τ for $0 \leq t, s, \tau \leq T$. Then problem (4.1) is well posed in $L_p(E_{\alpha,p})$ and the coercivity inequality holds:

$$||v'||_{L_p(E_{\alpha,p})} + \|A(.)v\|_{L_p(E_{\alpha,p})} \leq \frac{M}{\alpha(1-\alpha)}\|f\|_{L_p(E_{\alpha,p})} + M||A(0)v_0||_{E_{\alpha,p}},$$

where M does not depend on α, p, v_0 and $f(t)$.

Theorem 4.1.17. *Let $1 < p,\ q < \infty$. Suppose that assumptions (4.3) and (4.4) hold and $0 < \alpha \leq \varepsilon < 1$. Then problem (4.1) is well posed in $L_p(E_{\alpha,q})$ and the coercivity inequality holds:*

$$||v'||_{L_p(E_{\alpha,q})} + \|A(.)v\|_{L_p(E_{\alpha,q})} \leq \frac{M(q)}{\alpha(1-\alpha)}\|f\|_{L_p(E_{\alpha,q})} + M||A(0)v_0||_{E_{\alpha,q}},$$

where $M(q)$ and M does not depend on α, p, v_0 and $f(t)$.

Theorem 4.1.18. *Let $1 < p,\ q < \infty$. Suppose that assumption (4.3) holds. Suppose that the operator $A(t)A^{-1}(s)$ is Hölder continuous in t in the uniform operator topology for each fixed s, that is,*

$$||[A(t) - A(\tau)]A^{-1}(s)||_{E_{\alpha,q} \to E\alpha,q} \leq M|t-\tau|^{\varepsilon}, 0 < \varepsilon \leq 1,$$

where M and ε are positive constants independent of t, s and τ for $0 \leq t, s, \tau \leq T$. Then problem (4.1) is well posed in $L_p(E_{\alpha,q})$ and the coercivity inequality holds:

$$||v'||_{L_p(E_{\alpha,q})} + \|A(.)v\|_{L_p(E_{\alpha,q})} \leq \frac{M(q)}{\alpha(1-\alpha)}\|f\|_{L_p(E_{\alpha,q})} + M||A(0)v_0||_{E_{\alpha,q}},$$

where $M(q)$ and M does not depend on α, p, v_0 and $f(t)$.

Finally, we consider the applications of these results to the parabolic equations. First, we consider the mixed boundary-value problem for parabolic equation

$$\begin{cases} \dfrac{\partial u}{\partial t} - a(t,x)\dfrac{\partial^2 u}{\partial x^2} + \delta u = f(t,x), 0 < t \leq T, 0 < x < 1, \\ u(0,x) = \varphi(x),\ f(0,x) = 0,\ 0 \leq x \leq 1, \\ u(t,0) = u(t,1),\ u_x(t,0) = u_x(t,1),\ 0 \leq t \leq T, \end{cases} \tag{4.50}$$

where $a(t,x)$, $\varphi(x)$ and $f(t,x)$ are given sufficiently smooth functions and $a(t,x) > 0$, $\delta > 0$ is a sufficiently large number.

We introduce the Banach spaces $C^{\beta}[0,1]$ $(0 < \beta < 1)$ of all continuous functions $\varphi(x)$ satisfying a Hölder condition for which the following norms are finite:

$$\| \varphi \|_{C^{\beta}[0,1]} = \| \varphi \|_{C[0,1]} + \sup_{0 \le x < x+\tau \le 1} \frac{|\varphi(x+\tau) - \varphi(x)|}{\tau^{\beta}},$$

where $C[0,1]$ is the space of all continuous functions $\varphi(x)$ defined on [0,1] with the usual norm

$$\| \varphi \|_{C[0,1]} = \max_{0 \le x \le 1} |\varphi(x)|.$$

It is known that the differential expression

$$A^{t,x} v = -a(t,x) v''(t,x) + \delta v(t,x)$$

defines a positive operator $A^{t,x}$ acting in $C^{\beta}[0,1]$ with domain $C^{\beta+2}[0,1]$ and satisfying the conditions $v(t,0) = v(t,1)$, $v_x(t,0) = v_x(t,1)$.

Therefore, we can replace the mixed problem (4.50) by the abstract Cauchy problem (4.1). Using the results of Theorems 4.1.6, 4.1.8, 4.1.9, 4.1.10, 4.1.11, 4.1.12 and 4.1.13, we can obtain that

Theorem 4.1.19. *For the solution of the mixed problem* (4.50) *the following coercive inequalities are valid:*

$$\| u \|_{C_0^{1+\beta,\gamma}(C^{\mu}[0,1])} + \| u \|_{C_0^{\beta,\gamma}(C^{2+\mu}[0,1])}$$

$$\le \frac{M(\mu)}{(\beta-\gamma)(1-\beta)} \| f \|_{C_0^{\beta,\gamma}(C^{\mu}[0,1])} + M(\mu) \|\varphi\|_{C^{2+\mu+2(\beta-\gamma)}[0,1]},$$

$$0 \le \gamma < \beta < 1,\ 0 \le \mu + 2(\beta - \gamma) \le 1,$$

$$\| u \|_{C_0^{1+\beta,\gamma}(C^{2(\alpha-\beta)}[0,1])} + \| u \|_{C_0^{\beta,\gamma}(C^{2+2(\alpha-\beta)}[0,1])}$$

$$\le \frac{M(\alpha,\beta)}{\alpha(1-\alpha)} \| f \|_{C_0^{\beta,\gamma}(C^{2(\alpha-\beta)}[0,1])} + M(\alpha,\beta) \|\varphi\|_{C^{2+2(\alpha-\gamma)}[0,1]},$$

$$0 \le \gamma \le \beta \le \alpha, 0 < \alpha - \beta < \frac{1}{2}.$$

Here $M(\mu)$ and $M(\alpha,\beta)$ are independent of γ, $f(t,x)$, $\varphi(x)$.

Second, let Ω be the unit open cube in the n-dimensional Euclidean space $\mathbb{R}^n (0 < x_k < 1, 1 \le k \le n)$ with boundary S, $\overline{\Omega} = \Omega \cup S$. In $[0,T] \times \overline{\Omega}$ we consider the mixed boundary-value problem for the multidimensional parabolic equation

$$\begin{cases} \dfrac{\partial u(t,x)}{\partial t} - \displaystyle\sum_{r=1}^{n} \alpha_r(t,x) \frac{\partial^2 u(t,x)}{\partial x_r^2} + \delta u(t,x) = f(t,x), \\ x = (x_1, \ldots, x_n) \in \overline{\Omega}, 0 < t \le T, \\ u(0,x) = \varphi(x),\ f(0,x) = 0,\ x \in \overline{\Omega}, \\ u(t,x) = 0,\ x \in S, \end{cases} \tag{4.51}$$

where $\alpha_r(t,x)$, $f(t,x)$ $(t \in [0,T],\ x \in \overline{\Omega})$, $\varphi(x)(x \in \overline{\Omega})$ are given smooth functions and $\alpha_r(t,x) > 0$, $\delta > 0$ is a sufficiently large number.

We introduce the Banach spaces $C_{01}^{\beta}(\overline{\Omega})$ $(\beta = (\beta_1, \dots, \beta_n),\ 0 < x_k < 1,\ k = 1, \dots, n)$ of all continuous functions satisfying a Hölder condition with the indicator $\beta = (\beta_1, \dots, \beta_n), \beta_k \in (0,1), 1 \le k \le n$ and with weight $x_k^{\beta_k}(1 - x_k - h_k)^{\beta_k}$, $0 \le x_k < x_k + h_k \le 1, 1 \le k \le n$ which is equipped with the norm

$$\| f \|_{C_{01}^{\beta}(\overline{\Omega})} = \| f \|_{C(\overline{\Omega})}$$

$$+ \sup_{0 \le x_k < x_k + h_k \le 1, 1 \le k \le n} |f(x_1, \dots, x_n) - f(x_1 + h_1, \dots, x_n + h_n)|$$

$$\times \prod_{k=1}^{n} h_k^{-\beta_k} x_k^{\beta_k} (1 - x_k - h_k)^{\beta_k},$$

where $C(\overline{\Omega})$-is the space of all continuous functions defined on $\overline{\Omega}$, equipped with the norm

$$\| f \|_{C(\overline{\Omega})} = \max_{x \in \overline{\Omega}} |f(x)|.$$

It is known that the differential expression

$$A^{t,x} v = - \sum_{r=1}^{n} \alpha_r(t,x) \frac{\partial^2 v(t,x)}{\partial x^2} + \delta v(t,x)$$

defines a positive operator $A^{t,x}$ acting on $C_{01}^{\beta}(\overline{\Omega})$ with domain $D(A^{t,x}) \subset C_{01}^{2+\beta}(\overline{\Omega})$ and satisfying the condition $v = 0$ on S.

Therefore, we can replace the mixed problem (4.51) by the abstract Cauchy problem (4.1). Using the results of Theorems 4.1.6, 4.1.8, 4.1.9, 4.1.10, 4.1.11, 4.1.12 and 4.1.13, we can obtain that

Theorem 4.1.20. *For the solution of the mixed boundary-value problem* (4.51) *the following coercive inequality is valid:*

$$\| u \|_{C_{0T}^{1+\beta,\gamma}(C_{01}^{\mu}(\overline{\Omega}))} + \sum_{r=1}^{n} \| \frac{\partial^2 u}{\partial x_r^2} \|_{C_{0T}^{\beta,\gamma}(C_{01}^{\mu}(\overline{\Omega}))}$$

$$\le \frac{M(\mu)}{(\beta - \gamma)(1 - \beta)} \| f \|_{C_{0T}^{\beta,\gamma}(C_{01}^{\mu}(\overline{\Omega}))} + M(\mu) \sum_{r=1}^{n} \| \frac{\partial^2 \varphi}{\partial x_r^2} \|_{C_{01}^{\mu+2(\beta-\gamma)}(\overline{\Omega})},$$

$$0 \le \gamma < \beta < 1, \mu = \{\mu_1, \dots, \mu_n\},\ 0 < \mu_k + 2(\beta - \gamma) < 1, 1 \le k \le n,$$

where $M(\mu)$ is independent of β, γ and $f(t,x)$, $\varphi(x)$.

Third, we consider the Cauchy problem on the range $\{0 \le t \le T,\ x \in \mathbb{R}^n\}$ for the $2m$th-order multidimensional parabolic equation

$$\begin{cases} \dfrac{\partial u}{\partial t} + \displaystyle\sum_{|r|=2m} a_r(t,x) \frac{\partial^{|r|} u}{\partial x_1^{r_1} \cdots \partial x_n^{r_n}} + \delta u(t,x) = f(t,x), \\ 0 < t \le T,\ x, r \in \mathbb{R}^n, |r| = r_1 + \cdots + r_n, \\ u(0,x) = \varphi(x),\ f(0,x) = 0,\ x \in \mathbb{R}^n, \end{cases} \tag{4.52}$$

where $a_r(t,x)$ and $f(t,x)$, $\varphi(x)$ are given sufficiently smooth functions and $\delta > 0$ is the sufficiently large number.

Let us consider a differential operator with constant coefficients of the form

$$B = \sum_{|r|=2m} b_r \frac{\partial^{r_1 + \cdots + r_n}}{\partial_{x_1}^{r_1} \cdots \partial_{x_n}^{r_n}},$$

acting on functions defined on the entire space R^n. Here $r \in R^n$ is a vector with nonnegative integer components, $|r| = r_1 + \cdots + r_n$. If $\varphi(y)$ $(y = (y_1, \ldots, y_n) \in R^n)$ is an infinitely differentiable function that decays at infinity together with all its derivatives, then by means of the Fourier transformation one establishes the equality

$$F(B_\varphi)(\xi) = B(\xi) F(\varphi)(\xi).$$

Here the Fourier transform operator is defined by the rule

$$F(\varphi)(\xi) = (2\pi)^{-n/2} \int_{R^n} \exp\{-i(y,\xi)\} \varphi(y)\, dy,$$

$$(y,\xi) = y_1\xi_1 + \cdots + y_n\xi_n.$$

The function $B(\xi)$ is called the symbol of the operator B and is given by

$$B(\xi) = \sum_{|r|=2m} b_r (i\xi_1)^{r_1} \cdots (i\xi_n)^{r_n}.$$

We will assume that the symbol

$$B^{t,x}(\xi) = \sum_{|r|=2m} a_r(t,x) (i\xi_1)^{r_1} \cdots (i\xi_n)^{r_n}, \xi = (\xi_1, \ldots, \xi_n) \in R^n$$

of the differential operator of the form

$$B^{t,x} = \sum_{|r|=2m} a_r(t,x) \frac{\partial^{|r|}}{\partial x_1^{r_1} \ldots \partial x_n^{r_n}} \tag{4.53}$$

acting on functions defined on the space $\mathbb{R}^n$, satisfies the inequalities

$$0 < M_1|\xi|^{2m} \leq (-1)^m B^{t,x}(\xi) \leq M_2|\xi|^{2m} < \infty$$

for $\xi \neq 0$. Problem (4.52) has a unique smooth solution. This allows us to reduce problem (4.52) to the abstract Cauchy problem (4.1) in a Banach space $E = C^\mu(R^n)$ of all continuous bounded functions defined on $\mathbb{R}^n$satisfying a Hölder condition with the indicator $\mu \in (0,1)$ with a strongly positive operator $A^{t,x} = B^{t,x} + \delta I$ defined by (4.52).

Theorem 4.1.21. *For the solution of the Cauchy problem* (4.52) *the following coercivity inequalities are satisfied:*

$$\| u \|_{C_{0T}^{1+\beta,\gamma}(C^\mu(R^n))} + \sum_{|\tau|=2m} \| \frac{\partial^{|r|}u}{\partial x_1^{r_1}\dots\partial x_n^{r_n}} \|_{C_{0T}^{\beta,\gamma}(C^\mu(R^n))}$$

$$\leq \frac{M(\mu)}{(\beta-\gamma)(1-\beta)} \| f \|_{C_{0T}^{\beta,\gamma}(C^\mu(R^n))} + M(\mu) \sum_{|\tau|=2m} \| \frac{\partial^{|r|}\varphi}{\partial x_1^{r_1}\dots\partial x_n^{r_n}} \|_{C^{\mu+2m(\beta-\gamma)}(R^n)},$$

$$0 \leq \gamma < \beta < 1,\ 0 < \mu + 2m(\beta-\gamma) < 1,$$

$$\| u \|_{C_{0T}^{2+\beta,\gamma}(C^{2m(\alpha-\beta)}(R^n))} + \sum_{|\tau|=2m} \| \frac{\partial^{|r|}u}{\partial x_1^{r_1}\dots\partial x_n^{r_n}} \|_{C_{0T}^{\beta,\gamma}(C^{2m(\alpha-\beta)}(R^n))}$$

$$\leq M(\alpha,\beta) \| f \|_{C_{0T}^{\beta,\gamma}(C^{2m(\alpha-\beta)}(R^n))} + M(\alpha,\beta) \sum_{|\tau|=2m} \| \frac{\partial^{|r|}\varphi}{\partial x_1^{r_1}\dots\partial x_n^{r_n}} \|_{C^{\mu+2m(\alpha-\gamma)}(R^n)},$$

$$0 \leq \gamma \leq \beta,\ 0 < \mu + 2m(\alpha-\beta) < 1,$$

where $M(\mu)$ and $M(\alpha,\beta)$ does not depend on γ and $f(t,x), \varphi(x)$.

The proof of Theorem 4.1.21 is based on the abstract Theorems 4.1.6, 4.1.8, 4.1.9, 4.1.10, 4.1.11, 4.1.12 and 4.1.13, the coercivity inequality for an elliptic operator $A^{t,x}$ in $C^\mu(R^n)$ and on the following theorem on the structure of the fractional spaces $E_\alpha(C(R^n), A^{t,x})$.

Theorem 4.1.22. $E_\alpha(C(R^n), A^{t,x}) = C^{2m\alpha}(R^n)$ *for all* $0 < \alpha < \frac{1}{2m}$ *and* $0 \leq t \leq T$.

4.2 Difference Schemes Generated by an Exact Difference Scheme

In this section we consider the difference schemes of a high order accuracy for the approximate solutions of problem (4.1). On the segment $[0,T]$ we consider again a uniform grid space

$$[0,T]_\tau = \{t_k = k\tau, k = 0,1,\dots,N, N\tau = T\}$$

with step $\tau > 0$. Here N is a fixed positive integer. From formula (4.5) it follows that

$$\frac{1}{\tau}(v(t_k) - v(t_{k-1})) + \frac{1}{\tau}(I - v(t_k, t_{k-1}))v(t_{k-1})$$

$$= \frac{1}{\tau}\int_{t_{k-1}}^{t_k} v(t_k, s)f(s)ds, 1 \le k \le N, v(0) = v_0$$

or

$$\frac{1}{\tau}(u_k - u_{k-1}) + \frac{1}{\tau}(I - v(t_k, t_{k-1}))u_{k-1} \tag{4.54}$$

$$= f_k,\ f_k = \frac{1}{\tau}\int_{t_{k-1}}^{t_k} v(t_k, s)f(s)ds, 1 \le k \le N, u_0 = v_0.$$

The last system is called the single-step exact difference scheme for the solution of the initial-value problem (4.1).

Let us investigate the well-posedness of the exact single-step difference scheme (4.54). This problem is uniquely solvable and the following formula holds:

$$u_k = v(t_k, 0)u_0 + \sum_{j=1}^{k} v(t_k, t_j)\ f_j\tau. \tag{4.55}$$

Let us reduce the difference scheme (4.54) to an operator problem in the space $F_\tau(E)$ of vectors $v^\tau = \{v_k\}_{k=1}^N$. Recall that the space $F_\tau(E)$ can be equipped with various norms and thus become a normed space. Thus, for instance, the vector space $F_\tau(E)$ generates the normed space $C_\tau(E) = C([0,T]_\tau, E)$ with the norm

$$\| \varphi^\tau \|_{C_\tau(E)} = \max_{1 \le k \le N} \| \varphi_k \|_E,$$

the normed space $C_\tau^\beta(E) = C^\beta([0,T]_\tau, E)$, $0 < \beta < 1$, with the norm

$$\| \varphi^\tau \|_{C_\tau^\beta(E)} = \| \varphi^\tau \|_{C_\tau(E)} + \max_{1 \le k < k+r \le N} \| \varphi_{k+r} - \varphi_k \|_E \frac{1}{(r\tau)^\beta},$$

the weighted normed space $C_\tau^{\beta,\gamma}(E) = C^{\beta,\gamma}([0,T]_\tau, E), 0 < \gamma \le \beta,\ 0 < \beta < 1$, with the norm

$$\| \varphi^\tau \|_{C_\tau^{\beta,\gamma}(E)} = \| \varphi^\tau \|_{C_\tau(E)} + \max_{1 \le k < k+r \le N} \| \varphi_{k+r} - \varphi_k \|_E \frac{((k+r)\tau)^\gamma}{(r\tau)^\beta},$$

and the normed space $L_{p,\tau}(E) = L_p([0,T]_\tau, E), 1 \le p < \infty$, with the norm

$$\| \varphi^\tau \|_{L_{p,\tau}(E)} = (\sum_{k=1}^{N} \| \varphi_k \|_E^p \tau)^{\frac{1}{p}}.$$

First we will define the operator D_τ, acting from the space $E \times F_\tau(E)$ of vectors $w^\tau = \{w_k\}_{k=0}^N$ into the space $F_\tau(E)$ of vectors $v^\tau = \{v_k\}_{k=1}^N$ by the rule

$$v^\tau = D_\tau w^\tau, v_k = \frac{1}{\tau}(w_k - w_{k-1}), k = 1, \ldots, N.$$

Second we will define an operator A_τ from the space $E \times F_\tau(E)$ of vectors $w^\tau = \{w_k\}_{k=0}^N$ into the space $F_\tau(E)$ of vectors $v^\tau = \{v_k\}_{k=1}^N$ by the rule

$$v^\tau = A_\tau w^\tau, v_k = \tau^{-1}(I - v(t_k, t_{k-1})\,)w_{k-1},\ k = 1, \ldots, N.$$

Next, let us introduce the continuation operator $\Pi(u_0)$, which acts from $E \times F_\tau(E)$ to $F_\tau(E)$ according to the rule

$$\Pi(u_0)(u_1, \ldots, u_N) = (u_0, u_1, \ldots, u_N).$$

Then the difference schemes (4.54) can obviously be rewritten as the equivalent operator equation

$$D_\tau \Pi(u_0) u^\tau + A_\tau \Pi(u_0) u^\tau = f^\tau.$$

Here f^τ is defined by the formula

$$f^\tau = (f_1, \ldots, f_N).$$

The last operator problem will be considered in the space $F_\tau(E)$. From its unique solvability for any $u_0 \in E$ and $f^\tau \in F_\tau(E)$ it follows that its solution u^τ defines an additive and homogeneous operator $u^\tau(f^\tau, u_0)$.

The initial-value difference problem (4.54) is said to be stable in $F_\tau(E)$ if we have the stability inequality

$$\| u^\tau(f^\tau, u_0) \|_{F_\tau(E)} \leq M[\| f^\tau \|_{F_\tau(E)} + \| u_0 \|_E],$$

where M is independent not only of f^τ, u_0 but also of τ.

From formula (4.55) one derives the following results.

Theorem 4.2.1. *Suppose that assumptions* (4.3) *and* (4.4) *hold and* $0 < \varepsilon \leq 1$. *Then difference problem* (4.54) *is stable in* $C_\tau(E)$.

Theorem 4.2.2. *Suppose that assumptions* (4.3) *and* (4.4) *hold and* $0 < \alpha \leq \varepsilon \leq 1$. *Then difference problem* (4.54) *is stable in* $C_\tau^{\alpha,\alpha}(E)$.

Theorem 4.2.3. *Suppose that assumptions* (4.3) *and* (4.4) *hold and* $0 < \varepsilon \leq 1$. *Then difference problem* (4.54) *is stable in* $L_{p,\tau}(E)$, $1 \leq p < \infty$.

The initial-value difference problem (4.54) is said to be well posed (coercively stable) in $F_\tau(E)$ if we have the coercive inequality

$$\| \{\tau^{-1}(u_k - u_{k-1})\}_1^N \|_{F_\tau(E)} + \| A_\tau \Pi(u_0) u^\tau \|_{F_\tau(E)}$$
$$\leq M[\| f^\tau \|_{F_\tau(E)} + \| A(0) u_0 \|_E],$$

where M is independent not only of f^τ, u_0 but also of τ.

Since the initial-value differential problem (4.1) in $C(E)$ is not well posed for the unbounded operator $A(t)$ and space E, then the well-posedness of the difference initial-value in $C_\tau(E)$ norm does not take place uniformly with respect to $\tau > 0$. This means that the coercive norm

$$\| u^\tau \|_{K_\tau(E)} = \| \{\tau^{-1}(u_k - u_{k-1})\}_1^N \|_{C_\tau(E)} + \| A_\tau \Pi(u_0) u^\tau \|_{C_\tau(E)}$$

tends to ∞ as $\tau \to +0$. The investigation of the difference problem (4.54) permits us to establish the order of growth of this norm to ∞.

Theorem 4.2.4. *Suppose that assumptions* (4.3) *and* (4.4) *hold and* $0 < \varepsilon \leq 1$. *The solutions of the difference problem* (4.54) *in* $C_\tau(E)$ *obey the almost coercive inequality*

$$\| u^\tau \|_{K_\tau(E)} \leq M_1 [\| A(0) u_0 \|_E \tag{4.56}$$

$$+ \min\{\ln \frac{1}{\tau}, 1 + |\ln ||A(0)||_{E\to E}\} \| f^\tau \|_{C_\tau(E)}],$$

where M_1 *is independent not only of* f^τ, u_0, *but also of* τ.

Proof. Using formula (4.55), we can write

$$\tau^{-1}(I - v(t_k, t_{k-1})) u_{k-1} = \tau^{-1}(I - v(t_k, t_{k-1})) v(t_{k-1}, 0) u_0 \tag{4.57}$$

$$+ \sum_{j=1}^{k-1} \tau^{-1}(I - v(t_k, t_{k-1}) v(t_{k-1}, t_j) \ f_j \tau = J_k + G_k,$$

where

$$J_k = \tau^{-1}(I - v(t_k, t_{k-1})) v(t_{k-1}, 0) u_0,$$

$$G_k = \sum_{j=1}^{k-1} (I - v(t_k, t_{k-1})) v(t_{k-1}, t_j) \ f_j.$$

Let us estimate J_k and G_k separately. We start with J_k. Using property 4) of $v(t, s)$ and estimates (4.17), (4.18) and (4.21), we obtain

$$||\tau^{-1}(I - v(t_k, t_{k-1})) v(t_{k-1}, 0) A^{-1}(0)||_{E\to E} \leq M, 1 \leq k \leq N, \tag{4.58}$$

$$||(I - v(t_k, t_{k-1})) v(t_{k-1}, 0)||_{E\to E} \leq M \min\{\frac{1}{k}, \tau ||A(0)||_{E\to E}\}, 1 \leq k \leq N. \tag{4.59}$$

From estimate (4.58) it follows that

$$||J_k \|_E \leq \| \tau^{-1}(I - v(t_k, t_{k-1})) v(t_{k-1}, 0) A^{-1}(0)||_{E\to E} ||A(0) u_0 \|_E$$
$$\leq M ||A(0) u_0||_E$$

for all k, $k = 1, \dots, N$. Using the last inequality, we obtain

$$\| J^\tau \|_{C_\tau(E)} \leq M ||A(0) u_0||_E.$$

Now let us estimate G_k separately. From estimate (4.59) it follows that

$$||G_k \,||\,_E \leq \sum_{j=1}^{k-1} ||\,(1 - v(t_k, t_{k-1}))v(t_{k-1}, t_j)||_{E\to E}||f_j \,||_E$$

$$\leq M \sum_{j=1}^{N} \frac{1}{j} \| \, f^\tau \, \|_{C_\tau(E)} \; \leq M_1 \ln \frac{1}{\tau} \| \, f^\tau \, \|_{C_\tau(E)}$$

for all k, $k = 1, \ldots, N$. Using the last inequality, we obtain

$$\| \, G^\tau \, \|_{C_\tau(E)} \leq M_1 \ln \frac{1}{\tau} \| \, f^\tau \, \|_{C_\tau(E)}. \tag{4.60}$$

Further, using the triangle inequality, we obtain

$$||G_k \,||\,_E \leq \sum_{j=1}^{k-1} ||\,(I - v(t_k, t_{k-1}))v(t_{k-1}, t_j)||_{E\to E}||f_j \,||_E$$

$$\leq \sum_{j=1}^{k-1} ||\,(I - v(t_k, t_{k-1}))v(t_{k-1}, t_j)||_{E\to E}\| \, f^\tau \, \|_{C_\tau(E)}$$

for all k, $k = 1, \ldots, N$. Let us estimate

$$a_k = \sum_{j=1}^{k-1} ||\,(1 - v(t_k, t_{k-1}))v(t_{k-1}, t_j)||_{E\to E}.$$

If $||A(0)||_{E\to E} > N$, then

$$a_k \leq M \sum_{j=1}^{N-1} \frac{1}{j} \leq M \int\limits_1^N \frac{ds}{s} < M \int\limits_1^{||A(0)||_{E\to E}} \frac{ds}{s} \leq M |\ln ||A(0)||_{E\to E}|.$$

If $||A(0)||_{E\to E} < 1$, then

$$a_k \leq M \sum_{j=1}^{N-1} \tau ||A(0)||_{E\to E} \leq M_1.$$

Finally, if $1 \leq ||A(0)||_{E\to E} \leq N$, then

$$a_k \leq M[\sum_{j=1}^{[\frac{N}{||A(0)||_{E\to E}}]} \tau ||A(0)||_{E\to E} + \sum_{j=[\frac{N}{||A(0)||_{E\to E}}]+1}^{N-1} \frac{1}{j}]$$

$$\leq M(1 + \int\limits_{\frac{T}{||A(0)||_{E\to E}}}^{T} \frac{ds}{s}) \leq M(1 + \ln ||A(0)||_{E\to E}).$$

Thus, in all three cases we have the estimate

$$a_k \leq M(1 + |\ln \|A(0)\|_{E\to E}|).$$

Therefore,

$$\|G^\tau\|_{C_\tau(E)} \leq M_1(1 + |\ln \|A(0)\|_{E\to E})\| \| f^\tau \|_{C_\tau(E)}. \tag{4.61}$$

Finally, the estimate

$$\|G^\tau\|_{C_\tau(E)} \leq M_1 \min\{\ln \frac{1}{\tau}, 1 + |\ln \|A(0)\|_{E\to E}|\} \| f^\tau \|_{C_\tau(E)}$$

follows from (4.60) and (4.61). Combining the estimates for J^τ and G^τ, we obtain

$$\|\{\tau^{-1}(I - v(t_k, t_{k-1}))u_{k-1}\}_1^N\|_{C_\tau(E)} \leq M_1[\|A(0)u_0\|_E$$

$$+ \min\{\ln \frac{1}{\tau}, 1 + |\ln \|A(0)\|_{E\to E}|\} \| f^\tau \|_{C_\tau(E)}].$$

The estimate for $\|\{\tau^{-1}(u_k - u_{k-1})\}_1^N\|_{C_\tau(E)}$ follows from the triangle inequality and the last estimate. Theorem 4.2.4 is proved.

Now let us study the well-posedness of the difference problem (4.54) in the various Banach spaces.

Theorem 4.2.5. *Suppose that assumptions* (4.3) *and* (4.4) *hold and* $0<\alpha\leq\varepsilon\leq 1$. *The solutions of the difference problem* (4.54) *in* $C_\tau^{\alpha,\alpha}(E)$ *obey the coercivity inequality*

$$\|\{\tau^{-1}(u_k - u_{k-1})\}_1^N\|_{C_\tau^{\alpha,\alpha}(E)} + \| \{\tau^{-1}(I - v(t_k, t_{k-1}))u_{k-1}\}_1^N\|_{C_\tau^{\alpha,\alpha}(E)}$$

$$\leq M_1[\frac{1}{\alpha(1-\alpha)}\|f^\tau\|_{C_\tau^{\alpha,\alpha}(E)} + \|A(0)u_0\|_E],$$

where M_1 *is independent not only of* f^τ, u_0, α, *but also of* τ.

The proof of Theorem 4.2.5 follows the scheme of the proof of Theorem 4.1.6 and relies on estimates (4.3), (4.4), (4.17), (4.18), (4.19), (4.20), (4.21), (4.22), (4.23), (4.24) and on the formula

$$\tau^{-1}(I - v(t_k, t_{k-1}))u_{k-1} = \tau^{-1}(I - v(t_k, t_{k-1}))v(t_{k-1}, 0)u_0$$

$$+(I - \exp\{-t_{k-1}A(t_{k-1})\})f_k$$

$$+\sum_{j=1}^{k-1}[(I - v(t_k, t_{k-1})v(t_{k-1}, t_j) - (I - \exp\{-\tau A(t_{k-1})\}) \exp\{-t_{k-1}A(t_{k-1})\}]f_k$$

$$+\sum_{j=1}^{k-1}(I - v(t_k, t_{k-1})v(t_{k-1}, t_j)(f_j - f_k).$$

Theorem 4.2.6. *Suppose that assumptions* (4.3), (4.4) *hold and* $0<\alpha\leq\varepsilon\leq 1$. *The solutions of the difference problem* (4.54) *in* $C_\tau(E_\alpha)$ *obey the coercivity inequality*

$$||\{\tau^{-1}(u_k-u_{k-1})\}_1^N\|_{C_\tau(E_\alpha)}+\|\{\tau^{-1}(I-v(t_k,t_{k-1}))u_{k-1}\}_1^N\|_{C_\tau(E_\alpha)}$$

$$\leq M_1[\frac{1}{\alpha(1-\alpha)}\|f^\tau\|_{C_\tau(E_\alpha)}+\|A(0)u_0\|_{E_\alpha}],$$

where M_1 *is independent not only of* f^τ, u_0, α, *but also of* τ.

The proof of Theorem 4.2.6 follows the scheme of the proof of Theorem 4.1.8 and relies on estimates (4.3), (4.4), (4.17), (4.18), (4.19), (4.20), (4.21), (4.22), (4.23), (4.24) and on the formula

$$\tau^{-1}(I-v(t_k,t_{k-1}))u_{k-1}=\tau^{-1}(I-v(t_k,t_{k-1}))v(t_{k-1},0)u_0$$

$$+\sum_{j=1}^{k-1}(I-\exp\{-\tau A(t_{k-1})\})\exp\{-t_{k-1-j}A(t_{k-1})\}]f_s$$

$$+\sum_{j=1}^{k-1}[(I-v(t_k,t_{k-1})v(t_{k-1},t_j)-(I-\exp\{-\tau A(t_{k-1})\})\exp\{-t_{k-1-j}A(t_{k-1})\}]f_j.$$

Let us give, without proof, the following results.

Theorem 4.2.7. *Suppose that assumptions of Theorem* 4.1.11 *hold. Then for all* $0<\alpha<1$ *the solutions of the difference problem* (4.54) *in* $C_\tau(E_\alpha)$ *obey the coercivity inequality*

$$||\{\tau^{-1}(u_k-u_{k-1})\}_1^N\|_{C_\tau(E_\alpha)}+\|\{\tau^{-1}(I-v(t_k,t_{k-1}))u_{k-1}\}_1^N\|_{C_\tau(E_\alpha)}$$

$$\leq M_1[\frac{1}{\alpha(1-\alpha)}\|f^\tau\|_{C_\tau(E_\alpha)}+\|A(0)u_0\|_{E_\alpha}],$$

where M_1 *is independent not only of* f^τ, u_0, α, *but also of* τ.

Theorem 4.2.8. *Suppose that assumptions* (4.3) *and* (4.4) *hold and* $0<\beta\leq\varepsilon\leq 1$. *The solutions of the difference problem* (4.54) *in* $C_\tau^{\beta,\gamma}(E)$ $(0\leq\gamma\leq\beta, 0<\beta<1)$ *satisfy the following coercivity inequalities:*

$$\|\{\tau^{-1}(u_k-u_{k-1})\}_1^N\|_{C_\tau(E_{\beta-\gamma})}\leq\frac{M_1}{\beta(1-\beta)}\|f^\tau\|_{C_\tau^{\beta,\gamma}(E)}$$

$$+M_1\|(I-e^{-\tau A(0)})\tau^{-1}u_0-f_1\|_{E_{\beta-\gamma}},$$

$$\|\{\tau^{-1}(u_k-u_{k-1})\}_1^N\|_{C_\tau^{\beta,\gamma}(E)}+\|\{\tau^{-1}(I-v(t_k,t_{k-1}))u_{k-1}\}_1^N\|_{C_\tau^{\beta,\gamma}(E)}$$

$$\leq\frac{M_1}{\beta(1-\beta)}\|f^\tau\|_{C_\tau^{\beta,\gamma}(E)}+M_1|(I-e^{-\tau A(0)})\tau^{-1}u_0-f_1|_0^{\beta,\gamma},$$

where M_1 *is independent not only of* f^τ, u_0, β, γ *but also of* τ.

Note that the parameter γ can be chosen freely in $[0,\beta)$, which increases the number of spaces $C_\tau^{\beta,\gamma}(E)$, $(0 \le \gamma \le \beta, 0 < \beta < 1)$ of grid functions in which difference problem (4.54) is well posed.

Theorem 4.2.9. *Suppose that assumptions* (4.3) *and* (4.4) *hold. The solutions of difference problem* (4.54) *in* $C_\tau^{\beta,\gamma}(E_{\alpha-\beta})$, $(0 \le \gamma \le \beta \le \alpha \le \varepsilon \le 1)$ *satisfy the following coercivity inequalities:*

$$\| \{\tau^{-1}(u_k - u_{k-1})\}_1^N \|_{C_\tau^{\beta,\gamma}(E_{\alpha-\beta})} + \| \{\tau^{-1}(u_k - u_{k-1})\}_1^N \|_{C_\tau(\widetilde{E}_{\alpha-\beta}^{\beta,\gamma})}$$

$$+ \| \{\tau^{-1}(I - v(t_k, t_{k-1}))u_{k-1}\}_1^N \|_{C_\tau^{\beta,\gamma}(E_{\alpha-\beta})}$$

$$\le \frac{M_1}{\alpha(1-\alpha)} \| f^\tau \|_{C_\tau^{\beta,\gamma}(E_{\alpha-\beta})} + M_1 |(I - e^{-\tau A(0)})\tau^{-1}u_0 - f_1|_{\alpha-\beta}^{\beta,\gamma},$$

$$\| \{\tau^{-1}(u_k - u_{k-1})\}_1^N \|_{C_\tau^{\beta,\gamma}(E_{\alpha-\beta})} + \| \{\tau^{-1}(u_k - u_{k-1})\}_1^N \|_{C_\tau(E_{\alpha-\gamma})}$$

$$+ \| \{\tau^{-1}(I - v(t_k, t_{k-1}))u_{k-1}\}_1^N \|_{C_\tau^{\beta,\gamma}(E_{\alpha-\beta})}$$

$$\le \frac{M_1}{\alpha(1-\alpha)} \| f^\tau \|_{C_\tau^{\beta,\gamma}(E_{\alpha-\beta})} + M_1 \|(I - e^{-\tau A(0)})\tau^{-1}u_0 - f_1\|_{E_{\alpha-\gamma}},$$

where M_1 *is independent not only of* f^τ, u_0, α, β, γ *but also of* τ.

Note that the spaces $C_\tau^{\beta,\gamma}(E_{\alpha-\beta})$ of grid functions, in which coercive solvability has been established, depend on the parameters α, β and γ. However, the constants in the coercive inequalities depend only on α. Hence, we can choose the parameters β and γ freely, which increases the number of spaces of grid functions in which difference problem (4.54) is well posed.

Now let us consider the initial-value difference problem (4.54) in the spaces $L_{p,\tau}(E) = L_p([0,T]_\tau, E)$, $1 \le p < \infty$ of all grid functions. We have not been able to obtain the coercivity inequality

$$\| \{\tau^{-1}(u_k - u_{k-1})\}_1^N \|_{L_{p,\tau}(E)} + \| \{\tau^{-1}(I - v(t_k, t_{k-1}))u_{k-1}\}_1^N \|_{L_{p,\tau}(E)}$$

$$\le M[\| f^\tau \|_{L_{p,\tau}(E)} + \| A(0)u_0\|_E],$$

in the arbitrary Banach space E and for the general strong positive operator $A(0)$. Nevertheless, we have established the almost coercivity inequality.

Theorem 4.2.10. *Suppose that assumptions* (4.3) *and* (4.4) *hold. The solutions of the difference problem* (4.54) *in* $L_{p,\tau}(E)$ *obey the almost coercive inequality*

$$\| \{\tau^{-1}(u_k - u_{k-1})\}_1^N \|_{L_{p,\tau}(E)} + \| \{\tau^{-1}(I - v(t_k, t_{k-1}))u_{k-1}\}_1^N \|_{L_{p,\tau}(E)}$$

$$\le M_1[\| A(0)u_0\|_E + \min\left\{\ln\frac{1}{\tau}, 1 + |\ln \| A(0) \|_{E\to E}|\right\} \| f^\tau \|_{L_{p,\tau}(E)}],$$

where M_1 *does not depend on* f^τ, u_0, p *and* τ.

Note that if $A(0)$ is the unbounded positive operator in E, then

$$\min\left\{\ln\frac{1}{\tau}, 1+|\ln \| A(0) \|_{E\to E}|\right\} = \ln\frac{1}{\tau}.$$

Finally, let us give, without proof, the following results about well-posedness of the initial-value difference problem (4.54) in the spaces $L_{p,\tau}(E) = L_p([0,T]_\tau, E)$, $1 \le p < \infty$.

Theorem 4.2.11. *Suppose that assumptions* (4.3) *and* (4.4) *hold and* $0 < \varepsilon \le 1$. *Suppose that difference problem* (4.54) *is well posed in* $L_{p_0,\tau}(E)$ *for some* p_0, $1 < p_0 < \infty$. *Then it is well posed in* $L_{p,\tau}(E)$ *for all* p, $1 < p < \infty$ *and the following coercivity inequality holds:*

$$\| \{\tau^{-1}(u_k - u_{k-1})\}_1^N \|_{L_{p,\tau}(E)} + \| \{\tau^{-1}(I - v(t_k, t_{k-1}))u_{k-1}\}_1^N \|_{L_{p,\tau}(E)}$$

$$+\|u^\tau\|_{C_c(E_{1-1/p,p})} \le M(p_0)[\| u_0 \|_{E_{1-1/p,p}} + \frac{p^2}{p-1} \| f^\tau \|_{L_{p,\tau}(E)}],$$

where $M(p_0)$ *does not depend on* f^τ, u_0, p *and* τ.

Theorem 4.2.12. *Let* $1 \le p \le \infty$. *Suppose that assumptions* (4.3) *and* (4.4) *hold and* $0 < \alpha \le \varepsilon < 1$. *Then problem* (4.54) *is well posed in* $L_{p,\tau}(E_{\alpha,p})$ *and the following coercivity inequality holds:*

$$\| \{\tau^{-1}(u_k - u_{k-1})\}_1^N \|_{L_{p,\tau}(E_{\alpha,p})} + \| \{\tau^{-1}(I - v(t_k, t_{k-1}))u_{k-1}\}_1^N \|_{L_{p,\tau}(E_{\alpha,p})}$$

$$\le M[\| A(0)u_0\|_{E_{\alpha,p}} + \frac{1}{\alpha(1-\alpha)} \| f^\tau \|_{L_{p,\tau}(E_{\alpha,p})}],$$

where M *does not depend on* f^τ, u_0, p, α *and* τ.

Theorem 4.2.13. *Let* $1 \le p \le \infty$ *and* $0 < \alpha < 1$. *Suppose that the assumptions of Theorem* 4.1.18 *hold. Then problem* (4.54) *is well posed in* $L_{p,\tau}(E_{\alpha,p})$ *and the following coercivity inequality holds:*

$$\| \{\tau^{-1}(u_k - u_{k-1})\}_1^N \|_{L_{p,\tau}(E_{\alpha,p})} + \| \{\tau^{-1}(I - v(t_k, t_{k-1}))u_{k-1}\}_1^N \|_{L_{p,\tau}(E_{\alpha,p})}$$

$$\le M[\| A(0)u_0\|_{E_{\alpha,p}} + \frac{1}{\alpha(1-\alpha)} \| f^\tau \|_{L_{p,\tau}(E_{\alpha,p})}],$$

where M *does not depend on* f^τ, u_0, p, α *and* τ.

Theorem 4.2.14. *Let* $1 < p, q < \infty$. *Suppose that the assumptions* (4.3) *and* (4.4) *hold and* $0 < \alpha \le \varepsilon < 1$. *Then difference problem* (4.54) *is well posed in* $L_{p,\tau}(E_{\alpha,q})$ *and the following coercivity inequality holds:*

$$\| \{\tau^{-1}(u_k - u_{k-1})\}_1^N \|_{L_{p,\tau}(E_{\alpha,q})} + \| \{\tau^{-1}(I - v(t_k, t_{k-1}))u_{k-1}\}_1^N \|_{L_{p,\tau}(E_{\alpha,q})}$$

$$\le M(q)[\| A(0)u_0\|_{E_{\alpha,q}} + \frac{p^2}{p-1} \| f^\tau \|_{L_{p,\tau}(E_{\alpha,q})}],$$

where $M(q)$ *does not depend on* f^τ, u_0, p, α *and* τ.

Theorem 4.2.15. *Let $1 < p, q < \infty$ and $0 < \alpha < 1$. Suppose that the assumptions of Theorem 4.1.18 hold. Then difference problem (4.54) is well posed in $L_{p,\tau}(E_{\alpha,q})$ and the following coercivity inequality holds:*

$$\| \{\tau^{-1}(u_k-u_{k-1})\}_1^N \|_{L_{p,\tau}(E_{\alpha,q})} + \| \{\tau^{-1}(I-v(t_k,t_{k-1}))u_{k-1}\}_1^N \|_{L_{p,\tau}(E_{\alpha,q})}$$

$$\leq M(q)[\| A(0)u_0\|_{E_{\alpha,q}} + \frac{p^2}{p-1} \| f^\tau \|_{L_{p,\tau}(E_{\alpha,q})}],$$

where $M(q)$ does not depend on f^τ, u_0, p, α and τ.

Now, we will consider the applications of this exact difference scheme (4.54). From (4.54) it is clear that for the approximate solutions of problem (4.1) it is necessary to approximate the expressions

$$v(t_k,t_{k-1}) \text{ and } \frac{1}{\tau}\int_{t_{k-1}}^{t_k} v(t_k,s)f(s)ds.$$

The choice of $v(t_k,t_{k-1})$ and $\frac{1}{\tau}\int_{t_{k-1}}^{t_k} v(t_k,s)f(s)ds$ is not unique. In this section we consider two approaches. The first approach is carried out in three steps. Applying the definition of $v(t,s)$, we can write

$$v(t,s) = \sum_{i=0}^{\infty} v_i(t,s),\ v_0(t,s) = \exp\{-(t-s)A(s)\},$$

$$v_i(t,s) = \int_s^t v_{i-1}(t,z)[A(s)-A(z)]\exp\{-(z-s)A(s)\}dz, i \geq 1.$$

Therefore, in the first step we consider

$$u^\tau(k,k-1) = u_I^\tau(k,k-1) = \sum_{i=0}^{m} v_i(t_k,t_{k-1})$$

$$= \exp\{-\tau A(t_{k-1})\} + \sum_{i=1}^{\tilde{m}} \int_{t_{k-1}}^{t_k}\int_{z_1}^{t_k}\cdots\int_{z_{i-1}}^{t_k} \exp\{-(t_k-z_i)A(z_i)\}$$
$$\times(A(z_{i-1})-A(z_i))\exp\{-(z_i-z_{i-1})A(z_{i-1})\}dz_i$$
$$\times(A(z_{i-2})-A(z_{i-1}))\exp\{-(z_{i-1}-z_{i-2})A(z_{i-2})\}dz_{i-1}$$
$$\times\cdots(A(z_1)-A(z_2))\exp\{-(z_2-z_1)A(z_2)(A(z_1)-A(z_2))\exp\{-(z_2-z_1)A(z_2)\}$$
$$\times(A(t_{k-1})-A(z_1))\exp\{-(z_1-t_{k-1})A(t_{k-1})\}dz_1.$$

Suppose that the operator $A^r(t)A(z)A^{-(r+1)}(s)$ satisfies the condition

$$\left\|A^r(t)[A(z_2)-A(z_1)]A^{-(r+1)}(s)\right\|_{E\to E} \leq M|z_2-z_1|,\ r=0,1,\ldots, \tag{4.62}$$

where M is a positive constant independent of t, s, z_1 and z_2 for $0 \leq t, s, z_1, z_2 \leq T$.

Note that if $r=0$, then it follows that

$$\|v(t_k,t_{k-1})-u^\tau(k,k-1)\|_{E\to E}\le M\tau^{\widetilde{m}+1}.$$

If $0\le r\le\widetilde{m}$, then it follows that

$$\left\|[v(t_k,t_{k-1})-u^\tau(k,k-1)]A^{-\widetilde{m}}(0)\right\|_{E\to E}\le M\tau^{2\widetilde{m}+2}.$$

From the formula for $u_I^\tau(k,n)$ it is clear that for the approximate calculation of $u_I^\tau(k,n)$ it is necessary to approximate the expressions $A(t),\exp\{-sA(z)\}$ and their product. For an analytic semigroup we have the formula

$$\exp\{-(t-s)A\}-\exp\{-(t-s)B\}=\sum_{i=1}^{\infty}\psi_i(t,s;A,B),$$

where

$$\psi_i(t,s;A,B)=\int_s^t\psi_{i-1}(t,z;A,B)(A-B)\psi_0(t,s,A,B)dz,i=1,2,\cdots,$$

$$\psi_0(t,s,A,B)=\exp\{-(t-s)B\},0\le s\le t\le T.$$

Therefore, in the second step applying the last formula and Taylor's formula and neglecting the small terms, we can write

$$u^\tau(k,k-1)=\ u_{II}^\tau(k,k-1)=\exp\{-\tau A(t_{k-1})\}$$

$$+\sum_{r=1}^{\widetilde{m}}\int_{t_{k-1}}^{t_k}\int_{z_1}^{t_k}\cdots\int_{z_{r-1}}^{t_k}\sum_{a_1=0}^{\widetilde{m}-r}\psi_{r,a_1}[t_k,z_r;A(z_r),A(t_{k-1})]$$

$$\times(A(z_{r-1})-A(z_r))\sum_{a_2=0}^{\widetilde{m}-r-a_1}\psi_{a_1,a_2}[z_r,t_{k-1};A(z_{r-1}),A(t_{k-1})]dz_r$$

$$\times(A(z_{r-2})-A(z_{r-1}))\sum_{a_3=0}^{\widetilde{m}-r-a_1-a_2}\psi_{a_2,a_3}[z_{r-1},t_{k-1};A(z_{r-2}),A(t_{k-1})]dz_{r-1}$$

$$\times\cdots\times(A(z_1)-A(z_2))\sum_{a_r=0}^{\widetilde{m}-r-a_1-a_2-\cdots-a_{r-1}}\psi_{r,a_r}[z_2,z_1;A(z_1),A(t_{k-1})]dz_2$$

$$\times(A(t_{k-1})-A(z_1))\psi_{r,0}[z_1,t_{k-1};A(t_{k-1}),A(t_{k-1})]dz_1,$$

where

$$\psi_{r,0}[z_1,t_{k-1};A(t_{k-1}),A(t_{k-1})]$$

$$=\sum_{i=0}^{2\widetilde{m}-2r}\frac{(\tau-z_1+t_{k-1})^i}{i!}A^i(t_{k-1})\exp\{-\tau A(t_{k-1})\},$$

$$\psi_{r,a_n}[z,x_n;A(x_n),A(t_{k-1})] = \int_{x_n}^{z}\int_{s_1}^{z}\cdots\int_{s_{a_n-1}}^{z} \sum_{i=0}^{2\widetilde{m}-2\tau-2a_n} \frac{(\tau-z+s_1)^i}{i!}$$

$$\times A^i(t_{k-1})\exp\{-\tau A(t_{k-1})\}(A(t_{k-1})-A(x_n))$$

$$\times \sum_{i=0}^{2\widetilde{m}-2\tau-2a_n} \frac{(\tau-s_1+s_2)^i}{i!} A^i(t_{k-1})\exp\{-\tau A(t_{k-1})\}$$

$$\times\cdots\times(A(t_{k-1})-A(x_n)) \sum_{i=0}^{2\widetilde{m}-2\tau-2a_n} \frac{(\tau-s_{a_n}+x_n)^i}{i!} A^i(t_{k-1})$$

$$\times \exp\{-\tau A(t_{k-1})\}ds_1\cdots ds_{a_n},$$

$$a_n = 1,\ldots,\widetilde{m}; n=1,\ldots,r; r=1,\ldots,\widetilde{m}.$$

Finally, in the third step replacing the operator $\exp\{-\tau A(t_{k-1})\}$ by its Pade approximant $R_{j,l}(\tau A(t_{k-1}))$, we obtain

$$u^\tau(k,k-1) = u^\tau_{III}(k,k-1) = B^{l+j}_{0,0}[t_k,t_{k-1};A(t_k),A(t_{k-1})]$$

$$+\sum_{r=1}^{l}\int_{t_{k-1}}^{t_k}\int_{z_1}^{t_k}\cdots\int_{z_{r-1}}^{t_k}\sum_{a_1=0}^{l-r} B^{l+j-2r-2a_1}_{r,a_1}[t_k,z_r;A(z_r),A(t_{k-1})]$$

$$\times(A(z_{r-1})-A(z_r))\sum_{a_2=0}^{l-r-a_1} B^{l+j-2r-2a_2}_{a_1,a_2}[z\Upsilon,t_{k-1};A(z_{r-1}),A(t_{k-1})]dz_r$$

$$\times(A(z_{r-2})-A(z_{r-1}))\sum_{a_3=0}^{l-r-a_1-a_2} B^{l+j-2r-2a_3}_{a_2,a_3}[z_{r-1},t_{k-1};A(z_{r-2}),A(t_{k-1})]dz_{r-1}$$

$$\times\cdots\times(A(z_1)-A(z_2))\sum_{a_r=0}^{l-r-a_1-a_2-\cdots-a_{r-1}} B^{l+j-2r-2a_r}_{r,a_r}[z_2,z_1;A(z_1),A(t_{k-1})]dz_2$$

$$\times(A(t_{k-1})-A(z_1))B^{l+j-2r}_{r,0}[z_1,t_{k-1};A(t_{k-1}),A(t_{k-1})]\,dz_1,$$

where

$$B^{l+j}_{0,0}[t_k,t_{k-1};A(t_{k-1}),A(t_{k-1})] = R_{j,l}(\tau A(t_{k-1})),$$

$$B^{l+j-2r}_{n,0}[z_1,t_{k-1};A(t_{k-1}),A(t_{k-1})]$$

$$=\sum_{i=0}^{l+j}\frac{(\tau-z_1+t_{k-1})^i}{i!}A^i(t_{k-1})R_{j-r-[\frac{i}{2}],l-r+i+1}(\tau A(t_{k-1})),$$

$$B^{l+j-2r-2a_n}_{r,a_n}(z,x_n;A(x_n),A(t_{k-1}))$$

$$=\int_{x_n}^{z}\int_{s_1}^{z}\cdots\int_{s_{a_n-1}}^{z}\sum_{i=0}^{l+j-2r-a_n}\frac{(\tau-z+s_1)^i}{i!}A^i(t_{k-1})$$

$$\times R_{j-r-[\frac{i}{2}],l-r+a_n+i+1-[\frac{i}{2}]}(\tau A(t_{k-1}))(\tau A(t_{k-1}) - A(x_n))$$
$$\times \sum_{i=0}^{l+j-2r-a_n} \frac{(\tau - s_{a_n} + x_n)^i}{i!} A^i(t_{k-1})$$

$$\times R_{j-r-[\frac{i}{2}],l-r-a_n+i+1-[\frac{i}{2}]}(\tau A(t_{k-1}) \cdots (A(t_{k-1}) - A(x_n))$$
$$\times \sum_{i=0}^{l+j-2r-a_n} \frac{(\tau - s_{a_n} + x_n)^i}{i!} A^i(t_{k-1})$$
$$\times R_{j-r-[\frac{i}{2}],l-r+a_n+i+1-[\frac{i}{2}]}(\tau A(t_{k-1}))(\tau A(t_{k-1}))ds_1 \cdots ds_{a_n};$$
$$a_n = 1, \ldots, l; n = 1, \ldots, r; r = 1, \ldots, l.$$

Here $[a]$ denotes the integer part of $a > 0$. In exactly the same manner using the three steps, we can obtain the formula for $f_k^{j,l}$ such that

$$||f_k - f_k^{j,l}||_E \leq M\tau^{l+j},$$

where M does not depend on τ.

Thus, replacing the operator $v(t_k, t_{k-1})$ by $u^\tau(k, k-1))$ and the elements f_k by elements $f_k^{j,l}$, we obtain the difference schemes of approximation order $l + j$,

$$\tau^{-1}(u_k - u_{k-1}) + \tau^{-1}(I - u^\tau(k, k-1))u_{k-1} = f_k^{j,1}, \tag{4.63}$$
$$1 \leq k \leq N, \ u_0 = v_0$$

on the solutions $v(t)$ of the Cauchy problem (4.1).

Note that the difference schemes (4.63) for $j = l - 1$ and $j = l - 2$ include difference schemes of arbitrary high order of approximation. Moreover, the main corresponding functions $R_{j,l}(z)$ tend to 0 as $z \to \infty$ for $j = l - 2,\ l - 1$. Such difference schemes are simplest, in the sense that the degrees of the denominators of the corresponding Pade approximants of the function $\exp\{-z\}$ are minimal for a fixed order of approximation of the difference schemes.

Now let us give, without proof, the following estimates concerning $R_{j,l}^k(\tau A(t))$ for $j = l - 1$ and $j = l - 2$ which will be useful in the sequel.

Lemma 4.2.16. *Suppose that the assumption* (4.3) *holds. Then for any t, t, $\in [0, T]$ the following estimates are satisfied:*

$$||R_{j,l}^k(\tau A(t))||_{E\to E} \leq M, 1 \leq k \leq N, \tag{4.64}$$

$$||A(t)R_{j,l}^k(\tau A(t))||_{E\to E} \leq M(k\tau)^{-1}, 1 \leq k \leq N, \tag{4.65}$$

$$||\exp(-k\tau A(t)) - R_{j,l}^k(\tau A(t))||_{E\to E} \leq Mk^{-j-l}, 1 \leq k \leq N, \tag{4.66}$$

$$\left\| R_{j,l}^k(\tau A(t)) - R_{j,l}^{k+r}(\tau A(t)) \right\|_{E\to E} \leq M\frac{r^\alpha}{(k+r)^\alpha}, 1 \leq k < k + r \leq N. \tag{4.67}$$

Here M does not depend on k, r and τ.

Now we shall investigate the well-posedness of the single-step difference schemes (4.63). This problem is uniquely solvable and the formula

$$u_k = u^\tau(k,0)u_0 + \sum_{r=1}^{k} u^\tau(k,r) f_r^{j,l}\tau \tag{4.68}$$

holds, where

$$u^\tau(k,n) = \begin{cases} u^\tau(k,k-1)\cdots u^\tau(n+1,n), k > n, \\ I, k = n. \end{cases} \tag{4.69}$$

Let us reduce the difference scheme (4.63) to an operator problem in the space $F_\tau(E)$ of vectors $v^\tau = \{v_k\}_{k=1}^N$. Recall that the space $F_\tau(E)$ can be equipped with the various norms and thus become a normed space.

First we will define the operator D_τ, acting from the space $E \times F_\tau(E)$ of vectors $w^\tau = \{w_k\}_{k=0}^N$ into the space $F_\tau(E)$ of vectors $v^\tau = \{v_k\}_{k=1}^N$ by the rule

$$v^\tau = D_\tau w^\tau, v_k = \frac{1}{\tau}(w_k - w_{k-1}), k = 1, \ldots, N.$$

Second we will define an operator A_τ from the space $E \times F_\tau(E)$ of vectors $w^\tau = \{w_k\}_{k=0}^N$ into the space $F_\tau(E)$ of vectors $v^\tau = \{v_k\}_{k=1}^N$ by the rule

$$v^\tau = A_\tau w^\tau, v_k = \tau^{-1}(I - u^\tau(k,k-1))w_{k-1}, \ k = 1, \ldots, N.$$

Next, let us introduce the continuation operator $\Pi(u_0)$, which acts from $E \times F_\tau(E)$ to $F_\tau(E)$ according to the rule

$$\Pi(u_0)(u_1, \ldots, u_N) = (u_0, u_1, \ldots, u_N).$$

Then the difference schemes (4.63) can obviously be rewritten as the equivalent operator equation

$$D_\tau \Pi(u_0) u^\tau + A_\tau \Pi(u_0) u^\tau = f^\tau.$$

Here f^τ is defined by the formula

$$f^\tau = (f_1^{j,l}, \ldots, f_N^{j,l}).$$

The last operator problem will be considered in the space $F_\tau(E)$. From its unique solvability for any $u_0 \in E$ and $f^\tau \in F_\tau(E)$ it follows that its solution u^τ defines an additive and homogeneous operator $u^\tau(f^\tau, u_0)$ and is continuous.

The initial-value difference problem (4.63) is said to be stable in $F_\tau(E)$ if we have the stability inequality

$$\| u^\tau(f^\tau, u_0) \|_{F_\tau(E)} \leq M[\| f^\tau \|_{F_\tau(E)} + \| u_0 \|_E],$$

where M is independent not only of f^τ, u_0 but also of τ.

Let us prove a number of estimates for $u^\tau(k,n)$ that will be needed in the sequel. First we consider $\widetilde{u}^\tau(k,n)$ defined by

$$\widetilde{u}^\tau(k,n) = \begin{cases} \widetilde{u}^\tau(k,k-1)\cdots\widetilde{u}^\tau(n+1,n), k > n, \\ \\ I, k = n, \end{cases} \tag{4.70}$$

where $\widetilde{u}^\tau(k,k-1) = R_{j,l}(\tau A(t_{k-1}))$.

Lemma 4.2.17. *For any $0 \le n \le r \le k \le N$ the following identities hold:*

$$\widetilde{u}^\tau(k,n) = \widetilde{u}^\tau(k,r)\widetilde{u}^\tau(r,n), 0 \le n \le k \le k \le N, \tag{4.71}$$

$$\widetilde{u}^\tau(k,n) = R_{j,l}^{k-n}(\tau A(t_{n-1})) \tag{4.72}$$

$$+ \sum_{r=n+1}^{k} \widetilde{u}^\tau(k,r)[R_{j,l}(\tau A(t_{r-1})) - R_{j,l}(\tau A(t_{n-1}))]R_{j,l}^{r-n-1}(\tau A(t_{n-1})),$$

$$\widetilde{u}^\tau(k,n) = R_{j,l}^{k-n}(\tau A(t_{k-1})) \tag{4.73}$$

$$+ \sum_{r=n+1}^{k} R_{j,l}^{k-r}(\tau A(t_{r-1}))[R_{j,l}(\tau A(t_{k-1})) - R_{j,l}(\tau A(t_{r-1}))]\widetilde{u}^\tau(r-1,n).$$

Proof. Identity (4.71) follows from the definition of $\widetilde{u}^\tau(k,n)$. Using formula (4.70), we obtain

$$\tau^{-1}(\widetilde{u}^\tau(k,n) - \widetilde{u}^\tau(k-1,n)) + \tau^{-1}(I - R_{j,l}(\tau A(t_{k-1})))\widetilde{u}^\tau(k-1,n) = 0.$$

From this it follows that

$$\begin{aligned} &\tau^{-1}(\widetilde{u}^\tau(k,n) - \widetilde{u}^\tau(k-1,n)) + \tau^{-1}(I - R_{j,l}(\tau A(t_{q-1}))\widetilde{u}^\tau(k-1,n) \\ = \ &\tau^{-1}(R_{j,l}(\tau A(t_{q-1})) - R_{j,l}(\tau A(t_{k-1})))\widetilde{u}^\tau(k-1,n) \end{aligned}$$

for any $1 \le q \le N$. Therefore

$$\widetilde{u}^\tau(k,n) = R_{j,l}^{k-n}(\tau A(t_{q-1})) + \sum_{r=n+1}^{k} R_{j,l}^{k-r}(\tau A(t_{q-1}))$$

$$\times[R_{j,l}(\tau A(t_{q-1})) - R_{j,l}(\tau A(t_{r-1}))]\widetilde{u}^\tau(r-1,n).$$

Choosing $q = k$, we obtain the identity (4.72). In a similar manner one establishes the identity (4.72). Lemma 4.2.17 is proved.

Lemma 4.2.18. *Suppose that assumption* (4.62) *holds for $r = 0$. Then for any $0 \le n < k \le N$ the following estimates hold:*

$$\|\widetilde{u}^\tau(k,n)\|_{E\to E} \le M, \tag{4.74}$$

$$\left\|A(t_k)\widetilde{u}^\tau(k,n)A^{-1}(t_n)\right\|_{E\to E}\le M, \tag{4.75}$$

$$\|A(t_k)\widetilde{u}^\tau(k,n)\|_{E\to E}\le \frac{M}{(k-n)\tau}, \tag{4.76}$$

where $M\ge 0$ does not depend on τ, k, n.

Proof. Using identity (4.72) and estimates (4.64), (4.65) and

$$\left\|[R_{j,l}(\tau A(t_{t-1}))-R_{j,l}(\tau A(t_{n-1})]\,A^{-1}(t_{n-1})\right\|_{E\to E}\le M(r-n)\tau^2, \tag{4.77}$$

we obtain

$$\|\widetilde{u}^\tau(k,n)\|_{E\to E}\le \|R_{j,l}^{k-n}(\tau A(t_{n-1}))\|_{E\to E}$$

$$+\sum_{r=n+1}^{k}\|\widetilde{u}^\tau(k,r)\|_{E\to E}\|[R_{j,l}(\tau A(t_{r-1}))-R_{j,l}(\tau A(t_{n-1}))]A^{-1}(t_{n-1})\|_{E\to E}$$

$$\times\|A(t_{n-1})R_{j,l}^{r-n-1}(\tau A(t_{n-1}))\|_{E\to E}\le M+M\sum_{r=n+1}^{k}\|\widetilde{u}^\tau(k,r)\|_{E\to E}\,\tau.$$

Therefore,

$$\|\widetilde{u}^\tau(k,n)\|_{E\to E}\le M+M\sum_{r=n+1}^{k}\|\widetilde{u}^\tau(k,r)\|_{E\to E}\,\tau.$$

Now using the difference analogy of the integral inequality, we obtain (4.74). In a similar manner one establishes the inequality (4.75). Interpolating estimates (4.64) and (4.65), we obtain

$$\|A^{\frac12}(t)R_{j,l}^{k}(\tau A(t))\|_{E\to E}\le M(k\tau)^{-\frac12}, 1\le k\le N, \tag{4.78}$$

Using identity (4.72) and estimates (4.64), (4.65), (4.77) and (4.78), we obtain

$$\|A^{\frac12}(t_k)\widetilde{u}^\tau(k,n)\|_{E\to E}\le \|A^{\frac12}(t_k)R_{j,l}^{k-n}(\tau A(t_{n-1}))\|_{E\to E}$$

$$+\sum_{r=n+1}^{k}\|A^{\frac12}(t_k)\widetilde{u}^\tau(k,r)\|_{E\to E}\|[R_{j,l}(\tau A(t_{r-1}))-R_{j,l}(\tau A(t_{n-1}))]A^{-1}(t_{n-1})\|_{E\to E}$$

$$\times\|A(t_{n-1})R_{j,l}^{r-n-1}(\tau A(t_{n-1}))\|_{E\to E}$$

$$\le \frac{M}{((k-n)\tau)^{\frac12}}+M\sum_{r=n+1}^{k}\left\|A^{\frac12}(t_k)\widetilde{u}^\tau(k,r)\right\|_{E\to E}\tau.$$

Therefore,

$$\|A^{\frac12}(t_k)\widetilde{u}^\tau(k,n)\|_{E\to E}\le \frac{M}{((k-n)\tau)^{\frac12}}+M\sum_{r=n+1}^{k}\left\|A^{\frac12}(t_k)\widetilde{u}^\tau(k,r)\right\|_{E\to E}\tau.$$

Now using the difference analogy of the integral inequality, we obtain the estimate

$$\left\|A^{1/2}(t_k)\widetilde{u}^\tau(k,n)\right\|_{E\to E} \le \frac{M}{((k-n)\tau)^{1/2}}.$$

In a similar manner one establishes the estimate

$$\left\|A(t_k)\widetilde{u}^\tau(k,n)A^{-1/2}(t_n)\right\|_{E\to E} \le \frac{M}{((k-n)\tau)^{1/2}}.$$

From these and the identity

$$A(t_k)\widetilde{u}^\tau(k,n) = A(t_k)\widetilde{u}^\tau(k,i)A^{-1/2}(t_i)A^{1/2}(t_i)\widetilde{u}^\tau(i,n)$$

it follows that

$$\begin{aligned}\|A(t_k)\widetilde{u}^\tau(k,n)\|_{E\to E} &\le \left\|A(t_k)\widetilde{u}^\tau(k,i)A^{-1/2}(t_i)\|_{E\to E}\|A^{1/2}(t_i)\widetilde{u}^\tau(i,n)\right\|_{E\to E} \\ &\le \frac{M_1}{(k-n)\tau}.\end{aligned}$$

Lemma 4.2.18 is proved.

Lemma 4.2.19. *Suppose that the assumption* (4.62) *holds for* $r=0$. *Then for any* $0\le n<r\le k\le N$ *and* $0\le\alpha\le 1$ *the following estimates hold:*

$$\left\|R_{j,l}^{k+r-n}(\tau A(t_{k+r})) - R_{j,l}^{k-n}(\tau A(t_k))\right\|_{E\to E} \le M\Psi, \tag{4.79}$$

$$\left\|[A(t_{k+r})R_{j,l}^{k+r-n}(\tau A(t_{k+r})) - A(t_k)R_{j,l}^{k-n}(\tau A(t_k))]A^{-1}(t_n)\right\|_{E\to E} \le M\Psi, \tag{4.80}$$

$$\left\|A(t_{k+r})R_{j,l}^{k+r-n}(\tau A(t_{k+r})) - A(t_k)R_{j,l}^{k-n}(\tau A(t_k))\right\|_{E\to E} \le M\frac{\Psi}{(k-n)\tau}, \tag{4.81}$$

$$\|\widetilde{u}^\tau(k+r,n) - \widetilde{u}^\tau(k,n)\|_{E\to E} \le M\Psi, \tag{4.82}$$

$$\|[A(t_{k+r})\widetilde{u}^\tau(k+r,n) - A(t_k)\widetilde{u}^\tau(k,n)]A^{-1}(t_n)\|_{E\to E} \le M\Psi, \tag{4.83}$$

$$\|A(t_{k+r})\widetilde{u}^\tau(k+r,n) - A(t_k)\widetilde{u}^\tau(k,n)\|_{E\to E} \le \frac{M\Psi}{(k-n)\tau}, \tag{4.84}$$

where $M\ge 0$ *does not depend on* τ, k, r, n. *Here* $\Psi = r^\alpha(k-n)^{-\alpha}$.

Proof. We have that

$$\begin{aligned}R_{j,l}^{i+r}(\tau A(t_{k+r})) - R_{j,l}^{i}(\tau A(t_k)) &= R_{j,l}^{i+r}(\tau A(t_{k+r})) - R_{j,l}^{i+r}(\tau A(t_k)) \\ &+ R_{j,l}^{i+r}(\tau A(t_k)) - R_{j,l}^{i}(\tau A(t_k)) = J_1 + J_2,\end{aligned}$$

where

$$J_1 = R_{j,l}^{i+r}(\tau A(t_{k+r})) - R_{j,l}^{i+r}(\tau A(t_k)),$$
$$J_2 = R_{j,l}^{i+r}(\tau A(t_k)) - R_{j,l}^{i}(\tau A(t_k)).$$

From estimate (4.67) it follows that

$$\|J_2\|_{E\to E} = \left\|R_{j,l}^{i+r}(\tau A(t_k)) - R_{j,l}^{i}(\tau A(t_k))\right\|_{E\to E} \le M(r\tau)^{\alpha}(i\tau)^{-\alpha}.$$

Now let us estimate J_1. We have that

$$J_1 = \Big(\sum_{s=0}^{\left[\frac{i+r}{2}\right]} + \sum_{s=\left[\frac{i+r}{2}\right]+1}^{i+r-1}\Big) R_{j,l}^{i+r-s}(\tau A(t_{k+r}))$$

$$\times (R_{j,l}(\tau A(t_{k+r})) - R_{j,l}(\tau A(t_k)))R_{j,l}^{s}\,(\tau A(t_k)) = J_{11} + J_{12},$$

where

$$J_{11} = \sum_{s=0}^{\left[\frac{i+r}{2}\right]} R_{j,l}^{i+r-s}(\tau A(t_{k+r}))(R_{j,l}(\tau A(t_{k+r})) - R_{j,l}(\tau A(t_k)))R_{j,l}^{s}\,(\tau A(t_k)),$$

$$J_{12} = \sum_{s=\left[\frac{i+r}{2}\right]+1}^{i+r-1} R_{j,l}^{i+r-s}(\tau A(t_{k+r}))(R_{j,l}(\tau A(t_{k+r})) - R_{j,l}(\tau A(t_k)))R_{j,l}^{s}\,(\tau A(t_k)).$$

Using estimates (4.64), (4.65) and (4.77), we obtain

$$\|J_{12}\|_{E\to E} \le \sum_{s=\left[\frac{i+r}{2}\right]+1}^{i+r-1} \left\|R_{j,l}^{i+r-s}(\tau A(t_{k+r}))\right\|_{E\to E}$$

$$\times \left\|(R_{j,l}(\tau A(t_{k+r})) - R_{j,l}(\tau A(t_k)))A^{-1}(t_k)\right\|_{E\to E}$$

$$\times \left\|A(t_k)R_{j,l}^{s}\,(\tau A(t_k))\right\|_{E\to E} \le M \sum_{s=\left[\frac{i+r}{2}\right]+1}^{i+r-1} r\tau\frac{1}{s} \le M_1 r\tau \le M_1\Psi.$$

Applying Abel's formula, we can write

$$J_{11} = \sum_{s=0}^{\left[\frac{i+r}{2}\right]} R_{j,l}^{i+r-s}(\tau A(t_{k+r}))(R_{j,l}(\tau A(t_{k+r})) - R_{j,l}(\tau A(t_k)))(I - R_{j,l}(\tau A(t_k)))^{-1}$$

$$\times (R_{j,l}^{s}\,(\tau A(t_k)) - R_{j,l}^{s}\,(\tau A(t_k))) = \sum_{s=1}^{\left[\frac{i+r}{2}\right]} (R_{j,l}^{i+r-s+1}(\tau A(t_{k+r})) - R_{j,l}^{i+r-s}(\tau A(t_{k+r})))$$

$$\times\left(R_{j,l}(\tau A(t_{k+r}))-R_{j,l}(\tau A(t_k))\right)\left(I-R_{j,l}(\tau A(t_k))^{-1}R_{j,l}^{s}(\tau A(t_k)\right)$$

$$+R_{j,l}^{i+r-\left[\frac{i+r}{2}\right]}(\tau A(t_{k+r}))(R_{j,l}(\tau A(t_{k+r}))-R_{j,l}(\tau A(t_k)))R_{j,l}^{\left[\frac{i+r}{2}\right]+1}(\tau A(t_k))$$

$$-R_{j,l}^{i+r}(\tau A(t_{k+r}))\left[R_{j,l}(\tau A(t_{k+r}))-R_{j,l}(\tau A(t_k))\right].$$

Using the last formula and estimates (4.64), (4.65), (4.77) and

$$\left\|(\tau A(t_k))^{-1}(I-R_{j,l}(\tau A(t_k))\right\|_{E\to E}\le M,$$

$$\left\|\tau A(t_k)(I-R_{j,l}(\tau A(t_k)))^{-1}R_{j,l}(\tau A(t_k))\right\|_{E\to E}\le M,$$

$$\|R_{j,l}(\tau A(t_{k+r}))-R_{j,l}(\tau A(t_k))\|_{E\to E}\le Mr\tau,$$

we obtain

$$\|J_{11}\|_{E\to E}\le\sum_{s=1}^{\left[\frac{i+r}{2}\right]}\left\|R_{j,l}^{i+r-s}(\tau A(t_{k+r}))-R_{j,l}^{i+r-s}(\tau A(t_{k+r}))\right\|_{E\to E}$$

$$\times\|R_{j,l}(\tau A(t_{k+r}))-R_{j,l}(\tau A(t_k))\|_{E\to E}\left\|(I-R_{j,l}(\tau A(t_k))^{-1}R_{j,l}^{s}(\tau A(t_k))\right\|_{E\to E}$$

$$+\|R_{j,l}^{i+r-\left[\frac{i+r}{2}\right]}(\tau A(t_{k+r}))\|_{E\to E}\|R_{j,l}(\tau A(t_{k+r}))-R_{j,l}(\tau A(t_k))\|_{E\to E}$$

$$\times\|R_{j,l}^{\left[\frac{i+r}{2}\right]+1}(\tau A(t_k))\|_{E\to E}$$

$$+\left\|R_{j,l}^{i+r}(\tau A(t_{k+r}))\right\|_{E\to E}\|R_{j,l}(\tau A(t_{k+r}))-R_{j,l}(\tau A(t_k))\|_{E\to E}$$

$$\le Mr\tau+M\tau\sum_{s=1}^{\left[\frac{i+r}{2}\right]}\frac{1}{n+r-s}\le M_1r\tau\le M_1\Psi.$$

Combining the estimates for J_{11} and J_{12}, we obtain

$$\|J_{11}\|_{E\to E}\le M_1\Psi.$$

From this and the estimate for J_2 follows estimate (4.79). The proofs of estimates (4.80) and (4.81) follow the scheme of the proof of estimate (4.79). Now let us establish (4.83). Let us denote

$$w^{\tau}(k,n)=A(t_k)\widetilde{u}^{\tau}(k,n)A^{-1}(t_n).$$

By (4.73),

$$w^{\tau}(k,n)=A(t_k)R_{j,l}^{k-n}(\tau A(t_{k-1}))A^{-1}(t_n)$$

$$+\sum_{r=n+1}^{k}A(t_k)R_{j,l}^{k-r+1}(\tau A(t_{k-1}))(R_{j,l}(\tau A(t_{k-1}))$$

$$-R_{j,l}(\tau A(t_{t-1})))A^{-1}(t_{r-1})w^{\tau}(r-1,n)=J_1(k,n)+J_2(k,n),$$

where

$$J_1(k,n) = A(t_k)R_{j,l}^{k-n}(\tau A(t_{k-1}))A^{-1}(t_n),$$

$$J_2(k,n) = \sum_{r=n+1}^{k} A(t_k)R_{j,l}^{k-r+1}(\tau A(t_{k-1}))(R_{j,l}(\tau A(t_{k-1}))$$

$$-R_{j,l}(\tau A(t_{t-1})))A^{-1}(t_{r-1})w^\tau(r-1,n).$$

Applying the identity

$$J_1(k+r,n) - J_1(k,n) = A(t_{k+r})[R_{j,l}^{k+r-n}(\tau A(t_{k+r-1})) - R_{j,l}^{k-n}(\tau A(t_{k-1}))]A^{-1}(t_n)$$

$$+(A(t_{k+r}) - A(t_k))R_{j,l}^{k-n}(\tau A(t_k))A^{-1}(t_n)$$

and estimates (4.64), (4.65), (4.62), we obtain

$$\|J_1(k+r,n) - J_1(k,n)\|_{E\to E}$$

$$\leq \left\|A(t_{k+r})[R_{j,l}^{k+r-n}(\tau A(t_{k+r-1})) - R_{j,l}^{k-n}(\tau A(t_{k-1}))]A^{-1}(t_n)\right\|_{E\to E}$$

$$+ \left\|(A(t_{k+r}) - A(t_k))A^{-1}(t_n)\right\|_{E\to E} \left\|R_{j,l}^{k-n}(\tau A(t_k))\right\|_{E\to E} \leq M\Psi.$$

Now let us estimate $J_2(k+r,n)-J_2(k,n)$. Using estimates (4.65), (4.62), we obtain

$$\|J_2(k,n)\|_{E\to E} \leq M \sum_{r=n+1}^{k} \left\|A(t_k)R_{j,l}^{k-r+1}(\tau A(t_{k-1}))\right\|_{E\to E}$$

$$\times \left\|(R_{j,l}(\tau A(t_{k-1})) - R_{j,l}(\tau A(t_{t-1})))A^{-1}(t_{r-1})\right\|_{E\to E} \|w^\tau(r-1,n)\|_{E\to E}$$

$$\leq M \sum_{r=n+1}^{k} \frac{(k-r)\tau^2}{(k-r+1)\tau} \leq M(k-n)\tau.$$

From this and the triangle estimate it follows that

$$\|J_2(k+r,n) - J_2(k,n)\|_{E\to E} \leq Mr\tau$$

for all $k-n \leq 2r$. Now let $k-n > 2r$. We have that

$$J_2(k+r,n) - J_2(k,n) = \sum_{i=k-r+1}^{k+r} A(t_{k+r})R_{j,l}^{k+r-i+1}(\tau A(t_{k+r-1}))$$

$$\times(R_{j,l}(\tau A(t_{k+r-1})) - R_{j,l}(\tau A(t_{i-1})))A^{-1}(t_{i-1})w^\tau(i-1,n)$$

$$- \sum_{i=k-r+1}^{k} A(t_k)R_{j,l}^{k-i+1}(\tau A(t_{k-1}))(R_{j,l}(\tau A(t_{k-1}))$$

$$-R_{j,l}(\tau A(t_{i-1})))A^{-1}(t_{i-1})w^\tau(i-1,n)$$

$$+\sum_{i=n+1}^{k-r}(A(t_{k+r})R_{j,l}^{k+r-i+1}(\tau A(t_{k+r-1}))-A(t_{k-1})R_{j,l}^{k-i+1}(\tau A(t_{k-1})))$$
$$\times(R_{j,l}(\tau A(t_{k-1}))-R_{j,l}(\tau A(t_{i-1})))A^{-1}(t_{i-1})w^{\tau}(i-1,n)$$
$$+\sum_{i=n+1}^{k-r}A(t_{k+r})R_{j,l}^{k+r-i+1}(\tau A(t_{k+r-1}))(R_{j,l}(\tau A(t_{k+r-1}))$$
$$-R_{j,l}(\tau A(t_{k-1})))A^{-1}(t_{i-1})w^{\tau}(i-1,n)=\sum_{n=1}^{4}L_n,$$

where

$$L_1=\sum_{i=k-r+1}^{k+r}A(t_{k+r})R_{j,l}^{k+r-i+1}(\tau A(t_{k+r-1}))(R_{j,l}(\tau A(t_{k+r-1}))$$
$$-R_{j,l}(\tau A(t_{i-1})))A^{-1}(t_{i-1})w^{\tau}(i-1,n),$$
$$L_2=-\sum_{i=k-r+1}^{k}A(t_k)R_{j,l}^{k-i+1}(\tau A(t_{k-1}))(R_{j,l}(\tau A(t_{k-1}))$$
$$-R_{j,l}(\tau A(t_{i-1})))A^{-1}(t_{i-1})w^{\tau}(i-1,n),$$
$$L_3=\sum_{i=n+1}^{k-r}(A(t_{k+r})R_{j,l}^{k+r-i+1}(\tau A(t_{k+r-1}))-A(t_{k-1})R_{j,l}^{k-i+1}(\tau A(t_{k-1})))$$
$$\times(R_{j,l}(\tau A(t_{k-1}))-R_{j,l}(\tau A(t_{i-1})))A^{-1}(t_{i-1})w^{\tau}(i-1,n),$$
$$L_4=\sum_{i=n+1}^{k-r}A(t_{k+r})R_{j,l}^{k+r-i+1}(\tau A(t_{k+r-1}))(R_{j,l}(\tau A(t_{k+r-1}))$$
$$-R_{j,l}(\tau A(t_{k-1})))A^{-1}(t_{i-1})w^{\tau}(i-1,n).$$

Using estimates (4.65), (4.62) and (4.75), we obtain

$$\|L_1\|_{E\to E}\le M\sum_{i=k-r+1}^{k+r}\left\|A(t_{k+r})R_{j,l}^{k+r-i+1}(\tau A(t_{k+r-1}))\right\|_{E\to E}$$
$$\times\left\|(R_{j,l}(\tau A(t_{k+r-1}))-R_{j,l}(\tau A(t_{i-1})))A^{-1}(t_{i-1})\right\|_{E\to E}\|w^{\tau}(i-1,n)\|_{E\to E}$$
$$\le M\sum_{i=k-r+1}^{k+r}\frac{(k+r-i)\tau^2}{(k+r-i+1)\tau}\le M_1 r\tau.$$

In a similar manner one establishes the estimate

$$\|L_2\|_{E\to E}\le M_1 r\tau.$$

Now let us estimate L_3. Using estimates (4.82), (4.62) and (4.75), we obtain

$$\|L_3\|_{E\to E}$$

$$\leq \sum_{i=1}^{k-r} \left\|A(t_{k+r})R_{j,l}^{k+r-i+1}(\tau A(t_{k+r-1})) - A(t_{k-1})R_{j,l}^{k-i+1}(\tau A(t_{k-1}))\right\|_{E\to E}$$

$$\times \left\|(R_{j,l}(\tau A(t_{k-1})) - R_{j,l}(\tau A(t_{i-1})))A^{-1}(t_{i-1})\right\|_{E\to E} \|w^\tau(i-1,n)\|_{E\to E}$$

$$\leq M\sum_{i=1}^{k-r}(\frac{r\tau}{(k+r-i+1)\tau} + \frac{r\tau}{((k-i)\tau)^2})(k-i)\tau^2 \leq M_1 r\tau.$$

Finally let us estimate L_4. Using estimates (4.65), (4.62) and (4.75), we obtain

$$\|L_4\|_{E\to E} \leq \sum_{i=n+1}^{k-r} \left\|A(t_{k+r})R_{j,l}^{k+r-i+1}(\tau A(t_{k+r-1}))\right\|_{E\to E}$$

$$\times \left\|(R_{j,l}(\tau A(t_{k+r-1})) - R_{j,l}(\tau A(t_{k-1})))A^{-1}(t_{i-1})\right\|_{E\to E} \|w^\tau(i-1,n)\|_{E\to E}$$

$$\leq M \sum_{i=n+1}^{k-r} \frac{r\tau\tau}{(k+r-i+1)\tau} \leq M_1 r\tau(1 + |\ln(r\tau)|).$$

Thus, the operator L_4 has "worse" properties than the operators L_n for $n = 1, 2, 3$. Combining the estimates for L_n for $n = 1, 2, 3, 4$, we obtain

$$\|w^\tau(k+r,n) - w^\tau(k,n)\|_{E\to E} \leq M(\alpha)(r\tau)^\alpha((k-n)\tau)^{-\alpha}$$

for all $\alpha < 1$. Nevertheless this estimate permits us to establish the same estimate for L_4. Indeed, using the formula

$$L_4 = A(t_{k+r})(I - R_{j,l}(\tau A(t_{k+r-1})))^{-1}(R_{j,l}^{2r}(\tau A(t_{k+r-1}))$$

$$-R_{j,l}^{k+r-n}(\tau A(t_{k+r-1})))(R_{j,l}(\tau A(t_{k+r-1}))$$

$$-R_{j,l}(\tau A(t_{k-1})))A^{-1}(t_{k-1})w^\tau(k-1,n) + \sum_{i=n+1}^{k-r} A(t_{k+r})R_{j,l}^{k+r-i+1}(\tau A(t_{k+r-1}))$$

$$\times(R_{j,l}(\tau A(t_{k+r-1})) - R_{j,l}(\tau A(t_{k-1})))(A^{-1}(t_{i-1}) - A^{-1}(t_{k-1}))w^\tau(i-1,n)$$

$$+ \sum_{i=n+1}^{k-r} A(t_{k+r})R_{j,l}^{k+r-i+1}(\tau A(t_{k+r-1}))(R_{j,l}(\tau A(t_{k+r-1}))$$

$$-R_{j,l}(\tau A(t_{k-1})))A^{-1}(t_{k-1})(w^\tau(k-1,n) - w^\tau(i-1,n))$$

and estimates (4.65), (4.62) and (4.75), we obtain

$$\|L_4\|_{E\to E} \leq Mr\tau + M\sum_{i=n+1}^{k-r}\frac{r\tau}{(k+r-i)\tau}(k-i)\tau\tau$$

$$+M(\alpha)\sum_{i=n+1}^{k-r}\frac{r\tau}{(k+r-i)\tau}\frac{((k-i)\tau)^{\alpha}}{((i+1-n)\tau)^{\alpha}}\tau \leq M_2 r\tau.$$

Now combining the estimates for L_n for $n = 1,2,3,4$, we obtain estimate (4.83). In a similar manner one establishes estimate (4.82). Finally let us establish (4.84). Estimate (4.84) for $k-n \leq 2$ follows from estimates (4.76) and (4.77). Let $k-n > 2$. Then using the formula

$$A(t_{k+r})\widetilde{u}^{\tau}(k+r,n) - A(t_k)\widetilde{u}^{\tau}(k,n) = (w^{\tau}(k+r,[\frac{k-n}{2}])$$

$$-w^{\tau}(k,[\frac{k-n}{2}]))A([\frac{k-n}{2}]\tau)\widetilde{u}^{\tau}([\frac{k-n}{2}],n)$$

and estimates (4.82) and (4.76), we obtain estimate (4.84) for $k-n > 2$. Lemma 4.2.19 is proved.

Suppose that the assumption (4.62) holds for $r = 0$ and $r = 1$. Then

$$\|u^{\tau}(k,k-1) - R_{j,l}(\tau A(t_{k-1}))\|_{E\to E} \leq M\tau,$$

$$\left\|[u^{\tau}(k,k-1) - R_{j,l}(\tau A(t_{k-1}))]A^{-1}(t_{k-1})\right\|_{E\to E} \leq M\tau^2.$$

Now using the formulas

$$u^{\tau}(k,n) - \widetilde{u}^{\tau}(k,n) = \sum_{i=n}^{k-1} u^{\tau}(k,i+1)(u^{\tau}(i+1,i) - \widetilde{u}^{\tau}(i+1,i))\widetilde{u}^{\tau}(i,n),$$

$$u^{\tau}(k,n) - \widetilde{u}^{\tau}(k,n) = \sum_{i=n}^{k-1} \widetilde{u}^{\tau}(k,i+1)(u^{\tau}(i+1,i) - \widetilde{u}^{\tau}(i+1,i))u^{\tau}(i,n)$$

and the last estimates and estimates of Lemma 4.2.18 and Lemma 4.2.19, we can establish the estimates for $u^{\tau}(k,n)$.

Lemma 4.2.20. *Suppose that the assumption* (4.62) *holds for* $r = 0$ *and* $r = 1$. *Then for any* $0 \leq n < r \leq k \leq N$, $0 \leq \alpha \leq 1$ *the following estimates hold:*

$$\|u^{\tau}(k,n)\|_{E\to E} \leq M, \left\|A(t_k)u^{\tau}(k,n)A^{-1}(t_n)\right\|_{E\to E} \leq M,$$

$$\|A(t_k)u^{\tau}(k,n)\|_{E\to E} \leq \frac{M}{(k-n)\tau},$$

$$\|u^{\tau}(k+r,n)-u^{\tau}(k,n)\|_{E\to E}\leq M\Psi,$$

$$\left\|[A(t_{k+r})u^{\tau}(k+r,n)-A(t_k)u^{\tau}(k,n)]A^{-1}(t_n)\right\|_{E\to E}\leq M\Psi,$$

$$\|A(t_{k+r})u^{\tau}(k+r,n)-A(t_k)u^{\tau}(k,n)\|_{E\to E}\leq\frac{M\Psi}{(k-n)\tau},$$

where $M\geq 0$ does not depend on τ, k, r, n. Here $\Psi=r^{\alpha}(k-n)^{-\alpha}$.

From formula (4.68) and the estimates of Lemma 4.2.20 one derives the following results.

Theorem 4.2.21. *The difference problem* (4.63) *is stable in $C_{\tau}(E)$.*

Theorem 4.2.22. *The difference problem* (4.63) *is stable in $C_{\tau}^{\alpha,\alpha}(E)$.*

Theorem 4.2.23. *The difference problem* (4.63) *is stable in $L_{p,\tau}(E)$, $1\leq p<\infty$.*

Theorem 4.2.24. *The solutions of the difference problem* (4.63) *in $C_{\tau}(E)$ obey the almost coercive inequality*

$$\|\{\tau^{-1}(u_k-u_{k-1})\}_1^N\|_{C_{\tau}(E)}+\|\{\tau^{-1}(I-\widetilde{u}^{\tau}(k,k-1))u_{k-1}\}_1^N\|_{C_{\tau}(E)}$$

$$\leq M_1[\|A(0)u_0\|_E+\min\{\ln\frac{1}{\tau},1+|\ln\|A(0)\|_{E\to E}|\}\|f^{\tau}\|_{C_{\tau}(E)}],$$

where M_1 is independent not only of f^{τ}, u_0, but also of τ.

The proof of Theorem 4.2.24 follows the scheme of the proof of Theorem 4.2.4 and relies on the estimates of Lemma 4.2.20.

Now let us study the well-posedness of the difference problem (4.63) in the various Banach spaces.

Theorem 4.2.25. *The solutions of the difference problem* (4.63) *in $C_{\tau}^{\alpha,\alpha}(E)$ obey the coercivity inequality*

$$\|\{\tau^{-1}(u_k-u_{k-1})\}_1^N\|_{C_{\tau}^{\alpha,\alpha}(E)}+\|\{\tau^{-1}(I-u^{\tau}(k,k-1))u_{k-1}\}_1^N\|_{C_{\tau}^{\alpha,\alpha}(E)}$$

$$\leq M_1[\frac{1}{\alpha(1-\alpha)}\|f^{\tau}\|_{C_{\tau}^{\alpha,\alpha}(E)}+\|A(0)u_0\|_E],$$

where M_1 is independent not only of f^{τ}, u_0, α, but also of τ.

The proof of Theorem 4.2.25 follows the scheme of the proof of Theorem 4.2.5 and relies on the estimates of Lemma 4.2.20 and on the formula

$$\tau^{-1}(I-u^{\tau}(k,k-1))u_{k-1}=\tau^{-1}(I-u^{\tau}(k,k-1))u^{\tau}(k-1,0))u_0$$

$$+(I-R_{j,l}^{k-1}\{-\tau A(t_{k-1})\})f_k^{j,l}$$

$$+\sum_{i=1}^{k-1}[(I-u^{\tau}(k,k-1))u^{\tau}(k-1,i)-(I-R_{j,l}\{-\tau A(t_{k-1})\})R_{j,l}^{k-i-1}\{-\tau A(t_{k-1})\}]f_k^{j,l}$$

$$+\sum_{i=1}^{k-1}(I-u^{\tau}(k,k-1))u^{\tau}(k-1,i)(f_i^{j,l}-f_k^{j,l}).$$

Now let us give the definition of positive operators and introduce the fractional spaces that will be needed in the sequel.

The operator A is said to be strongly positive if its spectrum $\sigma(A)$ lies in the interior of the sector of angle ϕ, $0<2\phi<\pi$, symmetric with respect to the real axis and if on the edges of this sector, $S_1(\phi)=\{\rho e^{i\phi}:0\leq\rho\leq\infty\}$ and $S_2(\phi)=\{\rho e^{-i\phi}:0\leq\rho\leq\infty\}$ and outside of the resolvent $(\lambda-A)^{-1}$ is subject to the bound

$$\left\|(\lambda-A)^{-1}\right\|_{E\to E}\leq\frac{M(\phi)}{1+|\lambda|}. \tag{4.85}$$

The infimum of all such angles ϕ is called the spectral angle of the strongly positive operator A and is denoted by $\phi(A)=\phi(A,E)$. Since the spectrum $\sigma(A)$ is a closed set, it lies inside the sector formed by the rays $S_1(\phi(A))$ and $S_2(\phi(A))$ and some neighborhood of the apex of this sector does not intersect $\sigma(A)$. We shall consider contours $\Gamma=\Gamma(\phi,r)$ composed by the rays $S_1(\phi)$, $S_2(\phi)$ and an arc of circle of radius r centered at the origin; ϕ and r will be chosen so that $\sigma(A)<|\sigma|<\pi/2$ and the arc of circle of radius r lies in the resolvent set $\rho(A)$ of the operator A.

Let $f(z)$ be an analytic function on the set bounded by such a contour Γ and suppose that f satisfies estimate

$$|f(z)|\leq M|z|^{-\varepsilon}$$

for some $\varepsilon>0$. Then the operator Cauchy–Riesz integral

$$f(A)=\frac{1}{2\pi i}\int_{\Gamma}f(z)(z-A)^{-1}dz \tag{4.86}$$

converges in the operator norm and defines a bounded linear operator $f(A)$, a function of the strongly positive operator A. If $f(z)$ is continuous in a neighborhood of the origin, then in (4.86) we shall consider that $r=0$, i.e., $\Gamma=S_1(\phi)\cup S_2(\phi)$.

As in the case of a bounded operator A one shows that $f(A)$ does not depend on the choice of the contour Γ in the domain of analyticity of the function $f(z)$ and that the correspondence between the function $f(z)$ and the operator $f(A)$ is linear and multiplicative.

The function $f(z)=z^{-\alpha}$ defines a bounded operator $A^{-\alpha}$ whenever $\alpha>0$. Here the contour Γ is chosen with $r>0$. By the multiplicativity property, $A^{-(\alpha+\beta)}=A^{-\alpha}A^{-\beta}=A^{-\beta}A^{-\alpha}$ for any powers of the strongly positive operator A and not only for negative integer ones. From this identity it follows (when $\alpha+\beta$ is an integer) that the equation $A^{-\alpha}x=0$ has the unique solution $x=0$.

Hence, the positive powers $A^\alpha = \left(A^{-\alpha}\right)^{-1}$ of the strongly positive operator are defined. The operators A^α $(\alpha > 0)$ are unbounded if A is unbounded; they have dense domains $D(A^\alpha)$ and one has the continuous embeddings $D(A^\alpha) \subset D\left(A^\beta\right)$ if $\beta < \alpha$.

The theory of fractional powers of operators can be constructed for a wider class of positive operators. For such operators the estimate (4.85) is required to hold for some ϕ and not only from the interval $[0, \pi/2]$, but from the larger interval $[0, \pi)$.

Now let us consider the function $f(z) = e^{-tz}$. For any $t > 0$ this function tends to zero faster than any power $z^{-\alpha}$ as $|z| \to \infty$ and its values lie inside any sector bounded by a contour Γ. Therefore, formula (4.86) can be used to define the function $\exp\{-tA\}$ of the strongly positive operator A. By multiplicativity, the semigroup property holds:

$$\exp\{-(t_1 + t_2) A\} = \exp\{-t_1 A\} \exp\{-t_2 A\}, \; t_1, t_2 > 0.$$

Consider the function $\Psi(z) = z^\alpha e^{-tz}$ for some $\alpha > 0$ and $t > 0$. Since, obviously, $\Psi(z) \to 0$ faster than any negative power of z as $|z| \to \infty$, $\Psi(z)$ defines the operator function

$$\Psi(A) = \frac{1}{2\pi i} \int_\Gamma z^\alpha e^{-tz} (z - A)^{-1} dz. \tag{4.87}$$

Let us show that the operator $\exp\{-tA\}$ maps E into $D(A^\alpha)$ and $A^\alpha \exp\{-tA\}$ $= \Psi(A)$. Let x be an arbitrary element of E. By the multiplicativity property, (4.69) implies that

$$A^{-\alpha}\Psi(A)x = \frac{1}{2\pi i} \int_\Gamma e^{-tz} (z - A)^{-1} x dz = \exp\{-tA\} x,$$

which proves our assertion. Thus, we have the formula

$$A^\alpha \exp\{-tA\} = \frac{1}{2\pi i} \int_\Gamma z^\alpha e^{-tz} (z - A)^{-1} dz. \tag{4.88}$$

In the above argument we must assume that the contour Γ contains an arc of radius r, since we applied the operator $A^{-\alpha}$, which corresponds to the function $z^{-\alpha}$. The final formula (4.88) is valid for any (small) $r > 0$. Since the integrand in (4.88) is continuous at the point $z = 0$, letting $z \to 0$ we obtain the formula

$$A^\alpha \exp\{-tA\} = \frac{1}{2\pi i} [\int_\infty^0 \rho^\alpha e^{i\alpha\phi} e^{-t\rho e^{i\phi}} \left(\rho e^{i\phi} - A\right)^{-1} d\rho$$

$$+ \int_0^\infty \rho^\alpha e^{-i\alpha\phi} e^{-t\rho e^{-i\phi}} \left(\rho e^{-i\phi} - A\right)^{-1} d\rho]$$

for some $0 < \phi < \pi/2$. From this and estimate (4.85) it follows that

$$\|A^{\alpha}\exp\{-tA\}\|_{E\to E} \le \frac{M(\phi)}{\pi}\int_0^{\infty}\rho^{\alpha-1}e^{-t\rho\cos\phi}d\rho = \frac{M(\phi)\Gamma(\alpha)}{\pi(\cos\phi)^{\alpha}}t^{-\alpha}. \quad (4.89)$$

In particular, we have the estimate

$$\|\exp\{-tA\}\|_{E\to E} \le \frac{M(\phi)}{\pi}. \quad (4.90)$$

Let us show that the estimate (4.90) can be sharpened by a factor that decays exponentially when $t \to +\infty$.

Let A be a strongly positive operator. We claim that for sufficiently small $\delta > 0$ the operator $A - \delta$ is also strongly positive and $\phi(A-\delta) = \phi(A)$. Indeed, let $\lambda \in \Gamma(\phi)$. Consider the equation $\lambda x - (A-\delta)x = y$ for an arbitrary $y \in E$. The substitution $\lambda x - Ax = z$ yields the equation $z + \delta(\lambda - A)^{-1}z = y$. Since

$$\left\|\delta(\lambda - A)^{-1}\right\|_{E\to E} \le \delta M(\phi)$$

if $\lambda \in \Gamma(\phi)$, we see that for $\delta \le [2M(\phi)]^{-1}$the equation for z has a unique solution and $\|z\| \le 2\|y\|$. Consequently, the equation for x has a unique solution and

$$\|x\| \le M(\phi)[|\lambda|+1]^{-1}\|z\| \le 2M(\phi)[|\lambda|+1]^{-1}\|y\|.$$

This means that the operator $\lambda - (A+\delta)$ has a bounded inverse for $0 < \delta \le [2M(\phi)]^{-1}$ and

$$\left\|[\lambda - (A-\delta)]^{-1}\right\|_{E\to E} \le 2M(\phi)[|\lambda|+1]^{-1}.$$

Thus, we have shown that $A-\delta$ is a strongly positive operator. Hence, by (4.90), we have the estimate

$$\|\exp\{-(A-\delta)t\}\|_{E\to E} \le \frac{2M(\phi)}{\pi}.$$

This obviously yields

$$\|\exp\{-At\}\|_{E\to E} \le \frac{2M(\phi)}{\pi}e^{-\delta t}, \quad (4.91)$$

where we can put $\delta = [2M(\phi)]^{-1}$.

Let $t > 1$. Then, using the semigroup property, we can write

$$\exp\{-tA\} = \exp\{-A\}\exp\{-(t-1)A\}.$$

Next, applying estimates (4.89) with $t = 1$ and (4.91), we obtain

$$\|A^{\alpha}\exp\{-tA\}\|_{E\to E} \le \frac{M(\phi)\Gamma(\alpha)}{\pi(\cos\phi)^{\alpha}}\frac{2M(\phi)}{\pi}e^{-\delta(t-1)}.$$

Hence, the following estimate holds for $t > 1$:

$$\|A^{\alpha} \exp\{-tA\}\|_{E \to E} \leq M_1(\phi) e^{-\delta t}.$$

If $0 < t \leq 1$, then estimate (4.91) prevails. Combining these two estimates, we conclude that

$$\|A^{\alpha} \exp\{-tA\}\|_{E \to E} \leq \tilde{M}(\phi) e^{-\delta t} t^{-\alpha} \tag{4.92}$$

for some $\tilde{M}(\phi) > 0$ and $\delta > 0$.

Further, formula (4.86) allows us to establish that the operator-valued function $\exp\{-tA\}$ is differentiable in the operator norm for $t > 0$ and

$$\frac{d}{dt} \exp\{-tA\} = -A \exp\{-tA\}. \tag{4.93}$$

In particular, this implies that $\exp\{-tA\}$ is continuous in the operator norm. Using the semigroup property we deduce that the derivative of $\exp\{-tA\}$ is also continuous in the operator norm for $t > 0$. Finally, formula (4.93) shows that the operator-valued function $\exp\{-tA\}$ has derivatives of arbitrary order in the operator norm for $t > 0$.

Now let $x \in D(A)$. Then the (E-valued) function $\exp\{-tA\} x$ has a derivative for $t > 0$ and, by (4.93),

$$\frac{d}{dt} \exp\{-tA\} x = -\exp\{-tA\} Ax. \tag{4.94}$$

Next, for x as above we can write

$$(z - A)^{-1} x = z^{-1} x + z^{-1} (z - A)^{-1} Ax.$$

Using formula (4.86), we obtain

$$\exp\{-tA\} x = \frac{1}{2\pi i} \int_{\Gamma} e^{-tz} [z^{-1} x + z^{-1} (z - A)^{-1} Ax] dz.$$

Here the contour Γ has the form

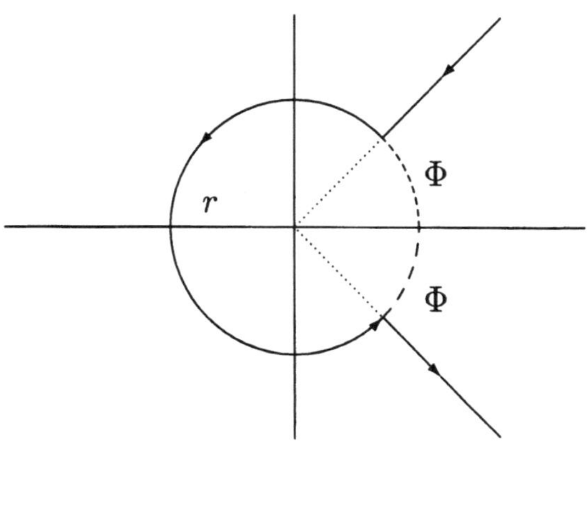

Using the Cauchy theorem, we get

$$\exp\{-tA\}x = \frac{1}{2\pi i}\int_\Gamma e^{-tz}z^{-1}(z-A)^{-1}Axdz + x.$$

Estimate (4.85) shows that in the last equality one can pass to the limit under the integral sign when $t \to +0$. Hence, the limit

$$\lim_{t\to+0}\exp\{-tA\}x = x + \frac{1}{2\pi i}\int_\Gamma z^{-1}(z-A)^{-1}Axdz$$

exists (in the norm of E). By Cauchy's theorem, the integral

$$\vartheta = \frac{1}{2\pi i}\int_\Gamma z^{-1}(z-A)^{-1}Axdz = \frac{1}{2\pi i}\int_{-\sigma-i\infty}^{-\sigma+i\infty} z^{-1}(z-A)^{-1}Axdz$$

for some $\sigma > 0$. Hence, by (4.85),

$$\|\vartheta\|_E \le \frac{M}{2\pi}\int_{-\infty}^{\infty}\frac{dt}{\sigma^2+t^2}\|Ax\|_E.$$

Since ϑ does not depend on σ, it follows that $\vartheta \equiv 0$. Hence, we proved that

$$\lim_{t\to+0}\exp\{-tA\}x = x \tag{4.95}$$

for any $x \in D(A)$. Since the norm $\|\exp\{-tA\}\|_{E\to E}$ is uniformly bounded for $t > 0$, the limit relation (4.95) holds for any $x \in E$.

Thus, if we extend the operator-valued function $U(t) = \exp\{-tA\}$, $t > 0$, at $t = 0$ by $U(0) = I$, we obtain a strongly continuous semigroup. From estimate (4.92) (with $\alpha = 0$) it follows that this semigroup is analytic. Finally, let us show that its generator is $U'(0) = -A$. From (4.94) and estimate (4.91) we derive the identity

$$U(t)x - x = -\int_0^t U(s)Axds$$

for $x \in D(A)$. Since $U(t)$ is strongly continuous to the left at the point $t = 0$, this implies that $x \in D(U'(0))$ and $U'(0)x = -Ax$. Hence, $U'(0)$ is an extension of the operator $-A$. By estimate (4.91), the operator $U'(0) + \lambda$ and $-A + \lambda$ have bounded inverses for any $\lambda < 0$. Therefore, $U'(0) = -A$.

We have shown that the operator-valued function $\exp\{-tA\}$ is an analytic semigroup with generator $-A$ and with an exponentially decaying norm. Operators $-A$ that generate such semigroups were called strongly positive operators.

Let A be a strongly positive operator. With the help of A we introduce the fractional space $E'_{\alpha,q}(E,A), 0 < \alpha < 1$, consisting of all $v \in E$ for which the following norms are finite:

$$\|v\|'_{\alpha,q} = \left(\int_0^\infty \left\|\lambda^\alpha A(\lambda+A)^{-1}v\right\|_E^q \frac{d\lambda}{\lambda}\right)^{\frac{1}{q}}, 1 \le q < \infty,$$

$$\|v\|'_\alpha = \|v\|'_{\alpha,\infty} = \sup_{\lambda>0}\left\|\lambda^\alpha A(\lambda+A)^{-1}v\right\|_E.$$

Recall that in Section 4.1 of Chapter 4 we introduced the fractional spaces $E_{\alpha,q}$ in which the norms are defined with the help of the semigroup $\exp\{-tA\}$ generated by a strongly positive operator A.

Theorem 4.2.26. *$E'_{\alpha,q}(E,A(t)) = E_{\alpha,q}(E,A(t))$ for all $0 \le t \le T, 0 < \alpha < 1$ and $1 \le q \le \infty$.*

Now let us consider the well-posedness of the difference problem (4.63) in the various Banach spaces.

Theorem 4.2.27. *The solutions of the difference problem (4.63) in $C_\tau(E'_\alpha)$ obey the coercivity inequality*

$$\|\{\tau^{-1}(u_k - u_{k-1})\}_1^N \|_{C_\tau(E'_\alpha)} + \| \{\tau^{-1}(I - u^\tau(k,k-1))u_{k-1}\}_1^N \|_{C_\tau(E'_\alpha)}$$

$$\le M_1[\frac{1}{\alpha(1-\alpha)}\| f^\tau \|_{C_\tau(E'_\alpha)} + \| A(0)u_0\|_{E'_\alpha}],$$

where M_1 is independent not only of f^τ, u_0, α, but also of τ.

The proof of Theorem 4.2.27 follows the scheme of the proof of Theorem 4.2.6 and relies on the estimates of Lemma 4.2.20 and on the formula

$$\tau^{-1}(I - u^\tau(k,k-1))u_{k-1} = \tau^{-1}(I - u^\tau(k,k-1))u^\tau(k-1,0)u_0$$

$$+\sum_{i=1}^{k-1}(I - R_{j,l}\{-\tau A(t_{k-1})\})R_{j,l}^{k-i-1}\{-\tau A(t_{k-1})\}]f_i^{j,l}$$

$$+\sum_{i=1}^{k-1}[(I - u^\tau(k,k-1))u^\tau(k-1,i) - (I - R_{j,l}\{-\tau A(t_{k-1})\})$$

$$\times R_{j,l}^{k-i-1}\{-\tau A(t_{k-1})\}]f_k^{j,l}.$$

Let us give, without proof, the following results.

Theorem 4.2.28. *The solutions of the difference problem (4.63) in $C_\tau^{\beta,\gamma}(E)$ $(0 \le \gamma \le \beta, 0 < \beta < 1)$ satisfy the following coercivity inequalities:*

$$\| \{\tau^{-1}(u_k - u_{k-1})\}_1^N \|_{C_\tau(E'_{\beta-\gamma})} \le \frac{M_1}{\beta(1-\beta)}\| f^\tau \|_{C_\tau^{\beta,\gamma}(E)}$$

$$+M_1\| (I - R_{j,l}(\tau A(0)))\tau^{-1}u_0 - f_1^{j,l} \|_{E'_{\beta-\gamma}},$$

$$\| \{\tau^{-1}(u_k - u_{k-1})\}_1^N \|_{C_\tau^{\beta,\gamma}(E)} + \| \{\tau^{-1}(I - u^\tau(k,k-1))u_{k-1}\}_1^N \|_{C_\tau^{\beta,\gamma}(E)}$$

$$\le \frac{M_1}{\beta(1-\beta)}\| f^\tau \|_{C_\tau^{\beta,\gamma}(E)} + M_1|(I - R_{j,l}(\tau A(0)))\tau^{-1}u_0 - f_1^{j,l}|_0^{\beta,\gamma},$$

where M_1 is independent not only of $f^\tau, u_0, \beta, \gamma$ but also of τ.

Here, $|w|_0^{\beta,\gamma}$ denotes the norm of the Banach space $E_0^{\beta,\gamma}$ and consists of those $w \in E$ for which the norm

$$|w|_0^{\beta,\gamma} = \max_{0\le i\le N} ||R_{j,l}^{i-1}(\tau A(t))w||_E$$

$$+ \sup_{0\le i<i+r\le N} (r\tau)^{-\beta}((i+r)\tau)^{\gamma}||(R_{j,l}^{i+r-1}(\tau A(t)) - R_{j,l}^{i-1}(\tau A(t)))w||_E$$

is finite.

Note that the parameter γ can be chosen freely in $[0,\beta)$, which increases the number of spaces $C_\tau^{\beta,\gamma}(E)(0 \le \gamma \le \beta, 0 < \beta < 1)$ of grid functions in which difference problem (4.63) is well posed.

Theorem 4.2.29. *The solutions of the difference problem* (4.63) *in* $C_\tau^{\beta,\gamma}(E'_{\alpha-\beta})$ $(0 \le \gamma \le \beta \le \alpha < 1)$ *satisfy the following coercivity inequalities:*

$$\| \{\tau^{-1}(u_k - u_{k-1})\}_1^N \|_{C_\tau^{\beta,\gamma}(E'_{\alpha-\beta})} + \| \{\tau^{-1}(u_k - u_{k-1})\}_1^N \|_{C_\tau(\tilde{E}^{\beta,\gamma}_{\alpha-\beta})}$$

$$+ \| \{\tau^{-1}(I - u^\tau(k,k-1))u_{k-1}\}_1^N \|_{C_\tau^{\beta,\gamma}(E'_{\alpha-\beta})}$$

$$\le \frac{M_1}{\alpha(1-\alpha)} \| f^\tau \|_{C_\tau^{\beta,\gamma}(E'_{\alpha-\beta})} + M_1|(I - R_{j,l}(\tau A(0)))\tau^{-1}u_0 - f_1^{j,l}|_{\alpha-\beta}^{\beta,\gamma},$$

$$\| \{\tau^{-1}(u_k - u_{k-1})\}_1^N \|_{C_\tau^{\beta,\gamma}(E'_{\alpha-\beta})} + \| \{\tau^{-1}(u_k - u_{k-1})\}_1^N \|_{C_\tau(E'_{\alpha-\gamma})}$$

$$+ \| \{\tau^{-1}(I - u^\tau(k,k-1))u_{k-1}\}_1^N \|_{C_\tau^{\beta,\gamma}(E'_{\alpha-\beta})}$$

$$\le \frac{M_1}{\alpha(1-\alpha)} \| f^\tau \|_{C_\tau^{\beta,\gamma}(E'_{\alpha-\beta})} + M_1||(I - R_{j,l}(\tau A(0)))\tau^{-1}u_0 - f_1^{j,l}||_{E'_{\alpha-\gamma}},$$

where M_1 is independent not only of f^τ, u_0, α, β, γ but also of τ.

Here, $|w|_{\alpha-\beta}^{\beta,\gamma}$ denotes that the norm of the Banach space $E_{\alpha-\beta}^{\beta,\gamma}$ consists of those $w \in E$ for which the norm

$$|w|_0^{\beta,\gamma} = \max_{0\le i\le N} ||R_{j,l}^{i-1}(\tau A(t))w||_{E_{\alpha-\beta}}$$

$$+ \sup_{0\le i<i+r\le N} (r\tau)^{-\beta}((i+r)\tau)^{\gamma}||(R_{j,l}^{i+r-1}(\tau A(t)) - R_{j,l}^{i-1}(\tau A(t)))w||_{E_{\alpha-\beta}}$$

is finite.

Note that the spaces $C_\tau^{\beta,\gamma}(E'_{\alpha-\beta})$ of grid functions, in which coercive solvability has been established, depend on the parameters α, β and γ. However, the constants in the coercive inequalities depend only on α. Hence, we can choose the parameters β and γ freely, which increases the number of spaces of grid functions in which difference problem (4.63) is well posed.

Now let us consider the initial-value difference problem (4.63) in the spaces $L_{p,\tau}(E) = L_p([0,T]_\tau, E)$, $1 \le p < \infty$ of all grid functions. We have not been able to obtain the coercivity inequality

$$\| \{\tau^{-1}(u_k - u_{k-1})\}_1^N \|_{L_{p,\tau}(E)} + \| \{\tau^{-1}(I - u^\tau(k,k-1))u_{k-1}\}_1^N \|_{L_{p,\tau}(E)} \le M[\| f^\tau \|_{L_{p,\tau}(E)} + \| A(0)u_0 \|_E],$$

in the arbitrary Banach space E and for the general strong positive operator $A(t)$. Nevertheless, we can establish the almost coercivity inequality.

Theorem 4.2.30. *The solutions of the difference problem* (4.63) *in* $L_{p,\tau}(E)$ *obey the almost coercive inequality*

$$\| \{\tau^{-1}(u_k - u_{k-1})\}_1^N \|_{L_{p,\tau}(E)} + \| \{\tau^{-1}(I - u^\tau(k,k-1))u_{k-1}\}_1^N \|_{L_{p,\tau}(E)}$$
$$\le M_1[\| A(0)u_0 \|_E + \min\left\{\ln\frac{1}{\tau}, 1 + |\ln \| A(0) \|_{E \to E}|\right\} \| f^\tau \|_{L_{p,\tau}(E)}],$$

where M_1 does not depend on f^τ, u_0, p and τ.

Note that if $A(0)$ is the unbounded positive operator in E, then

$$\min\left\{\ln\frac{1}{\tau}, 1 + |\ln \| A(0) \|_{E \to E}|\right\} = \ln\frac{1}{\tau}.$$

Finally, let us give, without proof, the following results about well-posedness of the initial-value difference problem (4.63) in the spaces $L_{p,\tau}(E) = L_p([0,T]_\tau, E)$, $1 \le p < \infty$.

Theorem 4.2.31. *Suppose that the difference problem* (4.63) *is well posed in* $L_{p_0,\tau}(E)$ *for some p_0, $1 < p_0 < \infty$. Then it is well posed in $L_{p,\tau}(E)$ for all $p, 1 < p < \infty$ and the following coercivity inequality holds:*

$$\| \{\tau^{-1}(u_k - u_{k-1})\}_1^N \|_{L_{p,\tau}(E)} + \| \{\tau^{-1}(I - u^\tau(k,k-1))u_{k-1}\}_1^N \|_{L_{p,\tau}(E)}$$
$$+ \|u^\tau\|_{C_\tau(E_{1-1/p,p})} \le M(p_0)[\| u_0 \|_{E_{1-1/p,p}} + \frac{p^2}{p-1} \| f^\tau \|],$$

where $M(p_0)$ does not depend on f^τ, u_0, p and τ.

Theorem 4.2.32. *Let $1 \le p \le \infty$ and $0 < \alpha < 1$. The problem* (4.63) *is well posed in $L_{p,\tau}(E'_{\alpha,p})$ and the following coercivity inequality holds:*

$$\| \{\tau^{-1}(u_k - u_{k-1})\}_1^N \|_{L_{p,\tau}(E'_{\alpha,p})} + \| \{\tau^{-1}(I - u^\tau(k,k-1))u_{k-1}\}_1^N \|_{L_{p,\tau}(E'_{\alpha,p})}$$
$$\le M[\| A(0)u_0 \|_{E'_{\alpha,p}} + \frac{1}{\alpha(1-\alpha)} \| f^\tau \|_{L_{p,\tau}(E'_{\alpha,p})}],$$

where M does not depend on f^τ, u_0, p, α and τ.

Theorem 4.2.33. *Let $1 < p, q < \infty$ and $0 < \alpha < 1$. The difference problem (4.63) is well posed in $L_{p,\tau}(E'_{\alpha,q})$ and the following coercivity inequality holds:*

$$\| \{\tau^{-1}(u_k - u_{k-1})\}_1^N \|_{L_{p,\tau}(E'_{\alpha,q})} + \| \{\tau^{-1}(I - u^\tau(k,k-1))u_{k-1}\}_1^N \|_{L_{p,\tau}(E'_{\alpha,q})}$$

$$\leq M(q)[\| A(0)u_0\|_{E'_{\alpha,q}} + \frac{p^2}{p-1} \| f^\tau \|_{L_{p,\tau}(E'_{\alpha,q})}],$$

where $M(q)$ does not depend on f^τ, u_0, p, $\alpha$and τ.

Now, the abstract theorems given above are applied in the investigation of difference schemes of higher order of accuracy with respect to the set of all variables for approximate solution of the Cauchy problem (4.52). The discretization of problem (4.52) is carried out in two steps. Let us define the grid space R_h^n $(0 < h \leq h_0)$ as the set of all points of the Euclidean space R^n whose coordinates are given by

$$x_k = s_k h, \qquad s_k = 0, \pm 1, \pm 2, \ldots, \; k = 1, \ldots, n.$$

The number h is called the step of the grid space. A function defined on R_h^n will be called a grid function. To the differential operator A with constant coefficients of the form

$$A = \sum_{|r|=2m} a_r \frac{\partial^{r_1 + \cdots + r_n}}{\partial_{x_1^{r_1}} \cdots \partial_{x_n^{r_n}}},$$

we assign the difference operator

$$A_h = h^{-m} \sum_{2m \leq |s| \leq S} b_s \Delta_{1-}^{s_1} \Delta_{1+}^{s_2} \cdots \Delta_{n-}^{s_{2n-1}} \Delta_{n+}^{s_{2n}}, \tag{4.96}$$

which acts on functions defined on the entire space R_h^n. Here $s \in R^{2n}$ is a vector with nonnegative integer coordinates,

$$\Delta_{k\pm} f^h(x) = \pm \left(f^h(x \pm e_k h) - f^h(x) \right),$$

and e_k is the unit vector of the axis x_k.

An infinitely differentiable function of the continuous argument $y \in R^n$ that is continuous and bounded together with all its derivatives is said to be smooth. Let $\varphi(y)$ be a smooth function on R^n. Using the Taylor expansion of $\varphi(y)$, one can show that

$$\sup_{x \in R_h^n} \left| h^{-1} \Delta_{k\pm} \varphi(x) - \frac{\partial}{\partial y_k} \varphi(x) \right| \leq M(\varphi) h.$$

Here the grid function $\varphi(x)$ and $\frac{\partial}{\partial y_k}\varphi(x)$ are the traces of the functions $\varphi(y)$ and $\frac{\partial}{\partial y_k}\varphi(y)$, respectively. The last inequality means that the difference operator $h^{-1}\Delta_{k\pm}$ is a first-order approximation for the differential operator $\frac{\partial}{\partial y_k}$.

We say that the difference operator A_h is a λth-order ($\lambda > 0$) approximation of the differential operator A if the inequality

$$\sup_{x\in R_h^n} |A_h\varphi(x) - A\varphi(x)| \leq M(\varphi)\, h^\lambda$$

holds for any smooth function $\varphi(y)$. We shall assume that the operator A_h approximates the differential operator A with any prescribed order.

A function of a continuous [resp., discrete] argument that decays at infinity faster than any negative power of $|y|$ [resp., $|x|$] is said to be rapidly decreasing. Let us define the Fourier transform of a grid function $f^h(x)$ by the formula

$$\tilde{f}(\xi) = (2\pi)^{-n} \sum_{x\in R_h^n} \exp\{-i(x,\xi)\} f^h(x)\, h^n, \xi \in R^n. \tag{4.97}$$

This formula defines a $2\pi h^{-1}$-periodic smooth function of the continuous argument ξ whenever $f^h(x)$ is a rapidly decreasing grid function. In this last case (4.97) is just a Fourier series expansion of the function $\tilde{f}(\xi)$ and the numbers $f^h(x)$ are the Fourier coefficients, given by the formula

$$f^h(x) = \int_{|\xi_1|\leq\pi h^{-1}} \cdots \int_{|\xi_n|\leq\pi h^{-1}} \exp\{i(x,\xi)\}\, \tilde{f}(\xi)\, d\xi_1 \cdots d\xi_n. \tag{4.98}$$

The inverse Fourier transform of a $2\pi h^{-1}$-periodic function $\varphi(\xi)$ is defined to be the grid function $\tilde{\varphi}^h(x)$ given by the formula

$$\tilde{\varphi}^h(x) = \int_{|\xi_1|\leq\pi h^{-1}} \cdots \int_{|\xi_n|\leq\pi h^{-1}} \exp\{i(x,\xi)\}\, \varphi(\xi)\, d\xi_1 \cdots d\xi_n. \tag{4.99}$$

Formulas (4.98) and (4.99) establish a one-to-one correspondence between rapidly decreasing grid functions of a continuous argument. In particular, if $f^h(x)$ is a rapidly decreasing grid function, then

$$\widehat{\left[\tilde{f}\right]}^h(x) = f^h(x).$$

If $f^h(x)$ is a rapidly decreasing grid function, then the grid function $A_h f^h(x)$ exists and is given by (4.97) and we have the equality

$$\tilde{A}_h f(\xi) = A(\xi h, h)\, \tilde{f}(\xi).$$

The function $A(\xi h, h)$ is obtained by replacing the operator $\Delta_{k\pm}$ in the right-hand side of equality (4.96) with the expression $\pm(\exp\{\pm i\xi_k h\} - 1)$, respectively and is called the symbol of the difference operator. Since $\exp\{\pm i\xi_k h\}$ are bounded analytic $2\pi h^{-1}$-periodic functions, the symbol $A(\xi h, h)$ is a bounded analytic $2\pi h^{-1}$-periodic function. It follows that for large $|\xi|$ one has the estimate

$$|A(\xi h, h)| \leq M(h)\, |\xi|^m, |\xi|^2 = |\xi_1|^2 + \cdots + |\xi_n|^2.$$

In the first step let us give the difference operator $A_h^{t,x}$ by the formula

$$A_h^{t,x} u_x^h = \sum_{2m \leq |r| \leq S} b_r^{t,x} D_h^r u_x^h + \delta u_x^h. \tag{4.100}$$

The coefficients are chosen in such a way that the operator $A_h^{t,x}$ approximates in a specified way the operator

$$\sum_{|r|=2m} a_r(t,x) \frac{\partial^{|r|}}{\partial x_1^{r_1} \cdots \partial x_n^{r_n}} + \delta.$$

We shall assume that for $|\xi_k h| \leq \pi$ and fixed t and x the symbol $A^{t,x}(\xi h, h)$ of the operator $A_h^{t,x} - \delta$ satisfies the inequalities

$$(-1)^m A^{t,x}(\xi h, h) \geq M_1 |\xi|^{2m}, |\arg A^{t,x}(\xi h, h)| \leq \phi < \phi_0 \leq \frac{\pi}{2}.$$

With the help of $A_h^{t,x}$ we arrive at the initial-value problem

$$\frac{dv^h(t,x)}{dt} + A_h^{t,x} v^h(t,x) = f^h(t,x), 0 \leq t \leq T, \tag{4.101}$$

$$v^h(0,x) = v_0^h(x), x \in \mathbb{R}_h^n,$$

for an infinite system of ordinary differential equations.

In the second step we replace problem (4.101) by the difference scheme

$$\frac{1}{\tau}(u_k^h(x) - u_{k-1}^h(x)) + \frac{1}{\tau}(I - R_{j,l}(\tau A_h^{t_{k-1},x})) u_{k-1}^h(x) \tag{4.102}$$

$$= f_k^{j,l}(x), 1 \leq k \leq N-1, u_0^h(x) = v_0^h(x), \ x \in R_h^n.$$

Let us give a number of corollaries of the abstract theorems given in the above. To formulate our result we need to introduce the space $C_h^\beta = C^\beta(R_h^n)$, $0 \leq \beta \leq 1$ of all bounded grid functions $u^h(x)$ defined on R_h^n, equipped with the norm

$$||u^h||_{C_h^\beta} = ||u^h||_{C_h} + \sup_{x,y \in R_h^n, x \neq y} \frac{|u^h(x) - u^h(y)|}{|x-y|^\beta}.$$

Here $C_h = C(R_h^n)$ denotes the Banach space of bounded grid functions $u^h(x)$ defined on R_h^n, equipped with the norm

$$||u^h||_{C_h} = \sup_{x \in R_h^n} |u^h(x)|.$$

Theorem 4.2.34. *The solutions of the difference schemes* (4.102) *satisfy the following stability estimates:*

$$\| u^{\tau,h} \|_{C_\tau^{\alpha,\alpha}(C_h^\beta)} \leq M[\|u_0^h\|_{C_h^\beta} + \| f^{\tau,h} \|_{C_\tau^{\alpha,\alpha}(C_h^\beta)}], 0 < \alpha < 1, 0 \leq \beta < 1,$$

$$\| u^{\tau,h} \|_{L_{p,\tau}(C_h^\beta)} \leq M[\|u_0^h\|_{C_h^\beta} + \| f^{\tau,h} \|_{L_{p,\tau}(C_h^\beta)}], 1 \leq p \leq \infty, 0 \leq \beta < 1,$$

where M does not depend on $f^{\tau,h}$, u_0^h, α, β, p, h and τ

The proof of Theorem 4.2.34 is based on the abstract Theorems 4.2.21, 4.2.22, 4.2.23 and on the following two theorems.

Theorem 4.2.35. *The operator $A_h^{t,x}$ is strongly positive in C_h.*

Theorem 4.2.36. *For any $0 < \beta < \frac{1}{2m}$ the norms in the spaces $E'_\beta(C_h, A_h^{t,x})$ and $C_h^{2m\beta}$ are equivalent uniformly in h.*

Theorem 4.2.37. *The solutions of the difference schemes* (4.102) *satisfy the following almost coercive stability estimates:*

$$\| \{\tau^{-1}(u_k^h - u_{k-1}^h)\}_1^{N-1} \|_{L_{p,\tau}(C_h)} \leq M[\ln \frac{1}{h} \sum_{|r|=2m} \|D_h^r u_0^h\|_{C_h}$$

$$+ \ln \frac{1}{\tau + h} \| f^{\tau,h} \|_{L_{p,\tau}(C_h)}], 1 \leq p \leq \infty,$$

where M does not depend on $f^{\tau,h}$, u_0^h, p, h and τ.

The proof of Theorem 4.2.37 is based on the abstract Theorems 4.2.24 and 4.2.30, the positivity of the operator A_h^x in C_h^β and on the almost coercivity inequality for an elliptic operator $A_h^{t,x}$ in C_h and on the estimate

$$\min\left\{\ln \frac{1}{\tau}, 1 + \left|\ln \| A_h^{t,x} \|_{C_h^\beta \to C_h^\beta}\right|\right\} \leq M \ln \frac{1}{\tau + h}.$$

Next, to formulate our result we need to introduce the space $W_{p,h}^\beta = W_p^\beta(R_h^n)$, $0 \leq \beta \leq 1, 1 \leq p < \infty$ of all bounded grid functions $u^h(x)$ defined on R_h^n, equipped with the norm

$$\|u^h\|_{W_{p,h}^\beta} = [\sum_{x \in R_h^n} \sum_{y \in R_h^n, y \neq 0} \frac{|u^h(x) - u^h(x+y)|^p}{|y|^{n+\beta p}} h^{2n} + \|u^h\|_{L_{p,h}}^p]^{\frac{1}{p}}.$$

Here $L_{p,h} = L_p(R_h^n)$ denotes the Banach space of bounded grid functions $u^h(x)$ defined on R_h^n, equipped with the norm

$$\|u^h\|_{L_{p,h}} = [\sum_{x \in R_h^n} |u^h(x)|^p h^n]^{\frac{1}{p}}.$$

Theorem 4.2.38. *The solutions of the difference schemes* (4.102) *satisfy the coercivity estimates:*

$$\| \{\tau^{-1}(u_k^h - u_{k-1}^h)\}_1^{N-1} \|_{C_\tau^{\alpha,\alpha}(C_h^\beta)}$$

$$\leq M(\alpha,\beta)[\sum_{|r|=2m} ||D_h^r u_0^h||_{C_h^\beta} + \| f^{\tau,h} \|_{C_\tau^{\alpha,\alpha}(C_h^\beta)}], 0 \leq \alpha < 1, 0 < \beta < 1,$$

$$\| \{\tau^{-1}(u_k^h - u_{k-1}^h)\}_1^{N-1} \|_{L_{p,\tau}(W_{q,h}^{m\alpha})}$$

$$\leq M(\alpha,p,q)[\sum_{|r|=2m} \|D_h^r u_0^h\|_{W_{q,h}^{m\alpha}} + \| f^{\tau,h} \|_{L_{p,\tau}(W_{q,h}^{m\alpha})}], 1 < p,q < \infty, 0 < \alpha < \frac{1}{m},$$

where $M(\alpha,\beta)$ *and* $M(\alpha,p,q)$ *do not depend on* $f^{\tau,h}$, u_0^h, h *and* τ

The proof of Theorem 4.2.38 is based on the abstract Theorems 4.2.25, 4.2.33 and Theorems 4.2.35, 4.2.36 and on the following two theorems.

Theorem 4.2.39. *The operator* $A_h^{t,x}$ *is strongly positive in* $W_{q,h}^\beta$.

Theorem 4.2.40. *For any* $0 < \beta < \frac{1}{2m}$ *the norms in the spaces* $E'_{\beta,q}(L_{q,h}, A_h^x)$ *and* $W_{q,h}^{2m\beta}$ *are equivalent uniformly in* h.

Note that in a similar way the difference schemes of the high order of accuracy with respect to one variable for approximate solutions of the mixed problems (4.50) and (4.51) can be constructed. Abstract theorems given from above permit us to obtain the stability, the almost stability and the coercive stability estimates for the solutions of these difference schemes.

Now, we will consider the second approach of choice $v(t_k, t_{k-1})$ and

$$\frac{1}{\tau}\int_{t_{k-1}}^{t_k} v(t_k,s)f(s)ds.$$

By (4.15) and (4.16), we have

$$v(t_k,t_{k-1}) = e^{-A(t_{k-1})\tau} \tag{4.103}$$

$$-\int_{t_{k-1}}^{t_k} v(t_k,z)[A(z)-A(t_{k-1})]e^{-A(t_{k-1})(z-t_{k-1})}dz,$$

$$v(t_k,t_{k-1}) = e^{-A(t_k)\tau} \tag{4.104}$$

$$+\int_{t_{k-1}}^{t_k} e^{-A(t_k)(t_k-s)}[A(t_k)-A(z)]v(z,t_{k-1})dz,$$

$$v(t_k,t_{k-1}) = e^{-A(t_{k-\frac{1}{2}})\tau} \tag{4.105}$$

$$-\int_{t_{k-1}}^{t_k} v(t_k,z)[A(z)-A(t_{k-\frac{1}{2}})]e^{-A(t_{k-\frac{1}{2}})(z-t_{k-1})}dz.$$

Using Taylor's formula, we obtain

$$\frac{1}{\tau}\int_{t_{k-1}}^{t_k} v(t_k,s)f(s)ds \tag{4.106}$$

$$=\sum_{m=0}^{j+l-1}\sum_{\lambda=0}^{m}\binom{m}{\lambda} v(t_k,t_{k-1})B_{m-\lambda}(t_{k-1})f^{(\lambda)}(t_{k-1})\frac{\tau^m}{(m+1)!}+o(\tau^{j+l}),$$

$$\frac{1}{\tau}\int_{t_{k-1}}^{t_k} v(t_k,s)f(s)ds \tag{4.107}$$

$$=\sum_{m=0}^{j+l-1}\sum_{\lambda=0}^{m}\binom{m}{\lambda} B_{m-\lambda}(t_k)f^{(\lambda)}(t_k)\frac{(-\tau)^m}{(m+1)!}+o(\tau^{j+l}),$$

$$\frac{1}{\tau}\int_{t_{k-1}}^{t_k} v(t_k,s)f(s)ds \tag{4.108}$$

$$=\sum_{m=0}^{l-1}\sum_{\lambda=0}^{2m}\binom{2m}{\lambda} v(t_k,t_{k-\frac{1}{2}})B_{2m-\lambda}(t_{k-\frac{1}{2}})f^{(\lambda)}(t_{k-\frac{1}{2}})\frac{\tau^{2m}}{2^{2m}(2m+1)!}+o(\tau^{2l}),$$

where

$$\begin{cases} B_i(t)=1, & i=0,\\ B_i(t)=B_{i-1}^{(\prime)}(t)+B_{i-1}(t)A(t), & 1\le i\le j+l. \end{cases}$$

Now, we will study the approximate formulas for the expressions $v(t_k,t_{k-1})$ and $v(t_k,t_{k-\frac{1}{2}})$. Using Taylor's formula and formulas (4.103), (4.104) and (4.105), we obtain

$$v(t_k,t_{k-1})=e^{-A(t_{k-1})\tau}-\tau(A(t_k)-A(t_{k-1}))e^{-A(t_{k-1})\tau}$$

$$+\sum_{n=1}^{j+l-1}\frac{(-\tau)^{n+1}}{(n+1)!}\sum_{i=0}^{n}\binom{n}{i} B_i(t_k)[(A(t_k)-A(t_{k-1}))(-A(t_{k-1}))^{n-i}$$

$$+\sum_{s=1}^{n-i}\binom{n-i}{s} A^{(s)}(t_k)(-A(t_{k-1}))^{n-i-s}]e^{-A(t_{k-1})\tau}+o(\tau^{j+l+1}),$$

$$v(t_k,t_{k-1})=e^{-A(t_k)\tau}+\tau(A(t_k)-A(t_{k-1}))e^{-A(t_k)\tau}$$

$$+\sum_{n=1}^{j+l-1}\frac{\tau^{n+1}}{(n+1)!}\sum_{i=0}^{n}\binom{n}{i} \widetilde{B}_i(t_{k-1})[(A(t_k)-A(t_{k-1}))(A(t_k))^{n-i}$$

$$-\sum_{s=1}^{n-i}\binom{n-i}{s} A^{(s)}(t_{k-1})(A(t_k))^{n-i-s}]e^{-A(t_k)\tau}+o(\tau^{j+l+1}),$$

$$v(t_k,t_{k-1})=e^{-A(t_{k-\frac{1}{2}})\tau}-\tau(A(t_k)-A(t_{k-\frac{1}{2}}))e^{-A(t_{k-\frac{1}{2}})\tau}$$
$$+\sum_{n=1}^{j+l-1}\frac{(-\tau)^{n+1}}{(n+1)!}\sum_{i=0}^{n}\binom{n}{i}B_i(t_k)[(A(t_k)-A(t_{k-\frac{1}{2}}))(-A(t_{k-\frac{1}{2}}))^{n-i}$$
$$\times\sum_{s=1}^{n-i}\binom{n-i}{s}A^{(s)}(t_k)(-A(t_{k-\frac{1}{2}}))^{n-i-s}]e^{-A(t_{k-\frac{1}{2}})\tau}+o(\tau^{j+l+1}),$$

where

$$\begin{cases}\widetilde{B}_i(t)=1, & i=0,\\ \widetilde{B}_i(t)=\widetilde{B}^{(\prime)}_{i-1}(t)-\widetilde{B}_{i-1}(t)A(t), & 1\le i\le j+l.\end{cases}$$

Further, using the last formulas and Pade fractions for the function e^{-z}, we can write

$$v(t_k,t_{k-1})=R_{j,l}(A(t_{k-1})\tau)-\tau(A(t_k)-A(t_{k-1}))R_{j-1,l-1}(A(t_{k-1})\tau) \tag{4.109}$$
$$+\sum_{n=1}^{j+l-1}\frac{(-\tau)^{n+1}}{(n+1)}\sum_{i=0}^{n}\binom{n}{i}B_i(t_k)[(A(t_k)-A(t_{k-1}))(-A(t_{k-1}))^{n-i}$$
$$+\sum_{s=1}^{n-i}\binom{n-i}{s}A^{(s)}(t_k)(-A(t_{k-1}))^{n-i-s}]R_{j-n+[\frac{n}{2}]-1,l-[\frac{n}{2}]}(A(t_{k-1})\tau)+o(\tau^{j+l+1}),$$

$$v(t_k,t_{k-1})=R_{j,l}(\tau A(t_k))+\tau(A(t_k)-A(t_{k-1}))R_{j-1,l-1}(A(t_k)\tau) \tag{4.110}$$
$$+\sum_{n=1}^{j+l-1}\frac{\tau^{n+1}}{(n+1)!}\sum_{i=0}^{n}\binom{n}{i}\widetilde{B}_i(t_{k-1})[(A(t_k)-A(t_{k-1}))(A(t_k))^{n-i}$$
$$-\sum_{s=1}^{n-i}\binom{n-i}{s}A^{(s)}(t_{k-1})(A(t_k))^{n-i-s}]R_{q-n+[\frac{n}{2}]-1,p-[\frac{n}{2}]}(\tau A(t_k))+o(\tau^{j+l+1}),$$

$$v(t_k,t_{k-1})=R_{j,l}(\tau A(t_{k-\frac{1}{2}}))-\tau(A(t_k)-A(t_{k-\frac{1}{2}}))R_{j-1,l-1}(\tau A(t_{k-\frac{1}{2}})) \tag{4.111}$$
$$+\sum_{n=1}^{j+l-1}\frac{(-\tau)^{n+1}}{(n+1)!}\sum_{i=0}^{n}\binom{n}{i}B_i(t_k)[(A(t_k)-A(t_{k-\frac{1}{2}}))(-A(t_{k-\frac{1}{2}}))^{n-i}$$
$$+\sum_{s=1}^{n-i}\binom{n-i}{s}A^{(s)}(t_k)(-A(t_{k-\frac{1}{2}}))^{n-i-s}]R_{j-n+[\frac{n}{2}]-1,l-[\frac{n}{2}]}(\tau A(t_{k-\frac{1}{2}}))+o(\tau^{j+l+1}).$$

The approximate formulas for the expression $v(t_k,t_{k-\frac{1}{2}})$ can be constructed using Pade's rational functions. We have that

$$v(t_k,t_{k-\frac{1}{2}})=e^{-A(t_k)\frac{\tau}{2}}+\int_{t_{k-\frac{1}{2}}}^{t_k}e^{-A(t_k)(t_k-z)}(A(t_k)-A(z))v(z,t_{k-\frac{1}{2}})dz,$$
$$v(t_k,t_{k-\frac{1}{2}})=e^{-A(t_{k-\frac{1}{2}})\frac{\tau}{2}}-\int_{t_{k-\frac{1}{2}}}^{t_k}v(t_k,z)(A(z)-A(t_{k-\frac{1}{2}}))e^{-A(t_{k-\frac{1}{2}})(z-t_{k-\frac{1}{2}})}dz.$$

Using Taylor's formula, we obtain

$$v(t_k, t_{k-\frac{1}{2}}) = e^{-A(t_k)\frac{\tau}{2}} + \frac{1}{2}\tau(A(t_k) - A(t_{k-\frac{1}{2}}))e^{-\frac{1}{2}A(t_k)\tau}$$

$$+ \sum_{n=1}^{j+l-1} \frac{\tau^{n+1}}{2^{n+1}(n+1)!} \sum_{i=0}^{n} \binom{n}{i} \widetilde{B}_i(t_{k-\frac{1}{2}})[(A(t_k) - A(t_{k-\frac{1}{2}}))(A(t_k))^{n-i}$$

$$-\sum_{s=1}^{n-i} \binom{n-i}{s} A^{(s)}(t_{k-\frac{1}{2}})(A(t_k))^{n-i-s}]e^{-A(t_k)\frac{\tau}{2}} + o(\tau^{j+l+1}),$$

$$v(t_k, t_{k-\frac{1}{2}}) = e^{-A(t_{k-\frac{1}{2}})\frac{\tau}{2}} - \frac{1}{2}\tau(A(t_k) - A(t_{k-\frac{1}{2}}))e^{-\frac{1}{2}A(t_{k-\frac{1}{2}})\tau}$$

$$+ \sum_{n=1}^{j+l-1} \frac{(-\tau)^{n+1}}{2^{n+1}(n+1)!} \sum_{i=0}^{n} \binom{n}{i} B_i(t_k)[(A(t_k) - A(t_{k-\frac{1}{2}}))(-A(t_{k-\frac{1}{2}}))^{n-i}$$

$$+\sum_{s=1}^{n-i} \binom{n-i}{s} A^{(s)}(t_k)(-A(t_{k-\frac{1}{2}}))^{n-i-s}]e^{-A(t_{k-\frac{1}{2}})\frac{\tau}{2}} + o(\tau^{j+l+1}).$$

Therefore, using the last formulas and Pade's approximants of the function e^{-z}, we can write

$$v(t_k, t_{k-\frac{1}{2}}) = R_{j,l}(\frac{\tau}{2}A(t_k)) + \frac{1}{2}\tau(A(t_k) - A(t_{k-\frac{1}{2}}))R_{j-1,l-1}(\frac{\tau}{2}A(t_k)) \quad (4.112)$$

$$+ \sum_{n=1}^{j+l-1} \frac{\tau^{n+1}}{2^{n+1}(n+1)!} \sum_{i=0}^{n} \binom{n}{i} \widetilde{B}_i(t_{k-\frac{1}{2}})[(A(t_k) - A(t_{k-\frac{1}{2}}))(A(t_k))^{n-i}$$

$$-\sum_{s=1}^{n-i} \binom{n-i}{s} A^{(s)}(t_{k-\frac{1}{2}})(A(t_k))^{n-i-s}]R_{j-n+[\frac{n}{2}]-1,l-[\frac{n}{2}]}(\frac{\tau}{2}A(t_k)) + o(\tau^{j+l+1}),$$

$$v(t_k, t_{k-\frac{1}{2}}) = R_{j,l}(\frac{\tau}{2}A(t_{k-\frac{1}{2}})) - \frac{1}{2}\tau(A(t_k) - A(t_{k-\frac{1}{2}}))R_{j-1,l-1}(\frac{\tau}{2}A(t_{k-\frac{1}{2}})) \quad (4.113)$$

$$+ \sum_{n=1}^{j+l-1} \frac{(-\tau)^{n+1}}{2^{n+1}(n+1)!} \sum_{i=0}^{n} \binom{n}{i} B_i(t_k)[(A(t_k) - A(t_{k-\frac{1}{2}}))(-A(t_{k-\frac{1}{2}}))^{n-i}$$

$$+\sum_{s=1}^{n-i} \binom{n-i}{s} A^{(s)}(t_k)(A(t_{k-\frac{1}{2}}))^{n-i-s}]R_{j-n+[\frac{n}{2}]-1,l-[\frac{n}{2}]}(\frac{\tau}{2}A(t_{k-\frac{1}{2}})) + o(\tau^{j+l+1}).$$

Finally, replacing the expression $v(t_k, t_{k-1})$ by its Pade approximants from (4.109) or (4.110) or (4.111) (neglecting the last terms, respectively) and the expression

$$\frac{1}{\tau}\int_{t_{k-1}}^{t_k} v(t_k, s)f(s)ds$$

by the expressions from (4.106) or (4.107) or (4.108) and replacing the expressions $v(t_k, t_{k-1})$ and $v(t_k, t_{k-\frac{1}{2}})$ by their Pade approximants from (4.109) or (4.110) or (4.111) or (4.112) or (4.113) (neglecting the last terms, respectively), we obtain the following difference schemes of $(j+l)$-th order of accuracy for the approximate solution of the initial-value problem (4.1):

$$\begin{aligned}
&\tau^{-1}(u_k - u_{k-1}) + \tau^{-1}[1 - R_{j,l}(A(t_{k-1})\tau) \\
&+\tau(A(t_k) - A(t_{k-1}))R_{j-1,l-1}(A(t_{k-1})\tau) \\
&-\sum_{n=1}^{j+l-1} \frac{(-\tau)^{n+1}}{(n+1)} \sum_{i=0}^{n} \binom{n}{i} B_i(t_k)[(A(t_k) - A(t_{k-1}))(-A(t_{k-1}))^{n-i} \\
&+\sum_{s=1}^{n-i} \binom{n-i}{s} A^{(s)}(t_k)(-A(t_{k-1}))^{n-i-s} \\
&\times R_{j-n+[\frac{n}{2}]-1,l-[\frac{n}{2}]}(A(t_{k-1})\tau)]]u_{k-1} = f_k^{j,l}, 1 \le k \le N, u_0 = y_0;
\end{aligned} \tag{4.114}$$

$$\begin{aligned}
&\tau^{-1}(u_k - u_{k-1}) + \tau^{-1}[1 - R_{j,l}(\tau A(t_k)) \\
&-\tau(A(t_k) - A(t_{k-1}))R_{j-1,l-1}(A(t_k)\tau) \\
&-\sum_{n=1}^{j+l-1} \frac{\tau^{n+1}}{(n+1)!} \sum_{i=0}^{n} \binom{n}{i} \widetilde{B}_i(t_{k-1})[(A(t_k) - A(t_{k-1}))(A(t_k))^{n-i} \\
&-\sum_{s=1}^{n-i} \binom{n-i}{s} A^{(s)}(t_{k-1})(A(t_k))^{n-i-s} \\
&\times R_{j-n+[\frac{n}{2}]-1,l-[\frac{n}{2}]}(\tau A(t_k))]]u_{k-1} = f_k^{j,l}, 1 \le k \le N, u_0 = y_0;
\end{aligned} \tag{4.115}$$

$$\begin{aligned}
&\tau^{-1}(u_k - u_{k-1}) + \tau^{-1}[1 - R_{j,l}(\tau A(t_{k-\frac{1}{2}})) \\
&+\tau(A(t_k) - A(t_{k-\frac{1}{2}}))R_{j-1,l-1}(\tau A(t_{k-\frac{1}{2}})) \\
&-\sum_{n=1}^{j+l-1} \frac{(-\tau)^{n+1}}{(n+1)!} \sum_{i=0}^{n} \binom{n}{i} B_i(t_k)[(A(t_k) - A(t_{k-\frac{1}{2}}))(-A(t_{k-\frac{1}{2}}))^{n-i} \\
&+\sum_{s=1}^{n-i} \binom{n-i}{s} A^{(s)}(t_k)(-A(t_{k-\frac{1}{2}}))^{n-i-s} \\
&\times R_{j-n+[\frac{n}{2}]-1,l-[\frac{n}{2}]}(\tau A(t_{k-\frac{1}{2}}))]]u_{k-1} = f_k^{j,l}, 1 \le k \le N, u_0 = y_0.
\end{aligned} \tag{4.116}$$

Here, $f_k^{j,l}$ can be defined by one of the following formulas:

$$f_k^{j,l} = \sum_{m=0}^{j+l-1} \sum_{\lambda=0}^{m} \binom{m}{\lambda} B_{m-\lambda}(t_k) f^{(\lambda)}(t_k) \frac{(-\tau)^m}{(m+1)!},$$

or

$$f_k^{j,l} = \sum_{m=0}^{j+l-1} \sum_{\lambda=0}^{m} \binom{m}{\lambda} B_{m-\lambda}(t_{k-1})$$

$$\times \frac{\tau^m}{(m+1)!}[R_{j-m-[\frac{m}{2}],l+[\frac{m}{2}]}(A(t_{k-1})\tau) - \tau(A(t_k)$$
$$-A(t_{k-1}))R_{q-m-[\frac{m}{2}]-1,p+[\frac{m}{2}]-1}(A(t_{k-1})\tau)$$

$$+\sum_{n=1}^{m-1} \frac{(-\tau)^{n+1}}{(n+1)} \sum_{i=0}^{n} \binom{n}{i} B_i(t_k)[(A(t_k) - A(t_{k-1}))(-A(t_{k-1}))^{n-i}$$

$$+\sum_{s=1}^{n-i} \binom{n-i}{s} A^{(s)}(t_k)(-A(t_{k-1}))^{n-i-s}]R_{j-n+[\frac{n}{2}]-1,l-[\frac{n}{2}]}(A(t_{k-1})\tau)]f^{(\lambda)}(t_{k-1}),$$

or

$$f_k^{j,l} = \sum_{m=0}^{j+l-1} \sum_{\lambda=0}^{m} \binom{m}{\lambda} B_{m-\lambda}(t_{k-1})$$

$$\times \frac{\tau^m}{(m+1)!}[R_{j-m-[\frac{m}{2}],l+[\frac{m}{2}]}(\tau A(t_k)) + \tau(A(t_k) - A(t_{k-1}))$$
$$\times R_{j-m-[\frac{m}{2}]-1,l+[\frac{m}{2}]-1}(A(t_k)\tau)$$

$$+\sum_{n=1}^{m-1} \frac{\tau^{n+1}}{(n+1)!} \sum_{i=0}^{n} \binom{n}{i} \widetilde{B}_i(t_{k-1})[(A(t_k) - A(t_{k-1}))(A(t_k))^{n-i}$$

$$-\sum_{s=1}^{n-i} \binom{n-i}{s} A^{(s)}(t_{k-1})(A(t_k))^{n-i-s} R_{j-n+[\frac{n}{2}]-1,l-[\frac{n}{2}]}(\tau A(t_k))]f^{(\lambda)}(t_{k-1}),$$

or

$$f_k^{j,l} = \sum_{m=0}^{j+l-1} \sum_{\lambda=0}^{m} \binom{m}{\lambda} B_{m-\lambda}(t_{k-1})$$

$$\times \frac{\tau^m}{(m+1)!}[R_{j-m-[\frac{m}{2}],l+[\frac{m}{2}]}(\tau A(t_{k-\frac{1}{2}})) - \tau(A(t_k) - A(t_{k-\frac{1}{2}}))$$
$$\times R_{j-m-[\frac{m}{2}]-1,l+[\frac{m}{2}]-1}(\tau A(t_{k-\frac{1}{2}}))$$

$$+\sum_{n=1}^{m-1} \frac{(-\tau)^{n+1}}{(n+1)!} \sum_{i=0}^{n} \binom{n}{i} B_i(t_k)[(A(t_k) - A(t_{k-\frac{1}{2}}))(-A(t_{k-\frac{1}{2}}))^{n-i}$$

$$+\sum_{s=1}^{n-i} \binom{n-i}{s} A^{(s)}(t_k)(-A(t_{k-\frac{1}{2}}))^{n-i-s} R_{j-n+[\frac{n}{2}]-1,l-[\frac{n}{2}]}(\tau A(t_{k-\frac{1}{2}}))]f^{(\lambda)}(t_{k-1}),$$

or

$$f_k^{j,l}=\sum_{m=0}^{l-1}\sum_{\lambda=0}^{2m}\binom{2m}{\lambda}B_{2m-\lambda}(t_{k-\frac{1}{2}})$$

$$\times\frac{\tau^{2m}}{2^{2m}(2m+1)!}[R_{j-m,l-m}(\frac{\tau}{2}A(t_k))+\frac{1}{2}\tau(A(t_k)-A(t_{k-\frac{1}{2}}))R_{j-m-1,l-m-1}(\frac{\tau}{2}A(t_k))$$

$$+\sum_{n=1}^{m-1}\frac{\tau^{n+1}}{2^{n+1}(n+1)!}\sum_{i=0}^{n}\binom{n}{i}\widetilde{B}_i(t_{k-\frac{1}{2}})[(A(t_k)-A(t_{k-\frac{1}{2}}))(A(t_k))^{n-i}$$

$$-\sum_{s=1}^{n-i}\binom{n-i}{s}A^{(s)}(t_{k-\frac{1}{2}})(A(t_k))^{n-i-s}R_{j-n+[\frac{n}{2}]-1,l-[\frac{n}{2}]}(\frac{\tau}{2}A(t_k))]f^{(\lambda)}(t_{k-\frac{1}{2}}),$$

or

$$f_k^{j,l}=\sum_{m=0}^{l-1}\sum_{\lambda=0}^{2m}\binom{2m}{\lambda}B_{2m-\lambda}(t_{k-\frac{1}{2}})$$

$$\times\frac{\tau^{2m}}{2^{2m}(2m+1)!}[R_{j-m,l-m}(\frac{\tau}{2}A(t_{k-\frac{1}{2}}))-\frac{1}{2}\tau(A(t_k)-A(t_{k-\frac{1}{2}}))R_{j-m-1,l-m-1}$$

$$\times(\frac{\tau}{2}A(t_{k-\frac{1}{2}}))$$

$$+\sum_{n=1}^{m-1}\frac{(-\tau)^{n+1}}{2^{n+1}(n+1)!}\sum_{i=0}^{n}\binom{n}{i}B_i(t_k)[(A(t_k)-A(t_{k-\frac{1}{2}}))(-A(t_{k-\frac{1}{2}}))^{n-i}$$

$$+\sum_{s=1}^{n-i}\binom{n-i}{s}A^{(s)}(t_k)(A(t_{k-\frac{1}{2}}))^{n-i-s}R_{j-n+[\frac{n}{2}]-1,l-[\frac{n}{2}]}(\frac{\tau}{2}A(t_{k-\frac{1}{2}}))]f^{(\lambda)}(t_{k-\frac{1}{2}}).$$

Note that in a similar manner with the difference schemes (4.63) one can investigate the stability, coercive stability of the difference schemes (4.114), (4.115) and (4.116) under the smooth assumptions for $A(t)$.

4.3 Single-Step Difference Schemes Generated by Taylor's Decomposition

We consider again the Cauchy problem (4.1). Let the function $v(t)$, for $(0\le t\le T)$ have $(l+j+1)$ continuous derivatives and $t_{k-1},t_k\in[0,T]_\tau$. Then by Theorem 2.2.1 we have the following Taylor decomposition on two points

$$v(t_k)-v(t_{k-1})+\sum_{i=1}^{l}\alpha_i v^{(i)}(t_k)\tau^i-\sum_{i=1}^{j}\beta_i v^{(i)}(t_{k-1})\tau^i \tag{4.117}$$

$$=\frac{(-1)^l}{(l+j)!}\int_{t_{k-1}}^{t_k}(t_k-s)^j(s-t_{k-1})^l v^{(l+j+1)}(s)ds,$$

where

$$\begin{cases} \alpha_i = \frac{(l+j-i)!l!(-1)^i}{(l+j)!i!(l-i)!} & \text{for any i}, 1 \leq i \leq l, \\ \beta_i = \frac{(l+j-i)!j!}{(l+j)!i!(j-i)!} & \text{for any i}, 1 \leq i \leq j. \end{cases} \tag{4.118}$$

Now, we will consider the applications of Taylor's decomposition of function on two points. From (4.117) it is clear that for the approximate solution of problem (4.1) it is necessary to find $v^{(i)}(t_k)$ for any $i, 1 \leq i \leq l$ and $v^{(i)}(t_{k-1})$ for any $i, 1 \leq i \leq j$. Using the equation

$$v\prime(t) = -A(t)v(t) + f(t),$$

we obtain

$$v^{(i)}(t) = -A_i(t)v(t) + f_i(t), \tag{4.119}$$

where

$$A_1(t) = -A(t),\ A_i(t) = A_{i-1}^{(1)}(t) - A_{i-1}(t)A(t) \text{ for any } i, 2 \leq i \leq l,$$

$$f_1(t) = f(t),\ f_i(t) = f_{i-1}^{(\prime)}(t) + A_{i-1}(t)f(t) \text{ for any } i,\ 2 \leq i \leq l.$$

From this formula for $A_i(t)$ it follows that

$$A_i\,(t) = (-1)^i A^{i+1}\,(t) + B_i(t), \tag{4.120}$$

where

$$B_0(t) = 0,\ B_i(t) = A_{i-1}^{(1)}\,(t) - B_{i-1}(t)A\,(t)\,,\ i \geq 1.$$

Therefore,

$$B_i(t) = \sum_{r=1}^{i}(-1)^{i-r}A_{r-1}^{(1)}\,(t)\,A^{i-r}\,(t)\,. \tag{4.121}$$

Replacing the $v^{(i)}(t)$ by (4.119) and neglecting the last term, we obtain the single-step difference schemes of $(l+j)$-th order of accuracy for the approximate solution of problem (4.1):

$$\tau^{-1}(u_k - u_{k-1}) - \sum_{i=1}^{l}\alpha_i\tau^{i-1}A_{i-1}\,(t_k)\,u_k \tag{4.122}$$

$$+\sum_{i=1}^{j}\beta_i\tau^{i-1}A_{i-1}\,(t_{k-1})\,u_{k-1} = f_k^{j,l},$$

$$f_k^{j,l} = -\sum_{i=1}^{l}\tau^{i-1}\alpha_i f_{i-1}\,(t_k) + \sum_{i=1}^{j}\tau^{i-1}\beta_i f_{i-1}\,(t_{k-1})\,, 1 \leq k \leq N,\ u_0 = y_0.$$

Suppose that the operator $A^r(t)A^{(i-r)}(t)A^{-(r+1)}(t)$ satisfies the condition

$$\left\|A^r(t)\,A^{(i-r)}(t)\,A^{-(1+r)}(t)\right\|_{E\to E}\le M, 1\le i\le l-r, 0\le r\le l-1, \tag{4.123}$$

where M is a positive constant independent of t, $0\le t\le T$. Then the following estimate holds:

$$\left\|A^{(1)}_{r-1}(t)\,A^{-r}(t)\right\|_{E\to E}\le M, 1\le r\le l.$$

From this it follows that

$$\left\|B_i(t)A^{-i}(t)\right\|_{E\to E}\le M, 1\le i\le l. \tag{4.124}$$

Using the identity

$$(I-\sum_{i=1}^{l}\alpha_i\tau^iA_{i-1}(t_k))^{-1}-Q_{j,l}^{-1}(\tau A(t_k))$$

$$=(I-\sum_{i=1}^{l}\alpha_i\tau^iA_{i-1}(t_k))^{-1}\tau\sum_{i=1}^{l}\alpha_i\tau^{i-1}B_{i-1}(t_k)\,Q_{j,l}^{-1}(\tau A(t_k)),$$

the estimate (4.124) and

$$\left\|Q_{j,l}^{-1}(\tau A(t_k))\right\|_{E\to E}\le M,$$

we obtain that the operator $I-\sum_{i=1}^{l}\alpha_i\tau^iA_{i-1}(t_k)$ has the bounded inverse $(I-\sum_{i=1}^{l}\alpha_i\tau^iA_{i-1}(t_k))^{-1}$. Therefore the difference problem (4.122) is uniquely solvable and the following formula holds:

$$u_k=u^\tau(k,0)u_0+\sum_{s=1}^{k}u^\tau(k,s)(I-\sum_{i=1}^{l}\alpha_i\tau^iA_{i-1}(t_s))^{-1}f_s^{j,l}\tau, \tag{4.125}$$

where

$$u^\tau(k,n)=\begin{cases}u^\tau(k,k-1)\cdots u^\tau(n+1,n), k>n,\\ I, k=n.\end{cases} \tag{4.126}$$

Here

$$u^\tau(k,k-1)=(I-\sum_{i=1}^{l}\alpha_i\tau^iA_{i-1}(t_k))^{-1}(I-\sum_{i=1}^{j}B_i\tau^iA_{i-1}(t_{k-1})). \tag{4.127}$$

Let us reduce the difference scheme (4.122) to an operator problem in the space $F_\tau(E)$ of vectors $v^\tau=\{v_k\}_{k=1}^N$. Recall that the space $F_\tau(E)$ can be equipped with the various norms and thus become a normed space.

First we will define the operator D_τ, acting from the space $E \times F_\tau(E)$ of vectors $w^\tau = \{w_k\}_{k=0}^N$ into the space $F_\tau(E)$ of vectors $v^\tau = \{v_k\}_{k=1}^N$ by the rule

$$v^\tau = D_\tau w^\tau, v_k = \frac{1}{\tau}(w_k - w_{k-1}), k = 1, \dots, N.$$

Second we will define an operator A_τ from the space $E \times F_\tau(E)$ of vectors $w^\tau = \{w_k\}_{k=0}^N$ into the space $F_\tau(E)$ of vectors $v^\tau = \{v_k\}_{k=1}^N$ by the rule

$$v^\tau = A_\tau w^\tau, v_k = -\sum_{i=1}^{l} \alpha_i \tau^{i-1} A_{i-1}(t_k) w_k + \sum_{i=1}^{j} \beta_i \tau^{i-1} A_{i-1}(t_{k-1}) w_{k-1},$$
$$k = 1, \dots, N.$$

Next, let us introduce the continuation operator $\Pi(u_0)$, which acts from $E \times F_\tau(E)$ to $F_\tau(E)$ according to the rule

$$\Pi(u_0)(u_1, \dots, u_N) = (u_0, u_1, \dots, u_N).$$

Then the difference schemes (4.122) can obviously be rewritten as the equivalent operator equation

$$D_\tau \Pi(u_0) u^\tau + A_\tau \Pi(u_0) u^\tau = f^\tau.$$

Here f^τ is defined by the formula

$$f^\tau = (f_1^{j,l}, \dots, f_N^{j,l}).$$

The last operator problem will be considered in the space $F_\tau(E)$. From its unique solvability for any $u_0 \in E$ and $f^\tau \in F_\tau(E)$ it follows that its solution u^τ defines an additive and homogeneous operator $u^\tau(f^\tau, u_0)$ that is continuous.

The initial-value difference problem (4.122) is said to be stable in $F_\tau(E)$ if we have the stability inequality

$$\| u^\tau(f^\tau, u_0) \|_{F_\tau(E)} \leq M[\| f^\tau \|_{F_\tau(E)} + \| u_0 \|_E],$$

where M is independent not only of f^τ, u_0 but also of τ.

Let us consider the operator $v^\tau(k, n)$ defined by

$$v^\tau(k,n) = \begin{cases} v^\tau(k, k-1) \cdots v^\tau(n+1, n), k > n, \\ I, k = n, \end{cases} \tag{4.128}$$

where

$$v^\tau(k, k-1) = (I - \sum_{i=1}^{j} \beta_i \tau^i A_{i-1}(t_{k-1}))(I - \sum_{i=1}^{l} \alpha_i \tau^i A_{i-1}(t_{k-1}))^{-1}. \tag{4.129}$$

Using formulas (4.126), (4.127), (4.128) and (4.129), we obtain

$$u^\tau(k,n)(I-\sum_{i=1}^{l}\alpha_i\tau^i A_{i-1}(t_n))^{-1}$$

$$=(I-\sum_{i=1}^{L}\alpha_i\tau^i A_{i-1}(t_k))^{-1}v^\tau(k-1,n), 1\le n\le k-1.$$

From this and formula (4.125) it follows that

$$u_k=(I-\sum_{i=1}^{l}\alpha_i\tau^i A_{i-1}(t_k))^{-1}v^\tau(k-1,k)(I-\sum_{i=1}^{j}\beta_i\tau^i A_{i-1}(0))u_0 \tag{4.130}$$

$$+\sum_{s=1}^{k}(I-\sum_{i=1}^{l}\alpha_i\tau^i A_{i-1}(t_k))^{-1}v^\tau(k-1,s)f_s^{j,l}\tau,\ k\ge 2,$$

$$u_1=(I-\sum_{i=1}^{l}\alpha_i\tau^i A_{i-1}(t_1))^{-1}(I-\sum_{i=1}^{j}\beta_i\tau^i A_{i-1}(0))u_0$$

$$+(I-\sum_{i=1}^{l}\alpha_i\tau^i A_{i-1}(t_1))^{-1}f_1^{j,l}\tau.$$

By formula (4.127),

$$v^\tau(k,k-1)-R_{j,l}(\tau A(t_k))=v^\tau(k,k-1)-(I-\sum_{i=1}^{j}\beta_i\tau^i A_{i-1}(t_k))Q_{j,l}^{-1}(\tau A(t_k))$$

$$+(I-\sum_{i=1}^{j}\beta_i\tau^i A_{i-1}(t_k)-P_{j,l}(\tau A(t_k)))Q_{j,l}^{-1}(\tau A(t_k))$$

$$=-v^\tau(k,k-1)\sum_{i=1}^{l}\beta_i\tau^i B_{i-1}(t_k)\,Q_{j,L}^{-1}(\tau A(t_k))-\sum_{i=1}^{j}\beta_i\tau^i B_{i-1}(t_k)\,Q_{j,L}^{-1}(\tau A(t_k)).$$

Using estimate (4.124), we obtain

$$\|v^\tau(k,k-1)-R_{j,l}(\tau A(t_k))\|_{E\to E}\le M\tau\,\|v^\tau(k,k-1)\|_{E\to E}+M\tau.$$

From this estimate it follows that

$$\|v^\tau(k,k-1)\|_{E\to E}\le M,\ \ \|v^\tau(k,k-1)-R_{j,l}(\tau A(t_k))\|_{E\to E}\le M\tau. \tag{4.131}$$

In a similar manner one establishes the estimate

$$\left\|[v^\tau(k,k-1)-R_{j,l}(\tau A(t_k))]\,A^{-1}(t_k)\right\|_{E\to E}\le M\tau^2. \tag{4.132}$$

Now we define the operator

$$\widetilde{v}^{\tau}(k,k-1) = R_{j,l}(\tau A(t_k)).$$

Since

$$\widetilde{v}^{\tau}(k,k-1) = \widetilde{u}^{\tau}(k+1,k),$$

we obviously have

$$\widetilde{v}^{\tau}(k,n) = \widetilde{u}^{\tau}(k+1,n+1).$$

Therefore, by Lemma 4.2.18 and Lemma 4.2.19, we have that

$$\|v^{\tau}(k,n)\|_{E\to E} \le M, \left\|A(t_k)v^{\tau}(k,n)A^{-1}(t_n)\right\|_{E\to E} \le M \tag{4.133}$$

for $1 \le n \le k \le N$,

$$\|A(t_k)v^{\tau}(k,n)\|_{E\to E} \le \frac{M}{(k-n)\tau}, 1 \le n < k \le N, \tag{4.134}$$

for $1 \le n < k \le N$,

$$\|A(t_{k+r})v^{\tau}(k+r,n) - A(t_k)v^{\tau}(k,n)\|_{E\to E} \le \frac{M\Psi}{(k-n)\tau}, \tag{4.135}$$

$$\left\|[A(t_{k+r})v^{\tau}(k+r,n) - A(t_k)v^{\tau}(k,n)]A^{-1}(t_n)\right\|_{E\to E} \le M\Psi, \tag{4.136}$$

for $1 \le n < k \le k+r \le N$. Here $\Psi = r^{\alpha}(k-n)^{-\alpha}, 0 \le \alpha \le 1$.

From formula (4.130) and estimates (4.133), (4.134), (4.135) and (4.136) one derives the following results.

Theorem 4.3.1. *The solutions of the difference problem* (4.122) *in* $C_{\tau}(E)$ *obey the stability inequality*

$$\|u^{\tau}\|_{C(\tau,E)} \le M[\|(I - \sum_{i=1}^{j} \beta_i \tau^i A_{i-1}(0))u_0\|_E + \|f^{\tau}\|_{C(\tau,E)}],$$

where M_1 *is independent not only of* f^{τ}, u_0, *but also of* τ.

Theorem 4.3.2. *The solutions of the difference problem* (4.122) *in* $C_{\tau}^{\alpha,\alpha}(E)$ *obey the stability inequality*

$$\|u^{\tau}\|_{C_0^{\alpha,\alpha}(\tau,E)} \le M[\|(I - \sum_{i=1}^{j} \beta_i \tau^i A_{i-1}(0))u_0\|_E + \|f^{\tau}\|_{C_0^{\alpha,\alpha}(\tau,E)}],$$

where M_1 *is independent not only of* f^{τ}, u_0, α, *but also of* τ.

Theorem 4.3.3. *The solutions of the difference problem* (4.122) *in* $L_{p,\tau}(E)$, $1 \leq p < \infty$ *obey the stability inequality*

$$\|u^\tau\|_{L_{p,\tau}(E)} \leq M[\|(I - \sum_{i=1}^{j} \beta_i \tau^i A_{i-1}(0))u_0\|_E + \|f^\tau\|_{L_{p,\tau}(E)}],$$

where M_1 *is independent not only of* f^τ, u_0, p. *but also of* τ.

Theorem 4.3.4. *The solutions of the difference problem* (4.122) *in* $C_\tau(E)$ *obey the almost coercive inequality*

$$\|\{\tau^{-1}(u_k - u_{k-1})\}_1^N\|_{C_\tau(E)}$$

$$+ \| \{-\sum_{i=1}^{l} \alpha_i \tau^{i-1} A_{i-1}(t_k) u_k + \sum_{i=1}^{j} \beta_i \tau^{i-1} A_{i-1}(t_{k-1}) u_{k-1}\}_1^N \|_{C_\tau(E)}$$

$$\leq M_1[\| A(0)(I - \sum_{i=1}^{j} \beta_i \tau^i A_{i-1}(0)) u_0\|_E$$

$$+ \min\{\ln \frac{1}{\tau}, 1 + |\ln \|A(0)\|_{E\to E}|\} \| f^\tau \|_{C_\tau(E)}],$$

where M_1 *is independent not only of* f^τ, u_0, *but also of* τ.

Theorem 4.3.5. *The solutions of the difference problem* (4.122) *in* $C_\tau^{\alpha,\alpha}(E)$ *obey the coercivity inequality*

$$\|\{\tau^{-1}(u_k - u_{k-1})\}_1^N\|_{C_\tau^{\alpha,\alpha}(E)}$$

$$+ \| \{-\sum_{i=1}^{l} \alpha_i \tau^{i-1} A_{i-1}(t_k) u_k + \sum_{i=1}^{j} \beta_i \tau^{i-1} A_{i-1}(t_{k-1}) u_{k-1}\}_1^N \|_{C_\tau^{\alpha,\alpha}(E)}$$

$$\leq M_1[\frac{1}{\alpha(1-\alpha)}\|f^\tau\|_{C_\tau^{\alpha,\alpha}(E)} + \|A(0)(I - \sum_{i=1}^{j} \beta_i \tau^i A_{i-1}(0)) u_0\|_E],$$

where M_1 *is independent not only of* f^τ, u_0, α, *but also of* τ.

Theorem 4.3.6. *The solutions of the difference problem* (4.122) *in* $C_\tau(E'_\alpha)$ *obey the coercivity inequality*

$$\|\{\tau^{-1}(u_k - u_{k-1})\}_1^N \|_{C_\tau(E'_\alpha)}$$

$$+ \| \{-\sum_{i=1}^{l} \alpha_i \tau^{i-1} A_{i-1}(t_k) u_k + \sum_{i=1}^{j} \beta_i \tau^{i-1} A_{i-1}(t_{k-1}) u_{k-1}\}_1^N \|_{C_\tau(E'_\alpha)}$$

$$\leq M_1[\frac{1}{\alpha(1-\alpha)}\| f^\tau \|_{C_\tau(E'_\alpha)} + \| A(0)(I - \sum_{i=1}^{j} \beta_i \tau^i A_{i-1}(0)) u_0\|_{E'_\alpha}],$$

where M_1 *is independent not only of* f^τ, u_0, α, *but also of* τ.

Theorem 4.3.7. *The solutions of the difference problem* (4.122) *in* $L_{p,\tau}(E)$ *obey the almost coercive inequality*

$$\| \{\tau^{-1}(u_k - u_{k-1})\}_1^N \|_{L_{p,\tau}(E)}$$

$$+\| \{-\sum_{i=1}^{l} \alpha_i \tau^{i-1} A_{i-1}(t_k) u_k + \sum_{i=1}^{j} \beta_i \tau^{i-1} A_{i-1}(t_{k-1}) u_{k-1}\}_1^N \|_{L_{p,\tau}(E)}$$

$$\leq M_1 [\| A(0)(I - \sum_{i=1}^{j} \beta_i \tau^i A_{i-1}(0)) u_0 \|_E$$

$$+ \min\left\{ \ln \frac{1}{\tau}, 1 + |\ln \| A(0) \|_{E \to E}| \right\} \| f^\tau \|_{L_{p,\tau}(E)}],$$

where M_1 *does not depend on* f^τ, u_0, p *and* τ.

Theorem 4.3.8. *Suppose that difference problem* (4.122) *is well posed in* $L_{p_0,\tau}(E)$ *for some* p_0, $1 < p_0 < \infty$. *Then it is well posed in* $L_{p,\tau}(E)$ *for all* p, $1 < p < \infty$ *and the following coercivity inequality holds:*

$$\| \{\tau^{-1}(u_k - u_{k-1})\}_1^N \|_{L_{p,\tau}(E)}$$

$$+ \| \{-\sum_{i=1}^{l} \alpha_i \tau^{i-1} A_{i-1}(t_k) u_k + \sum_{i=1}^{j} \beta_i \tau^{i-1} A_{i-1}(t_{k-1}) u_{k-1}\}_1^N \|_{L_{p,\tau}(E)}$$
$$+ \| u^\tau \|_{C_\tau(E_{1-1/p,p})}$$

$$\leq M(p_0) [\| (I - \sum_{i=1}^{j} \beta_i \tau^i A_{i-1}(0)) u_0 \|_{E_{1-1/p,p}} + \frac{p^2}{p-1} \| f^\tau \|_{L_{p,\tau}(E)}],$$

where $M(p_0)$ *does not depend on* f^τ, u_0, p *and* τ.

Theorem 4.3.9. *Let* $1 \leq p \leq \infty$ *and* $0 < \alpha < 1$. *Problem* (4.122) *is well posed in* $L_{p,\tau}(E'_{\alpha,p})$ *and the following coercivity inequality holds:*

$$\| \{\tau^{-1}(u_k - u_{k-1})\}_1^N \|_{L_{p,\tau}(E'_{\alpha,p})}$$

$$+\| \{-\sum_{i=1}^{l} \alpha_i \tau^{i-1} A_{i-1}(t_k) u_k + \sum_{i=1}^{j} \beta_i \tau^{i-1} A_{i-1}(t_{k-1}) u_{k-1}\}_1^N \|_{L_{p,\tau}(E'_{\alpha,p})}$$

$$\leq M [\| A(0)(I - \sum_{i=1}^{j} \beta_i \tau^i A_{i-1}(0)) u_0 \|_{E'_{\alpha,p}} + \frac{1}{\alpha(1-\alpha)} \| f^\tau \|_{L_{p,\tau}(E'_{\alpha,p})}],$$

where M *does not depend on* f^τ, u_0, p, α *and* τ.

Theorem 4.3.10. *Let $1 < p, q < \infty$ and $0 < \alpha < 1$. The difference problem* (4.122) *is well posed in $L_{p,\tau}(E'_{\alpha,q})$ and the following coercivity inequality holds:*

$$\| \{\tau^{-1}(u_k - u_{k-1})\}_1^N \|_{L_{p,\tau}(E'_{\alpha,q})}$$

$$+\| \{-\sum_{i=1}^{l} \alpha_i \tau^{i-1} A_{i-1}(t_k) u_k + \sum_{i=1}^{j} \beta_i \tau^{i-1} A_{i-1}(t_{k-1}) u_{k-1}\}_1^N \|_{L_{p,\tau}(E'_{\alpha,q})}$$

$$\leq M(q)[\| A(0)(I - \sum_{i=1}^{j} \beta_i \tau^i A_{i-1}(0)) u_0 \|_{E'_{\alpha,q}} + \frac{p^2}{p-1} \| f^\tau \|_{L_{p,\tau}(E'_{\alpha,q})}],$$

where $M(q)$ does not depend on f^τ, u_0, p, α and τ.

Note that the difference schemes of the high order of accuracy with respect to one variable for approximate solutions of the mixed problems (4.50), (4.51) and (4.52) generated by (4.122) can be constructed. Abstract theorems given above permit us to establish the stability, almost stability and coercive stability of these difference schemes.

Chapter 5

Partial Differential Equations of Elliptic Type

In the present chapter we consider the well-posedness of an abstract boundary-value problem for differential equations of elliptic type

$$-v''(t) + Av(t) = f(t) \quad (0 \le t \le T), v(0) = v_0, v(T) = v_T$$

in an arbitrary Banach space E with the positive operator A. The high order of accuracy two-step difference schemes generated by an exact difference scheme or by the Taylor decomposition on three points for the numerical solutions of this problem are presented. The well-posedness of these difference schemes in various Banach spaces are studied. The stability and coercive stability estimates in Hölder norms for solutions of the high order of accuracy difference schemes of the mixed type boundary-value problems for elliptic equations are obtained.

5.1 A Boundary-Value Problem. Well-posedness

We consider the boundary-value problem for differential equation

$$-v''(t) + Av(t) = f(t) \quad (0 \le t \le T), v(0) = v_0, v(T) = v_T \tag{5.1}$$

in an arbitrary Banach space E with the linear positive operator A.

A function $v(t)$ is called a solution of problem (5.1) if the following conditions are satisfied:

i. $v(t)$ is twice continuously differentiable on the segment $[0, T]$. The derivatives at the endpoints of the segment are understood as the appropriate unilateral derivatives.

ii. The element $v(t)$ belongs to $D(A)$ for all $t \in [0, T]$, and the function $Av(t)$ is continuous on the segment $[0, T]$.

iii. $v(t)$ satisfies the equation and boundary conditions (5.1).

A solution of problem (5.1) defined in this manner will from now on be referred to as a solution of problem (5.1) in the space $C(E) = C([0,T],E)$. Here $C(E)$ stands for the Banach space of all continuous functions $\varphi(t)$ defined on $[0,T]$ with values in E equipped with the norm

$$||\varphi||_{C(E)} = \max_{0\leq t\leq T} \|\varphi(t)\|_E.$$

If $v(t)$ is a solution in $C(E)$ of problem (5.1), then the data of the problem satisfy the following conditions:

a) $f(t)$ belongs to $C(E)$;

b) v_0 and v_T belongs to $D(A)$.

We say that problem (5.1) is well posed in $C(E)$, if the following two conditions are satisfied:

1. For any $f \in C(E)$ and v_0, $v_T \in D(A)$ there exists the unique solution $v(t) = v(t; f(t), v_0, v_T)$ in $C(E)$ of problem (5.1). This means that an additive and homogeneous operator $v(t; f(t), v_0, v_T)$ is defined which acts from $C(E) \times D(A) \times D(A)$ to $C(E)$ and gives the solution of (5.1) in $C(E)$.
2. $v(t; f(t), v_0, v_T)$, regarded as an operator from $C(E) \times D(A) \times D(A)$ to $C(E)$, is continuous. Here $C(E) \times D(A) \times D(A)$ is understood as the normed space of the triplets $(f(t), v_0, v_T)$, $f(t) \in C(E)$ and v_0, $v_T \in D(A)$, equipped with the norm

$$||(f(t), v_0, v_T)||_{C(E)\times D(A)\times D(A)} = ||f||_{C(E)} + ||v_0||_{D(A)} + ||v_T||_{D(A)}.$$

By Banach's theorem in $C(E)$ and these properties one has coercive inequality

$$\|v''\|_{C(E)} + \|Av\|_{C(E)} \leq M_C[\|f\|_{C(E)} + ||Av_0||_E + ||Av_T||_E], \tag{5.2}$$

where M_C $(1 \leq M_C < +\infty)$ does not depend on v_0, v_T and $f(t)$. Inequality (5.2) permits us to investigate the spectral properties of operator A generated by the well-posedness in $C(E)$ of problem (5.1). For any $u \in D(A)$ and $\lambda \geq 0$ we will put

$$\psi = \lambda u + Au.$$

Then, evidently, the function $e^{i\sqrt{\lambda}t}u(i = \sqrt{-1})$ is a solution in $C(E)$ of equation

$$-v''(t) + Av(t) = f(t)$$

with $f(t) = e^{i\sqrt{\lambda}t}\psi$. Let $w(t)$ be a smooth function defined on $[0,T]$ by the formula

$$w(t) = \begin{cases} 0, & \text{for any } t \in [0, \frac{T}{8}] \cup [\frac{7T}{8}, T], \\ a(t), & \text{for any } t \in [\frac{T}{8}, \frac{T}{4}] \cup [\frac{3T}{4}, \frac{7T}{8}], \\ 1, & \text{for any } t \in [\frac{T}{4}, \frac{3T}{4}], \end{cases}$$

where $a(t)$ be some smooth function and $|a(t)| \leq 1$. Then, the function $v(t) = w(t)e^{i\sqrt{\lambda}t}u$ is a solution in $C(E)$ of problem (5.1) with $v(0) = v(T) = 0$ and

$$f(t) = \begin{cases} 0, & \text{for any } t \in [0, \frac{T}{8}] \cup [\frac{7T}{8}, T], \\ -(a(t)e^{i\sqrt{\lambda}t})_{tt}u + a(t)e^{i\sqrt{\lambda}t}Au, & \text{for any } t \in [\frac{T}{8}, \frac{T}{4}] \cup [\frac{3T}{4}, \frac{7T}{8}], \\ e^{i\sqrt{\lambda}t}\psi, & \text{for any } t \in [\frac{T}{4}, \frac{3T}{4}]. \end{cases}$$

Therefore from the coercive inequality (5.2) it follows that

$$\begin{aligned} \lambda \|u\|_E + \|Au\|_E &= \max_{t\in[\frac{T}{4},\frac{3T}{4}]} \|v''(t)\|_E + \max_{t\in[\frac{T}{4},\frac{3T}{4}]} \|Av(t)\|_E \\ &\leq \max_{t\in[\frac{T}{8},\frac{7T}{8}]} \|v''(t)\|_E + \max_{t\in[\frac{T}{8},\frac{7T}{8}]} \|Av(t)\|_E \\ &= \|v''\|_{C(E)} + \|Av\|_{C(E)} \leq M_C \|f\|_{C(E)} \\ = M_C \max\{\|\psi\|_E, &\max_{t\in[\frac{T}{8},\frac{T}{4}]\cup[\frac{3T}{4},\frac{7T}{8}]} \| - (a(t)e^{i\sqrt{\lambda}t})_{tt}u + a(t)e^{i\sqrt{\lambda}t}Au\|_E\}. \end{aligned}$$

Since

$$-(a(t)e^{i\sqrt{\lambda}t})_{tt}u + a(t)e^{i\sqrt{\lambda}t}Au = a(t)e^{i\sqrt{\lambda}t}\psi - 2a'(t)i\sqrt{\lambda}e^{i\sqrt{\lambda}t}u - a''(t)e^{i\sqrt{\lambda}t}u,$$

we have that

$$\begin{aligned} &\| - (a(t)e^{i\sqrt{\lambda}t})_{tt}u + a(t)e^{i\sqrt{\lambda}t}Au\|_E \\ \leq\ &|a(t)|\|\psi\|_E + 2|a'(t)||\frac{d}{dt}(e^{i\sqrt{\lambda}t})|||u||_E + |a''(t)|||u||_E \\ \leq\ &\|\psi\|_E + 2M_1\sqrt{\lambda}||u||_E + M_2||u||_E, \end{aligned}$$

where

$$M_1 = \max_{t\in[\frac{T}{8},\frac{T}{4}]\cup[\frac{3T}{4},\frac{7T}{8}]} |a'(t)|, M_2 = \max_{t\in[\frac{T}{8},\frac{T}{4}]\cup[\frac{3T}{4},\frac{7T}{8}]} |a''(t)|.$$

Therefore

$$\lambda \|u\|_E + \|Au\|_E \leq M_C[\|\psi\|_E + 2M_1\sqrt{\lambda}||u||_E + M_2||u||_E].$$

Using the inequality

$$0 < 2\sqrt{\lambda} < \varepsilon\lambda + \varepsilon^{-1}$$

for $\varepsilon > 0$, we obtain that

$$[\lambda(1 - \varepsilon M_1 M_C) - (\varepsilon^{-1}M_1M_C + M_2M_C)]\, \|u\|_E + \|Au\|_E \leq M_C\|\psi\|_E.$$

Let

$$0 < \varepsilon < (M_1M_C)^{-1}, \delta = \min_{0<\varepsilon<(M_1M_C)^{-1}} (\varepsilon^{-1}M_1M_C + M_2M_C)(1 - \varepsilon M_1M_C)^{-1}$$

and $\lambda \geq \delta + \eta$ for some $\eta > 0$. Then the inequality

$$(\lambda + 1)\|u\|_E + \|Au\|_E \leq M_p(\eta)\|\psi\|_E$$

holds for some $0 \leq M_p(\eta) < \infty$. This means that there exists the inverse operator $(\lambda I + A)^{-1}$, which is defined on the range $R(\lambda I + A)$ of operator $\lambda I + A$. We will prove that $R(\lambda I + A) = E$ for sufficiently large λ, $\lambda > 0$.

Then it follows that there exists the inverse operator $(\lambda I + A)^{-1}$ and the inequality

$$(\lambda + 1)\left\|(\lambda I + A)^{-1}\psi\right\|_E + \left\|A(\lambda I + A)^{-1}\psi\right\|_E \leq M_p(\eta)\|\psi\|_E$$

holds for all λ, $\lambda \geq \delta + \eta$ and $\psi \in E$. Now we will prove that $R(\lambda I + A) = E$ for all λ, $\lambda > 0$. We consider the boundary-value problem (5.1). Making the substitution $t = \alpha\tau$ for $\alpha > 0$, we obtain the well-posed in $C([0, \frac{T}{\alpha}], E)$ boundary-value problem

$$-\frac{1}{\alpha^2}\widetilde{v}''(\tau) + A\widetilde{v}(\tau) = \widetilde{f}(\tau)(0 \leq \tau \leq \frac{T}{\alpha}), \widetilde{v}(0) = v_0, \widetilde{v}(\frac{T}{\alpha}) = v_T.$$

Let $\widetilde{v}(\tau)$ be the solution of this problem for $v_0 = v_T = 0$ and $\widetilde{f}(\tau) = f_0$, where f_0 is an arbitrary element of E. Since f_0 and A are not dependent in τ, we have that $\widetilde{v}(\tau) = \widetilde{v}(\frac{T}{2\alpha} - \tau)$ for all $0 \leq \tau \leq \frac{T}{2\alpha}$. Then, in particular, it follows that $\widetilde{v}'(0) = -\widetilde{v}'(\frac{T}{\alpha})$. Finally, choosing $\sqrt{\lambda}\frac{T}{\alpha} = (2m+1)\pi$ for some $m = 1, 2, \ldots$ and integrating by parts, we obtain

$$-\int_0^{\frac{T}{\alpha}} \widetilde{v}''(\tau)e^{i\sqrt{\lambda}\tau}d\tau = [-\widetilde{v}'(\frac{T}{\alpha})e^{i\sqrt{\lambda}\frac{T}{\alpha}} - \widetilde{v}'(0)] + i\sqrt{\lambda}[\widetilde{v}(\frac{T}{\alpha})e^{i\sqrt{\lambda}\frac{T}{\alpha}} - \widetilde{v}(0)]$$

$$-(i\sqrt{\lambda})^2\int_0^{\frac{T}{\alpha}} \widetilde{v}(\tau)e^{i\sqrt{\lambda}\tau}d\tau = \lambda\int_0^{\frac{T}{\alpha}} \widetilde{v}(\tau)e^{i\sqrt{\lambda}\tau}d\tau.$$

Therefore,

$$-\int_0^{\frac{T}{\alpha}} \widetilde{v}''(\tau)e^{i\sqrt{\lambda}\tau}d\tau = [-\widetilde{v}'(\frac{T}{\alpha})e^{i\sqrt{\lambda}\frac{T}{\alpha}} - \widetilde{v}'(0)] + i\sqrt{\lambda}[\widetilde{v}(\frac{T}{\alpha})e^{i\sqrt{\lambda}\frac{T}{\alpha}} - \widetilde{v}(0)]$$

$$\frac{\lambda}{\alpha^2}\int_0^{\frac{T}{\alpha}} \widetilde{v}(\tau)e^{i\sqrt{\lambda}\tau}d\tau + A\int_0^{\frac{T}{\alpha}} \widetilde{v}(\tau)e^{i\sqrt{\lambda}\tau}d\tau = \int_0^{\frac{T}{\alpha}} f_0e^{i\sqrt{\lambda}\tau}d\tau = -\frac{2}{i\sqrt{\lambda}}f_0$$

for all λ, $\lambda > 0$ and $f_0 \in E$. This means that $R(\lambda I + A) = E$ for all $\sqrt{\lambda}\frac{T}{\alpha} = (2m+1)\pi$ and $m = 1, 2, \ldots$. From that it follows that there exists the bounded inverse $(\lambda I + A)^{-1}$ for all $\lambda \geq \delta + \eta$.

An operator A acting in a Banach space E and having a dense domain $D(A)$ is called positive if the operator $\lambda I + A$ has a bounded-in-E inverse for all λ, $\lambda \geq 0$, and the estimate

$$\left\|(\lambda I + A)^{-1}\right\|_{E \to E} \leq M_p(\lambda + 1)^{-1} \tag{5.3}$$

is satisfied for some $1 \leq M_p < +\infty$. Note that we are not able to obtain the estimate (5.3) for all λ, $\lambda \geq 0$.

Is the positivity of operator A a sufficient condition for the well-posedness of (5.1) in $C(E)$? As it turns out, problem (5.1) is not well posed in $C(E)$ for all such operators. Let us give an example.

Let A be the operator acting in $E = C[0,1]$ defined by the formula $Av(x) = -v''(x) + v(x)$, with the domain

$$D(A) = \{v(x) : v''(x) \in C[0,1], v(0) = v(1) = 0\}.$$

Evidently, the operator $\lambda I + A$ has a bounded inverse for any $\lambda \geq 0$, and formula

$$[(\lambda I + A)^{-1} v](x) = -\frac{1}{2\sqrt{1+\lambda}} \left(1 - e^{-2\sqrt{1+\lambda}}\right)^{-1} [e^{-(1-x)\sqrt{1+\lambda}} - e^{-(1+x)\sqrt{1+\lambda}}]$$

$$\times \int_0^1 [e^{-(1-\tau)\sqrt{1+\lambda}} - e^{-(1+\tau)\sqrt{1+\lambda}}] v(\tau) d\tau$$

$$+ \frac{1}{2\sqrt{1+\lambda}} \int_0^1 [e^{-|x-\tau|\sqrt{1+\lambda}} - e^{-|x+\tau|\sqrt{1+\lambda}}] v(\tau) d\tau$$

holds. From this formula it follows that A is a positive operator in the Banach space $E = C[0,1]$. Now let us show that in this case the boundary-value problem (5.1) is not well posed in $C(E) = C([0,1], C[0,1])$. Then (5.1) turns into the boundary-value problem

$$\begin{cases} -\dfrac{\partial^2 u}{\partial t^2} - \dfrac{\partial^2 u}{\partial x^2} + u = f(t,x), \quad 0 < t < 1, \ 0 < x < 1, \\ u(0,x) = u(1,x) = 0, \ 0 \leq x \leq 1, \\ u(t,0) = u(t,1) = 0, \ 0 \leq t \leq 1. \end{cases} \tag{5.4}$$

However, it is well known that problem (5.4) is not well posed in

$$C(E) = C([0,1], C[0,1]).$$

The corresponding counterexample can be given by

$$u(t,x) = [x^2(1-x)^2 - t^2(1-t)^2] \ln \frac{1}{x^2(1-x)^2 + t^2(1-t)^2}.$$

It means that $u(t,x)$, $u_t(t,x)$, $u_x(t,x) \in C([0,1], C[0,1])$ and

$$u_{tt}(t,x) = -2 \ln \frac{1}{x^2(1-x)^2 + t^2(1-t)^2} + \psi(t,x),$$

$$u_{xx}(t,x) = 2 \ln \frac{1}{x^2(1-x)^2 + t^2(1-t)^2} + \varphi(t,x)$$

for some continuous functions $\psi(t,x)$ and $\varphi(t,x)$. Therefore, evidently, the boundary-value problem (5.4) is ill posed in $C([0,1],C[0,1])$. It means that coercive inequality (5.2) is not true for any solution in $C(E)$ of the boundary-value problem (5.1).

Note that from the estimate (5.3), evidently, it follows that the operator $\lambda I + A$ has bounded inverse for all complex numbers

$$\lambda = \sigma + i\tau \in G_\varepsilon^+ = G_\varepsilon^+(M_p) \quad (0 < \varepsilon < 1),$$

such that

$$|\tau| \leq \frac{1-\varepsilon}{M_p}(1+\sigma), \text{ for } \sigma \geq 0$$

or

$$(\sigma^2+\tau^2)^{1/2} \leq \frac{1-\varepsilon}{M_p}\ , \text{ for } \sigma \leq 0$$

and the estimate

$$\left\|(\lambda I + A)^{-1}\right\|_{E\to E} \leq M_1\varepsilon^{-1}(1+|\lambda|)^{-1}$$

holds for some $M_1 \in [1,+\infty)$ but does not depend on $\varepsilon \in (0,1)$. It means that the spectrum $\sigma(A)$ of A is a subset of set $G_\varepsilon^- = -G_\varepsilon^+$ and inside of G_ε^- and on its boundary ∂G_ε^- the resolvent $(\lambda I - A)^{-1}$ is subject to the bound

$$\left\|(\lambda I - A)^{-1}\right\|_{E\to E} \leq M_1\varepsilon^{-1}(1+|\lambda|)^{-1}. \tag{5.5}$$

Let $\psi(z)$ be an analytic function on a neighborhood of $\sigma(A)$, and suppose that ψ satisfies the estimate

$$(1+|z|)^\alpha\,|\psi(z)| \leq M_2$$

for some $0<\alpha<+\infty$, $1 \leq M_2 < +\infty$. Then the Cauchy–Riesz integral operator

$$\psi(A) = \frac{1}{2\pi i}\int\limits_{\partial G_\varepsilon^-} \psi(z)(zI-A)^{-1}dz \tag{5.6}$$

converges in the operator norm and defines a bounded linear operator $\psi(A)$, a function of A. In particular, the function $\psi(A) = z^{-\alpha}$ defines a bounded operator $A^{-\alpha}$ whenever $\alpha > 0$. The positive powers $A^\alpha = (A^{-\alpha})^{-1}(\alpha > 0)$ of the operator A are defined. The operators A^α $(\alpha > 0)$ are unbounded, and their domains $D(A^a)$ are dense in E if A is unbounded. One has the continuous embeddings $D(A^\alpha) \subset D(A^\beta)$ if $\beta < \alpha$. The following moment inequality holds:

$$\|A^\alpha u\|_E \leq M(\alpha,\beta)\left\|A^\beta u\right\|_E^{\alpha/\beta}\|u\|_E^{1-\alpha/\beta}\,(0\leq\alpha\leq\beta<+\infty, u\in D(A^\beta)) \tag{5.7}$$

for some $M(\alpha,\beta) \in [1,+\infty)$ that does not depend on $u \in D(A^\beta)$. The operator $A^{1/2}$ has better spectral properties than the positive operator A. Indeed, using the identity

$$\lambda I + A = (\sqrt{\lambda}I - A^{\frac{1}{2}})(\sqrt{\lambda}I + A^{\frac{1}{2}}),$$

inequality (5.7) for $\alpha = \frac{1}{2}$, $\beta = 1$ and estimate (5.5) we obtain

$$\left\|(\sqrt{\lambda}I - A^{\frac{1}{2}})^{-1}\right\|_{E\to E} = \left\|(\sqrt{\lambda}I + A^{\frac{1}{2}})(\lambda I + A)^{-1}\right\|_{E\to E}$$

$$\leq \sqrt{|\lambda|}\left\|(\lambda I + A)^{-1}\right\|_{E\to E} + M(\alpha,\beta)\left\|A(\lambda I + A)^{-1}\right\|^{\frac{1}{2}}_{E\to E}\left\|(\lambda I + A)^{-1}\right\|^{\frac{1}{2}}_{E\to E}$$

$$\leq \sqrt{|\lambda|}M_1\varepsilon^{-1}(1+|\lambda|)^{-1}$$
$$+M(\alpha,\beta)(1 + M_1\varepsilon^{-1}|\lambda|(1+|\lambda|)^{-1})(M_1\varepsilon^{-1}(1+|\lambda|)^{-1})^{\frac{1}{2}}.$$

Thus, the operator $\sqrt{\lambda}I - A^{\frac{1}{2}}$ has a bounded inverse for $\lambda \in G_\varepsilon^-$, and the estimate

$$\left\|(\sqrt{\lambda}I - A^{\frac{1}{2}})^{-1}\right\|_{E\to E} \leq M_3\varepsilon^{-1}(1+\sqrt{|\lambda|})^{-1}$$

holds for some $M_3 \in [1,+\infty)$ and does not depend on $\in$ and λ. This means that $B = A^{\frac{1}{2}}$ is a strongly positive operator in a Banach space E (see Chapter 4). Therefore, the operator $-B$ is a generator of an analytic semigroup $\exp\{-tB\}$ $(t \geq 0)$ with exponentially decreasing norm, when $t \longrightarrow +\infty$, i.e., the estimates

$$\|\exp(-tB)\|_{E\to E}, \|tB\exp(-tB)\|_{E\to E} \leq M(B)e^{-\alpha(B)t}(t>0) \qquad (5.8)$$

hold for some $M(B) \in [1,+\infty)$, $a(B) \in (0,+\infty)$. From that it follows that the operator $I - e^{-2TB}$ has a bounded inverse and the following estimate holds:

$$\|(I - e^{-2TB})^{-1}\|_{E\to E} \leq M(B)(1 - e^{-2T\alpha(B)})^{-1}. \qquad (5.9)$$

First of all let us prove a lemma that will be needed in the sequel.

Lemma 5.1.1. *For any $0 < t < t+\tau < T$ and $0 \leq \alpha \leq 1$ one has the inequality*

$$\|\exp(-tB) - \exp\{-(t+\tau)B\}\|_{E\to E} \leq M\frac{\tau^\alpha}{(t+\tau)^\alpha}, \qquad (5.10)$$

where M does not depend on α, t, and τ.

Proof. Using the formula

$$\exp(-tB) - \exp\{-(t+\tau)B\} = \int_t^{t+\tau} B\exp(-sB)ds$$

and estimate (5.8), we obtain that

$$\|\exp(-tB) - \exp\{-(t+\tau)B\}\|_{E\to E} \leq M_1\frac{\tau}{t}.$$

The inequality

$$\|\exp(-tB) - \exp\{-(t+\tau)B\}\|_{E\to E} \leq M_1$$

follows from (5.8) and the triangle inequality. Therefore

$$\|\exp(-tB)-\exp\{-(t+\tau)B\}\|_{E\to E}\leq M_1\min\{1,\frac{\tau}{t}\}.$$

From this it follows that

$$\|\exp(-tB)-\exp\{-(t+\tau)B\}\|_{E\to E}\leq M_1\frac{2\tau}{t+\tau}.$$

Interpolating the last inequality and

$$\|\exp(-tB)-\exp\{-(t+\tau)B\}\|_{E\to E}\leq M_1$$

we obtain (5.10). Lemma 5.1.1 is proved.

Now, we consider the boundary-value problem (5.1). It can be obviously rewritten as the equivalent boundary-value problem for the system of first-order linear differential equations

$$\begin{cases} v'(t)+Bv(t)=z(t)(0\leq t\leq T), v(0)=v_0, v(T)=v_T, \\ \\ -z'(t)+Bz(t)=f(t). \end{cases}$$

Integrating these, we at once obtain

$$\begin{cases} v(t)=e^{-tB}v_0+\int_0^t e^{-(t-s)B}z(s)ds, \\ z(t)=e^{-(T-t)B}z(T)+\int_t^T e^{-(s-t)B}f(s)ds. \end{cases}$$

From these formulas and the condition $v(T)=v_T$ it follows that

$$v(T)=e^{-TB}v_0+\int_0^T e^{-2(T-s)B}dsz(T)$$

$$+\int_0^T\int_s^T e^{-(T+y-2s)B}f(y)dyds.$$

Since $\int_0^T e^{-2(T-s)B}ds=(2B)^{-1}(I-e^{-2TB})$, we have that

$$z(T)=(I-e^{-2TB})^{-1}2B[v_T-e^{-TB}v_0$$

$$-\int_0^T\int_s^T e^{-(T+y-2s)B}f(y)dyds]$$

$$=(I-e^{-2TB})^{-1}2B[v_T-e^{-TB}v_0$$

$$-(2B)^{-1}\int_0^T(e^{-(T-y)B}-e^{-(T+y)B})f(y)dy].$$

So,

$$z(t) = e^{-(T-t)B}(I - e^{-2TB})^{-1}2B[v_T - e^{-TB}v_0$$

$$-(2B)^{-1}\int_0^T (e^{-(T-y)B} - e^{-(T+y)B})f(y)dy] + \int_t^T e^{-(s-t)B}f(s)ds$$

and

$$v(t) = e^{-tB}v_0 + \int_0^t e^{-(T+t-2s)B}ds(I - e^{-2TB})^{-1}2B[v_T - e^{-TB}v_0$$

$$-(2B)^{-1}\int_0^T (e^{-(T-y)B} - e^{-(T+y)B})f(y)dy]$$

$$+\int_0^t \int_y^T e^{-(t+s-2y)B}f(s)dyds.$$

By an interchange of the order of integration, we obtain

$$v(t) = e^{-tB}v_0 + (e^{-(T-t)B} - e^{-(T+t)B})(I - e^{-2TB})^{-1}[v_T - e^{-TB}v_0$$

$$-(2B)^{-1}\int_0^T (e^{-(T-y)B} - e^{-(T+y)B})f(y)dy]$$

$$+\int_0^t \int_0^y e^{-(t+s-2y)B}f(s)dsdy + \int_t^T \int_0^t e^{-(t+s-2y)B}f(s)dyds$$

$$= e^{-tB}v_0 + (e^{-(T-t)B} - e^{-(T+t)B})(I - e^{-2TB})^{-1}[v_T - e^{-TB}v_0$$

$$-(2B)^{-1}\int_0^T (e^{-(T-s)B} - e^{-(T+s)B})f(s)ds]$$

$$+(2B)^{-1}\int_0^t (e^{-(t-s)B} - e^{-(t+s)B})f(s)ds$$

$$+(2B)^{-1}\int_t^T (e^{-(s-t)B} - e^{-(t+s)B})f(s)ds$$

$$= (I - e^{-2TB})^{-1}\{(e^{-tB} - e^{-(2T-t)B})v_0 + (e^{-(T-t)B} - e^{-(T+t)B})v_T$$

$$-(e^{-(T-t)B} - e^{-(T+t)B})(2B)^{-1}\int_0^T (e^{-(T-s)B} - e^{-(T+s)B})f(s)ds\}$$

$$+(2B)^{-1}\int_0^T (e^{-|t-s|B} - e^{-(t+s)B})f(s)ds.$$

Thus,

$$v(t) = (I - e^{-2TB})^{-1}\{(e^{-tB} - e^{-(2T-t)B})v_0 + (e^{-(T-t)B} - e^{-(T+t)B})v_T \quad (5.11)$$

$$-(e^{-(T-t)B} - e^{-(T+t)B})(2B)^{-1} \int_0^T (e^{-(T-s)B} - e^{-(T+s)B}) f(s)ds\}$$

$$+(2B)^{-1} \int_0^T (e^{-|t-s|B} - e^{-(t+s)B}) f(s)ds.$$

It is easy to show that $v(t)$ defined on $[0,T]$ by formula (5.11) is a unique solution in $C(E)$ of problem (5.1) if, for example, $v_0,\ v_T \in D(A^2)$ and $Af(t) \in C(E)$ or $f'(t) \in C(E)$. Sufficient conditions for the well-posedness of the boundary-value problem (5.1) can be established if one considers this problem in certain spaces $F(E)$ of smooth E-valued functions on $[0,T]$.

A function $v(t)$ is said to be a solution of problem (5.1) in $F(E)$ if it is a solution of this problem in $C(E)$ and the functions $v''(t)$ and $Av(t)$ belong to $F(E)$.

As in the case of the space $C(E)$, we say that problem (5.1) is well posed in $F(E)$, if the following two conditions are satisfied:

1. For any $f \in F(E)$ and $v_0, v_T \in D(A)$ there exists a unique solution $v(t) = v(t; f(t), v_0, v_T)$ in $F(E)$ of problem (5.1). This means that an additive and homogeneous operator $v(t; f(t), v_0, v_T)$ is defined which acts from $F(E) \times D(A) \times D(A)$ to $F(E)$ and gives the solution of (5.1) in $F(E)$.

2. $v(t; f(t), v_0, v_T)$, regarded as an operator from $F(E) \times D(A) \times D(A)$ to $F(E)$, is continuous. Here $F(E) \times D(A) \times D(A)$ is understood as the normed space of the triplets $(f(t), v_0, v_T)$, $f(t) \in F(E)$ and $v_0, v_T \in D(A)$, equipped with the norm

$$||(f(t), v_0, v_T)||_{F(E)\times D(A)\times D(A)} = ||f||_{F(E)} + ||v_0||_{D(A)} + ||v_T||_{D(A)}.$$

First, we set $F(E)$ equal to $C_{0T}^{\alpha,\alpha}(E) = C_{0T}^{\alpha,\alpha}([0,T], E), 0 < \alpha < 1$, the Banach space obtained by completion of the set of smooth E-valued functions $\varphi(t)$ on $[0,T]$ in the norm

$$\| \varphi \|_{C_{0T}^{\alpha,\alpha}(E)} = \|\varphi\|_{C(E)} + \sup_{0\le t<t+\tau\le T} \frac{\|\varphi(t+\tau) - \varphi(t)\|_E}{\tau^\alpha} (T-t)^\alpha (t+\tau)^\alpha.$$

As in the case of the space $C(E)$, from the well-posedness of the boundary-value problem (5.1) one derives the coercivity inequality

$$\|v''\|_{C_{0T}^{\alpha,\alpha}(E)} + \|Av\|_{C_{0T}^{\alpha,\alpha}(E)} \le M_C(\alpha)[\|f\|_{C_{0T}^{\alpha,\alpha}(E)} + ||Av_0||_E + ||Av_T||_E] \quad (5.12)$$

where $M_C\ (\alpha)(1 \le M_C(\alpha) < +\infty)$ does not depend on v_0, v_T and $f(t)$.

Theorem 5.1.2. *Suppose $v_0, v_T \in D(A), f(t) \in C_{0T}^{\alpha,\alpha}(E)(0 < \alpha < 1)$. Then the boundary-value problem (5.1) is well posed in the Hölder space $C_{0T}^{\alpha,\alpha}(E)$, if A is* *the positive operator in the Banach space E. For the solution $v(t)$ in $C_{0T}^{\alpha,\alpha}(E)$ of the boundary-value problem the coercive inequality*

$$\|v''\|_{C_{0T}^{\alpha,\alpha}(E)} + \|Av\|_{C_{0T}^{\alpha,\alpha}(E)} \leq \frac{M}{\alpha(1-\alpha)} \| f \|_{C_{0T}^{\alpha,\alpha}(E)} + M[\| Av_0\|_E + \|Av_T\|_E]$$

holds, where M does not depend on α, v_0, v_T and $f(t)$.

Proof. If $v(t)$ is a solution in $C_{0T}^{\alpha,\alpha}(E)$ of problem (5.1), then it is a solution in $C(E)$ of this problem. Hence, by (5.11), we have the representation

$$v(t) = (I - e^{-2TB})^{-1}\{(e^{-tB} - e^{-(2T-t)B})v_0 + (e^{-(T-t)B} - e^{-(T+t)B})v_T$$
$$-(e^{-(T-t)B} - e^{-(T+t)B})(2B)^{-1} \int_0^T (e^{-(T-s)B} - e^{-(T+s)B})f(s)ds\}$$
$$+(2B)^{-1} \int_0^T (e^{-|t-s|B} - e^{-(t+s)B})f(s)ds.$$

To prove the theorem we must show that the functions $v''(t)$ and $Av(t)$ belong to $C_{0T}^{\alpha,\alpha}(E)$, and finally obtain estimates for the norms of these functions. Here we address only the last task. Using formula (5.11), we obtain

$$Av(t) = (I - e^{-2TB})^{-1}\{(e^{-tB} - e^{-(2T-t)B})Av_0 \tag{5.13}$$
$$+(e^{-(T-t)B} - e^{-(T+t)B})Av_T\}$$
$$-(e^{-(T-t)B} - e^{-(T+t)B})\frac{1}{2}B \int_0^T (e^{-(T-s)B} - e^{-(T+s)B})ds f(t)\}$$
$$+\frac{1}{2}B \int_0^T (e^{-|t-s|B} - e^{-(t+s)B})ds f(t)$$
$$-(e^{-(T-t)B} - e^{-(T+t)B})\frac{1}{2}B \int_0^T (e^{-(T-s)B} - e^{-(T+s)B})(f(s) - f(t))ds\}$$
$$+\frac{1}{2}B \int_0^T (e^{-|t-s|B} - e^{-(t+s)B})(f(s) - f(t))ds$$
$$= (I - e^{-2TB})^{-1}(e^{-tB} - e^{-(2T-t)B})Av_0 + (I - e^{-2TB})^{-1}(e^{-(T-t)B} - e^{-(T+t)B})Av_T$$
$$+(I - e^{-2TB})^{-1}(I - e^{-tB})(I - e^{-(T-t)B})(I - e^{-TB})f(t)$$
$$+(I - e^{-2TB})^{-1}\frac{B}{2}(I - e^{-(2T-2t)B}) \int_0^t e^{-(t-s)B}(I - e^{-2sB})(f(s) - f(t))ds$$
$$+(I - e^{-2TB})^{-1}\frac{B}{2}(I - e^{-2tB}) \int_t^T e^{-(s-t)B}(I - e^{-(2T-2s)B})(f(s) - f(t))ds$$
$$= \sum_{k=1}^{5} J_k(t).$$

Let us estimate $J_k(t)$ for any $k = 1, 2, 3, 4, 5$ separately. We start with $J_1(t)$. Using estimates (5.8)and (5.9) we show that, for $0 \le t \le T$,

$$||J_1(t)||_E \le ||(I - e^{-2TB})^{-1}||_{E\to E}||e^{-tB} - e^{-(2T-t)B}||_{E\to E}||Av_0||_E \le M||Av_0||_E.$$

Further, applying (5.9) and (5.10) we show that, for $0 < t < t+\tau < T$,

$$||J_1(t+\tau) - J_1(t)||_E \le ||(I - e^{-2TB})^{-1}||_{E\to E}[||e^{-tB} - e^{-(t+\tau)B}||_{E\to E}$$

$$+||e^{-(2T-t)B} - e^{-(2T-t-\tau)B}||_{E\to E}]||Av_0||_E \le M[\frac{\tau^\alpha}{(t+\tau)^\alpha} + \frac{\tau^\alpha}{(T-t)^\alpha}]||Av_0||_E.$$

Thus, we have proved that

$$||J_1||_{C_{0T}^{\alpha,\alpha}(E)} \le M||Av_0||_E. \tag{5.14}$$

In a similar manner we can show that

$$||J_2||_{C_{0T}^{\alpha,\alpha}(E)} \le M||Av_T||_E. \tag{5.15}$$

Now let us estimate $J_3(t)$. Using estimates (5.8) and (5.9), we obtain that

$$\| J_3(t) \|_E \le ||(I - e^{-2TB})^{-1}||_{E\to E}||I - e^{-tB}||_{E\to E}$$

$$\times||I - e^{-(T-t)B}||_{E\to E}||I - e^{-B}||_{E\to E}\|f(t)\|_E \le M\|f\|_{C_{0T}^{\alpha,\alpha}(E)}$$

for $0 \le t \le T$. From this inequality it follows that

$$\max_{0\le t\le T} \| J_3(t) \|_E \le M \| f \|_{C_{0T}^{\alpha,\alpha}(E)}\ .$$

Since

$$J_3(t+\tau) - J_3(t) = (I-e^{-2TB})^{-1}(I-e^{-tB})(I-e^{-(T-t)B})(I-e^{-TB})(f(t+\tau)-f(t))$$

$$+(I - e^{-2TB})^{-1}\{(e^{-tB} - e^{-(t+\tau)B})(I - e^{-(T-t)B})$$
$$+(I - e^{-tB})(e^{-(T-t)B} - e^{-(T-t-\tau)B})\}(I - e^{-TB})f(t)$$

it follows that

$$\| J_3(t+\tau) - J_3(t) \|_E \le ||(I - e^{-2TB})^{-1}||_{E\to E}$$

$$\times||I - e^{-tB}||_{E\to E}||I - e^{-(T-t)B}||_{E\to E}||I - e^{-TB}||_{E\to E}\|f(t+\tau) - f(t)\|_E$$

$$+||(I - e^{-2TB})^{-1}||_{E\to E}\{||e^{-tB} - e^{-(t+\tau)B}||_{E\to E}||I - e^{-(T-t)B}||_{E\to E}$$
$$+||I - e^{-tB}||_{E\to E}||e^{-(T-t)B} - e^{-(T-t-\tau)B}||_{E\to E}\}||I - e^{-TB}||_{E\to E}\|f(t)\|_E$$

for $0 < t < t + \tau < T$. Using estimates (5.8), (5.9) and (5.10), we obtain

$$\| J_3(t+\tau) - J_3(t) \|_E \le M[\frac{\tau^\alpha}{(t+\tau)^\alpha} + \frac{\tau^\alpha}{(T-t)^\alpha}] \| f \|_{C_{0T}^{\alpha,\alpha}(E)}$$

for $0 < t < t + \tau < T$. From this it follows that

$$\| J_3 \|_{C_{0T}^{\alpha,\alpha}(E)} \le M_1 \| f \|_{C_{0T}^{\alpha,\alpha}(E)} . \tag{5.16}$$

Now let us estimate $J_4(t)$. We have to consider two cases: $0 \le t \le \frac{T}{2}$ and $\frac{T}{2} \le t \le T$. In the first case, using estimates (5.8) and (5.9), we obtain that

$$\| J_4(t) \|_E \le \|(I - e^{-2TB})^{-1}\|_{E\to E} \frac{1}{2} \int_0^t \|Be^{-(t-s)B}(I - e^{-(2T-2t)B}) \|_{E\to E}$$

$$\times \|I - e^{-2sB} \|_{E\to E} \|f(s) - f(t) \|_E \, ds \le M \int_0^t \frac{ds}{(t-s)^{1-\alpha} t^\alpha (T-s)^\alpha} \| f \|_{C_{0T}^{\alpha,\alpha}(E)}$$

$$\le \frac{M}{\alpha}(T-t)^{-\alpha} \| f \|_{C_{0T}^{\alpha,\alpha}(E)} \le \frac{M2^\alpha}{\alpha T^\alpha} \| f \|_{C_{0T}^{\alpha,\alpha}(E)} .$$

In the second case, using estimates (5.8), (5.9) and (5.10), we obtain that

$$\| J_4(t) \|_E \le \|(I - e^{-2TB})^{-1}\|_{E\to E} \frac{1}{2} \int_0^t \|Be^{-(t-s)B}(I - e^{-(2T-2t)B}) \|_{E\to E}$$

$$\times \|I - e^{-2sB} \|_{E\to E} \|f(s) - f(t) \|_E \, ds$$

$$\le M \int_0^t \frac{2(T-t)ds}{(t-s)^{1-\alpha}(2T-t-s)t^\alpha (T-s)^\alpha} \| f \|_{C_{0T}^{\alpha,\alpha}(E)} .$$

Since

$$\int_0^t \frac{2(T-t)ds}{(t-s)^{1-\alpha}(2T-t-s)t^\alpha (T-s)^\alpha} \le \frac{1}{t^\alpha (T-t)^\alpha} \int_0^t \frac{2(T-t)ds}{(t-s)^{1-\alpha}(2T-t-s)}$$

$$= \frac{1}{t^\alpha (T-t)^\alpha} \int_0^t \frac{2(T-t)dz}{z^{1-\alpha}(2T-2t+z)} \le \frac{1}{t^\alpha (T-t)^\alpha (2T-2t)^{1-\alpha}} \int_0^\infty \frac{2(T-t)dp}{p^{1-\alpha}(1+p)}$$

$$= \frac{2^\alpha}{t^\alpha} \int_0^\infty \frac{dp}{p^{1-\alpha}(1+p)} \le \frac{4^\alpha}{T^\alpha} \int_0^\infty \frac{dp}{p^{1-\alpha}(1+p)},$$

it follows that

$$\| J_4(t) \|_E \leq \frac{M}{\alpha(1-\alpha)} \| f \|_{C_{0T}^{\alpha,\alpha}(E)} .$$

Combining the estimates for $\| J_4(t) \|_E$, we obtain

$$\max_{0\leq t\leq T} \| J_4(t) \|_E \leq \frac{M}{\alpha(1-\alpha)} \| f \| . \tag{5.17}$$

Now let us estimate the difference $J_4(t+\tau) - J_4(t)$ for $0 < t < t+\tau < T$. We will consider separately the cases $t \leq \tau$, $T-t \leq 2\tau$ and $t > \tau$, $T-t > 2\tau$. If $t \leq \tau$, then (5.17) yields

$$\| J_4(t+\tau) - J_4(t) \|_E \leq \| J_4(t+\tau) \|_E + \| J_4(t) \|_E$$

$$\leq \frac{M2^{1+\alpha}}{\alpha(1-\alpha)} \| f \|_{C_{0T}^{\alpha,\alpha}(E)} \tau^\alpha (t+\tau)^{-\alpha} = \frac{M_1}{\alpha(1-\alpha)} \| f \|_{C_{0T}^{\alpha,\alpha}(E)} \tau^\alpha (t+\tau)^{-\alpha}.$$

If $T-t \leq 2\tau$, then (5.17) yields

$$\| J_4(t+\tau) - J_4(t) \|_E \leq \| J_4(t+\tau) \|_E + \| J_4(t) \|_E$$

$$\leq \frac{M2^{1+\alpha}}{\alpha(1-\alpha)} \| f \|_{C_{0T}^{\alpha,\alpha}(E)} \tau^\alpha (T-t)^{-\alpha} = \frac{M_1}{\alpha(1-\alpha)} \| f \|_{C_{0T}^{\alpha,\alpha}(E)} \tau^\alpha (T-t)^{-\alpha}.$$

Now let $t > \tau$ and $T-t > 2\tau$. Let us represent the difference $J_4(t+\tau) - J_4(t)$ as the sum of the following integrals:

$$J_4(t+\tau) - J_4(t) = (I-e^{-2TB})^{-1} \frac{B}{2} (I - e^{-(2T-2t-2\tau)B})$$

$$\times \int_{t-\tau}^{t+\tau} e^{-(t+\tau-s)B} (I - e^{-2sB})(f(s) - f(t+\tau))ds$$

$$-(I-e^{-2TB})^{-1} \frac{B}{2} (I - e^{-(2T-2t)B}) \int_{t-\tau}^{t} e^{-(t-s)B} (I - e^{-2sB})(f(s) - f(t))ds$$

$$+(I-e^{-2TB})^{-1} \frac{B}{2} (I - e^{-(2T-2t-2\tau)B}) \int_{0}^{t-\tau} e^{-(t+\tau-s)B} (I - e^{-2sB})(f(t) - f(t+\tau))ds$$

$$+(I-e^{-2TB})^{-1} \int_{0}^{t-\tau} \frac{B}{2} [(e^{-(2T-2t)B} - e^{-(2T-2t-2\tau)B}) e^{-(t-s)B}$$

$$+(I - e^{-(2T-2t)B})(e^{-(t+\tau-s)B} - e^{-(t-s)B})](I - e^{-2sB})(f(s) - f(t))ds$$

$$= P_1 + P_2 + P_3 + P_4.$$

Now let us estimate P_1. Using estimates (5.8) and (5.9), we obtain that

$$\| P_1 \|_E \leq \|(I - e^{-2TB})^{-1}\|_{E\to E}\frac{1}{2}\int_{t-\tau}^{t+\tau} \|Be^{-(t+\tau-s)B}(I - e^{-(2T-2t-2\tau)B}) \|_{E\to E}$$

$$\times\|I - e^{-2sB} \|_{E\to E} \|f(s) - f(t+\tau) \|_E \, ds$$

$$\leq M\int_{t-\tau}^{t+\tau} \frac{ds}{(t+\tau-s)^{1-\alpha}(t+\tau)^{\alpha}(T-s)^{\alpha}} \| f \|_{C_{0T}^{\alpha,\alpha}(E)}$$

$$\begin{aligned} &\leq \frac{M}{\alpha}(2\tau)^{\alpha}(T-t-\tau)^{-\alpha}(t+\tau)^{-\alpha} \| f \|_{C_{0T}^{\alpha,\alpha}(E)} \\ &\leq \frac{2^{2\alpha}M}{\alpha}\tau^{\alpha}(T-t)^{-\alpha}(t+\tau)^{-\alpha} \| f \|_{C_{0T}^{\alpha,\alpha}(E)} \\ &= \frac{M_1}{\alpha}\tau^{\alpha}(T-t)^{-\alpha}(t+\tau)^{-\alpha} \| f \|_{C_{0T}^{\alpha,\alpha}(E)} . \end{aligned}$$

In exactly the same manner one establishes the estimate

$$\| P_2 \|_E \leq \frac{M_1}{\alpha}\tau^{\alpha}(T-t)^{-\alpha}(t+\tau)^{-\alpha} \| f \|_{C_{0T}^{\alpha,\alpha}(E)} .$$

Since

$$P_3 = (I - e^{-2TB})^{-1}\frac{B}{2}(I - e^{-(2T-2t-2\tau)B})$$

$$\times\int_0^{t-\tau} e^{-(t+\tau-s)B}(I - e^{-2sB})ds(f(t) - f(t+\tau))$$

$$= (I - e^{-2TB})^{-1}\frac{1}{2}(I - e^{-(2T-2t-2\tau)B})$$

$$\times[e^{-2\tau B} - 2e^{-(t+\tau)B} + e^{-2tB}](f(t) - f(t+\tau)),$$

it follows that

$$\| P_3 \|_E \leq \|(I - e^{-2TB})^{-1}\|_{E\to E}\frac{1}{2}\|I - e^{-(2T-2t-2\tau)B}\|_{E\to E}$$

$$\times\|e^{-2\tau B} - 2e^{-(t+\tau)B} + e^{-2tB}\|_{E\to E}\|f(t) - f(t+\tau)\|_E.$$

Using estimates (5.8) and (5.9), we obtain

$$\| P_3 \|_E \leq M_1\tau^{\alpha}(T-t)^{-\alpha}(t+\tau)^{-\alpha} \| f \|_{C_{0T}^{\alpha,\alpha}(E)} .$$

Now let us estimate P_4. Using estimates (5.8), (5.9) and (5.10), we obtain

$$\| P_3 \|_E \leq \|(I - e^{-2TB})^{-1}\|_{E\to E}\frac{1}{2}\int_0^{t-\tau} \{\| B(e^{-(2T-2t)B} - e^{-(2T-2t-2\tau)B})e^{-(t-s)B} \|_{E\to E}$$

$$+ \| B(I - e^{-(2T-2t)B})(e^{-(t+\tau-s)B} - e^{-(t-s)B}) \|_{E\to E} \}$$

$$\times\|I - e^{-2sB}\|_{E\to E} \| f(t) - f(s) \|_E \, ds$$

$$\leq M \int_0^{t-\tau} \frac{\tau ds}{(t-s)^{2-\alpha}t^\alpha (T-s)^\alpha} \| f \|_{C_{0T}^{\alpha,\alpha}(E)} .$$

Since for $t > \tau$ and $T - t > 2\tau$, we have the bound

$$\int_0^{t-\tau} \frac{\tau ds}{(t-s)^{2-\alpha}t^\alpha(T-s)^\alpha} \leq \frac{\tau}{t^\alpha(T-t+\tau)^\alpha}\int_0^{t-\tau}\frac{ds}{(t-s)^{2-\alpha}}$$

$$\leq \frac{\tau}{(1-\alpha)t^\alpha(T-t+\tau)^\alpha}\tau^{-1+\alpha} \leq \frac{2^\alpha\tau^\alpha}{(1-\alpha)(t+\tau)^\alpha(T-t)^\alpha}.$$

Therefore, we conclude that

$$\| P_4 \|_E \leq \frac{M_1}{1-\alpha}\tau^\alpha(T-t)^{-\alpha}(t+\tau)^{-\alpha} \| f \|_{C_{0T}^{\alpha,\alpha}(E)} .$$

Thus, we have shown that for any $0 < t < t+\tau < T$ the following inequality holds:

$$\| J_4(t+\tau) - J_4(t) \|_E \leq \frac{M_1}{\alpha(1-\alpha)}\tau^\alpha(T-t)^{-\alpha}(t+\tau)^{-\alpha} \| f \|_{C_{0T}^{\alpha,\alpha}(E)} . \quad (5.18)$$

Estimates (5.17) and (5.18) give

$$\| J_4 \|_{C_{0T}^{\alpha,\alpha}(E)} \leq \frac{M}{\alpha(1-\alpha)} \| f \|_{C_{0T}^{\alpha,\alpha}(E)} . \quad (5.19)$$

In exactly the same manner one establishes the inequality

$$\| J_5 \|_{C_{0T}^{\alpha,\alpha}(E)} \leq \frac{M}{\alpha(1-\alpha)} \| f \|_{C_{0T}^{\alpha,\alpha}(E)} . \quad (5.20)$$

Finally, applying estimates (5.14), (5.15), (5.16), (5.19) and (5.20) we obtain the estimate

$$\| Av \|_{C_{0T}^{\alpha,\alpha}(E)} \leq \frac{M}{\alpha(1-\alpha)} \| f \|_{C_{0T}^{\alpha,\alpha}(E)} + M[\| Av_0\|_E + \|Av_T\|_E].$$

By the triangle inequality, this last estimate and equation (5.1) yields

$$\| v''(t) \|_{C_{0T}^{\alpha,\alpha}(E)} \leq \frac{M}{\alpha(1-\alpha)} \| f \|_{C_{0T}^{\alpha,\alpha}(E)} + M[\| Av_0 \|_E + \|Av_T\|_E].$$

Theorem 5.1.2 is proved.

We denote by $\tilde{C}_{0T}^{\alpha,\alpha}(E) = \tilde{C}_{0T}^{\alpha,\alpha}([0,T],E)$, $0 < \alpha < 1$, the Banach space obtained by completion of the set of smooth E-valued functions $\varphi(t)$ on $[0,T]$ in the norm

$$\| \varphi \|_{\tilde{C}_{0T}^{\alpha,\alpha}(E)} = \|\varphi\|_{C(E)} + \sup_{0 \leq t < t+\tau \leq T} \frac{\|\varphi(t+\tau) - \varphi(t)\|_E}{\tau^\alpha} \min\{(T-t)^\alpha, (t+\tau)^\alpha\}.$$

It is easy to show that

$$\| \varphi \|_{C_{0T}^{\alpha,\alpha}(E)} \leq T^\alpha \| \varphi \|_{\tilde{C}_{0T}^{\alpha,\alpha}(E)} .$$

By the way, the norms of the spaces $C_{0T}^{\alpha,\alpha}(E)$ and $\tilde{C}_{0T}^{\alpha,\alpha}(E)$ are not equivalent. Nevertheless let us give, without proof, the following result.

Theorem 5.1.3. *Suppose v_0, $v_T \in D(A), f(t) \in \tilde{C}_{0T}^{\alpha,\alpha}(E)(0 < \alpha < 1)$. Then the boundary-value problem (5.1) is well posed in Hölder space $\tilde{C}_{0T}^{\alpha,\alpha}(E)$, if A is the positive operator in the Banach space E. For the solution $v(t)$ in $\tilde{C}_{0T}^{\alpha,\alpha}(E)$ of the boundary-value problem the coercive inequality*

$$\|v''\|_{\tilde{C}_{0T}^{\alpha,\alpha}(E)} + \|Av\|_{\tilde{C}_{0T}^{\alpha,\alpha}(E)} \leq \frac{M}{\alpha(1-\alpha)} \| f \|_{\tilde{C}_{0T}^{\alpha,\alpha}(E)} + M[\| Av_0\|_E + \|Av_T\|_E]$$

holds, where M does not depend on α, v_0, v_T and $f(t)$.

Note that the coercivity inequality

$$\|v''\|_{F(E)} + \|Av\|_{F(E)} \leq M_C[\|f\|_{F(E)} + \|Av_0\|_E + \|Av_T\|_E]$$

fails if we set $F(E)$ equal to $C^\alpha(E), (0 < \alpha < 1)$, the Banach space obtained by completion of the set of all smooth E-valued functions $\varphi(t)$ on [0, T] in the norm

$$\| \varphi \|_{C^\alpha(E)} = \max_{0 \leq t \leq T} \|\varphi(t)\|_E + \sup_{0 \leq t < t+\tau \leq T} \frac{\| \varphi(t+\tau) - \varphi(t) \|_E}{\tau^\alpha}.$$

Nevertheless, we have the following result.

Theorem 5.1.4. *Suppose v_0'', $v_T'' \in E_\alpha$, $f(t) \in C^\alpha(E)(0 < \alpha < 1)$. Then the boundary-value problem (5.1) is well posed in the Hölder space $C^\alpha(E)$, if A is the positive operator in the Banach space E. For the solution $v(t)$ in $C^\alpha(E)$ of the boundary-value problem the coercive inequality*

$$\|v''\|_{C^\alpha(E)} + \|Av\|_{C^\alpha(E)} + \|v''\|_{C(E_\alpha)} \leq \frac{M}{\alpha(1-\alpha)} \| f \|_{C^\alpha(E)}$$
$$+ \frac{M}{\alpha}[\| v_0''\|_{E_\alpha} + \|v_T''\|_{E_\alpha}]$$

holds, where M does not depend on α, v_0, v_T and $f(t)$. Here, the Banach space $E_\alpha = E_\alpha(E, B)$ $(0 < \alpha < 1)$ consists of those $v \in E$ for which the norm

$$\| v \|_{E_\alpha} = \sup_{z>0} z^{1-\alpha} \| B exp\{-zB\} v \|_E + \| v \|_E$$

is finite.

Proof. By formula (5.11), we have

$$\begin{aligned}
v''(t) &= (I - e^{-2TB})^{-1}(e^{-tB} - e^{-(2T-t)B})v_0'' \\
&+(I - e^{-2TB})^{-1}(e^{-(T-t)B} - e^{-(T+t)B})v_T'' \\
&+(I - e^{-2TB})^{-1}(e^{-(t+T)B} - e^{-(2T-t)B})f(0) \\
&+(I - e^{-2TB})^{-1}(e^{-(2T-t)B} - e^{-(T+t)B})f(T) \\
&-(I - e^{-2TB})^{-1}e^{-tB}(I - e^{-TB})(f(t) - f(0)) \\
&-(I - e^{-2TB})^{-1}e^{-(T-t)B}(I - e^{-TB})(f(t) - f(T)) \\
&+(I - e^{-2TB})^{-1}\frac{B}{2}(I - e^{-(2T-2t)B}) \int_0^t e^{-(t-s)B}(I - e^{-2sB})(f(s) - f(t))ds \\
&+(I - e^{-2TB})^{-1}\frac{B}{2}(I - e^{-2tB}) \int_t^T e^{-(s-t)B}(I - e^{-(2T-2s)B})(f(s) - f(t))ds \\
&= \sum_{k=1}^{8} G_k(t).
\end{aligned}$$

To prove the theorem it suffices to establish the estimates for $v''(t)$ in $C^\alpha(E)$ and in $C(E_\alpha)$. Let us estimate $G_k(t)$ for any $k = 1, 2, 3, 4, 5, 6, 7, 8$ separately. We start with $G_1(t)$. Using estimates (5.8), (5.9) and the definition of the spaces E_α we show that, for $0 \le t \le T$,

$$\begin{aligned}
||G_1(t)||_E &\le ||(I - e^{-2TB})^{-1}||_{E\to E} \int_t^{2T-t} ||B\exp(-sB)v_0''||_E ds \\
&\le M \int_t^{2T-t} \frac{ds}{s^{1-\alpha}} ||v_0''||_{E_\alpha} \le \frac{M_1}{\alpha}||v_0''||_{E_\alpha}.
\end{aligned}$$

Further, applying (5.8), (5.9) and the definition of the spaces E_α we show that, for $0 < t < t + \tau < T$,

$$\begin{aligned}
||G_1(t+\tau) - G_1(t)||_E &\le ||(I - e^{-2TB})^{-1}||_{E\to E}[\int_t^{t+\tau} ||B\exp(-sB)v_0''||_E ds \\
&+ \int_{2T-t-\tau}^{2T-t} ||B\exp(-sB)v_0''||_E ds]
\end{aligned}$$

$$\leq M[\int_{t}^{t+\tau} \frac{ds}{s^{1-\alpha}} + \int_{2T-t-\tau}^{2T-t} \frac{ds}{s^{1-\alpha}}]||v_0''||_{E_\alpha} \leq \tau^\alpha \frac{M_1}{\alpha}||v_0''||_{E_\alpha}.$$

Thus, we have proved that

$$||G_1||_{C^\alpha(E)} \leq \frac{M_1}{\alpha}||v_0''||_{E_\alpha}.$$

In a similar manner we can show that

$$||G_2||_{C^\alpha(E)} \leq \frac{M_1}{\alpha}||v_T''||_{E_\alpha}.$$

Using estimates (5.8), (5.9) and the definition of the spaces E_α we show that, for $0 \leq t \leq T$,

$$||G_1(t)||_{E_\alpha} \leq ||(I - e^{-2TB})^{-1}||_{E\to E}||e^{-tB} - e^{-(2T-t)B}||_{E\to E}||v_0''||_{E_\alpha} \leq M||v_0''||_{E_\alpha}.$$

Thus, we have proved that

$$||G_1||_{C(E_\alpha)} \leq M||v_0''||_{E_\alpha}.$$

In a similar manner we can show that

$$||G_2||_{C(E_\alpha)} \leq M||v_T''||_{E_\alpha}.$$

Now let us estimate $G_3(t)$. Using estimates (5.8) and (5.9) we show that, for $0 \leq t \leq T$,

$$\begin{aligned}||G_3(t)||_E &\leq ||(I - e^{-2TB})^{-1}||_{E\to E}||e^{-(t+T)B} - e^{-(2T-t)B}||_{E\to E}||f(0)||_E \\ &\leq M\|f\|_{C^\alpha(E)}.\end{aligned}$$

Further, applying (5.8), (5.9) and (5.10) we show that, for $0 < t < t+\tau < T$,

$$\| G_3(t+\tau) - G_3(t) \|_E \leq ||(I - e^{-2TB})^{-1}||_{E\to E}$$

$$\times\{||e^{-(t+T)B} - e^{-(t+T+\tau)B}||_{E\to E} + ||e^{-(2T-t)B} - e^{-(2T-t-\tau)B}||_{E\to E}\}\|f(0)\|_E$$

$$\leq M[\frac{\tau^\alpha}{(t+T+\tau)^\alpha} + \frac{\tau^\alpha}{(2T-t)^\alpha}] \parallel f \parallel_{C^\alpha(E)} \leq M_1\tau^\alpha \parallel f \parallel_{C^\alpha(E)} .$$

Thus, we have proved that

$$||G_3||_{C^\alpha(E)} \leq M_1 \parallel f \parallel_{C^\alpha(E)} .$$

In a similar manner we can show that

$$||G_4||_{C^\alpha(E)} \leq M_1 \parallel f \parallel_{C^\alpha(E)} .$$

By the definition of the spaces E_α and using estimates (5.8), (5.9), we obtain that

$$\| \lambda^{1-\alpha} B e^{-\lambda B} G_3(t) \|_E \leq \lambda^{1-\alpha} \|(I - e^{-2TB})^{-1}\|_{E\to E}$$

$$\times \|Be^{-\lambda B}(e^{-(t+T)B} - e^{-(2T-t)B})f(0)\|_E$$

$$\leq M[\frac{\lambda^{1-\alpha}}{\lambda+t+T} + \frac{\lambda^{1-\alpha}}{\lambda-t+2T}]\|f(0)\|_E$$

$$\leq M[\frac{1}{(t+T)^\alpha} + \frac{1}{(2T-t)^\alpha}]\|f(0)\|_E \leq M_1 \| f \|_{C^\alpha(E)}$$

for all $\lambda, \lambda > 0$ and for $0 \leq t \leq T$. So, we have established that

$$\|G_3\|_{C(E_\alpha)} \leq M_1 \| f \|_{C^\alpha(E)} .$$

In a similar manner we can show that

$$\|G_4\|_{C(E_\alpha)} \leq M_1 \| f \|_{C^\alpha(E)} .$$

Now let us estimate $G_5(t)$. Using estimates (5.8) and (5.9) we obtain that

$$\| G_5(t) \|_E \leq \|(I - e^{-2TB})^{-1}\|_{E\to E}\|I - e^{-TB}\|_{E\to E}$$

$$\times \|e^{-tB}\|_{E\to E}\|f(t) - f(0)\|_E \leq M\|f\|_{C^\alpha(E)} \leq Mt^\alpha\|f\|_{C^\alpha(E)}$$

for $0 \leq t \leq T$. From this inequality it follows that

$$\max_{0\leq t\leq T} \| G_5(t) \|_E \leq M_1 \| f \|_{C^\alpha(E)} .$$

Since

$$G_5(t+\tau) - G_5(t) = (I - e^{-2TB})^{-1}e^{-(t+\tau)B}(I - e^{-TB})(f(t+\tau) - f(t))$$

$$-(I - e^{-2TB})^{-1}\{(e^{-tB} - e^{-(t+\tau)B})(I - e^{-TB})(f(t) - f(0)),$$

it follows that

$$\| G_5(t+\tau) - G_5(t) \|_E \leq \|(I - e^{-2TB})^{-1}\|_{E\to E}$$

$$\times \|e^{-(t+\tau)B}\|_{E\to E}\|I - e^{-TB}\|_{E\to E}\|f(t+\tau) - f(t)\|_E$$

$$+\|(I - e^{-2TB})^{-1}\|_{E\to E}\|e^{-tB} - e^{-(t+\tau)B}\|_{E\to E}\|I - e^{-TB}\|_{E\to E}\|f(t) - f(0)\|_E$$

$$\leq M(\tau^\alpha + \frac{\tau^\alpha}{t^\alpha}t^\alpha) \| f \|_{C^\alpha(E)} = M_1\tau^\alpha \| f \|_{C^\alpha(E)}$$

for $0 < t < t+\tau < T$. From this it follows that

$$\| G_5 \|_{C^\alpha(E)} \leq M_1 \| f \|_{C^\alpha(E)} .$$

In a similar manner we can show that

$$\|G_6\|_{C^\alpha(E)} \le M_1 \parallel f \parallel_{C^\alpha(E)} .$$

By the definition of the spaces E_α and using estimates (5.8), (5.9), we obtain that

$$\| \lambda^{1-\alpha} Be^{-\lambda B} G_5(t) \|_E \le \lambda^{1-\alpha} \|(I - e^{-2TB})^{-1}\|_{E\to E}$$

$$\times \|I - e^{-TB}\|_{E\to E} \|Be^{-(\lambda+t)B}(f(t) - f(0))\|_E$$

$$\le M \frac{\lambda^{1-\alpha} t^\alpha}{\lambda + t} \parallel f \parallel_{C^\alpha(E)} \le M \parallel f \parallel_{C^\alpha(E)}$$

for all λ, $\lambda > 0$ and for $0 \le t \le T$. So, we have established that

$$\|G_5\|_{C(E_\alpha)} \le M \parallel f \parallel_{C^\alpha(E)} .$$

In a similar manner we can show that

$$\|G_6\|_{C(E_\alpha)} \le M_1 \parallel f \parallel_{C^\alpha(E)} .$$

Now let us estimate $G_7(t)$. Using estimates (5.8) and (5.9) we obtain that

$$\| G_7(t) \|_E \le \|(I - e^{-2TB})^{-1}\|_{E\to E} \frac{1}{2} \int_0^t \|Be^{-(t-s)B}(I - e^{-(2T-2t)B}) \|_{E\to E} \tag{5.21}$$

$$\times \|I - e^{-2sB} \|_{E\to E} \|f(s) - f(t) \|_E \, ds$$

$$\le M \int_0^t \frac{ds}{(t-s)^{1-\alpha}} \parallel f \parallel_{C^\alpha(E)} = \frac{M}{\alpha} t^\alpha \parallel f \parallel_{C^\alpha(E)}$$

for $0 \le t \le T$. From this inequality it follows that

$$\max_{0\le t\le T} \| G_7(t) \|_E \le \frac{M}{\alpha} \parallel f \parallel_{C^\alpha(E)} . \tag{5.22}$$

Now let us estimate the difference $G_7(t+\tau) - G_7(t)$ for $0 < t < t+\tau < T$. We will consider separately the cases $t \le \tau$ and $t > \tau$. If $t \le \tau$, then (5.21) yields

$$\| G_7(t+\tau) - G_7(t) \|_E \le \| G_7(t+\tau) \|_E + \| G_7(t) \|_E$$

$$\le \frac{M}{\alpha}(t^\alpha + (t+\tau)^\alpha) \parallel f \parallel_{C^\alpha(E)} \le \frac{M_1 \tau^\alpha}{\alpha} \parallel f \parallel_{C^\alpha(E)} .$$

Now let $t > \tau$. Let us represent the difference $G_7(t+\tau) - G_7(t)$ as the sum of the following integrals:

$$G_7(t+\tau) - G_7(t) = (I - e^{-2TB})^{-1} \frac{B}{2} (I - e^{-(2T-2t-2\tau)B})$$

$$\times \int_{t-\tau}^{t+\tau} e^{-(t+\tau-s)B}(I - e^{-2sB})(f(s) - f(t+\tau))ds$$

$$-(I-e^{-2TB})^{-1}\frac{B}{2}(I-e^{-(2T-2t)B})\int\limits_{t-\tau}^{t} e^{-(t-s)B}(I-e^{-2sB})(f(s)-f(t))ds$$

$$+(I-e^{-2TB})^{-1}\frac{B}{2}(I-e^{-(2T-2t-2\tau)B})$$

$$\times\int\limits_{0}^{t-\tau} e^{-(t+\tau-s)B}(I-e^{-2sB})(f(t)-f(t+\tau))ds$$

$$+(I-e^{-2TB})^{-1}\int\limits_{0}^{t-\tau}\frac{B}{2}[(e^{-(2T-2t)B}-e^{-(2T-2t-2\tau)B})e^{-(t-s)B}$$

$$+(I-e^{-(2T-2t)B})(e^{-(t+\tau-s)B}-e^{-(t-s)B})](I-e^{-2sB})(f(s)-f(t))ds$$

$$=P_1+P_2+P_3+P_4.$$

Now let us estimate P_1. Using estimates (5.8) and (5.9) we obtain that

$$\| P_1 \|_E \leq \|(I-e^{-2TB})^{-1}\|_{E\to E}\frac{1}{2}\int\limits_{t-\tau}^{t+\tau}\|Be^{-(t+\tau-s)B}(I-e^{-(2T-2t-2\tau)B})\|_{E\to E}$$

$$\times\|I-e^{-2sB}\|_{E\to E}\|f(s)-f(t+\tau)\|_E\, ds$$

$$\leq M\int\limits_{t-\tau}^{t+\tau}\frac{ds}{(t+\tau-s)^{1-\alpha}}\| f \|_{C^\alpha(E)}\leq\frac{M}{\alpha}(2\tau)^\alpha \| f \|_{C^\alpha(E)}\,.$$

In exactly the same manner one establishes the estimate

$$\| P_2 \|_E\leq\frac{M_1}{\alpha}\tau^\alpha \| f \|_{C^\alpha(E)}\,.$$

Since

$$P_3=(I-e^{-2TB})^{-1}\frac{B}{2}(I-e^{-(2T-2t-2\tau)B})$$

$$\times\int\limits_{0}^{t-\tau} e^{-(t+\tau-s)B}(I-e^{-2sB})ds(f(t)-f(t+\tau))$$

$$=(I-e^{-2TB})^{-1}\frac{1}{2}(I-e^{-(2T-2t-2\tau)B})$$

$$\times[e^{-2\tau B}-2e^{-(t+\tau)B}+e^{-2tB}](f(t)-f(t+\tau)),$$

it follows that

$$\| P_3 \|_E\leq\|(I-e^{-2TB})^{-1}\|_{E\to E}\frac{1}{2}\|I-e^{-(2T-2t-2\tau)B}\|_{E\to E}$$

$$\times\|e^{-2\tau B}-2e^{-(t+\tau)B}+e^{-2tB}\|_{E\to E}\|f(t)-f(t+\tau)\|_E.$$

Using estimates (5.8) and (5.9) we obtain

$$\| P_3 \|_E \leq M_1 \tau^\alpha \| f \|_{C^\alpha(E)} .$$

Now let us estimate P_4. Using estimates (5.8), (5.9) and (5.10), we obtain

$$\| P_3 \|_E \leq \|(I - e^{-2TB})^{-1}\|_{E\to E}$$

$$\times \frac{1}{2} \int_0^{t-\tau} \{\| B(e^{-(2T-2t)B} - e^{-(2T-2t-2\tau)B}) e^{-(t-s)B} \|_{E\to E}$$

$$+ \| B(I - e^{-(2T-2t)B})(e^{-(t+\tau-s)B} - e^{-(t-s)B}) \|_{E\to E} \}$$

$$\times \|I - e^{-2sB}\|_{E\to E} \| f(t) - f(s) \|_E \, ds$$

$$\leq M \int_0^{t-\tau} \frac{\tau ds}{(t-s)^{2-\alpha}} \| f \|_{C^\alpha(E)} .$$

Since for $t > \tau$, we have the bound

$$\int_0^{t-\tau} \frac{\tau ds}{(t-s)^{2-\alpha}} \leq \frac{\tau}{t^\alpha (T-t+\tau)^\alpha} \int_0^{t-\tau} \frac{ds}{(t-s)^{2-\alpha}} \leq \frac{\tau}{1-\alpha} \tau^{-1+\alpha} = \frac{\tau^\alpha}{1-\alpha}$$

$$\leq \frac{\tau}{(1-\alpha)} \tau^{-1+\alpha} \leq \frac{2^\alpha \tau^\alpha}{(1-\alpha)(t+\tau)^\alpha (T-t)^\alpha} .$$

Therefore, we conclude that

$$\| P_4 \|_E \leq \frac{M_1}{1-\alpha} \tau^\alpha \| f \|_{C^\alpha(E)} .$$

Thus, we have shown that for any $0 < t < t+\tau < T$ the following inequality holds:

$$\| G_7(t+\tau) - G_7(t) \|_E \leq \frac{M_1}{\alpha(1-\alpha)} \tau^\alpha \| f \|_{C^\alpha(E)} . \tag{5.23}$$

Estimates (5.22) and (5.23) give

$$\| G_7 \|_{C^\alpha(E)} \leq \frac{M_1}{\alpha(1-\alpha)} \| f \|_{C^\alpha(E)} .$$

In exactly the same manner one establishes the inequality

$$\| G_8 \|_{C^\alpha(E)} \leq \frac{M}{\alpha(1-\alpha)} \| f \|_{C^\alpha(E)} .$$

By the definition of the spaces E_α and using estimates (5.8), (5.9), we obtain that

$$\| \lambda^{1-\alpha} B e^{-\lambda B} G_7(t) \|_E \le \lambda^{1-\alpha} \|(I - e^{-2TB})^{-1}\|_{E\to E}$$

$$\times \frac{1}{2} \int_0^t \|B^2 e^{-(\lambda+t-s)B} (I - e^{-(2T-2t)B}) \|_{E\to E}$$

$$\times \|I - e^{-2sB} \|_{E\to E} \|f(s) - f(t) \|_E \, ds$$

$$\le M \int_0^t \frac{\lambda^{1-\alpha} ds}{(\lambda + t - s)^{2-\alpha}} \| f \|_{C^\alpha(E)} \le \frac{M}{1-\alpha} \| f \|_{C^\alpha(E)}$$

for all $\lambda, \lambda > 0$ and for $0 \le t \le T$. So, we have established that

$$\|G_7\|_{C(E_\alpha)} \le \frac{M}{1-\alpha} \| f \|_{C^\alpha(E)} \, .$$

In a similar manner we can show that

$$\|G_8\|_{C(E_\alpha)} \le \frac{M}{1-\alpha} \| f \|_{C^\alpha(E)} \, .$$

Combining the estimates for $G_k(t)$, $k = 1, 2, 3, 4, 5, 6, 7, 8$ we obtain the estimates

$$\|v''\|_{C^\alpha(E)} \le \frac{M}{\alpha(1-\alpha)} \| f \|_{C^\alpha(E)} + \frac{M}{\alpha}[\| v_0''\|_{E_\alpha} + \|v_T''\|_{E_\alpha}],$$

$$\|v''\|_{C(E_\alpha)} \le \frac{M}{\alpha(1-\alpha)} \| f \|_{C^\alpha(E)} + \frac{M}{\alpha}[\| v_0''\|_{E_\alpha} + \|v_T''\|_{E_\alpha}].$$

The estimate for $Av(t)$ in the norm of $C^\alpha(E)$ follows from the triangle inequality. Theorem 5.1.4 is proved.

Second, one can enrich the family of spaces of smooth functions in which the boundary-value problem (5.1) is well posed by introducing the spaces $C_{0T}^{\beta,\gamma}(E)$, $(0 \le \gamma \le \beta, 0 < \beta < 1)$, obtained by completion of the set of all smooth E-valued functions $\varphi(t)$ on [0, T] with respect to the norm

$$\| \varphi \|_{C_{0T}^{\beta,\gamma}(E)} = \max_{0\le t\le T} \|\varphi(t)\|_E + \sup_{0\le t<t+\tau\le T} \frac{(t+\tau)^\gamma (T-t)^\gamma \| \varphi(t+\tau) - \varphi(t) \|_E}{\tau^\beta}.$$

Let us give, without proof, the following result.

Theorem 5.1.5. *Let A be the positive operator in a Banach space E and $f(t) \in C_{0T}^{\beta,\gamma}(E)$ $(0 \le \gamma \le \beta, 0 < \beta < 1)$. Then for the solution $v(t)$ in $C_{0T}^{\beta,\gamma}(E)$ of the boundary-value problem* (5.1) *the coercive inequalities*

$$\| v'' \|_{C(E_{\beta-\gamma})} \le M[\| v_0'' \|_{E_{\beta-\gamma}} + \| v_T'' \|_{E_{\beta-\gamma}} + \beta^{-1}(1-\beta)^{-1} \| f \|_{C_{0T}^{\beta,\gamma}(E)}],$$

$$\| v'' \|_{C_{0T}^{\beta,\gamma}(E)} + \| Av \|_{C_{0T}^{\beta,\gamma}(E)}$$

$$\le M[|v_0''|_0^{\beta,\gamma} + |v_T''|_0^{\beta,\gamma} + \beta^{-1}(1-\beta)^{-1} \| f \|_{C_{0T}^{\beta,\gamma}(E)}]$$

hold, where M does not depend on $\beta, \gamma, v_0'', v_T''$ and $f(t)$. Here, $|w|_0^{\beta,\gamma}$ denotes that the norm of the Banach space $E_0^{\beta,\gamma}$ consists of those $w \in E$ for which the norm

$$|w|_0^{\beta,\gamma} = \max_{0\le z\le T} ||e^{-zB}w||_E + \sup_{0\le z<z+\tau\le T} \tau^{-\beta}(z+\tau)^{\gamma}(T-z)^{\gamma}||(e^{-(z+\tau)B}-e^{-zB})w||_E$$

is finite.

Note that the parameter γ can be chosen freely in $[0,\beta)$, which increases the number of function spaces in which problem (5.1) is well posed. In particular, it is important that problem (5.1) is well posed in the Hölder space without a weight $(\gamma = 0)$. Recall that the positivity of A is a necessary condition for well-posedness of problem (5.1) in $C(E)$. However, problem (5.1) is not well posed in $C(E)$ for all positive operators. It turns out that a Banach space E can be restricted to a Banach space E' in such a manner that the restricted problem (5.1) in E' will be well posed in $C(E')$. The role of E' will be played here by the fractional spaces $E_\alpha = E_\alpha(E,B)(0<\alpha<1)$.

Theorem 5.1.6. *Let A be the positive operator in a Banach space E and $f(t) \in C(E_\alpha)$, $(0<\alpha<1)$. Then for the solution $v(t)$ in $C(E_\alpha)$ of the boundary-value problem (5.1) the coercive inequality*

$$\| v'' \|_{C(E_\alpha)} + \| Av \|_{C(E_\alpha)}$$

$$\le M[\| Av_0 \|_{E_\alpha} + \| Av_T \|_{E_\alpha} + \alpha^{-1}(1-\alpha)^{-1} \| f \|_{C(E_\alpha)}] \tag{5.24}$$

holds, where M does not depend on α, v_0'', v_T'' and $f(t)$.

Proof. Using formula (5.7), we obtain

$$Av(t) = (I-e^{-2TB})^{-1}\{(e^{-tB}-e^{-(2T-t)B})Av_0 + (e^{-(T-t)B}-e^{-(T+t)B})Av_T\}$$

$$-(I-e^{-2TB})^{-1}(e^{-(T-t)B}-e^{-(T+t)B})\int_0^T \frac{B}{2}e^{-(T-s)B}f(s)ds$$

$$+(I-e^{-2TB})^{-1}(e^{-(2T-t)B}-e^{-tB})\int_0^T \frac{B}{2}e^{-sB}f(s)ds$$

$$+\int_0^t \frac{B}{2}e^{-(t-s)B}f(s)ds + \int_t^T \frac{B}{2}e^{-(s-t)B}f(s)ds = \sum_{k=1}^{6} I_k(t).$$

Let us estimate $I_k(t)$ for any $k = 1,2,3,\ldots,6$ separately. We start with $I_1(t)$. Using estimates (5.8), (5.9) and the definition of the spaces E_α we show that, for $0 \le t \le T$,

$$||I_1(t)||_{E_\alpha} \le ||(I-e^{-2TB})^{-1}||_{E\to E}||e^{-tB}-e^{-(2T-t)B}||_{E\to E}||Av_0||_{E_\alpha}$$

$$\le M||Av_0||_{E_\alpha}.$$

Thus, we have proved that

$$||I_1||_{C(E_\alpha)} \leq M||Av_0||_{E_\alpha}.$$

In a similar manner we can show that

$$||I_2||_{C(E_\alpha)} \leq M||Av_T||_{E_\alpha}.$$

By the definition of the spaces E_α and using estimates (5.8), (5.9), we obtain that

$$\| \lambda^{1-\alpha} B e^{-\lambda B} I_3(t) \|_E \leq ||(I-e^{-2TB})^{-1}||_{E\to E} ||e^{-(T-t)B} - e^{-(T+t)B}||_{E\to E}$$

$$\times \int_0^T \frac{\lambda^{1-\alpha}}{2} ||B^2 e^{-(T-s+\lambda)B} f(s)||_E ds$$

$$\leq M \int_0^T \frac{\lambda^{1-\alpha} ds \, \| f \|_{C(E_\alpha)}}{(T-s+\lambda)(T-s)^{1-\alpha}} \leq M \int_0^\infty \frac{dz \, \| f \|_{C(E_\alpha)}}{(1+z)z^{1-\alpha}} \leq \frac{M_1}{\alpha(1-\alpha)} \| f \|_{C(E_\alpha)}$$

for all $\lambda, \lambda > 0$ and for $0 \leq t \leq T$. So, we have established that

$$||I_3||_{C(E_\alpha)} \leq \frac{M_1}{\alpha(1-\alpha)} \| f \|_{C(E_\alpha)} .$$

In a similar manner we can show that

$$||I_4||_{C(E_\alpha)} \leq \frac{M_1}{\alpha(1-\alpha)} \| f \|_{C(E_\alpha)} .$$

Next, using estimates (5.8), we obtain

$$\| \lambda^{1-\alpha} B e^{-\lambda B} I_5 \|_E \leq \frac{1}{2} \int_0^t \lambda^{1-\alpha} \| B^2 e^{-(s+\lambda)B} f(s) \|_E \, ds$$

$$\leq M \int_0^\infty \frac{dz \, \| f \|_{C(E_\alpha)}}{(1+z)z^{1-\alpha}} \leq \frac{M_1}{\alpha(1-\alpha)} \| f \|_{C(E_\alpha)}$$

for all λ, $\lambda > 0$ and for $0 \leq t \leq T$. So, we have established that

$$||I_5||_{C(E_\alpha)} \leq \frac{M_1}{\alpha(1-\alpha)} \| f \|_{C(E_\alpha)} .$$

In a similar manner we can show that

$$||I_6||_{C(E_\alpha)} \leq \frac{M_1}{\alpha(1-\alpha)} \| f \|_{C(E_\alpha)} .$$

Combining the estimates for I_1, I_2, I_3, I_4, I_5, I_6 we obtain estimate (5.24). Theorem 5.1.4 is proved.

Third, let us study now the boundary-value problem (5.1) in the spaces $C_{0T}^{\beta,\gamma}(E_{\alpha-\beta})$, $(0 \leq \gamma \leq \beta \leq \alpha, 0 < \alpha < 1)$. To these there correspond the spaces of traces $E^{\beta,\gamma}_{\alpha-\beta}$, which consist of elements $w \in E$ for which the norm

$$
\begin{aligned}
|w|^{\beta,\gamma}_{\alpha-\beta} &= \max_{0\le z\le T} \|e^{-zB}w\|_{E_{\alpha-\beta}} \\
&+ \sup_{0\le z<z+\tau\le T} \tau^{-\beta}(z+\tau)^{\gamma}(1-z)^{\gamma}\|(e^{-(z+\tau)B}-e^{-zB})w\|_{E_{\alpha-\beta}}
\end{aligned}
$$

is finite. Let us give, without proof, the following results.

Theorem 5.1.7. *Let A be the positive operator in a Banach space E, $v''_0 \in E^{\beta,\gamma}_{\alpha-\beta}$, $v''_T \in E^{\beta,\gamma}_{\alpha-\beta}$ and $f(t) \in C^{\beta,\gamma}_{0T}(E_{\alpha-\beta})$ $(0 \le \gamma \le \beta \le \alpha,\ 0<\alpha<1)$. Then for the solution $v(t)$ in $C^{\beta,\gamma}_{0T}(E_{\alpha-\beta})$ of the boundary-value problem* (5.1) *the coercive inequality*

$$
\begin{aligned}
&\| v'' \|_{C^{\beta,\gamma}_{0T}(E_{\alpha-\beta})} + \| Av \|_{C^{\beta,\gamma}_{0T}(E_{\alpha-\beta})} + \| v'' \|_{C(E^{\beta,\gamma}_{\alpha-\beta})} \\
&\le M[|v''_0|^{\beta,\gamma}_{\alpha-\beta} + |v''_T|^{\beta,\gamma}_{\alpha-\beta} + \alpha^{-1}(1-\alpha)^{-1} \| f \|_{C^{\beta,\gamma}_{0T}(E_{\alpha-\beta})}]
\end{aligned}
$$

holds, where M does not depend on α, β, γ, v''_0, v''_T and $f(t)$.

Theorem 5.1.8. *Let A be the positive operator in a Banach space E and $v''_0 \in E_{\alpha-\gamma}$, $v''_T \in E_{\alpha-\gamma}$ and $f(t) \in C^{\beta,\gamma}_{0T}(E_{\alpha-\gamma})$ $(0 \le \gamma \le \beta \le \alpha,\ 0<\alpha<1)$. Then for the solution $v(t)$ in $C^{\beta,\gamma}_{0T}(E_{\alpha-\beta})$ of the boundary-value problem* (5.1) *the coercive inequality*

$$
\begin{aligned}
&\| v'' \|_{C^{\beta,\gamma}_{0T}(E_{\alpha-\beta})} + \| Av \|_{C^{\beta,\gamma}_{0T}(E_{\alpha-\beta})} + \| v'' \|_{C(E_{\alpha-\gamma})} \\
&\le M\alpha^{-1}(1-\alpha)^{-1}[\| v''_0 \|_{E_{\alpha-\gamma}} + \| v''_T \|_{E_{\alpha-\gamma}} + \| f \|_{C^{\beta,\gamma}_{0T}(E_{\alpha-\beta})}]
\end{aligned}
$$

holds, where M does not depend on α, β, γ, v''_0, v''_T and $f(t)$.

Note that the spaces of smooth functions $C^{\beta,\gamma}_{0T}(E_{\alpha-\beta})$, in which coercive solvability has been established, depend on the parameters α, β and γ. However, the constants in the coercive inequalities depend only on α. Hence, we can choose the parameters β and γ freely, which increases the number of function spaces in which problem (5.1) is well posed. In particular, Theorem 5.1.6 implies the well-posedness Theorem 5.1.4 in $C(E_\alpha)$ established in the above.

Fourth, let us study now the boundary-value problem (5.1) in the spaces $L_p(E) = L_p([0,T],E)$ $(1 \le p < \infty)$ of all strongly measurable E-valued functions $v(t)$ on $[0,T]$ for which the norm

$$
\| v \|_{L_p(E)} = \Big(\int_0^T \| v(t) \|^p_E \, dt\Big)^{\frac{1}{p}}
$$

is finite.

Recall that a function $v(t)$ is said to be absolutely continuous if it has a derivative $v'(t)$ for almost every t such that $v'(t) \in L_1(E)$, and if the Newton–

Leibniz formula

$$v(t) - v(\tau) = \int_{\tau}^{t} v'(s)ds$$

holds for all t, $\tau \in [0, T]$. Here the integral is understood in the sense of Bochner.

A function $v(t)$ is said to be a solution of problem (5.1) in $L_p(E)$ if it is absolutely continuous, the functions $v''(t)$ and $Av(t)$ belong to $L_p(E)$, equation (5.1) is satisfied for almost every t, and $v(0) = v_0$, $v(T) = v_T$. From this definition it follows that a necessary condition for the solvability of problem (5.1) in $L_p(E)$ is that $f(t) \in L_p(E)$. It will be shown that in certain cases this condition is also sufficient for the solvability of problem (5.1). As concerns the boundary elements, in contrast to the situation considered earlier, from the solvability of problem (5.1) in $L_p(E)$ it follows only that v_0, $v_T \in E$. In the case of an unbounded operator A this does not allow us to prove the solvability of problem (5.1).

From the unique solvability of (5.1) it follows that the operator $v(t; f(t), v_0, v_T)$ is bounded in $L_p(E)$ and one has coercive inequality

$$\|v''\|_{L_p(E)} + \|Av(t)\|_{L_p(E)} \leq M_C[\|f\|_{L_p(E)} + ||Av_0||_E + ||Av_T||_E] ,$$

where M_C $(1 \leq M_C < +\infty)$ does not depend on v_0, v_T and $f(t)$. From that we can obtain the positivity of A under the stronger assumption that the operator A^{-1} is compact in E. Let us give, without proof, the following results.

Theorem 5.1.9. *Let A be the positive operator in a Banach space E. Suppose problem (5.1) is well posed in $L_{p_0}(E)$ for some p_0, $1 < p_0 < \infty$. Then it is well posed in $L_p(E)$ for any p, $1 < p < \infty$ and the coercivity inequality holds:*

$$\|v''\|_{L_p(E)} + \|Av\|_{L_p(E)} + \|v\|_{C(E_{1-1/p,p})}$$

$$\leq \frac{M(p_0)p^2}{p-1}\|f\|_{L_p(E)} + M(||v_0||_{E_{1-1/p,p}} + ||v_T||_{E_{1-1/p,p}}),$$

where $M(p_0)$ and M does not depend on p, v_0, v_T and $f(t)$. Here the Banach space $E_{1-1/p,p} = E_{1-1/p,p}(D(A^{1/2}), A^{1/2})$ consists of those $v \in E$ for which norm

$$\|v\|_{E_{1-1/p,p}} = \left(\sum_{0}^{T} \| \exp\{-zA^{1/2}\}v(z)\|_E^p dA\right)^{1/p} \quad (1 \leq p < \infty)$$

is finite.

Theorem 5.1.10. *Let $1 < p < \infty$ and $0 < \alpha < 1$. Suppose that A is the positive operator in a Banach space E. Then problem (5.1) is well posed in $L_p(E_{\alpha,p})$ and the coercivity inequality holds:*

$$||v''||_{L_p(E_{\alpha,p})} + \|Av\|_{L_p(E_{\alpha,p})} \leq \frac{M}{\alpha(1-\alpha)}\|f\|_{L_p(E_{\alpha,p})} + M(||Av_0||_{E_{\alpha,p}} + ||Av_T||_{E_{\alpha,p}}),$$

where M does not depend on α, p, v_0, v_T and $f(t)$.

Theorem 5.1.11. *Let $1 < p,\ q < \infty$ and $0 < \alpha < 1$. Suppose that A is the positive operator in a Banach space E. Then problem (5.1) is well posed in $L_p(E_{\alpha,q})$ and the coercivity inequality holds:*

$$||v''||_{L_p(E_{\alpha,q})} + \|Av\|_{L_p(E_{\alpha,q})} \leq \frac{M(q)}{\alpha(1-\alpha)}\|f\|_{L_p(E_{\alpha,q})} + M(||Av_0||_{E_{\alpha,q}} + ||Av_T||_{E_{\alpha,q}}),$$

where $M(q)$ and M does not depend on α, p, v_0, v_T and $f(t)$.

Here, the Banach space $E_{\alpha,q} = E_{\alpha,q}(B,E)(0 < \alpha < 1, 1 < q < \infty)$ consists of those $v \in E$ for which the norm

$$\| v \|_{E_{\alpha,q}} = \left(\int_0^\infty \lambda^{1-\alpha} \| Bexp\{-\lambda B\}v \|_E^q \frac{d\lambda}{\lambda}\right)^{1/q}$$

is finite.

Finally, we consider the applications of these results to the elliptic equations. First, we consider the boundary-value problems for two-dimensional elliptic equations

$$-\frac{\partial^2 u}{\partial y^2} - a(x)\frac{\partial^2 u}{\partial x^2} + \delta u = f(y,x), 0 < y < T, 0 < x < 1, \tag{5.25}$$

$$u(0,x) = \varphi(x), u(T,x) = \psi(x),\ f(0,x) = f(\tau,x) = 0,\ 0 \leq x \leq 1,$$
$$u(y,0) = u(y,1),\ \ u_x(y,0) = u_x(y,1),\ \ 0 \leq y \leq T,$$

where $a(x)$, $\varphi(x)$, $\psi(x)$ and $f(y,x)$ are given sufficiently smooth functions and $a(x) > 0$. $\delta > 0$ is a sufficiently large number.

We introduce the Banach spaces $C^\beta[0,1]$ $(0 < \beta < 1)$ of all continuous functions $\varphi(x)$ satisfying a Hölder condition for which the following norms are finite:

$$\| \varphi \|_{C^\beta[0,1]} = \| \varphi \|_{C[0,1]} + \sup_{0 \leq x < x+\tau \leq 1} \frac{|\varphi(x+\tau) - \varphi(x)|}{\tau^\beta},$$

where $C[0,1]$ is the space of all continuous functions $\varphi(x)$ defined on [0, 1] with the usual norm

$$\| \varphi \|_{C[0,1]} = \max_{0 \leq x \leq 1} |\varphi(x)|.$$

It is known that the differential expression

$$A^x v = -a(x)v''(x) + \delta v(x)$$

defines a positive operator A^x acting in $C^\beta[0,1]$ with domain $C^{\beta+2}[0,1]$ and satisfying the conditions $v(0) = v(1)$, $v_x(0) = v_x(1)$.

Therefore, we can replace boundary-value problem (5.25) by the abstract boundary-value problem (5.1). Using the results of Theorems 5.1.2, 5.1.5 and 5.1.8, we can obtain that

Theorem 5.1.12. *For the solution of the boundary-value problem* (5.25) *the following coercive inequalities are valid:*

$$\| u \|_{C_{0T}^{2+\beta,\gamma}(C^{\mu}[0,1])} + \| u \|_{C_{0T}^{\beta,\gamma}(C^{2+\mu}[0,1])}$$

$$\leq \frac{M(\mu)}{(\beta-\gamma)(1-\beta)} \| f \|_{C_{0T}^{\beta,\gamma}(C^{\mu}[0,1])} + M(\mu)(\|\varphi\|_{C^{2+\mu+\beta-\gamma}[0,1]} + \|\psi\|_{C^{2+\mu+\beta-\gamma}[0,1]}),$$

$$0 \leq \gamma < \beta < 1,\ 0 \leq \mu + \beta - \gamma < 1,$$

$$\| u \|_{C_{0T}^{2+\beta,\gamma}(C^{\alpha-\beta}[0,1])} + \| u \|_{C_{0T}^{\beta,\gamma}(C^{2+\alpha-\beta}[0,1])}$$

$$\leq \frac{M(\alpha,\beta)}{\alpha(1-\alpha)} \| f \|_{C_{0T}^{\beta,\gamma}(C^{\alpha-\beta}[0,1])} + M(\mu)(\|\varphi\|_{C^{2+\alpha-\gamma}[0,1]} + \|\psi\|_{C^{2+\alpha-\gamma}[0,1]}),$$

$$0 \leq \gamma \leq \beta \leq \alpha, 0 < \alpha - \beta < \frac{1}{2}.$$

Here $M(\mu)$ *and* $M(\alpha,\beta)$ *are independent of* γ, $f(y,x)$, $\varphi(x)$, $\psi(x)$.

Second, let Ω be the unit open cube in the n-dimensional Euclidean space $\mathbb{R}^n$ $(0 < x_k < 1, 1 \leq k \leq n)$ with boundary S, $\overline{\Omega} = \Omega \cup S$. In $[0,T] \times \Omega$ we consider the mixed boundary-value problem for the multidimensional elliptic equation

$$\begin{cases} -\dfrac{\partial^2 u(y,x)}{\partial y^2} - \displaystyle\sum_{r=1}^{n} \alpha_r(x) \dfrac{\partial^2 u(y,x)}{\partial x_r^2} + \delta u(y,x) = f(y,x), \\ x = (x_1, \ldots, x_n) \in \Omega, 0 < y < T, \\ u(0,x) = \varphi(x),\ u(T,x) = \psi(x),\ f(1,x) = f(T,x) = 0\ x \in \overline{\Omega}, \\ u(y,x) = 0,\ x \in S, \end{cases} \tag{5.26}$$

where $\alpha_r(x)$ $(x \in \Omega)$ and $f(y,x)$ $(y \in (0,T),\ x \in \Omega)$, $\varphi(x)$, $\psi(x) (x \in \overline{\Omega})$ are given smooth functions and $\alpha_r(x) > 0$, $\delta > 0$ is a sufficiently large number.

We introduce the Banach spaces $C_{01}^{\beta}(\overline{\Omega})$ $(\beta = (\beta_1, \ldots, \beta_n), 0 < x_k < 1$, $k = 1, \ldots, n)$ of all continuous functions satisfying a Hölder condition with the indicator $\beta = (\beta_1, \ldots, \beta_n), \beta_k \in (0,1), 1 \leq k \leq n$ and with weight $x_k^{\beta_k}(1 - x_k - h_k)^{\beta_k}, 0 \leq x_k < x_k + h_k \leq 1, 1 \leq k \leq n$ which is equipped with the norm

$$\| f \|_{C_{01}^{\beta}(\overline{\Omega})} = \| f \|_{C(\overline{\Omega})}$$

$$+ \sup_{0 \leq x_k < x_k + h_k \leq 1, 1 \leq k \leq n} |f(x_1, \ldots, x_n) - f(x_1 + h_1, \ldots, x_n + h_n)|$$

$$\times \prod_{k=1}^{n} h_k^{-\beta_k} x_k^{\beta_k} (1 - x_k - h_k)^{\beta_k},$$

where $C(\overline{\Omega})$ stands for the Banach space of all continuous functions defined on $\overline{\Omega}$, equipped with the norm

$$\| f \|_{C(\overline{\Omega})} = \max_{x \in \overline{\Omega}} |f(x)|.$$

It is known that the differential expression

$$A^x v = -\sum_{r=1}^{n} \alpha_r(x) \frac{\partial^2 v(y,x)}{\partial x^2} + \delta v(y,x)$$

defines a positive operator A^x acting on $C_{01}^{\beta}(\overline{\Omega})$ with domain $D(A^x) \subset C_{01}^{2+\beta}(\overline{\Omega})$ and satisfying the condition $v = 0$ on S.

Therefore, we can replace boundary-value problems (5.26) by the abstract boundary-value problems (5.1). Using the results of Theorems 5.1.2 and 5.1.5, we can obtain that

Theorem 5.1.13. *For the solution of the boundary-value problem* (5.26) *the following coercive inequality is valid:*

$$\| u \|_{C_{0T}^{2+\beta,\gamma}(C_{01}^{\mu}(\overline{\Omega}))} + \sum_{r=1}^{n} \| \frac{\partial^2 u}{\partial x_r^2} \|_{C_{0T}^{\beta,\gamma}(C_{01}^{\mu}(\overline{\Omega}))}$$

$$\leq \frac{M(\mu)}{(\beta-\gamma)(1-\beta)} \| f \|_{C_{0T}^{\beta,\gamma}(C_{01}^{\mu}(\overline{\Omega}))}$$

$$+M(\mu)\Big(\sum_{r=1}^{n} \| \frac{\partial^2 \varphi}{\partial x_r^2} \|_{C_{01}^{\mu+\beta-\gamma}(\overline{\Omega})} + \sum_{r=1}^{n} \| \frac{\partial^2 \psi}{\partial x_r^2} \|_{C_{01}^{\mu+\beta-\gamma}(\overline{\Omega})}\Big),$$

$$0 \leq \gamma < \beta < 1, \mu = \{\mu_1, \dots, \mu_n\}, 0 < \mu_k + \beta - \gamma < 1, 1 \leq k \leq n,$$

where $M(\mu)$ is independent of β, γ and $f(y,x)$, $\varphi(x)$, $\psi(x)$.

Third, we consider the boundary-value problem on the range $\{0 \leq y \leq T,\ x \in \mathbb{R}^n\}$ for $2m$-order multidimensional elliptic equations

$$\begin{cases} -\dfrac{\partial^2 u}{\partial y^2} + \sum\limits_{|r|=2m} a_r(x) \dfrac{\partial^{|r|} u}{\partial x_1^{r_1} \cdots \partial x_n^{r_n}} + \delta u(y,x) = f(y,x), \\ 0 < y < T,\ x, r \in \mathbb{R}^n, |r| = r_1 + \cdots + r_n, \\ u(0,x) = \varphi(x),\ u(T,x) = \psi(x),\ f(1,x) = f(T,x) = 0,\ x \in \mathbb{R}^n, \end{cases} \tag{5.27}$$

where $a_r(x)$ and $f(y,x)$, $\varphi(x)$, $\psi(x)$ are given sufficiently smooth functions and $\alpha_r(x) > 0$, $\delta > 0$ is a sufficiently large number.

We will assume that the symbol

$$B^x(\xi) = \sum_{|r|=2m} a_r(x) \left(i\xi_1\right)^{r_1} \cdots \left(i\xi_n\right)^{r_n}, \xi = (\xi_1, \dots, \xi_n) \in R^n$$

of the differential operator of the form

$$B^x = \sum_{|r|=2m} a_r(x) \frac{\partial^{|r|}}{\partial x_1^{r_1} \dots \partial x_n^{r_n}} \tag{5.28}$$

acting on functions defined on the space $\mathbb{R}^n$, satisfies the inequalities

$$0 < M_1|\xi|^{2m} \le (-1)^m B^x(\xi) \le M_2|\xi|^{2m} < \infty$$

for $\xi \neq 0$. Problem (5.48) has a unique smooth solution. This allows us to reduce the boundary-value problem (5.48) to the boundary-value problem (5.1) with a strongly positive operator $A^x = B^x + \delta I$ defined by (5.27) in a Banach space $E = C^\mu(R^n)$ of all continuous bounded functions defined on $\mathbb{R}^n$ satisfying a Hölder condition with the indicator $\mu \in (0,1)$.

Theorem 5.1.14. *For the solution of the boundary-value problem* (5.27) *the following coercivity inequalities are satisfied:*

$$\| u \|_{C_{0T}^{2+\beta,\gamma}(C^\mu(R^n))} + \sum_{|\tau|=2m} \| \frac{\partial^{|r|} u}{\partial x_1^{r_1} \ldots \partial x_n^{r_n}} \|_{C_{0T}^{\beta,\gamma}(C^\mu(R^n))}$$

$$\le \frac{M(\mu)}{(\beta-\gamma)(1-\beta)} \| f \|_{C_{0T}^{\beta,\gamma}(C^\mu(R^n))}$$

$$+M(\mu)\Big(\sum_{|\tau|=2m} \| \frac{\partial^{|r|}\varphi}{\partial x_1^{r_1}\ldots\partial x_n^{r_n}} \|_{C^{\mu+m(\beta-\gamma)}(R^n)} + \sum_{|\tau|=2m} \| \frac{\partial^{|r|}\psi}{\partial x_1^{r_1}\ldots\partial x_n^{r_n}} \|_{C^{\mu+m(\beta-\gamma)}(R^n)}\Big),$$

$$0 \le \gamma < \beta < 1,\ 0 < \mu + m(\beta-\gamma) < 1,$$

$$\| u \|_{C_{0T}^{2+\beta,\gamma}(C^{m(\alpha-\beta)}(R^n))} + \sum_{|\tau|=2m} \| \frac{\partial^{|r|} u}{\partial x_1^{r_1} \ldots \partial x_n^{r_n}} \|_{C_{0T}^{\beta,\gamma}(C^{m(\alpha-\beta)}(R^n))}$$

$$\le M(\alpha,\beta) \| f \|_{C_{0T}^{\beta,\gamma}(C^{m(\alpha-\beta)}(R^n))}$$

$$+M(\alpha,\beta)\Big(\sum_{|\tau|=2m} \| \frac{\partial^{|r|}\varphi}{\partial x_1^{r_1}\ldots\partial x_n^{r_n}} \|_{C^{m(\alpha-\gamma)}(R^n)} + \sum_{|\tau|=2m} \| \frac{\partial^{|r|}\psi}{\partial x_1^{r_1}\ldots\partial x_n^{r_n}} \|_{C^{m(\alpha-\gamma)}(R^n)}\Big),$$

$$0 \le \gamma \le \beta,\ 0 < m(\alpha-\beta) < 1,$$

where $M(\mu)$ *and* $M(\alpha,\beta)$ *does not depend on* γ *and* $f(y,x), \varphi(x), \psi(x)$.

The proof of Theorem 4.1.8 is based on the abstract Theorems 2.1.1 and 3.1.1, the positivity of the operator A^x in $C^\mu(R^n)$, the structure of the fractional spaces $E_\alpha(C(R^n),(A^x)^{\frac{1}{2}})$ and the coercivity inequality for an elliptic operator A^x in $C^\mu(R^n)$.

5.2 Difference Schemes Generated by an Exact Difference Scheme

We consider the boundary-value problem (5.1). For the construction of the two-step difference schemes of an arbitrary high order of accuracy for the approximate solutions of the boundary-value problem (5.1) we consider the uniform grid space

$$[0,T]_\tau = \{t_k = k\tau, k = 0,1,\ldots,N, N\tau = T\}.$$

By Theorem 3.1.2 we have the following two-step difference scheme for the solution of the boundary-value problem (5.1):

$$\begin{aligned}
&-\frac{1}{\tau^2}(v(t_{k+1}) - 2v(t_k) + v(t_{k-1}))\\
&+\frac{1}{\tau^2}[(I - (2B\tau)^{-1}\{I - \exp(-2\tau B)\}\exp(-\tau B))v(t_{k-1})\\
&\quad +(-2I + (2B\tau)^{-1}\{I - \exp(-2\tau B)\}\exp(-2\tau B)\\
&\quad +(2B\tau)^{-1}\{I - \exp(-2\tau B)\})v(t_k)\\
&+(I - (2B\tau)^{-1}\{I - \exp(-2\tau B)\}\exp(-\tau B))v(t_{k+1})]\\
= {}& \frac{1}{\tau^2}[(2B\tau)^{-1}\{I - \exp(-2\tau B)\}\int_{t_k}^{t_{k+1}}(2B)^{-1}\{\exp(-(z - t_k)B)\\
&- \exp(-(t_{k+2} - z)B))\}f(z)dz\\
&+(2B\tau)^{-1}\{I - \exp(-2\tau B)\}\int_{t_{k-1}}^{t_k}(2B)^{-1}\{\exp(-(t_k - z)B)\\
&- \exp(-(z - t_{k-2})B))\}f(z)dz],
\end{aligned}$$

$$1 \le k \le N-1, v(0) = v_0, v(T) = v_T.$$

It can obviously be rewritten as the equivalent boundary-value problem for the second-order difference equations

$$\begin{aligned}
&-(2B\tau)^{-1}\{I - \exp(-2\tau B)\}\exp(-\tau B)v(t_{k-1})\\
&+(2B\tau)^{-1}\{I - \exp(-2\tau B)\}\{I + \exp(-2\tau B)\})v(t_k)\\
&-(2B\tau)^{-1}\{I - \exp(-2\tau B)\}\exp(-\tau B)v(t_{k+1})\\
= {}& (2B\tau)^{-1}\{I - \exp(-2\tau B)\}[\int_{t_k}^{t_{k+1}}(2B)^{-1}\{\exp(-(z - t_k)B)\\
&- \exp(-(t_{k+2} - z)B))\}f(z)dz
\end{aligned}$$

$$+\int_{t_{k-1}}^{t_k}(2B)^{-1}\{\exp(-(t_k-z)B)-\exp(-(z-t_{k-2})B))\}f(z)dz],$$

$$1\le k\le N-1, v(0)=v_0, v(T)=v_T,$$

or

$$-\frac{1}{\tau^2}(u_{k+1}-2u_k+u_{k-1})+\frac{1}{\tau^2}[(I-\exp(-\tau B))(u_{k-1}+u_{k+1}) \tag{5.29}$$

$$+(\exp(-2\tau B)-I)u_k]=f_k,$$

$$f_k=(2B\tau)^{-1}(f_{1,k}+f_{2,k+1})-(2B\tau)^{-1}\exp(-\tau B)(f_{1,k+1}+f_{2,k}),$$

$$f_{1,k}=\frac{1}{\tau}\int_{t_{k-1}}^{t_k}\exp(-(t_k-z)B))f(z)dz,$$

$$f_{2,k}=\frac{1}{\tau}\int_{t_{k-1}}^{t_k}\exp(-(z-t_{k-1})B)f(z)dz,$$

$$1\le k\le N-1, u_k=v(t_k), 0\le k\le N, u_0=v(0), u_N=v(T).$$

The latter will be referred to as the exact two-step difference scheme for the boundary-value problem (5.1). Let us investigate the well-posedness of the exact two-step difference scheme (5.29). This problem is uniquely solvable, and the following formula holds:

$$u_k=(I-e^{-2t_NB})^{-1}\{(e^{-t_kB}-e^{-(2t_N-t_k)B})u_0 \tag{5.30}$$

$$+(e^{-(t_N-t_k)B}-e^{-(t_N+t_k)B})u_N$$

$$-(e^{-(t_N-t_k)B}-e^{-(t_N+t_k)B})\tau^2\sum_{i=1}^{N-1}(e^{-(t_N-t_i)B}-e^{-(t_N+t_i)B})(I-e^{-2\tau B})^{-1}f_i\}$$

$$+\tau^2\sum_{i=1}^{N-1}(e^{-|t_k-t_i|B}-e^{-(t_k+t_i)B})(I-e^{-2\tau B})^{-1}f_i, 1\le k\le N-1.$$

Really, problem (5.29) can be rewritten as the equivalent nonlocal boundary-value problem for first-order linear difference equations

$$\begin{cases}\tau^{-1}(u_k-u_{k-1})+\tau^{-1}(I-e^{-\tau B})u_{k-1}=z_k,\\ 1\le k\le N,\ u_0=v(0), u_N=v(T),\\ \\ -\tau^{-1}(z_{k+1}-z_k)+\tau^{-1}(I-e^{-\tau B})z_{k+1}=f_k, 1\le k\le N-1.\end{cases}$$

From this there follows the system of recursion formulas

$$\begin{cases}u_k=e^{-\tau B}u_{k-1}+\tau z_k, 1\le k\le N,\\ \\ z_k=e^{-\tau B}z_{k+1}+\tau f_k, 1\le k\le N-1.\end{cases}$$

Hence

$$\begin{cases} u_k = e^{-k\tau B}u_0 + \sum\limits_{i=1}^{k} e^{-(k-i)\tau B}\tau z_i, 1 \leq k \leq N, \\ \\ z_k = e^{-(N-k)\tau B}z_N + \sum\limits_{j=k}^{N-1} e^{-(j-k)\tau B}\tau f_j, 1 \leq k \leq N-1. \end{cases}$$

From this it follows that

$$u_N = e^{-N\tau B}u_0 + \sum_{i=1}^{N-1} e^{-(N-i)\tau B}\tau\{e^{-(N-i)\tau B}z_N + \sum_{j=i}^{N-1} e^{-(j-i)\tau B}\tau f_j\} + \tau z_N.$$

Since

$$\sum_{i=1}^{N-1} e^{-(N-i)\tau B}\tau e^{-(N-i)\tau B} + \tau I = \tau(I - e^{-2\tau B})^{-1}\sum_{i=1}^{N} e^{-2(N-i)\tau B}(I - e^{-2\tau B})$$

$$= \tau(I - e^{-2\tau B})^{-1}(I - e^{-2N\tau B}),$$

$$\sum_{i=1}^{N-1} e^{-(N-i)\tau B}\tau \sum_{j=i}^{N-1} e^{-(j-i)\tau B}\tau f_j = \sum_{j=1}^{N-1} \tau \sum_{i=1}^{j} e^{-(N+j-2i)\tau B}\tau f_j$$

$$= \sum_{j=1}^{N-1} \tau(I - e^{-2\tau B})^{-1}(e^{-(N-j)\tau B} - e^{-(N+j)\tau B})\tau f_j,$$

we have that

$$\begin{aligned} z_N &= \tau^{-1}(I - e^{-2\tau B})(I - e^{-2N\tau B})^{-1}\{u_N - e^{-N\tau B}u_0 \\ &\quad - \sum_{j=1}^{N-1} \tau(I - e^{-2\tau B})^{-1}(e^{-(N-j)\tau B} - e^{-(N+j)\tau B})\tau f_j\} \\ \\ &= \tau^{-1}(I - e^{-2\tau B})(I - e^{-2N\tau B})^{-1}\{u_N - e^{-N\tau B}u_0\} \\ &\quad -(I - e^{-2N\tau B})^{-1}\sum_{j=1}^{N-1}(e^{-(N-j)\tau B} - e^{-(N+j)\tau B})\tau f_j. \end{aligned}$$

Therefore

$$u_k = e^{-k\tau B}u_0 + \sum_{i=1}^{k} e^{-(k-i)\tau B}\tau\{e^{-(N-i)\tau B}z_N + \sum_{j=i}^{N-1} e^{-(j-i)\tau B}\tau f_j\}$$

$$= e^{-k\tau B}u_0 + \sum_{i=1}^{k} e^{-(k-i)\tau B}\tau e^{-(N-i)\tau B}z_N + \sum_{i=1}^{k} e^{-(k-i)\tau B}\tau \sum_{j=i}^{N-1} e^{-(j-i)\tau B}\tau f_j.$$

Since

$$\sum_{i=1}^{k} e^{-(k-i)\tau B}\tau e^{-(N-i)\tau B} = \tau(I-e^{-2\tau B})^{-1}\sum_{i=1}^{k} e^{-(N+k-2i)\tau B}(I-e^{-2\tau B})$$

$$= \tau(I-e^{-2\tau B})^{-1}(e^{-(N-k)\tau B} - e^{-(N+k)\tau B}),$$

$$\sum_{i=1}^{k} e^{-(k-i)\tau B}\tau \sum_{j=i}^{N-1} e^{-(j-i)\tau B}\tau f_j$$

$$= \sum_{j=1}^{k}\tau\sum_{i=1}^{j} e^{-(j+k-2i)\tau B}\tau f_j + \sum_{j=k+1}^{N-1}\tau\sum_{i=1}^{k} e^{-(k+j-2i)\tau B}\tau f_j$$

$$= \tau(I-e^{-2\tau B})^{-1}\sum_{j=1}^{k}(e^{-(k-j)\tau B} - e^{-(k+j)\tau B})\tau f_j$$

$$+\tau(I-e^{-2\tau B})^{-1}\sum_{j=k+1}^{N-1}(e^{-(j-k)\tau B} - e^{-(k+j)\tau B})\tau f_j$$

$$= \tau(I-e^{-2\tau B})^{-1}\sum_{j=1}^{N-1}(e^{-|k-j|\tau B} - e^{-(k+j)\tau B})\tau f_j,$$

we have that

$$\begin{aligned} u_k &= e^{-k\tau B}u_0 + \tau(I-e^{-2\tau B})^{-1}(e^{-(N-k)\tau B} - e^{-(N+k)\tau B})z_N \\ &\quad +\tau(I-e^{-2\tau B})^{-1}\sum_{j=1}^{N-1}(e^{-|k-j|\tau B} - e^{-(k+j)\tau B})\tau f_j. \end{aligned}$$

From this formula and using the formula for z_N, we can obtain formula (5.30).

Let us denote by $F_\tau(E) = F([0,T]_\tau, E)$ the space of grid functions $\varphi^\tau = \{\varphi_k\}_{k=1}^{N-1}$ for fixed $\tau = \frac{T}{N}$. Thus, $F_\tau(E)$ is the vector space whose elements are ordered $(N-1)$-tuples of elements of E. The space $F_\tau(E)$ can be equipped with various norms and thus become a normed space. Thus, for instance, the vector space $F_\tau(E)$ generates the normed space $C_\tau(E) = C([0,T]_\tau, E)$ with the norm

$$\| \varphi^\tau \|_{C_\tau(E)} = \max_{1\le k\le N-1} \| \varphi_k \|_E,$$

the normed space $C_\tau^\beta(E) = C^\beta([0,T]_\tau, E)$ $0<\beta<1$, with the norm

$$\| \varphi^\tau \|_{C_\tau^\beta(E)} = \| \varphi^\tau \|_{C_\tau(E)} + \max_{1\le k<k+r\le N-1} \| \varphi_{k+r} - \varphi_k \|_E \frac{1}{(r\tau)^\beta},$$

the weighted normed space $C_\tau^{\beta,\gamma}(E) = C^{\beta,\gamma}([0,T]_\tau, E), 0 < \gamma \leq \beta,\ 0 < \beta < 1$, with the norm

$$\| \varphi^\tau \|_{C_\tau^{\beta,\gamma}(E)} = \| \varphi^\tau \|_{C_\tau(E)}$$

$$+ \max_{1 \leq k < k+r \leq N-1} \| \varphi_{k+r} - \varphi_k \|_E \frac{((k+r)\tau)^\gamma((N-k)\tau)^\gamma}{(r\tau)^\beta},$$

and the normed space $L_{p,\tau}(E) = L_p([0,T]_\tau, E), 1 \leq p < \infty$, with the norm

$$\| \varphi^\tau \|_{L_{p,\tau}(E)} = (\sum_{k=1}^{N-1} \| \varphi_k \|_E^p \tau)^{\frac{1}{p}}.$$

Let us consider the difference problem (5.29). We will reduce it to an operator problem in the space $F_\tau(E)$. First we shall define an operator D_τ^2 acting from the space $E \times F_\tau(E) \times E$ of vectors $w^\tau = \{w_k\}_{k=0}^N$ into the space $F_\tau(E)$ of vectors $v^\tau = \{v_k\}_{k=1}^{N-1}$ according to the rule

$$v^\tau = D_\tau^2 u^\tau, v_k = \frac{1}{\tau^2}(w_{k+1} - 2w_k + w_{k-1}), k = 1, \dots, N-1.$$

Second we shall define an operator A_τ acting from the space $E \times F_\tau(E) \times E$ of vectors $w^\tau = \{w_k\}_{k=0}^N$ into the space $F_\tau(E)$ of vectors $v^\tau = \{v_k\}_{k=1}^{N-1}$ according to the rule

$$v^\tau = A_\tau u^\tau, v_k = \frac{1}{\tau^2}[(I - \exp(-\tau B))(w_{k-1} + w_{k+1})$$

$$+(\exp(-2\tau B) - I)w_k], k = 1, \dots, N-1.$$

Next, let us introduce the continuation operator $\Pi(u_0, u_N)$, which acts from $E \times F_\tau(E) \times E$ to $F_\tau(E)$ according to the rule

$$\Pi(u_0, u_N)(u_1, \dots, u_{N-1}) = (u_0, u_1, \dots, u_{N-1}, u_N).$$

Then, clearly, the difference problem (5.29) is equivalent to the operator problem

$$-D_\tau^2 \Pi(u_0, u_N)u^\tau + A_\tau \Pi(u_0, u_N)u^\tau = f^\tau.$$

Here f^τ is defined by the formula

$$f^\tau = (f_1, \dots, f_{N-1}).$$

The last operator problem will be considered in the space $F_\tau(E)$. From its unique solvability for any u_0, $u_N \in E$ and $f^\tau \in F_\tau(E)$ it follows that its solution u^τ defines a continuous additive and homogeneous operator $u^\tau(f^\tau, u_0, u_N)$.

The boundary-value problem (5.29) is said to be stable in $F_\tau(E)$ if we have the inequality

$$\| u^\tau(f^\tau, u_0, u_N) \|_{F_\tau(E)} \leq M[\| f^\tau \|_{F_\tau(E)} + \| u_0 \|_E + \|u_N\|_E],$$

where M is independent not only of f^τ, u_0, u_N, but also of τ.

From formula (5.30) one derives the following results.

Theorem 5.2.1. *Let A be the positive operator in a Banach space E. Then difference problem (5.29) is stable in $C_\tau(E)$.*

The proof of this theorem is based on formula (5.11) and the estimate

$$\|\exp(-k\tau B)\|_{E\to E} \le M\ e^{-\delta k\tau}, \delta > 0, \tag{5.31}$$

for all $k, k \ge 1$ and all $\tau > 0$,where M does not depend on k and τ. Estimate (5.31) follows from (5.8) for $t = t_k = k\tau$ and $a(B) = \delta$.

Theorem 5.2.2. *Let A be the positive operator in a Banach space E. Then difference problem (5.29) is stable in $C_\tau^{\alpha,\alpha}(E)$, $0 < \alpha < 1$.*

The proof of this theorem is based on formula (5.11), estimate (5.31) and

$$\|\exp(-k\tau B) - \exp\{-(k+r)\tau B\}\|_{E\to E} \le M\frac{r^\alpha}{(k+r)^\alpha}, \tag{5.32}$$

for all k, r, $1 \le k < k+r \le N-1$ and all $\tau > 0$, where M does not depend on k, r and τ. Estimate (5.32) follows from (5.10).

Theorem 5.2.3. *Let A be the positive operator in a Banach space E. Then difference problem (5.29) is stable in $L_{p,\tau}(E)$, $1 \le p < \infty$.*

The proof of this theorem is based on formula (5.11) and estimate (5.31).

The boundary-value difference problem (5.29) is said to be well posed (coercively stable) in $F_\tau(E)$ if we have the coercive inequality

$$\| \{\tau^{-2}(u_{k+1} - 2u_k + u_{k-1})\}_1^{N-1} \|_{F_\tau(E)}$$

$$+ \| \{\tau^{-2}[(I - \exp(-\tau B))(u_{k-1} + u_{k+1}) + (\exp(-2\tau B) - I)u_k]\}_1^{N-1} \|_{F_\tau(E)}$$

$$\le M[\| f^\tau \|_{F_\tau(E)} + \| Au_0\|_E + \|Au_N\|_E],$$

where M is independent not only of f^τ, u_0, u_N, but also of τ. Since the boundary-value differential problem (5.1) in $C(E)$ is not well posed for the general positive operator A and space E, then the well-posedness of the difference boundary-value in $C_\tau(E)$ norm does not take place uniformly with respect to $\tau > 0$. This means that the coercive norm

$$\| u^\tau \|_{K_\tau(E)} = \| \{\tau^{-2}(u_{k+1} - 2u_k + u_{k-1})\}_1^{N-1} \|_{C_\tau(E)}$$

$$+ \| \{\tau^{-2}[(I - \exp(-\tau B))(u_{k-1} + u_{k+1}) + (\exp(-2\tau B) - I)u_k]\}_1^{N-1} \|_{C_\tau(E)}$$

tends to ∞ as $\tau \to +0$. The investigation of the difference problem (5.29) permits us to establish the order of growth of this norm to ∞. First, let us prove a lemma that will be needed in the sequel.

Lemma 5.2.4. *Suppose A is the positive operator in a Banach space E. Then the following estimate holds:*

$$\sum_{j=1}^{N-1} ||(I - e^{-\tau B})e^{-t_{j-1}B}||_{E\to E} \leq M \min\left(\ln(1/\tau), 1 + \left|\ln \|B\|\right|\right), \tag{5.33}$$

where M does not depend on τ.

Proof. Using estimate (5.10), we obtain

$$\sum_{j=1}^{N-1} ||(I - e^{-\tau B})e^{-t_{j-1}B}||_{E\to E} \leq M \sum_{j=1}^{N-1} \frac{1}{j} \leq M \int_1^N \frac{ds}{s} = M \ \ln N \leq M_1 \ln(1/\tau). \tag{5.34}$$

Further, using estimate (5.10), we obtain

$$||(I - e^{-\tau B})e^{-t_{j-1}B}||_{E\to E} \leq M \min\{\frac{1}{j}, \tau||B||_{E\to E}\}.$$

If $||B||_{E\to E} > N$, then

$$\sum_{j=1}^{N-1} ||(I - e^{-\tau B})e^{-t_{j-1}B}||_{E\to E} \leq M \sum_{j=1}^{N-1} \frac{1}{j}$$

$$\leq M \int_1^N \frac{ds}{s} < M \int_1^{||B||_{E\to E}} \frac{ds}{s} \leq M|\ln ||B||_{E\to E}|.$$

If $||B||_{E\to E} < 1$, then

$$\sum_{j=1}^{N-1} ||(I - e^{-\tau B})e^{-t_{j-1}B}||_{E\to E} \leq M \sum_{j=1}^{N-1} \tau||B||_{E\to E} \leq M_1.$$

Finally, if $1 \leq ||B||_{E\to E} \leq N$, then

$$\sum_{j=1}^{N-1} ||(I - e^{-\tau B})e^{-t_{j-1}B}||_{E\to E} \leq \sum_{j=1}^{[\frac{N}{||B||_{E\to E}}]} \tau||B||_{E\to E} + \sum_{j=[\frac{N}{||B||_{E\to E}}]+1}^{N-1} \frac{1}{j}$$

$$\leq M(1 + \int_{\frac{T}{||B||_{E\to E}}}^{T} \frac{ds}{s}) \leq M(1 + \ln ||B||_{E\to E}).$$

Thus, in all three cases we have the estimate

$$\sum_{j=1}^{N-1} \|(I-e^{-\tau B})e^{-t_{j-1}B}\|_{E\to E} \le M(1+\left|\ln\|B\|_{E\to E}\right|). \tag{5.35}$$

Finally, estimate (5.33) follows from (5.34) and (5.35). Lemma 5.2.4 is proved.

Theorem 5.2.5. *Let A be the positive operator in a Banach space E and $u_0 \in D(A)$, $u_N \in D(A)$. Suppose the operator $I+\exp(-\tau B)$ has an inverse $(I+\exp(-\tau B))^{-1}$ and*

$$\|(I+\exp(-\tau B))^{-1}\|_{E\to E} \le M, \tag{5.36}$$

where M does not depend on τ. Then the solutions of the difference problem (5.29) in $C_\tau(E)$ obey the almost coercive inequality

$$\| u^\tau \|_{K_\tau(E)} \le M_1[\| Au_0\|_E + \|Au_N\|_E$$

$$+ \min\left\{\ln\frac{1}{\tau}, 1+|\ln\| B \|_{E\to E}|\right\} \| f^\tau \|_{C_\tau(E)}],$$

where M_1 is independent not only of f^τ, u_0, u_N, but also of τ.

Proof. Using formula (5.30), we obtain

$$\tau^{-2}(u_{k+1}-2u_k+u_{k-1})$$

$$= \tau^{-2}(I-e^{-2t_NB})^{-1}(I-e^{-\tau B})^2\{(e^{-t_{k-1}B}-e^{-(2t_N-t_{k+1})B})u_0$$
$$+(e^{-(t_N-t_{k+1})B}-e^{-(t_N+t_{k-1})B})u_N$$
$$-(e^{-(t_N-t_{k+1})B}-e^{-(t_N+t_{k-1})B})\tau^2\sum_{i=1}^{N-1}(e^{-(t_N-t_i)B}-e^{-(t_N+t_i)B})(I-e^{-2\tau B})^{-1}f_i\}$$
$$+\tau^{-2}\tau^2(I-e^{-\tau B})^2\sum_{i=1}^{k-1}e^{-(t_{k-1}-t_i)B}(I-e^{-2\tau B})^{-1}f_i$$
$$+\tau^{-2}\tau^2(I-e^{-\tau B})^2\sum_{i=k+1}^{N-1}e^{-(t_i-t_{k+1})B}(I-e^{-2\tau B})^{-1}f_i-\tau^{-2}\tau^2 2(I-e^{-\tau B})$$
$$\times(I-e^{-2\tau B})^{-1}f_k-\tau^{-2}\tau^2(I-e^{-\tau B})^2\sum_{i=1}^{N-1}e^{-(t_{k-1}+t_i)B})(I-e^{-2\tau B})^{-1}f_i$$

$$= (I-e^{-2t_NB})^{-1}(I-e^{-\tau B})^2\tau^{-2}\{(e^{-t_{k-1}B}-e^{-(2t_N-t_{k+1})B})u_0 \tag{5.37}$$
$$+(e^{-(t_N-t_{k+1})B}-e^{-(t_N+t_{k-1})B})u_N\}$$
$$-(I-e^{-2t_NB})^{-1}(I-e^{-\tau B})(e^{-(t_N-t_{k+1})B}-e^{-(t_N+t_{k-1})B})$$

$$\times \sum_{i=1}^{N-1}(e^{-(t_N-t_i)B}-e^{-(t_N+t_i)B})(I+e^{-\tau B})^{-1}f_i$$

$$+(I-e^{-\tau B})\sum_{i=1}^{k-1}e^{-(t_{k-1}-t_i)B}(I+e^{-\tau B})^{-1}f_i$$

$$+(I-e^{-\tau B})\sum_{i=k+1}^{N-1}e^{-(t_i-t_{k+1})B}(I+e^{-\tau B})^{-1}f_i-2(I+e^{-\tau B})^{-1}f_k$$

$$-(I-e^{-\tau B})\sum_{i=1}^{N-1}e^{-(t_{k-1}+t_i)B})(I+e^{-\tau B})^{-1}f_i=g_k+p_k,$$

where

$$g_k=\tau^{-2}(I-e^{-2t_NB})^{-1}(I-e^{-\tau B})^2\{(e^{-t_{k-1}B}-e^{-(2t_N-t_{k+1})B})u_0$$
$$+(e^{-(t_N-t_{k+1})B}-e^{-(t_N+t_{k-1})B})u_N\},$$
$$p_k=-(I-e^{-2t_NB})^{-1}(I-e^{-\tau B})(e^{-(t_N-t_{k+1})B}-e^{-(t_N+t_{k-1})B})$$
$$\times \sum_{i=1}^{N-1}(e^{-(t_N-t_i)B}-e^{-(t_N+t_i)B})(I+e^{-\tau B})^{-1}f_i$$
$$+(I-e^{-\tau B})\sum_{i=1}^{k-1}e^{-(t_{k-1}-t_i)B}(I+e^{-\tau B})^{-1}f_i$$
$$+(I-e^{-\tau B})\sum_{i=k+1}^{N-1}e^{-(t_i-t_{k+1})B}(I+e^{-\tau B})^{-1}f_i-2(I+e^{-\tau B})^{-1}f_k$$
$$-(I-e^{-\tau B})\sum_{i=1}^{N-1}e^{-(t_{k-1}+t_i)B})(I+e^{-\tau B})^{-1}f_i.$$

To this end it suffices to show that

$$\| g^\tau\|_{C_\tau(E)}\leq M_1[\| Au_0\|_E+\|Au_N\|_E], \tag{5.38}$$

$$\| p^\tau \|_{C_\tau(E)}\leq M_1\min\left\{\ln\frac{1}{\tau},1+|\ln\| B\|_{E\to E}|\right\}\| f^\tau \|_{C_\tau(E)}. \tag{5.39}$$

Estimate (5.38) follows from the estimate

$$\|(I-e^{-\tau B})^2(\tau B)^{-2}\|_{E\to E}\leq M$$

and the fact that the homogeneous difference equation ($f^\tau=0$) is stable in $C_\tau(E)$ uniformly in τ. Using estimates (5.10), (5.11) and (5.36), we obtain

$$\|p_k\|_E\leq\|(I-e^{-2t_NB})^{-1}\|_{E\to E}\|e^{-(t_N-t_{k+1})B}-e^{-(t_N+t_{k-1})B}\|_{E\to E}$$
$$\times\sum_{i=1}^{N-1}\|e^{-\tau B}-e^{-(t_{N+1}+t_i)B}\|_{E\to E}\|(I-e^{-\tau B})e^{-(t_{N-1}-t_i)B}\|_{E\to E}$$

$$\times\|(I+e^{-\tau B})^{-1}\|_{E\to E}\|f_i\|_E$$

$$+\sum_{i=1}^{k-1}\|(I-e^{-\tau B})e^{-(t_{k-1}-t_i)B}\|_{E\to E}\|(I+e^{-\tau B})^{-1}\|_{E\to E}\|f_i\|_E$$

$$+\sum_{i=k+1}^{N-1}\|(I-e^{-\tau B})e^{-(t_i-t_{k+1})B}\|_{E\to E}\|(I+e^{-\tau B})^{-1}\|_{E\to E}\|f_i\|_E$$

$$+2\|(I+e^{-\tau B})^{-1}\|_{E\to E}\|f_k\|_E$$

$$+\sum_{i=1}^{N-1}\|e^{-t_k B}\|_{E\to E}\|(I-e^{-\tau B})e^{-t_{i-1}B}\|_{E\to E}\|(I+e^{-\tau B})^{-1}\|_{E\to E}\|f_i\|_E$$

$$\leq M\sum_{j=1}^{N-1}\|(I-e^{-\tau B})e^{-t_{j-1}B}\|_{E\to E}\| f^\tau \|_{C_\tau(E)}.$$

From the last estimate and estimate (5.33) follows estimate (5.39). Theorem 5.2.5 is proved.

Note that estimate (5.37) is established for arbitrary positive operator A. In contrast, estimate (5.38) is established under the assumption (5.36). When is the last condition satisfied? We have an answer in the case where $E=H$ is a Hilbert space. Furthermore, the method of proof of Theorem 5.2.5 enables us to establish the following fact.

Theorem 5.2.6. *Let A be the positive operator in a Banach space E and $u_0,u_N\in D(A)$. Suppose the operator $I+\exp(-\tau B)$ has an inverse $(I+\exp(-\tau B))^{-1}$ and $f^\tau\in D((I+\exp(-\tau B))^{-1})$. Then the solutions of the difference problem* (5.29) *in $C_\tau(E)$ obey the almost coercive inequality*

$$\| u^\tau \|_{K_\tau(E)}\leq M_1[\| Au_0\|_E+\|Au_N\|_E$$

$$+\min\left\{\ln\frac{1}{\tau},1+|\ln\| B \|_{E\to E}|\right\}\| (I+\exp(-\tau B))^{-1}f^\tau \|_{C_\tau(E)}],$$

where M_1 is independent not only of f^τ, u_0, u_N, but also of τ.

Since

$$B^{-1}(I+\exp(-\tau B))^{-1}=\int_0^\tau\exp(-sB)ds(\,I-\exp(-2\tau B))^{-1},$$

using estimates (5.9) and (5.10), we obtain

$$\|B^{-1}(I+\exp(-\tau B))^{-1}\|_{E\to E} \tag{5.40}$$

$$\leq\int_0^\tau\|\exp(-sB)\|_{E\to E}ds\times\|(I-\exp(-2\tau B))^{-1}\|_{E\to E}\leq\frac{M(1-e^{-\delta\tau})}{\delta(1-e^{-2\delta\tau})}\leq\frac{M}{\delta}.$$

This yields the following result.

Theorem 5.2.7. *Suppose A is the positive operator in a Banach space E and u_0, $u_N \in D(A)$, $f^\tau \in D(B)$. Then the solutions of the difference problem (5.29) in $C_\tau(E)$ obey the almost coercive inequality*

$$\| u^\tau \|_{K_\tau(E)} \leq M_1[\| Au_0\|_E + \|Au_N\|_E$$

$$+ \min\left\{\ln\frac{1}{\tau}, 1 + |\ln \| B \|_{E\to E}|\right\} \| Bf^\tau \|_{C_\tau(E)}],$$

where M_1 is independent not only of f^τ, u_0, u_N, but also of τ.

Now let us study the well-posedness of the difference problem (5.29) in various Banach spaces.

Theorem 5.2.8. *Suppose that the assumptions of Theorem 5.17 hold. Then the solutions of the difference problem (5.29) in $C_\tau^{\alpha,\alpha}(E)$ obey the coercivity inequality*

$$\| \{\tau^{-2}(u_{k+1} - 2u_k + u_{k-1})\}_1^{N-1} \|_{C_\tau^{\alpha,\alpha}(E)}$$

$$+ \| \{\tau^{-2}[(I - \exp(-\tau B))(u_{k-1} + u_{k+1}) + (\exp(-2\tau B) - I)u_k]\}_1^{N-1} \|_{C_\tau^{\alpha,\alpha}(E)}$$

$$\leq M_1[\frac{1}{\alpha(1-\alpha)}\| f^\tau \|_{C_\tau^{\alpha,\alpha}(E)} + \| Au_0\|_E + \|Au_N\|_E],$$

where M_1 is independent not only of f^τ, u_0, u_N, α, but also of τ.

Proof. Using formula (5.37), we obtain

$$\tau^{-2}(u_{k+1} - 2u_k + u_{k-1})$$

$$= (I - e^{-2t_N B})^{-1}(I - e^{-\tau B})^2\tau^{-2}(e^{-t_{k-1}B} - e^{-(2t_N - t_{k+1})B})u_0$$

$$+(I - e^{-2t_N B})^{-1}(I - e^{-\tau B})^2\tau^{-2}(e^{-(t_N - t_{k+1})B} - e^{-(t_N + t_{k-1})B})u_N$$

$$+\{-(I - e^{-2t_N B})^{-1}(I - e^{-\tau B})(e^{-(t_N - t_{k+1})B} - e^{-(t_N + t_{k-1})B})$$

$$\times \sum_{i=1}^{N-1}(e^{-(t_N - t_i)B} - e^{-(t_N + t_i)B})(I + e^{-\tau B})^{-1}$$

$$+(I - e^{-\tau B})\sum_{i=1}^{k-1} e^{-(t_{k-1} - t_i)B}(I + e^{-\tau B})^{-1}$$

$$+(I - e^{-\tau B})\sum_{i=k+1}^{N-1} e^{-(t_i - t_{k+1})B}(I + e^{-\tau B})^{-1} - 2(I + e^{-\tau B})^{-1}$$

$$-(I - e^{-\tau B})\sum_{i=1}^{N-1} e^{-(t_{k-1} + t_i)B})(I + e^{-\tau B})^{-1}\}f_k$$

$$-(I-e^{-2t_NB})^{-1}(I-e^{-\tau B})(e^{-(t_N-t_{k+1})B}-e^{-(t_N+t_{k-1})B})$$
$$\times\sum_{i=1}^{N-1}(e^{-(t_N-t_i)B}-e^{-(t_N+t_i)B})(I+e^{-\tau B})^{-1}(f_i-f_k)$$
$$+(I-e^{-\tau B})\sum_{i=1}^{k-1}e^{-(t_{k-1}-t_i)B}(I+e^{-\tau B})^{-1}(f_i-f_k)$$
$$+(I-e^{-\tau B})\sum_{i=k+1}^{N-1}e^{-(t_i-t_{k+1})B}(I+e^{-\tau B})^{-1}(f_i-f_k)$$
$$-(I-e^{-\tau B})\sum_{i=1}^{N-1}e^{-(t_{k-1}+t_i)B})(I+e^{-\tau B})^{-1}(f_i-f_k)$$
$$=(I-e^{-2t_NB})^{-1}(I-e^{-\tau B})^2\tau^{-2}(e^{-t_{k-1}B}-e^{-(2t_N-t_{k+1})B})u_0 \tag{5.41}$$
$$+(I-e^{-2t_NB})^{-1}(I-e^{-\tau B})^2\tau^{-2}(e^{-(t_N-t_{k+1})B}-e^{-(t_N+t_{k-1})B})u_N$$
$$+\{-I+(I-e^{-2t_NB})^{-1}(I-e^{-t_NB})[I-e^{-t_{k-1}B}-e^{-t_{N-k-1}B}+e^{-t_NB}]\}f_k$$
$$+(I-e^{-2t_NB})^{-1}(I-e^{-2t_{N-k}B})(I-e^{-\tau B})$$
$$\times\sum_{i=1}^{k-1}e^{-(t_{k-1}-t_i)B}(I-e^{-2t_iB})(I+e^{-\tau B})^{-1}(f_i-f_k)$$
$$+(I-e^{-2t_NB})^{-1}(I-e^{-2t_kB})(I-e^{-\tau B})$$
$$\times\sum_{i=k+1}^{N-1}e^{-(t_i-t_{k+1})B}(I-e^{-2t_{N-i}B})(I+e^{-\tau B})^{-1}(f_i-f_k)$$
$$=\sum_{m=1}^{5}J_k^m,$$

where

$$J_k^1=(I-e^{-2t_NB})^{-1}(I-e^{-\tau B})^2\tau^{-2}(e^{-t_{k-1}B}-e^{-(2t_N-t_{k+1})B})u_0,$$
$$J_k^2=(I-e^{-2t_NB})^{-1}(I-e^{-\tau B})^2\tau^{-2}(e^{-(t_N-t_{k+1})B}-e^{-(t_N+t_{k-1})B})u_N,$$
$$J_k^3=\{-I+(I-e^{-2t_NB})^{-1}(I-e^{-t_NB})[I-e^{-t_{k-1}B}-e^{-t_{N-k-1}B}+e^{-t_NB}]\}f_k,$$
$$J_k^4=(I-e^{-2t_NB})^{-1}(I-e^{-2t_{N-k}B})(I-e^{-\tau B})$$
$$\times\sum_{i=1}^{k-1}e^{-(t_{k-1}-t_i)B}(I-e^{-2t_iB})(I+e^{-\tau B})^{-1}(f_i-f_k),$$
$$J_k^5=(I-e^{-2t_NB})^{-1}(I-e^{-2t_kB})(I-e^{-\tau B})$$
$$\times\sum_{i=k+1}^{N-1}e^{-(t_i-t_{k+1})B}(I-e^{-2t_{N-i}B})(I+e^{-\tau B})^{-1}(f_i-f_k).$$

Let us estimate $J_m^\tau = \{J_k^m\}_{k=1}^{N-1}$ for any $m = 1,2,3,4,5$ separately. We start with J_1^τ. Using estimates (5.8) and (5.9) we show that, for $1 \le k \le N-1$,

$$||J_k^1||_E \le ||(I-e^{-2t_N B})^{-1}||_{E\to E}||(I-e^{-\tau B})(\tau B)^{-1}||_{E\to E}^2$$
$$\times||e^{-t_{k-1}B} - e^{-(2t_N - t_{k+1})B}||_{E\to E}||Au_0||_E \le M||Au_0||_E.$$

From this it follows that

$$||J_1^1 - J_{1+r}^1||_E \le ||J_1^1||_E + ||J_{1+r}^1||_E \le 2M||Au_0||_E \le 2M\frac{(2r)^\alpha}{(1+r)^\alpha}||Au_0||_E$$
$$\le M_1[\frac{t_r^\alpha}{t_{r+1}^\alpha} + \frac{t_r^\alpha}{(t_N - t_1)^\alpha}]||Au_0||_E,$$

for $1 < r+1 \le N-1$. Further, applying (5.9) and (5.10) we show that, for $2 \le k < k+r \le N-1$,

$$||J_k^1 - J_{k+r}^1||_E \le ||(I-e^{-2t_N B})^{-1}||_{E\to E}||(I-e^{-\tau B})(\tau B)^{-1}||_{E\to E}^2$$
$$\times[||e^{-t_{k-1}B} - e^{-t_{k+r-1})B}||_{E\to E}$$
$$+||e^{-(2t_N - t_{k+1})B} - e^{-(2t_N - t_{k+r+1})B}||_{E\to E}]||Av_0||_E$$
$$\le M[\frac{t_r^\alpha}{t_{k+r-1}^\alpha} + \frac{t_r^\alpha}{(2t_N - t_{k+1})^\alpha}]||Av_0||_E \le M_1[\frac{t_r^\alpha}{t_{k+r}^\alpha} + \frac{t_r^\alpha}{(t_N - t_k)^\alpha}]||Au_0||_E.$$

Thus, we have proved that

$$||J_1^\tau||_{C_\tau^{\alpha,\alpha}(E)} \le M_1||Au_0||_E. \tag{5.42}$$

In a similar manner we can show that

$$||J_2^\tau||_{C_\tau^{\alpha,\alpha}(E)} \le M_1||Au_N||_E. \tag{5.43}$$

Now let us estimate J_3^τ. Using estimates (5.8) and (5.9) we obtain that

$$\| J_k^3 \|_E \le [1 + ||(I-e^{-2t_N B})^{-1}||_{E\to E}||I - e^{-t_N B}||_{E\to E}$$
$$\times[1 + ||e^{-t_{k-1}B}||_{E\to E} + ||e^{-t_{N-k-1}B}||_{E\to E} + ||e^{-t_N B}||_{E\to E}]]\|f_k\|_E$$
$$\le M_1\| f^\tau \|_{C_\tau^{\alpha,\alpha}(E)}$$

for $1 \le k \le N-1$. From this inequality it follows that

$$\max_{1\le k\le N-1} \| J_k^3 \|_E \le M_1\| f^\tau \|_{C_\tau^{\alpha,\alpha}(E)}.$$

From this it follows that

$$||J_1^3 - J_{1+r}^3||_E \le ||J_1^3||_E + ||J_{1+r}^3||_E \le 2M_1\| f^\tau \|_{C_\tau^{\alpha,\alpha}(E)}$$
$$\le 2M_1\frac{(2r)^\alpha}{(1+r)^\alpha}\| f^\tau \|_{C_\tau^{\alpha,\alpha}(E)} \le M_2[\frac{t_r^\alpha}{t_{r+1}^\alpha} + \frac{t_r^\alpha}{(t_N - t_1)^\alpha}]\| f^\tau \|_{C_\tau^{\alpha,\alpha}(E)},$$

for $1 < r+1 \leq N-1$ and

$$||J_k^3 - J_{k+r}^3||_E \leq ||J_k^3||_E + ||J_{k+r}^3||_E \leq 2M_1 \| f^\tau \|_{C_\tau^{\alpha,\alpha}(E)}$$

$$\leq 2M_1 \frac{(2r)^\alpha}{(1+r)^\alpha} \| f^\tau \|_{C_\tau^{\alpha,\alpha}(E)}$$

$$= 2M_1 \frac{(2r)^\alpha}{(1+r+k-k)^\alpha} \| f^\tau \|_{C_\tau^{\alpha,\alpha}(E)} \leq M_2 [\frac{t_r^\alpha}{t_{k+r}^\alpha} + \frac{t_r^\alpha}{(t_N - t_k)^\alpha}] \| f^\tau \|_{C_\tau^{\alpha,\alpha}(E)},$$

for $1 < k < r+k = N-1$. Since

$$J_{k+r}^3 - J_k^3 = \{-I + (I - e^{-2t_N B})^{-1}(I - e^{-t_N B})[I - e^{-t_{k+r-1}B}$$

$$-e^{-t_{N-k-r-1}B} + e^{-t_N B}]\}(f_{k+r} - f_k)$$

$$+(I - e^{-2t_N B})^{-1}(I - e^{-t_N B})[e^{-t_{k-1}B} - e^{-t_{k+r-1}B} + e^{-t_{N-k-1}B} - e^{-t_{N-k-r-1}B}]f_k,$$

it follows that

$$\| J_{k+r}^3 - J_k^3 \|_E \leq [1 + ||(I - e^{-t_N B})^{-1}||_{E\to E}$$

$$\times ||I - e^{-t_N B}||_{E\to E}[1 + ||e^{-t_{k-1}B}||_{E\to E} + ||e^{-t_{k+r-1}B}||_{E\to E}$$

$$+||e^{-t_{N-k-1}B}||_{E\to E} + ||e^{-t_{N-k-r-1}B}||_{E\to E}]]\|f_{k+r} - f_k\|_E$$

$$+||(I - e^{-2t_N B})^{-1}||_{E\to E}\{||e^{-t_{k-1}B} - e^{-t_{k+r-1}B}||_{E\to E}$$

$$+||e^{-t_{N-k-1}B} - e^{-t_{N-k-r-1}B}||_{E\to E}\}||I - e^{-t_N B}||_{E\to E}\|f_k\|_E$$

for $2 \leq k < k+r \leq N-2$. Using estimates (5.8), (5.9) and (5.10), we obtain

$$\| J_{k+r}^3 - J_k^3 \|_E \leq M_1 [\frac{t_r^\alpha}{(t_k + t_r)^\alpha} + \frac{t_r^\alpha}{(t_N - t_k)^\alpha}] \| f^\tau \|_{C_\tau^{\alpha,\alpha}(E)}$$

for $2 \leq k < k+r \leq N-2$. From this it follows that

$$\| J_3^\tau \|_{C_\tau^{\alpha,\alpha}(E} \leq M_1 \| f^\tau \|_{C_\tau^{\alpha,\alpha}(E)}. \tag{5.44}$$

Now let us estimate J_4^τ. By the definition $J_1^4 = 0$. Therefore, we have to consider two cases: $2 \leq k \leq \frac{N}{2}$ and $\frac{N}{2} \leq k \leq N-1$. In the first case, using estimates (5.8) and (5.9) we obtain that

$$\| J_k^4 \|_E \leq ||(I - e^{-2t_N B})^{-1}||_{E\to E} \sum_{i=1}^{k-1} ||e^{-(t_{k-1}-t_i)B}(I - e^{-2t_{N-k}B})(I - e^{-\tau B}) \|_{E\to E}$$

$$\times ||I - e^{-2t_i B} \|_{E\to E} \; ||(I + e^{-\tau B})^{-1}||_{E\to E}\|f_i - f_k \|_E$$

$$\leq M \sum_{i=1}^{k-1} \frac{\tau}{(t_k - t_i)^{1-\alpha} t_k^\alpha (t_N - t_i)^\alpha} \| f^\tau \|_{C_\tau^{\alpha,\alpha}(E)}$$

$$\leq \frac{M}{t_k^\alpha (t_N - t_k)^\alpha} [\sum_{i=1}^{k-1} \frac{\tau}{(t_k - t_i)^{1-\alpha}}] \| f^\tau \|_{C_\tau^{\alpha,\alpha}(E)}.$$

The sum enclosed in the right-hand side square brackets is the lower Darboux integral sum for the integral

$$\int_0^{t_k} \frac{ds}{(t_k - s)^\alpha} = \frac{t_k^\alpha}{\alpha}.$$

Thus, for $2 \le k \le \frac{N}{2}$ we have established the bound

$$\| J_k^4 \|_E \le \frac{M_1}{\alpha}(t_N - t_k)^{-\alpha} \| f^\tau \|_{C_\tau^{\alpha,\alpha}(E)} \le \frac{M_1 2^\alpha}{\alpha T^\alpha} \| f^\tau \|_{C_\tau^{\alpha,\alpha}(E)}.$$

In the second case, using estimates (5.8), (5.9) and (5.10) we obtain that

$$\| J_k^4 \|_E \le \|(I - e^{-2t_N B})^{-1}\|_{E \to E} \sum_{i=1}^{k-1} \|e^{-(t_{k-1}-t_i)B}(I - e^{-2t_{N-k}B})(I - e^{-\tau B}) \|_{E \to E}$$

$$\times \|I - e^{-2t_i B} \|_{E \to E} \|(I + e^{-\tau B})^{-1}\|_{E \to E} \| f_i - f_k \|_E$$

$$\le M \sum_{i=1}^{k-1} \frac{2(t_N - t_k)\tau}{(t_k - t_i)^{1-\alpha}(2t_N - t_k - t_i) t_k^\alpha (t_N - t_i)^\alpha} \| f^\tau \|_{C_\tau^{\alpha,\alpha}(E)}$$

$$\le \frac{2M(t_N - t_k)}{t_k^\alpha (t_N - t_k)^\alpha} [\sum_{i=1}^{k-1} \frac{\tau}{(t_k - t_i)^{1-\alpha}(2t_N - t_k - t_i)}] \| f^\tau \|_{C_\tau^{\alpha,\alpha}(E)}.$$

The sum enclosed in the right-hand side square brackets is the lower Darboux integral sum for the integral

$$\int_0^{t_k} \frac{ds}{(t_k - s)^{1-\alpha}(2t_N - t_k - s)}.$$

Since

$$\int_0^{t_k} \frac{ds}{(t_k - s)^{1-\alpha}(2t_N - t_k - s)} = \int_0^{t_k} \frac{dz}{z^{1-\alpha}(2t_N - 2t_k + z)}$$

$$\le \frac{1}{(2t_N - 2t_k)^{1-\alpha}} \int_0^\infty \frac{dp}{p^{1-\alpha}(1+p)} \le \frac{M}{(2t_N - 2t_k)^{1-\alpha}\alpha(1-\alpha)},$$

it follows that

$$\| J_k^4 \|_E \le \frac{2M(t_N - t_k)}{t_k^\alpha (t_N - t_k)^\alpha} \frac{M}{(2t_N - 2t_k)^{1-\alpha}\alpha(1-\alpha)} \| f^\tau \|_{C_\tau^{\alpha,\alpha}(E)}.$$

Thus, for $\frac{N}{2} \leq k \leq N-1$ we have established the bound

$$\| J_k^4 \|_E \leq \frac{M_1}{\alpha(1-\alpha)t_k^\alpha} \| f^\tau \|_{C_\tau^{\alpha,\alpha}(E)} \leq \frac{M_1 2^\alpha}{\alpha T^\alpha} \| f^\tau \|_{C_\tau^{\alpha,\alpha}(E)}.$$

Combining the estimates for $\| J_k^4 \|_E$, we obtain

$$\max_{1\leq k\leq N-1} \| J_k^4 \|_E \leq \frac{M_1}{\alpha(1-\alpha)} \| f^\tau \|_{C_\tau^{\alpha,\alpha}(E)}. \tag{5.45}$$

Now let us estimate the difference $J_{k+r}^4 - J_k^4$ for $2 \leq k < k+r \leq N-1$. We will consider separately the cases $k \leq r$, $N-k \leq 2r$ and $k > r$, $N-k > 2r$. If $k \leq r$, then (5.17) yields

$$\| J_{k+r}^4 - J_k^4 \|_E \leq \| J_{k+r}^4 \|_E + \| J_k^4 \|_E$$

$$\leq \frac{M_1 2^{1+\alpha}}{\alpha(1-\alpha)} \| f^\tau \|_{C_\tau^{\alpha,\alpha}(E)} r^\alpha (k+r)^{-\alpha} = \frac{M_2}{\alpha(1-\alpha)} \| f^\tau \|_{C_\tau^{\alpha,\alpha}(E)} t_r^\alpha (t_k + t_r)^{-\alpha}.$$

If $N-k \leq 2r$, then (5.17) yields

$$\| J_{k+r}^4 - J_k^4 \|_E \leq \| J_{k+r}^4 \|_E + \| J_k^4 \|_E$$

$$\leq \frac{M_1 2^{1+\alpha}}{\alpha(1-\alpha)} \| f^\tau \|_{C_\tau^{\alpha,\alpha}(E)} r^\alpha (N-k)^{-\alpha} = \frac{M_2}{\alpha(1-\alpha)} \| f^\tau \|_{C_\tau^{\alpha,\alpha}(E)} t_r^\alpha (t_N - t_k)^{-\alpha}.$$

Now let $k > r$ and $N-k > 2r$. Let us represent the difference $J_{k+r}^4 - J_k^4$ as the following sums:

$$J_{k+r}^4 - J_k^4 = (I - e^{-2t_N B})^{-1} (I - e^{-\tau B})$$

$$\times \{ \sum_{i=1}^{k+r-1} e^{-(t_{k+r-1}-t_i)B} (I - e^{-2t_i B})(I - e^{-2t_{N-k-r} B})(I + e^{-\tau B})^{-1} (f_i - f_{k+r})$$

$$- \sum_{i=1}^{k-1} e^{-(t_{k-1}-t_i)B} (I - e^{-2t_{N-k} B})(I - e^{-2t_i B})(I + e^{-\tau B})^{-1} (f_i - f_k) \}$$

$$= (I - e^{-2t_N B})^{-1} (I - e^{-\tau B}) \sum_{i=k-r}^{k+r-1} e^{-(t_{k+r-1}-t_i)B}$$

$$\times (I - e^{-2t_i B})(I - e^{-2t_{N-k-r} B})(I + e^{-\tau B})^{-1} (f_i - f_{k+r})$$

$$-(I - e^{-2t_N B})^{-1} (I - e^{-\tau B}) \sum_{i=k-r}^{k-1} e^{-(t_{k-1}-t_i)B}$$

$$\times (I - e^{-2t_{N-k} B})(I - e^{-2t_i B})(I + e^{-\tau B})^{-1} (f_i - f_k)$$

$$+(I-e^{-2t_NB})^{-1}(I-e^{-\tau B})\sum_{i=1}^{k-r-1}e^{-(t_{k+r-1}-t_i)B}$$

$$\times(I-e^{-2t_iB})(I-e^{-2t_{N-k-r}B})(I+e^{-\tau B})^{-1}(f_k-f_{k+r})$$

$$+(I-e^{-2t_NB})^{-1}(I-e^{-\tau B})\sum_{i=1}^{k-r-1}\{e^{-(t_{k+r-1}-t_i)B}(I-e^{-2t_{N-k-r}B})$$

$$-e^{-(t_{k-1}-t_i)B}(I-e^{-2t_{N-k}B})\}(I-e^{-2t_iB})(I+e^{-\tau B})^{-1}(f_i-f_k)$$

$$=P_k^1+P_k^2+P_k^3+P_k^4.$$

Now let us estimate P_k^1. Using estimates (5.8) and (5.9), we obtain that

$$\| P_k^1 \|_E\leq ||(I-e^{-2t_NB})^{-1}||_{E\to E}$$

$$\times\sum_{i=k-r}^{k+r-1}||(I-e^{-\tau B})e^{-(t_{k+r-1}-t_i)B}(I-e^{-2t_{N-k-r}B})||_{E\to E}$$

$$\times||I-e^{-2t_iB}\|_{E\to E}\|(I+e^{-\tau B})^{-1}||_{E\to E}||f_i-f_{k+r}\|_E$$

$$\leq\frac{M}{(t_k+t_r)^\alpha}[\frac{1}{\tau^\alpha(t_N-t_{k+r-1})^\alpha}+\sum_{i=k-r}^{k+r-2}\frac{\tau}{(t_{k+r}-t_i)^{1-\alpha}(t_N-t_i)^\alpha}]\| f^\tau \|_{C_\tau^{\alpha,\alpha}(E)}$$

$$\leq\frac{M_1}{(t_k+t_r)^\alpha}[\sum_{i=k-r}^{k+r-1}\frac{\tau}{(t_{k+r}-t_i)^{1-\alpha}(t_N-t_i)^\alpha}]\| f^\tau \|_{C_\tau^{\alpha,\alpha}(E)}.$$

The sum enclosed in the right-hand side square brackets is the lower Darboux integral sum for the integral

$$\int_{t_k-t_r}^{t_k+t_r}\frac{ds}{(t_k+t_r-s)^{1-\alpha}(t_N-s)^\alpha}.$$

Since

$$\int_{t_k-t_r}^{t_k+t_r}\frac{ds}{(t_k+t_r-s)^{1-\alpha}(t_N-s)^\alpha}\leq\frac{M}{\alpha}(2t_r)^\alpha(t_N-t_k-t_r)^{-\alpha},$$

it follows that

$$\| P_k^1 \|_E\leq\frac{M_1M}{\alpha}(2t_r)^\alpha(t_N-t_k-t_r)^{-\alpha}(t_k+t_r)^{-\alpha}\| f^\tau \|_{C_\tau^{\alpha,\alpha}(E)}.$$

Thus, we have established the bound

$$\| P_k^1 \|_E\leq\frac{M_2}{\alpha}t_r^\alpha(t_N-t_k)^{-\alpha}(t_k+t_r)^{-\alpha}\| f^\tau \|_{C_\tau^{\alpha,\alpha}(E)}.$$

In a similar manner we can show that

$$\| P_k^2 \|_E \leq \frac{M_2}{\alpha} t_r^\alpha (t_N - t_k)^{-\alpha} (t_k + t_r)^{-\alpha} \| f^\tau \|_{C_\tau^{\alpha,\alpha}(E)}.$$

Since

$$P_k^3 = (I - e^{-2t_N B})^{-1} e^{-2t_{2r} B} (I - e^{-t_{k-r-1} B})$$
$$\times (I - e^{-2t_i B})(I - e^{-2t_{N-k-r} B})(I + e^{-\tau B})^{-1}(f_k - f_{k+r}),$$

it follows that

$$\| P_k^3 \|_E \leq \|(I - e^{-2t_N B})^{-1}\|_{E\to E} \|e^{-2t_{2r} B}(I - e^{-t_{k-r-1} B})\|_{E\to E}$$
$$\times \|I - e^{-2t_i B}\|_{E\to E} \|I - e^{-2t_{N-k-r} B}\|_{E\to E} \|(I + e^{-\tau B})^{-1}\|_{E\to E} \|f_k - f_{k+r}\|_E.$$

Using estimates (5.8) and (5.9), we obtain

$$\| P_k^3 \|_E \leq M t_r^\alpha (t_N - t_k)^{-\alpha} (t_k + t_r)^{-\alpha} \| f^\tau \|_{C_\tau^{\alpha,\alpha}(E)}.$$

Now let us estimate P_k^4. Using estimates (5.8), (5.9) and (5.10), we obtain

$$\| P_k^4 \|_E \leq \|(I - e^{-2t_N B})^{-1}\|_{E\to E}$$
$$\times \sum_{i=1}^{k-r-1} \{ \| (I - e^{-\tau B})(e^{-(2t_N - 2t_k)B} - e^{-(2t_N - 2t_k - 2t_r)B}) e^{-(t_{k-1} - t_i)B} \|_{E\to E}$$
$$+ \| (I - e^{-\tau B})(I - e^{-t_r)B}) e^{-(t_{k-1} - t_i)B} \|_{E\to E} \}$$
$$\times \|I - e^{-2t_i B}\|_{E\to E} \|(I + e^{-\tau B})^{-1}\|_{E\to E} \| f_i - f_k \|_E$$
$$\leq M \frac{t_r}{t_k^\alpha} [\sum_{i=1}^{k-r-1} \frac{\tau}{(t_k - t_i)^{2-\alpha}(t_N - t_i)^\alpha}] \| f^\tau \|_{C_\tau^{\alpha,\alpha}(E)}.$$

The sum enclosed in the right-hand side square brackets is the lower Darboux integral sum for the integral

$$\int_0^{t_k - t_r} \frac{ds}{(t_k - s)^{2-\alpha}(t_N - s)^\alpha}.$$

Since for $k > r$ and $N - k > 2r$ we have the bound

$$\int_0^{t_k - t_r} \frac{ds}{(t_k - s)^{2-\alpha}(t_N - s)^\alpha} \leq \frac{1}{(t_N - t_k + t_r)^\alpha} \int_0^{t_k - t_r} \frac{ds}{(t_k - s)^{2-\alpha}}$$
$$\leq \frac{1}{(1-\alpha)(t_N - t_k + t_r)^\alpha} t_r^{-1+\alpha} \leq \frac{1}{(1-\alpha)(t_N - t_k)^\alpha} t_r^{-1+\alpha},$$

it follows that

$$\| P_k^4 \|_E \leq \frac{M}{1-\alpha} t_\tau^\alpha (t_N - t_k)^{-\alpha} t_k^{-\alpha} \| f^\tau \|_{C_\tau^{\alpha,\alpha}(E)}.$$

Thus, we have established the bound

$$\| P_k^4 \|_E \leq \frac{M 2^\alpha}{1-\alpha} t_\tau^\alpha (t_N - t_k)^{-\alpha} (t_k + t_r)^{-\alpha} \| f^\tau \|_{C_\tau^{\alpha,\alpha}(E)}.$$

Thus, we have shown that for any $2 \leq k < k+r \leq N-1$ the following inequality holds:

$$\| J_{k+r}^4 - J_k^4 \|_E \leq \frac{M_1}{\alpha(1-\alpha)} t_r^\alpha (t_N - t_k)^{-\alpha} (t_k + t_r)^{-\alpha} \| f^\tau \|_{C_\tau^{\alpha,\alpha}(E)}. \tag{5.46}$$

Estimates (5.45) and (5.46) give

$$\| J_4^\tau \|_{C_\tau^{\alpha,\alpha}(E)} \leq \frac{M_1}{\alpha(1-\alpha)} \| f^\tau \|_{C_\tau^{\alpha,\alpha}(E)} . \tag{5.47}$$

In exactly the same manner one establishes the inequality

$$\| J_5^\tau \|_{C_\tau^{\alpha,\alpha}(E)} \leq \frac{M_1}{\alpha(1-\alpha)} \| f^\tau \|_{C_\tau^{\alpha,\alpha}(E)} . \tag{5.48}$$

Finally, applying estimates (5.42), (5.43), (5.44), (5.47) and (5.48) we obtain the estimate

$$\| \{\tau^{-2}(u_{k+1} - 2u_k + u_{k-1})\}_1^{N-1} \|_{C_\tau^{\alpha,\alpha}(E)}$$
$$\leq M_1 [\frac{1}{\alpha(1-\alpha)} \| f^\tau \|_{C_\tau^{\alpha,\alpha}(E)} + \| A u_0 \|_E + \| A u_N \|_E].$$

By the triangle inequality, this last estimate and difference equation (5.29) yields

$$\| \{\tau^{-2}[(I - \exp(-\tau B))(u_{k-1} + u_{k+1}) + (\exp(-2\tau B) - I)u_k]\}_1^{N-1} \|_{C_\tau^{\alpha,\alpha}(E)}$$
$$\leq M_1 [\frac{1}{\alpha(1-\alpha)} \| f^\tau \|_{C_\tau^{\alpha,\alpha}(E)} + \| A u_0 \|_E + \| A u_N \|_E].$$

Theorem 5.2.8 is proved.

Furthermore, the method of proof of Theorem 5.2.8 enables us to establish the following fact.

Theorem 5.2.9. *Suppose that the assumptions of Theorem* 5.2.6 *hold. Then the solutions of the difference problem* (5.29) *in* $C_\tau^{\alpha,\alpha}(E)$ *obey the coercivity inequality*

$$\| \{\tau^{-2}(u_{k+1} - 2u_k + u_{k-1})\}_1^{N-1} \|_{C_\tau^{\alpha,\alpha}(E)}$$
$$+ \| \{\tau^{-2}[(I - \exp(-\tau B))(u_{k-1} + u_{k+1}) + (\exp(-2\tau B) - I)u_k]\}_1^{N-1} \|_{C_\tau^{\alpha,\alpha}(E)}$$
$$\leq M_1 [\frac{1}{\alpha(1-\alpha)} \| (I + \exp(-\tau B))^{-1} f^\tau \|_{C_\tau^{\alpha,\alpha}(E)} + \| A u_0 \|_E + \| A u_N \|_E],$$

where M_1 *is independent not only of* f^τ, u_0, u_N, α, *but also of* τ.

This yields the following result

Theorem 5.2.10. *Suppose that the assumptions of Theorem* 5.19 *hold. Then the solutions of the difference problem* (5.29) *in* $C_\tau^{\alpha,\alpha}(E)$ *obey the coercivity inequality*

$$\| \{\tau^{-2}(u_{k+1} - 2u_k + u_{k-1})\}_1^{N-1} \|_{C_\tau^{\alpha,\alpha}(E)}$$

$$+ \| \{\tau^{-2}[(I - \exp(-\tau B))(u_{k-1} + u_{k+1}) + (\exp(-2\tau B) - I)u_k]\}_1^{N-1} \|_{C_\tau^{\alpha,\alpha}(E)}$$

$$\leq M_1[\frac{1}{\alpha(1-\alpha)} \| Bf^\tau \|_{C_\tau^{\alpha,\alpha}(E)} + \| Au_0\|_E + ||Au_N\|_E],$$

where M_1 *is independent not only of* f^τ, u_0, u_N, α, *but also of* τ.

Note that the coercivity inequality

$$\| \{\tau^{-2}(u_{k+1} - 2u_k + u_{k-1})\}_1^{N-1} \|_{C_\tau^{\alpha}(E)}$$

$$+ \| \{\tau^{-2}[(I - \exp(-\tau B))(u_{k-1} + u_{k+1}) + (\exp(-2\tau B) - I)u_k]\}_1^{N-1} \|_{C_\tau^{\alpha}(E)}$$

$$\leq M_1[\frac{1}{\alpha(1-\alpha)} \| f^\tau \|_{C_\tau^{\alpha}(E)} + \| Au_0\|_E + ||Au_N\|_E]$$

fails. Nevertheless, we have the following result.

Theorem 5.2.11. *Suppose that the assumptions of Theorem* 5.17 *hold. Then the solutions of the difference problem* (5.29) *in* $C_\tau^{\alpha}(E)$ *obey the coercivity inequality*

$$\| \{\tau^{-2}(u_{k+1} - 2u_k + u_{k-1})\}_1^{N-1} \|_{C_\tau^{\alpha}(E)}$$

$$+ \| \{\tau^{-2}[(I - \exp(-\tau B))(u_{k-1} + u_{k+1}) + (\exp(-2\tau B) - I)u_k]\}_1^{N-1} \|_{C_\tau^{\alpha}(E)}$$

$$+ \| \{\tau^{-2}(u_{k+1} - 2u_k + u_{k-1})\}_1^{N-1} \|_{C_\tau(E_\alpha)}$$

$$\leq \frac{M_1}{\alpha(1-\alpha)} \| f^\tau \|_{C_\tau^{\alpha}(E)} + \frac{M_1}{\alpha}[\| (I - e^{-\tau B})^2 \tau^{-2} u_0 - f_1 \|_{E_\alpha}$$

$$+ ||(I - e^{-\tau B})^2 \tau^{-2} u_N - f_{N-1}\|_{E_\alpha}],$$

where M_1 *is independent not only of* f^τ, u_0, u_N, α, *but also of* τ.

Proof. To prove the theorem it suffices to establish the estimates for

$$\{\tau^{-2}(u_{k+1} - 2u_k + u_{k-1})\}_1^{N-1}$$

in $C_\tau^{\alpha}(E)$ and in $C_\tau(E_\alpha)$. By formula (5.41), we have

$$\tau^{-2}(u_{k+1} - 2u_k + u_{k-1})$$

$$= (I - e^{-2t_N B})^{-1}(e^{-t_{k-1}B} - e^{-(2t_N - t_{k+1})B})[(I - e^{-\tau B})^2 \tau^{-2} u_0 - f_1]$$

$$+(I-e^{-2t_NB})^{-1}(e^{-(t_N-t_{k+1})B}-e^{-(t_N+t_{k-1})B})[(I-e^{-\tau B})^2\tau^{-2}u_N-f_{N-1}]$$
$$+(I-e^{-2t_NB})^{-1}(e^{-t_{k-1}B}-e^{-(2t_N-t_{k+1})B})[f_1-f_k]$$
$$+(I-e^{-2t_NB})^{-1}(e^{-(t_N-t_{k+1})B}-e^{-(t_N+t_{k-1})B})[f_k-f_{N-1}]$$
$$+(I-e^{-2t_NB})^{-1}(I-e^{-2t_{N-k}B})(I-e^{-\tau B})$$
$$\times\sum_{i=1}^{k-1}e^{-(t_{k-1}-t_i)B}(I-e^{-2t_iB})(I+e^{-\tau B})^{-1}(f_i-f_k)$$
$$+(I-e^{-2t_NB})^{-1}(I-e^{-2t_kB})(I-e^{-\tau B})$$
$$\times\sum_{i=k+1}^{N-1}e^{-(t_i-t_{k+1})B}(I-e^{-2t_{N-i}B})(I+e^{-\tau B})^{-1}(f_i-f_k)$$
$$=\sum_{m=1}^{6}G_k^m,$$

where

$$G_k^1=(I-e^{-2t_NB})^{-1}(e^{-t_{k-1}B}-e^{-(2t_N-t_{k+1})B})[(I-e^{-\tau B})^2\tau^{-2}u_0-f_1],$$
$$G_k^2=(I-e^{-2t_NB})^{-1}(e^{-(t_N-t_{k+1})B}-e^{-(t_N+t_{k-1})B})[(I-e^{-\tau B})^2\tau^{-2}u_N-f_{N-1}],$$
$$G_k^3=(I-e^{-2t_NB})^{-1}(e^{-t_{k-1}B}-e^{-(2t_N-t_{k+1})B})[f_1-f_k],$$
$$G_k^4=(I-e^{-2t_NB})^{-1}(e^{-(t_N-t_{k+1})B}-e^{-(t_N+t_{k-1})B})[f_k-f_{N-1}],$$
$$G_k^5=(I-e^{-2t_NB})^{-1}(I-e^{-2t_{N-k}B})(I-e^{-\tau B})$$
$$\times\sum_{i=1}^{k-1}e^{-(t_{k-1}-t_i)B}(I-e^{-2t_iB})(I+e^{-\tau B})^{-1}(f_i-f_k),$$
$$G_k^6=(I-e^{-2t_NB})^{-1}(I-e^{-2t_kB})(I-e^{-\tau B})$$
$$\times\sum_{i=k+1}^{N-1}e^{-(t_i-t_{k+1})B}(I-e^{-2t_{N-i}B})(I+e^{-\tau B})^{-1}(f_i-f_k).$$

Let us estimate $G_m^\tau=\{G_k^m\}_{k=1}^{N-1}$ for any $m=1,2,3,4,5,6$ separately. We start with G_1^τ. Using estimates (5.8), (5.9) and the definition of the spaces E_α, we show that, for $1\le k\le N-1$,

$$||G_k^1||_E\le||(I-e^{-2t_NB})^{-1}||_{E\to E}$$
$$\times\int_{t_{k-1}}^{2t_N-t_{k+1}}||B\exp(-sB)[(I-e^{-\tau B})^2\tau^{-2}u_0-f_1]||_E ds$$
$$\le M\int_{t_{k-1}}^{2t_N-t_{k+1}}\frac{ds}{s^{1-\alpha}}||[(I-e^{-\tau B})^2\tau^{-2}u_0-f_1]||_{E_\alpha}\le\frac{M_1}{\alpha}||(I-e^{-\tau B})^2\tau^{-2}u_0-f_1||_{E_\alpha}.$$

Further, applying (5.8), (5.9) and the definition of the spaces E_α we show that, for $1 \leq k < k+r \leq N-1$,

$$||G_k^1 - G_{k+r}^1||_E \leq ||(I - e^{-2t_N B})^{-1}||_{E \to E}$$

$$\times [\int_{t_{k-1}}^{t_{k+r-1}} ||B \exp(-sB)[(I - e^{-\tau B})^2 \tau^{-2} u_0 - f_1]||_E ds$$

$$+ \int_{2t_N - t_{k+r+1}}^{2t_N - t_{k+1}} ||B \exp(-sB)[(I - e^{-\tau B})^2 \tau^{-2} u_0 - f_1]||_E ds]$$

$$\leq M[\int_{t_{k-1}}^{t_{k+r-1}} \frac{ds}{s^{1-\alpha}} + \int_{2t_N - t_{k+r+1}}^{2t_N - t_{k+1}} \frac{ds}{s^{1-\alpha}}]||(I - e^{-\tau B})^2 \tau^{-2} u_0 - f_1||_{E_\alpha}$$

$$\leq t_r^\alpha \frac{M_1}{\alpha} ||(I - e^{-\tau B})^2 \tau^{-2} u_0 - f_1||_{E_\alpha}.$$

Thus, we have proved that

$$||G_1^\tau||_{C_\tau^\alpha(E)} \leq \frac{M_1}{\alpha} ||(I - e^{-\tau B})^2 \tau^{-2} u_0 - f_1||_{E_\alpha}.$$

In a similar manner we can show that

$$||G_2^\tau||_{C_\tau^\alpha(E)} \leq \frac{M_1}{\alpha} ||(I - e^{-\tau B})^2 \tau^{-2} u_N - f_{N-1}||_{E_\alpha}.$$

Using estimates (5.8), (5.9) and the definition of the spaces E_α we show that, for $1 \leq k \leq N-1$,

$$||G_k^1||_{E_\alpha} \leq ||(I - e^{-2t_N B})^{-1}||_{E \to E} ||e^{-t_{k-1} B} - e^{-(2t_N - t_{k+1})B}||_{E \to E}$$

$$\times ||(I - e^{-\tau B})^2 \tau^{-2} u_0 - f_1||_{E_\alpha}$$

$$\leq M ||(I - e^{-\tau B})^2 \tau^{-2} u_0 - f_1||_{E_\alpha}.$$

Thus, we have proved that

$$||G_1^\tau||_{C_\tau(E_\alpha)} \leq M ||(I - e^{-\tau B})^2 \tau^{-2} u_0 - f_1||_{E_\alpha}.$$

In a similar manner we can show that

$$||G_2^\tau||_{C_\tau(E_\alpha)} \leq M ||(I - e^{-\tau B})^2 \tau^{-2} u_N - f_{N-1}||_{E_\alpha}.$$

Now let us estimate G_3^τ. Using estimates (5.8) and (5.9) we obtain that, for $1 \leq k \leq N-1$,

$$||G_k^3||_E \leq ||(I - e^{-2t_N B})^{-1}||_{E \to E} ||e^{-t_{k-1} B} - e^{-(2t_N - t_{k+1})B}||_{E \to E} ||f_1 - f_k||_E$$

$$\leq M ||f^\tau||_{C_\tau^\alpha(E)}.$$

Further, applying (5.8), (5.9) and (5.10) we show that, for $1 \leq k < k+r \leq N-1$,

$$\| G_{k+r}^3 - G_k^3 \|_E \leq ||(I - e^{-2t_N B})^{-1}||_{E \to E}$$

$$\times \{||e^{-t_{k-1}B} - e^{-t_{k+r-1}B}||_{E \to E} + ||e^{-(2t_N - t_{k+1})B} - e^{-(2t_N - t_{k+r+1})B}||_{E \to E}\} \|f_1 - f_k\|_E$$

$$+ ||(I - e^{-2t_N B})^{-1}||_{E \to E} ||e^{-t_{k+r-1}B} - e^{-(2t_N - t_{k+r+1})B}||_{E \to E} ||f_{k+r} - f_k||_E$$

$$\leq M t_r^\alpha \|f^\tau\|_{C_\tau^\alpha(E)}.$$

Thus, we have proved that

$$||G_3^\tau||_{C_\tau^\alpha(E)} \leq M_1 \|f^\tau\|_{C_\tau^\alpha(E)}.$$

In a similar manner we can show that

$$||G_4^\tau||_{C_\tau^\alpha(E)} \leq M_1 \|f^\tau\|_{C_\tau^\alpha(E)}.$$

By the definition of the spaces E_α and using estimates (5.8), (5.9), we obtain that

$$\| \lambda^{1-\alpha} B e^{-\lambda B} G_k^3 \|_E \leq \lambda^{1-\alpha} ||(I - e^{-2t_N B})^{-1}||_{E \to E}$$

$$\times ||B e^{-\lambda B} (e^{-t_{k-1}B} - e^{-(2t_N - t_{k+1})B})(f_1 - f_k)||_E$$

$$\leq M \frac{\lambda^{1-\alpha} t_{k-1}^\alpha}{\lambda + t_{k-1}} \|f^\tau\|_{C_\tau^\alpha(E)} \leq M \|f^\tau\|_{C_\tau^\alpha(E)}$$

for all λ, $\lambda > 0$ and for $1 \leq k \leq N-1$. So, we have established that

$$||G_3^\tau||_{C(E_\alpha)} \leq M_1 \| f^\tau \|_{C_\tau^\alpha(E)} .$$

In a similar manner we can show that

$$||G_4^\tau||_{C(E_\alpha)} \leq M_1 \| f^\tau \|_{C_\tau^\alpha(E)} .$$

Now let us estimate G_5^τ. By the definition $G_1^5 = 0$. Therefore, we have to consider the cases $2 \leq k \leq N-1$. Using estimates (5.8) and (5.9) we obtain that

$$\| G_k^5 \|_E \leq ||(I - e^{-2t_N B})^{-1}||_{E \to E} \sum_{i=1}^{k-1} ||e^{-(t_{k-1} - t_i)B}(I - e^{-2t_{N-k}B})(I - e^{-\tau B}) \|_{E \to E}$$

$$\times ||I - e^{-2t_i B} \|_{E \to E} ||(I + e^{-\tau B})^{-1}||_{E \to E} \|f_i - f_k \|_E$$

$$\leq M [\sum_{i=1}^{k-1} \frac{\tau}{(t_k - t_i)^{1-\alpha}}] \| f^\tau \|_{C_\tau^\alpha(E)}.$$

The sum enclosed in the right-hand side square brackets is the lower Darboux integral sum for the integral

$$\int_0^{t_k} \frac{ds}{(t_k - s)^\alpha} = \frac{t_k^\alpha}{\alpha}.$$

Thus, for $2 \leq k \leq N$ we have established the bound

$$\| G_k^5 \|_E \leq \frac{M_1}{\alpha} t_k{}^\alpha \| f^\tau \|_{C_\tau^\alpha(E)}. \tag{5.49}$$

From this inequality it follows that

$$\max_{1 \leq k \leq N-1} \| G_k^5 \|_E \leq \frac{M_1}{\alpha} T^\alpha \| f^\tau \|_{C_\tau^\alpha(E)}. \tag{5.50}$$

Now let us estimate the difference $G_{k+r}^5 - G_k^5$ for $2 \leq k < k+r \leq N-1$. We will consider separately the cases $k \leq r$ and $k > r$. If $k \leq r$, then (5.49) yields

$$\begin{aligned} \| G_{k+r}^5 - G_k^5 \|_E &\leq \| G_{k+r}^5 \|_E + \| G_k^5 \|_E \\ &\leq \frac{M_1}{\alpha} t_k{}^\alpha \| f^\tau \|_{C_\tau^\alpha(E)} + \frac{M_1}{\alpha} t_k{}^\alpha \| f^\tau \|_{C_\tau^\alpha(E)} \\ &\leq \frac{M_1(1+2^\alpha)}{\alpha} t_r{}^\alpha \| f^\tau \|_{C_\tau^\alpha(E)} = \frac{M_2}{\alpha} t_r{}^\alpha \| f^\tau \|_{C_\tau^\alpha(E)}. \end{aligned}$$

Now let $k > r$. Let us represent the difference $G_{k+r}^5 - G_k^5$ as the following sums:

$$G_{k+r}^5 - G_k^5 = (I - e^{-2t_N B})^{-1}(I - e^{-\tau B})$$

$$\times \{ \sum_{i=1}^{k+r-1} e^{-(t_{k+r-1}-t_i)B}(I - e^{-2t_i B})(I - e^{-2t_{N-k-r}B})(I + e^{-\tau B})^{-1}(f_i - f_{k+r})$$

$$- \sum_{i=1}^{k-1} e^{-(t_{k-1}-t_i)B}(I - e^{-2t_{N-k}B})(I - e^{-2t_i B})(I + e^{-\tau B})^{-1}(f_i - f_k)\}$$

$$= (I - e^{-2t_N B})^{-1}(I - e^{-\tau B}) \sum_{i=k-r}^{k+r-1} e^{-(t_{k+r-1}-t_i)B}$$

$$\times (I - e^{-2t_i B})(I - e^{-2t_{N-k-r}B})(I + e^{-\tau B})^{-1}(f_i - f_{k+r})$$

$$-(I - e^{-2t_N B})^{-1}(I - e^{-\tau B}) \sum_{i=k-r}^{k-1} e^{-(t_{k-1}-t_i)B}$$

$$\times (I - e^{-2t_{N-k}B})(I - e^{-2t_i B})(I + e^{-\tau B})^{-1}(f_i - f_k)$$

$$+(I - e^{-2t_N B})^{-1}(I - e^{-\tau B}) \sum_{i=1}^{k-r-1} e^{-(t_{k+r-1}-t_i)B}$$

$$\times (I - e^{-2t_i B})(I - e^{-2t_{N-k-r}B})(I + e^{-\tau B})^{-1}(f_k - f_{k+r})$$

$$+(I - e^{-2t_N B})^{-1}(I - e^{-\tau B}) \sum_{i=1}^{k-r-1} \{e^{-(t_{k+r-1}-t_i)B}(I - e^{-2t_{N-k-r}B})$$

$$-e^{-(t_{k-1}-t_i)B}(I - e^{-2t_{N-k}B})\}(I - e^{-2t_i B})(I + e^{-\tau B})^{-1}(f_i - f_k)$$

$$= P_k^1 + P_k^2 + P_k^3 + P_k^4.$$

Now let us estimate P_k^1. Using estimates (5.8) and (5.9), we obtain that

$$\| P_k^1 \|_E \leq \|(I - e^{-2t_N B})^{-1}\|_{E\to E}$$

$$\times \sum_{i=k-r}^{k+r-1} \|(I - e^{-\tau B})e^{-(t_{k+r-1}-t_i)B}(I - e^{-2t_{N-k-r}B})\|_{E\to E}$$

$$\times\|I - e^{-2t_i B}\|_{E\to E} \|(I + e^{-\tau B})^{-1}\|_{E\to E}\|f_i - f_{k+r}\|_E$$

$$\leq M[\frac{1}{\tau^\alpha} + \sum_{i=k-r}^{k+r-2} \frac{\tau}{(t_{k+r} - t_i)^{1-\alpha}}]\| f^\tau \|_{C_\tau^\alpha(E)}$$

$$\leq M_1[\sum_{i=k-r}^{k+r-1} \frac{\tau}{(t_{k+r} - t_i)^{1-\alpha}}]\| f^\tau \|_{C_\tau^\alpha(E)}.$$

The sum enclosed in the right-hand side square brackets is the lower Darboux integral sum for the integral

$$\int_{t_k-t_r}^{t_k+t_r} \frac{ds}{(t_k + t_r - s)^{1-\alpha}}.$$

Since

$$\int_{t_k-t_r}^{t_k+t_r} \frac{ds}{(t_k + t_r - s)^{1-\alpha}} = \frac{1}{\alpha}(2t_r)^\alpha,$$

it follows that

$$\| P_k^1 \|_E \leq \frac{M_1}{\alpha}(2t_r)^\alpha \| f^\tau \|_{C_\tau^\alpha(E)}.$$

Thus, we have established the bound

$$\| P_k^1 \|_E \leq \frac{M_2}{\alpha} t_r^\alpha \| f^\tau \|_{C_\tau^\alpha(E)}.$$

In a similar manner we can show that

$$\| P_k^2 \|_E \leq \frac{M_2}{\alpha} t_r^\alpha \| f^\tau \|_{C_\tau^\alpha(E)}.$$

Since

$$P_k^3 = (I - e^{-2t_N B})^{-1} e^{-2t_{2r} B}(I - e^{-t_{k-r-1}B})$$

$$\times(I - e^{-2t_i B})(I - e^{-2t_{N-k-r}B})(I + e^{-\tau B})^{-1}(f_k - f_{k+r}),$$

it follows that

$$\| P_k^3 \|_E \leq \|(I - e^{-2t_N B})^{-1}\|_{E\to E}\|e^{-2t_{2r}B}(I - e^{-t_{k-r-1}B})\|_{E\to E}$$

$$\times\|I - e^{-2t_i B}\|_{E\to E}\|I - e^{-2t_{N-k-r}B}\|_{E\to E}\|(I + e^{-\tau B})^{-1}\|_{E\to E}\|f_k - f_{k+r}\|_E.$$

Using estimates (5.8) and (5.9), we obtain

$$\| P_k^3 \|_E \leq M t_r^\alpha \| f^\tau \|_{C_\tau^\alpha(E)}.$$

Now let us estimate P_k^4. Using estimates (5.8), (5.9) and (5.10), we obtain

$$\| P_k^4 \|_E \leq \| (I - e^{-2t_N B})^{-1} \|_{E \to E}$$

$$\times \sum_{i=1}^{k-r-1} \{ \| (I - e^{-\tau B})(e^{-(2t_N - 2t_k)B} - e^{-(2t_N - 2t_k - 2t_r)B}) e^{-(t_{k-1} - t_i)B} \|_{E \to E}$$

$$+ \| (I - e^{-\tau B})(I - e^{-t_r)B}) e^{-(t_{k-1} - t_i)B} \|_{E \to E} \}$$

$$\times \| I - e^{-2t_i B} \|_{E \to E} \| (I + e^{-\tau B})^{-1} \|_{E \to E} \| f_i - f_k \|_E$$

$$\leq M t_r [\sum_{i=1}^{k-r-1} \frac{\tau}{(t_k - t_i)^{2-\alpha}}] \| f^\tau \|_{C_\tau^\alpha(E)}.$$

The sum enclosed in the right-hand side square brackets is the lower Darboux integral sum for the integral

$$\int_0^{t_k - t_r} \frac{ds}{(t_k - s)^{2-\alpha}}.$$

Since for $k > r$ we have the bound

$$\int_0^{t_k - t_r} \frac{ds}{(t_k - s)^{2-\alpha}} \leq \frac{1}{1-\alpha} t_r^{-1+\alpha},$$

it follows that

$$\| P_k^4 \|_E \leq \frac{M}{1-\alpha} t_r^\alpha \| f^\tau \|_{C_\tau^\alpha(E)}.$$

Thus, we have established the bound

$$\| P_k^4 \|_E \leq \frac{M}{1-\alpha} t_r^\alpha \| f^\tau \|_{C_\tau^\alpha(E)}.$$

Thus, we have shown that for any $2 \leq k < k + r \leq N - 1$ the following inequality holds:

$$\| G_{k+r}^5 - G_k^5 \|_E \leq \frac{M_1}{\alpha(1-\alpha)} t_r^\alpha \| f^\tau \|_{C_\tau^\alpha(E)}. \tag{5.51}$$

Estimates (5.50) and (5.51) give

$$\| G_5^\tau \|_{C_\tau^\alpha(E)} \leq \frac{M_2}{\alpha(1-\alpha)} \| f^\tau \|_{C_\tau^\alpha(E)} .$$

In exactly the same manner one establishes the inequality

$$\| G_6^\tau \|_{C_\tau^\alpha(E)} \leq \frac{M_2}{\alpha(1-\alpha)} \| f^\tau \|_{C_\tau^\alpha(E)} .$$

By the definition of the spaces E_α and using estimates (5.8), (5.9), we obtain that

$$\| \lambda^{1-\alpha} B e^{-\lambda B} G_k^5 \|_E \leq \lambda^{1-\alpha} \|(I - e^{-2t_N B})^{-1}\|_{E\to E}$$

$$\times \sum_{i=1}^{k-1} \|I - e^{-2t_{N-k}B} \|_{E\to E}$$

$$\times \|I - e^{-2t_i B} \|_{E\to E} \|(I + e^{-\tau B})^{-1}\|_{E\to E} \|(I - e^{-\tau B}) B e^{-(t_{k-1}-t_i+\lambda)B}(f_i - f_k) \|_E$$

$$\leq M\lambda^{1-\alpha} [\sum_{i=1}^{k-1} \frac{\tau}{(t_k - t_i + \lambda)^{2-\alpha}}] \| f^\tau \|_{C_\tau^\alpha(E)} .$$

The sum enclosed in the right-hand side square brackets is the lower Darboux integral sum for the integral

$$\int_0^{t_k} \frac{ds}{(t_k - s + \lambda)^{2-\alpha}} .$$

Since

$$\int_0^{t_k} \frac{ds}{(t_k - s + \lambda)^{2-\alpha}} \leq \frac{1}{1-\alpha} \lambda^{-1+\alpha},$$

it follows that

$$\| \lambda^{1-\alpha} B e^{-\lambda B} G_k^5 \|_E \leq \frac{M}{1-\alpha} \| f^\tau \|_{C_\tau^\alpha(E)}$$

for all λ, $\lambda > 0$ and for $2 \leq k \leq N-1$. So, we have established that

$$\|G_5^\tau\|_{C_\tau(E_\alpha)} \leq \frac{M}{1-\alpha} \| f^\tau \|_{C_\tau^\alpha(E)} .$$

In a similar manner we can show that

$$\|G_5^\tau\|_{C_\tau(E_\alpha)} \leq \frac{M}{1-\alpha} \| f^\tau \|_{C_\tau^\alpha(E)} .$$

Combining the estimates for G_m^τ, $m = 1, 2, 3, 4, 5, 6$ we obtain the estimates

$$\| \{\tau^{-2}(u_{k+1} - 2u_k + u_{k-1})\}_1^{N-1} \|_{C_\tau^\alpha(E)} \leq \frac{M_1}{\alpha(1-\alpha)} \| f^\tau \|_{C_\tau^\alpha(E)}$$

$$+ \frac{M_1}{\alpha} [\| (I - e^{-\tau B})^2 \tau^{-2} u_0 - f_1 \|_{E_\alpha} + \|(I - e^{-\tau B})^2 \tau^{-2} u_N - f_{N-1}\|_{E_\alpha}],$$

$$\| \{\tau^{-2}(u_{k+1} - 2u_k + u_{k-1})\}_1^{N-1} \|_{C_\tau(E_\alpha)} \leq \frac{M_1}{\alpha(1-\alpha)} \| f^\tau \|_{C_\tau^\alpha(E)}$$

$$+ \frac{M_1}{\alpha} [\| (I - e^{-\tau B})^2 \tau^{-2} u_0 - f_1 \|_{E_\alpha} + \|(I - e^{-\tau B})^2 \tau^{-2} u_N - f_{N-1}\|_{E_\alpha}],$$

The estimate for $\{\tau^{-2}[(I-\exp(-\tau B))(u_{k-1}+u_{k+1})+(\exp(-2\tau B)-I)u_k]\}_1^{N-1}$ in the norm of $C_\tau(E_\alpha)$ follows from the triangle inequality. Theorem 5.2.11 is proved.

Furthermore, the method of proof of Theorem 5.2.11 enables us to establish the following fact.

Theorem 5.2.12. *Suppose that the assumptions of Theorem* 5.2.6 *hold. Then the solutions of the difference problem* (5.29) *in* $C_\tau^\alpha(E)$ *obey the coercivity inequality*

$$\| \{\tau^{-2}(u_{k+1}-2u_k+u_{k-1})\}_1^{N-1} \|_{C_\tau^\alpha(E)}$$

$$+ \| \{\tau^{-2}[(I-\exp(-\tau B))(u_{k-1}+u_{k+1})+(\exp(-2\tau B)-I)u_k]\}_1^{N-1} \|_{C_\tau^\alpha(E)}$$

$$+ \| \{\tau^{-2}(u_{k+1}-2u_k+u_{k-1})\}_1^{N-1} \|_{C_\tau(E_\alpha)}$$

$$\leq \frac{M_1}{\alpha(1-\alpha)} \| (I+\exp(-\tau B))^{-1} f^\tau \|_{C_\tau^\alpha(E)}$$

$$+\frac{M_1}{\alpha}[\| (I-e^{-\tau B})^2\tau^{-2}u_0 - f_1 \|_{E_\alpha} + \|(I-e^{-\tau B})^2\tau^{-2}u_N - f_{N-1}\|_{E_\alpha}],$$

where M_1 *is independent not only of* f^τ, u_0, u_N, α, *but also of* τ.

This yields the following result.

Theorem 5.2.13. *Suppose that the assumptions of Theorem* 5.19 *hold. Then the solutions of the difference problem* (5.29) *in* $C_\tau^\alpha(E)$ *obey the coercivity inequality*

$$\| \{\tau^{-2}(u_{k+1}-2u_k+u_{k-1})\}_1^{N-1} \|_{C_\tau^\alpha(E)}$$

$$+ \| \{\tau^{-2}[(I-\exp(-\tau B))(u_{k-1}+u_{k+1})+(\exp(-2\tau B)-I)u_k]\}_1^{N-1} \|_{C_\tau^\alpha(E)}$$

$$+ \| \{\tau^{-2}(u_{k+1}-2u_k+u_{k-1})\}_1^{N-1} \|_{C_\tau(E_\alpha)}$$

$$\leq \frac{M_1}{\alpha(1-\alpha)} \| Bf^\tau \|_{C_\tau^\alpha(E)} + \frac{M_1}{\alpha}[\| (I-e^{-\tau B})^2\tau^{-2}u_0 - f_1 \|_{E_\alpha}$$

$$+\|(I-e^{-\tau B})^2\tau^{-2}u_N - f_{N-1}\|_{E_\alpha}],$$

where M_1 *is independent not only of* f^τ, u_0, u_N, α, *but also of* τ.

Let us give, without proof, the following results.

Theorem 5.2.14. *Suppose that the assumptions of Theorem* 5.2.5 *hold. Then the solutions of the difference problem* (5.29) *in* $C_\tau^{\beta,\gamma}(E)$ $(0\leq\gamma\leq\beta, 0<\beta<1)$ *obey the coercivity inequalities*

$$\| \{\tau^{-2}(u_{k+1}-2u_k+u_{k-1})\}_1^{N-1} \|_{C_\tau(E_{\beta-\gamma})} \leq \frac{M_1}{\beta(1-\beta)} \| f^\tau \|_{C_\tau^{\beta,\gamma}(E)}$$

$$+M_1[\| (I-e^{-\tau B})^2\tau^{-2}u_0 - f_1 \|_{E_{\beta-\gamma}} + \|(I-e^{-\tau B})^2\tau^{-2}u_N - f_{N-1}\|_{E_{\beta-\gamma}}],$$

$$\| \{\tau^{-2}(u_{k+1} - 2u_k + u_{k-1})\}_1^{N-1} \|_{C_\tau^{\beta,\gamma}(E)}$$
$$+ \| \{\tau^{-2}[(I - \exp(-\tau B))(u_{k-1} + u_{k+1}) + (\exp(-2\tau B) - I)u_k]\}_1^{N-1} \|_{C_\tau^{\beta,\gamma}(E)}$$
$$\leq \frac{M_1}{\beta(1-\beta)} \| f^\tau \|_{C_\tau^{\beta,\gamma}(E)} + M_1[|(I - e^{-\tau B})^2\tau^{-2}u_0 - f_1|_0^{\beta,\gamma}$$
$$+|(I - e^{-\tau B})^2\tau^{-2}u_N - f_{N-1}|_0^{\beta,\gamma}],$$

where M_1 is independent not only of f^τ, u_0, u_N, β, γ but also of τ.

Theorem 5.2.15. *Suppose that the assumptions of Theorem 5.2.6 hold. Then the solutions of the difference problem (5.29) in $C_\tau^{\beta,\gamma}(E)$ $(0 \leq \gamma \leq \beta, 0 < \beta < 1)$ obey the coercivity inequalities*

$$\| \{\tau^{-2}(u_{k+1} - 2u_k + u_{k-1})\}_1^{N-1} \|_{C_\tau(E_{\beta-\gamma})}$$
$$\leq \frac{M_1}{\beta(1-\beta)} \| (I + \exp(-\tau B))^{-1} f^\tau \|_{C_\tau^{\beta,\gamma}(E)}$$
$$+M_1[\| (I - e^{-\tau B})^2\tau^{-2}u_0 - f_1 \|_{E_{\beta-\gamma}} + \|(I - e^{-\tau B})^2\tau^{-2}u_N - f_{N-1}\|_{E_{\beta-\gamma}}],$$
$$\| \{\tau^{-2}(u_{k+1} - 2u_k + u_{k-1})\}_1^{N-1} \|_{C_\tau^{\beta,\gamma}(E)}$$
$$+ \| \{\tau^{-2}[(I - \exp(-\tau B))(u_{k-1} + u_{k+1}) + (\exp(-2\tau B) - I)u_k]\}_1^{N-1} \|_{C_\tau^{\beta,\gamma}(E)}$$
$$\leq \frac{M_1}{\beta(1-\beta)} \| (I + \exp(-\tau B))^{-1} f^\tau \|_{C_\tau^{\beta,\gamma}(E)}$$
$$+M_1[|(I - e^{-\tau B})^2\tau^{-2}u_0 - f_1|_0^{\beta,\gamma} + |(I - e^{-\tau B})^2\tau^{-2}u_N - f_{N-1}|_0^{\beta,\gamma}],$$

where M_1 is independent not only of f^τ, u_0, u_N, β, γ but also of τ.

Theorem 5.2.16. *Suppose that the assumptions of Theorem 5.2.7 hold. Then the solutions of the difference problem (5.29) in $C_\tau^{\beta,\gamma}(E)(0 \leq \gamma \leq \beta, 0 < \beta < 1)$ obey the coercivity inequalities*

$$\| \{\tau^{-2}(u_{k+1} - 2u_k + u_{k-1})\}_1^{N-1} \|_{C_\tau(E_{\beta-\gamma})} \leq \frac{M_1}{\beta(1-\beta)} \| Bf^\tau \|_{C_\tau^{\beta,\gamma}(E)}$$
$$+M_1[\| (I - e^{-\tau B})^2\tau^{-2}u_0 - f_1 \|_{E_{\beta-\gamma}} + \|(I - e^{-\tau B})^2\tau^{-2}u_N - f_{N-1}\|_{E_{\beta-\gamma}}],$$
$$\| \{\tau^{-2}(u_{k+1} - 2u_k + u_{k-1})\}_1^{N-1} \|_{C_\tau^{\beta,\gamma}(E)}$$
$$+ \| \{\tau^{-2}[(I - \exp(-\tau B))(u_{k-1} + u_{k+1}) + (\exp(-2\tau B) - I)u_k]\}_1^{N-1} \|_{C_\tau^{\beta,\gamma}(E)}$$
$$\leq \frac{M_1}{\beta(1-\beta)} \| Bf^\tau \|_{C_\tau^{\beta,\gamma}(E)} + M_1[|(I - e^{-\tau B})^2\tau^{-2}u_0 - f_1|_0^{\beta,\gamma}$$
$$+|(I - e^{-\tau B})^2\tau^{-2}u_N - f_{N-1}|_0^{\beta,\gamma}],$$

where M_1 is independent not only of f^τ, u_0, u_N, β, γ but also of τ.

Here, $|w|_0^{\beta,\gamma}$ denotes that the norm of the Banach space $\widetilde{E}_0^{\beta,\gamma}$ consists of those $v \in E$ for which the norm

$$|w|_0^{\beta,\gamma} = \max_{1\leq n\leq N-1} \|e^{-t_nB}w\|_E + \sup_{1\leq n<n+r\leq N-1} t_r^{-\beta}(t_n+t_r)^{\gamma}(t_N-t_n)^{\gamma}\|(e^{-(t_n+t_r)B}-e^{-t_nB})w\|_E$$

is finite.

Note that the parameter γ can be chosen freely in $[0,\beta)$, which increases the number of spaces of grid functions in which difference problem (5.29) is well posed.

Recall that the difference problem (5.29) is not well posed in $C_\tau(E)$ for all positive operators. It turns out that a Banach space E can be restricted to a Banach space E' in such a manner that the restricted problem (5.29) in E' will be well posed in $C_\tau(E')$. The role of E'will be played here by the fractional spaces $E_\alpha = E_\alpha(E,B)$ $(0<\alpha<1)$.

Theorem 5.2.17. *Suppose that the assumptions of Theorem* 5.2.5 *hold. Then the solutions of the difference problem* (5.29) *in* $C_\tau(E_\alpha)$ *obey the coercivity inequality*

$$\| \{\tau^{-2}(u_{k+1}-2u_k+u_{k-1})\}_1^{N-1} \|_{C_\tau(E_\alpha)}$$

$$+\| \{\tau^{-2}[(I-\exp(-\tau B))(u_{k-1}+u_{k+1})+(\exp(-2\tau B)-I)u_k]\}_1^{N-1} \|_{C_\tau(E_\alpha)}$$

$$\leq M_1[\frac{1}{\alpha(1-\alpha)}\| f^\tau \|_{C_\tau(E_\alpha)} + \| Au_0\|_{E_\alpha} + \|Au_N\|_{E_\alpha}],$$

where M_1 *is independent not only of* f^τ, u_0, u_N, α, *but also of* τ.

Proof. Using formula (5.37), we obtain

$$\tau^{-2}(u_{k+1}-2u_k+u_{k-1})$$

$$= (I-e^{-2t_NB})^{-1}(I-e^{-\tau B})^2\tau^{-2}\{(e^{-t_{k-1}B}-e^{-(2t_N-t_{k+1})B})u_0 + (e^{-(t_N-t_{k+1})B}-e^{-(t_N+t_{k-1})B})u_N\}$$

$$-(I-e^{-2t_NB})^{-1}(I-e^{-\tau B})(e^{-(t_N-t_{k+1})B}-e^{-(t_N+t_{k-1})B})$$

$$\times\sum_{i=1}^{N-1}(e^{-(t_N-t_i)B}-e^{-(t_N+t_i)B})(I+e^{-\tau B})^{-1}f_i$$

$$+(I-e^{-\tau B})\sum_{i=1}^{k-1}e^{-(t_{k-1}-t_i)B}(I+e^{-\tau B})^{-1}f_i$$

$$+(I-e^{-\tau B})\sum_{i=k+1}^{N-1}e^{-(t_i-t_{k+1})B}(I+e^{-\tau B})^{-1}f_i - 2(I+e^{-\tau B})^{-1}f_k$$

$$-(I-e^{-\tau B})\sum_{i=1}^{N-1}e^{-(t_{k-1}+t_i)B})(I+e^{-\tau B})^{-1}f_i = g_k+p_k,$$

where

$$g_k = \tau^{-2}(I - e^{-2t_N B})^{-1}(I - e^{-\tau B})^2\{(e^{-t_{k-1}B} - e^{-(2t_N - t_{k+1})B})u_0$$
$$+(e^{-(t_N - t_{k+1})B} - e^{-(t_N + t_{k-1})B})u_N\},$$
$$p_k = -(I - e^{-2t_N B})^{-1}(I - e^{-\tau B})(e^{-(t_N - t_{k+1})B} - e^{-(t_N + t_{k-1})B})$$
$$\times \sum_{i=1}^{N-1}(e^{-(t_N - t_i)B} - e^{-(t_N + t_i)B})(I + e^{-\tau B})^{-1} f_i$$
$$+(I - e^{-\tau B})\sum_{i=1}^{k-1} e^{-(t_{k-1} - t_i)B}(I + e^{-\tau B})^{-1} f_i$$
$$+(I - e^{-\tau B})\sum_{i=k+1}^{N-1} e^{-(t_i - t_{k+1})B}(I + e^{-\tau B})^{-1} f_i - 2(I + e^{-\tau B})^{-1} f_k$$
$$-(I - e^{-\tau B})\sum_{i=1}^{N-1} e^{-(t_{k-1} + t_i)B})(I + e^{-\tau B})^{-1} f_i.$$

To this end it suffices to show that

$$\| g^\tau \|_{C_\tau(E_\alpha)} \le M_1[\| Au_0\|_{E\alpha} + Au_N\|_{E_\alpha}], \tag{5.52}$$

$$\| p^\tau \|_{C_\tau(E_\alpha)} \le M_1 \alpha^{-1}(1-\alpha)^{-1}\| f^\tau \|_{C_\tau(E_\alpha)}. \tag{5.53}$$

Estimate (5.52) follows from the estimate

$$\|(I - e^{-\tau B})^2(\tau B)^{-2}\|_{E\to E} \le M$$

and the fact that the homogeneous difference equation ($f^\tau = 0$) is stable in $C_\tau(E_\alpha)$ uniformly in τ. By the definition of the spaces E_α and using estimates (5.10), (5.11) and (5.36), we obtain

$$\| \lambda^{1-\alpha} B e^{-\lambda B} p_k \|_E \le \lambda^{1-\alpha}\|(I - e^{-2t_N B})^{-1}\|_{E\to E}$$
$$\times \|e^{-(t_N - t_{k+1})B} - e^{-(t_N + t_{k-1})B}\|_{E\to E}$$
$$\times \sum_{i=1}^{N-1} \|e^{-\tau B} - e^{-(t_{N+1} + t_i)B}\|_{E\to E}$$
$$\times\|(I + e^{-\tau B})^{-1}\|_{E\to E}\|B(I - e^{-\tau B})e^{-(t_{N-1} - t_i + \lambda)B} f_i\|_E$$
$$+\lambda^{1-\alpha}\sum_{i=1}^{k-1}\|(I + e^{-\tau B})^{-1}\|_{E\to E}\|B(I - e^{-\tau B})e^{-(t_{k-1} - t_i + \lambda)B} f_i\|_E$$

$$+\lambda^{1-\alpha}\sum_{i=k+1}^{N-1}||(I+e^{-\tau B})^{-1}||_{E\to E}||B(I-e^{-\tau B})e^{-(t_i-t_{k+1})B}f_i||_E$$

$$+\lambda^{1-\alpha}2||(I+e^{-\tau B})^{-1}||_{E\to E}||Be^{-\lambda B}f_k||_E$$

$$+\lambda^{1-\alpha}\sum_{i=1}^{N-1}||e^{-t_kB}||_{E\to E}||(I+e^{-\tau B})^{-1}||_{E\to E}||B(I-e^{-\tau B})e^{-(t_{i-1}+\lambda)B}f_i||_E$$

$$\leq M(1+\lambda^{1-\alpha}[\sum_{j=1}^{N-1}\frac{\tau}{(t_{j-1}+\lambda)\lambda^{1-\alpha}}])\|\, f^\tau\, \|_{C_\tau(E_\alpha)}.$$

The sum enclosed in the right-hand side square brackets is the lower Darboux integral sum for the integral

$$\int_0^T \frac{ds}{(s+\lambda)s^{1-\alpha}}.$$

Since

$$\int_0^T \frac{ds}{(s+\lambda)s^{1-\alpha}} \leq \frac{1}{\lambda^{1-\alpha}}\int_0^\infty \frac{dr}{(1+r)r^{1-\alpha}},$$

it follows that

$$\|\, \lambda^{1-\alpha}Be^{-\lambda B}p_k\, \|_E \leq \frac{M_1}{\alpha(1-\alpha)}\|\, f^\tau\, \|_{C_\tau(E_\alpha)}$$

for all λ, $\lambda > 0$ and for $1 \leq k \leq N-1$. From the last estimate follows estimate (5.53). Theorem 5.2.17 is proved.

Furthermore, the method of proof of Theorem 5.2.17 enables us to establish the following fact.

Theorem 5.2.18. *Suppose that the assumptions of Theorem* 5.2.6 *hold. Then the solutions of the difference problem* (5.29) *in* $C_\tau(E_\alpha)$ *obey the coercivity inequality*

$$\|\, \{\tau^{-2}(u_{k+1}-2u_k+u_{k-1})\}_1^{N-1}\, \|_{C_\tau(E_\alpha)}$$

$$+\, \|\, \{\tau^{-2}[(I-\exp(-\tau B))(u_{k-1}+u_{k+1})+(\exp(-2\tau B)-I)u_k]\}_1^{N-1}\, \|_{C_\tau(E_\alpha)}$$

$$\leq M_1[\frac{1}{\alpha(1-\alpha)}\|\, (I+\exp(-\tau B))^{-1}f^\tau\, \|_{C_\tau(E_\alpha)}+\|\, Au_0\|_{E_\alpha}+||Au_N||_{E_\alpha}],$$

where M_1 *is independent not only of* f^τ, u_0, u_N, α, *but also of* τ.

This yields the following result.

Theorem 5.2.19. *Suppose that the assumptions of Theorem 5.2.7 hold. Then the solutions of the difference problem (5.29) in $C_\tau(E_\alpha)$ obey the coercivity inequality*

$$\| \{\tau^{-2}(u_{k+1} - 2u_k + u_{k-1})\}_1^{N-1} \|_{C_\tau(E_\alpha)}$$

$$+ \| \{\tau^{-2}[(I - \exp(-\tau B))(u_{k-1} + u_{k+1}) + (\exp(-2\tau B) - I)u_k]\}_1^{N-1} \|_{C_\tau(E_\alpha)}$$

$$\leq M_1[\frac{1}{\alpha(1-\alpha)} \| Bf^\tau \|_{C_\tau(E_\alpha)} + \| Au_0\|_{E_\alpha} + \|Au_N\|_{E_\alpha}],$$

where M_1 is independent not only of f^τ, u_0, u_N, α, but also of τ.

Let us give, without proof, the following results.

Theorem 5.2.20. *Suppose that the assumptions of Theorem 5.2.5 hold. Then the solutions of the difference problem (5.29) in $C_\tau^{\beta,\gamma}(E_{\alpha-\beta})$ $(0 \leq \gamma \leq \beta \leq \alpha, 0 < \alpha < 1)$ obey the coercivity inequalities*

$$\| \{\tau^{-2}(u_{k+1} - 2u_k + u_{k-1})\}_1^{N-1} \|_{C_\tau^{\beta,\gamma}(E_{\alpha-\beta})}$$

$$+ \| \{\tau^{-2}[(I - \exp(-\tau B))(u_{k-1} + u_{k+1}) + (\exp(-2\tau B) - I)u_k]\}_1^{N-1} \|_{C_\tau^{\beta,\gamma}(E_{\alpha-\beta})}$$

$$+ \| \{\tau^{-2}(u_{k+1} - 2u_k + u_{k-1})\}_1^{N-1} \|_{C_\tau(\widetilde{E}_{\alpha-\beta}^{\beta,\gamma})}$$

$$\leq \frac{M_1}{\alpha(1-\alpha)} \| f^\tau \|_{C_\tau^{\beta,\gamma}(E_{\alpha-\beta})}$$

$$+M_1[|(I - e^{-\tau B})^2\tau^{-2}u_0 - f_1|_{\alpha-\beta}^{\beta,\gamma} + |(I - e^{-\tau B})^2\tau^{-2}u_N - f_{N-1}|_{\alpha-\beta}^{\beta,\gamma}],$$

$$\| \{\tau^{-2}(u_{k+1} - 2u_k + u_{k-1})\}_1^{N-1} \|_{C_\tau^{\beta,\gamma}(E_{\alpha-\beta})}$$

$$+ \| \{\tau^{-2}[(I - \exp(-\tau B))(u_{k-1} + u_{k+1}) + (\exp(-2\tau B) - I)u_k]\}_1^{N-1} \|_{C_\tau^{\beta,\gamma}(E_{\alpha-\beta})}$$

$$+ \| \{\tau^{-2}(u_{k+1} - 2u_k + u_{k-1})\}_1^{N-1} \|_{C_\tau(E_{\alpha-\gamma})}$$

$$\leq \frac{M_1}{\alpha(1-\alpha)}[\| f^\tau \|_{C_\tau^{\beta,\gamma}(E_{\alpha-\beta})}$$

$$+M_1[\|(I - e^{-\tau B})^2\tau^{-2}u_0 - f_1\|_{E_{\alpha-\gamma}} + \|(I - e^{-\tau B})^2\tau^{-2}u_N - f_{N-1}\|_{E_{\alpha-\gamma}}]],$$

where M_1 is independent not only of f^τ, u_0, u_N, α, β, γ but also of τ.

Theorem 5.2.21. *Suppose that the assumptions of Theorem 5.2.6 hold. Then the solutions of the difference problem (5.29) in $C_\tau^{\beta,\gamma}(E_{\alpha-\beta})$ $(0 \leq \gamma \leq \beta \leq \alpha, 0 < \alpha < 1)$ obey the coercivity inequalities*

$$\| \{\tau^{-2}(u_{k+1} - 2u_k + u_{k-1})\}_1^{N-1} \|_{C_\tau^{\beta,\gamma}(E_{\alpha-\beta})}$$

$$+ \| \{\tau^{-2}[(I - \exp(-\tau B))(u_{k-1} + u_{k+1}) + (\exp(-2\tau B) - I)u_k]\}_1^{N-1} \|_{C_\tau^{\beta,\gamma}(E_{\alpha-\beta})}$$

$$+ \| \{\tau^{-2}(u_{k+1} - 2u_k + u_{k-1})\}_1^{N-1} \|_{C_\tau(\widetilde{E}_{\alpha-\beta}^{\beta,\gamma})}$$

$$\leq \frac{M_1}{\alpha(1-\alpha)} \| (I+\exp(-\tau B))^{-1} f^\tau \|_{C_\tau^{\beta,\gamma}(E_{\alpha-\beta})}$$

$$+M_1[|(I-e^{-\tau B})^2\tau^{-2}u_0 - f_1|_{\alpha-\beta}^{\beta,\gamma} + |(I-e^{-\tau B})^2\tau^{-2}u_N - f_{N-1}|_{\alpha-\beta}^{\beta,\gamma}],$$

$$\| \{\tau^{-2}(u_{k+1}-2u_k+u_{k-1})\}_1^{N-1} \|_{C_\tau^{\beta,\gamma}(E_{\alpha-\beta})}$$

$$+ \| \{\tau^{-2}[(I-\exp(-\tau B))(u_{k-1}+u_{k+1}) + (\exp(-2\tau B)-I)u_k]\}_1^{N-1} \|_{C_\tau^{\beta,\gamma}(E_{\alpha-\beta})}$$

$$+ \| \{\tau^{-2}(u_{k+1}-2u_k+u_{k-1})\}_1^{N-1} \|_{C_\tau(E_{\alpha-\gamma})}$$

$$\leq \frac{M_1}{\alpha(1-\alpha)} [\| (I+\exp(-\tau B))^{-1} f^\tau \|_{C_\tau^{\beta,\gamma}(E_{\alpha-\beta})}$$

$$+M_1[\|(I-e^{-\tau B})^2\tau^{-2}u_0 - f_1 \|_{C_\tau(E_{\alpha-\gamma})} + \|(I-e^{-\tau B})^2\tau^{-2}u_N - f_{N-1}\|_{C_\tau(E_{\alpha-\gamma})}]],$$

where M_1 is independent not only of f^τ, u_0, u_N, α, β, γ but also of τ.

Theorem 5.2.22. *Suppose that the assumptions of Theorem 5.19 hold. Then the solutions of the difference problem (5.29) in $C_\tau^{\beta,\gamma}(E_{\alpha-\beta})$ $(0 \leq \gamma \leq \beta \leq \alpha, 0 < \alpha < 1)$ obey the coercivity inequalities*

$$\| \{\tau^{-2}(u_{k+1}-2u_k+u_{k-1})\}_1^{N-1} \|_{C_\tau^{\beta,\gamma}(E_{\alpha-\beta})}$$

$$+ \| \{\tau^{-2}[(I-\exp(-\tau B))(u_{k-1}+u_{k+1}) + (\exp(-2\tau B)-I)u_k]\}_1^{N-1} \|_{C_\tau^{\beta,\gamma}(E_{\alpha-\beta})}$$

$$+ \| \{\tau^{-2}(u_{k+1}-2u_k+u_{k-1})\}_1^{N-1} \|_{C_\tau(\tilde{E}_{\alpha-\beta}^{\beta,\gamma})}$$

$$\leq \frac{M_1}{\alpha(1-\alpha)} \| Bf^\tau \|_{C_\tau^{\beta,\gamma}(E_{\alpha-\beta})}$$

$$+M_1[|(I-e^{-\tau B})^2\tau^{-2}u_0 - f_1|_{\alpha-\beta}^{\beta,\gamma} + |(I-e^{-\tau B})^2\tau^{-2}u_N - f_{N-1}|_{\alpha-\beta}^{\beta,\gamma}],$$

$$\| \{\tau^{-2}(u_{k+1}-2u_k+u_{k-1})\}_1^{N-1} \|_{C_\tau^{\beta,\gamma}(E_{\alpha-\beta})}$$

$$+ \| \{\tau^{-2}[(I-\exp(-\tau B))(u_{k-1}+u_{k+1}) + (\exp(-2\tau B)-I)u_k]\}_1^{N-1} \|_{C_\tau^{\beta,\gamma}(E_{\alpha-\beta})}$$

$$+ \| \{\tau^{-2}(u_{k+1}-2u_k+u_{k-1})\}_1^{N-1} \|_{C_\tau(E_{\alpha-\gamma})}$$

$$\leq \frac{M_1}{\alpha(1-\alpha)} [\| Bf^\tau \|_{C_\tau^{\beta,\gamma}(E_{\alpha-\beta})}$$

$$+M_1[\|(I-e^{-\tau B})^2\tau^{-2}u_0 - f_1\|_{C_\tau(E_{\alpha-\gamma})} + \|(I-e^{-\tau B})^2\tau^{-2}u_N - f_{N-1}\|_{C_\tau(E_{\alpha-\gamma})}]],$$

where M_1 is independent not only of f^τ, u_0, u_N, α, β, γ but also of τ.

Here, $|w|_{\alpha-\beta}^{\beta,\gamma}$ denotes that the norm of the Banach space $\widetilde{E}_{\alpha-\beta}^{\beta,\gamma}$ consists of those $v \in E$ for which the norm

$$|w|_{\alpha-\beta}^{\beta,\gamma} = \max_{1\le n\le N-1} \|e^{-t_n B} w\|_{E_{\alpha-\beta}}$$

$$+ \sup_{1\le n<n+r\le N-1} t_r^{-\beta}(t_n+t_r)^{\gamma}(t_N-t_n)^{\gamma}\|(e^{-(t_n+t_r)B} - e^{-t_n B})w\|_{E_{\alpha-\beta}}$$

is finite.

Note that the spaces of grid functions $C_\tau^{\beta,\gamma}(E_{\alpha-\beta})$, in which coercive solvability has been established, depend on the parameters α, β and γ. However, the constants in the coercive inequalities depend only on α. Hence, we can choose the parameters β and γ freely, which increases the number of spaces of grid functions in which difference problem (5.29) is well posed.

Let us consider now the boundary-value difference problem (5.29) in the spaces $L_{p,\tau}(E) = L_p([0,T]_\tau, E)$, $1 \le p < \infty$ of all grid functions. We have not been able to obtain the coercivity inequality

$$\| \{\tau^{-2}(u_{k+1} - 2u_k + u_{k-1})\}_1^{N-1} \|_{L_{p,\tau}(E)}$$

$$+ \| \{\tau^{-2}[(I - \exp(-\tau B))(u_{k-1} + u_{k+1}) + (\exp(-2\tau B) - I)u_k]\}_1^{N-1} \|_{L_{p,\tau}(E)}$$

$$\le M[\| f^\tau \|_{L_{p,\tau}(E)} + \| Au_0\|_E + \|Au_N\|_E],$$

in the arbitrary Banach space E and for the general positive operator A. Nevertheless, we can establish the almost coercivity inequality.

Theorem 5.2.23. *Suppose that the assumptions of Theorem* 5.2.5 *hold. Then the solutions of the difference problem* (5.29) *in* $L_{p,\tau}(E)$ *obey the almost coercive inequality*

$$\| \{\tau^{-2}(u_{k+1} - 2u_k + u_{k-1})\}_1^{N-1} \|_{L_{p,\tau}(E)}$$

$$+ \| \{\tau^{-2}[(I - \exp(-\tau B))(u_{k-1} + u_{k+1}) + (\exp(-2\tau B) - I)u_k]\}_1^{N-1} \|_{L_{p,\tau}(E)}$$

$$\le M_1[\| Au_0\|_E + \|Au_N\|_E$$

$$+ \min\left\{\ln\frac{1}{\tau}, 1 + |\ln \| B \|_{E\to E}|\right\} \| f^\tau \|_{L_{p,\tau}(E)}],$$

where M_1 *does not depend on* f^τ, u_0, u_N, p *and* τ.

Proof. The proof of this theorem in the case $p = 1$ is carried out according to the same scheme of Theorem 5.2.5 and is based on estimate (5.33). So, we will consider the proof of this theorem for any p, $1 < p < \infty$. By formula (5.37) we have that

$$\tau^{-2}(u_{k+1} - 2u_k + u_{k-1}) = g_k + p_k.$$

Therefore, to prove the theorem it suffices to show that

$$\| g^\tau \|_{L_{p,\tau}(E)} \le M_1[\| Au_0\|_E + \|Au_N\|_E], \tag{5.54}$$

$$\| p^\tau \|_{L_{p,\tau}(E)} \le M_1 \min\left\{\ln\frac{1}{\tau}, 1 + |\ln \| B \|_{E\to E}|\right\} \| f^\tau \|_{L_{p,\tau}(E)}. \tag{5.55}$$

Estimate (5.54) follows from the estimate

$$||(I - e^{-\tau B})^2(\tau B)^{-2}||_{E\to E} \le M$$

and the fact that the homogeneous difference equation ($f^\tau = 0$) is stable in $L_{p,\tau}(E)$ uniformly in τ. We have that

$$p_k = p_k^1 + p_k^2 + p_k^3,$$

where

$$p_k^1 = (I - e^{-2t_N B})^{-1}(I - e^{-\tau B})\sum_{i=1}^{k-1} e^{-(t_{k-1}-t_i)B}$$

$$\times(I - e^{-2t_i B})(I - e^{-2t_{N-k}B})(I + e^{-\tau B})^{-1} f_i, 2 \le k \le N-1, p_1^1 = 0,$$

$$p_k^2 = -(I - e^{-2t_N B})^{-1}(I - e^{-\tau B})(e^{-(t_N - t_{k+1})B} - e^{-(t_N + t_{k-1})B})$$

$$\times(e^{-(t_N-t_k)B} - e^{-(t_N+t_k)B})(I + e^{-\tau B})^{-1} f_k - 2(I + e^{-\tau B})^{-1} f_k$$

$$-(I - e^{-\tau B})e^{-(t_{k-1}+t_k)B})(I + e^{-\tau B})^{-1} f_k,$$

$$p_k^3 = (I - e^{-2t_N B})^{-1}(I - e^{-\tau B})\sum_{i=k+1}^{N-1} e^{-(t_i - t_{k+1})B}$$

$$\times(I - e^{-2t_k B})(I - e^{-2t_{N-i}B})(I + e^{-\tau B})^{-1} f_i, 1 \le k \le N-2, p_{N-1}^3 = 0.$$

Let us estimate $\{p_k^m\}_{k=1}^{N-1}$ for any $m = 1, 2, 3$ separately. Using estimates (5.10), (5.11) and (5.36), we obtain

$$||p_k^2||_E \le ||(I - e^{-2t_N B})^{-1}||_{E\to E}||I - e^{-2t_{N-k}B}||_{E\to E}$$

$$\times\sum_{i=1}^{k-1}||I - e^{-2t_i B}||_{E\to E}||(I - e^{-\tau B})e^{-(t_{k-1}-t_i)B}||_{E\to E}$$

$$\times||(I + e^{-\tau B})^{-1}||_{E\to E}||f_i||_E \le M\sum_{i=1}^{k-1}||(I - e^{-\tau B})e^{-(t_{k-1}-t_i)B}||_{E\to E}||f_i||_E$$

$$= M\sum_{j=1}^{k-1}||(I - e^{-\tau B})e^{-t_{j-1}B}||_{E\to E}||f_{k-j}||_E, 2 \le k \le N-1.$$

Set $\varphi_i = f_i$ if $i = 1, \ldots, k-1$, and $\varphi_i = 0$ otherwise. We have that

$$||p_k^1||_E \le M\sum_{j=1}^{N-1}||(I - e^{-\tau B})e^{-t_{j-1}B}||_{E\to E}||\varphi_{k-j}||_E, 2 \le k \le N-1.$$

Using this estimate and Minkowski sum inequality it follows that

$$\| \{p_k^1\}_{k=1}^{N-1} \|_{L_{p,\tau}(E)} = (\sum_{k=2}^{N-1} \| p_k^1 \|_E^p \tau)^{\frac{1}{p}}$$

$$\leq \sum_{j=1}^{N-1} (\sum_{k=2}^{N-1} (M||(I - e^{-\tau B})e^{-t_{j-1}B}||_{E\to E}||\varphi_{k-j}||_E)^p \tau)^{\frac{1}{p}}$$

$$\leq \sum_{j=1}^{N-1} M||(I - e^{-\tau B})e^{-t_{j-1}B}||_{E\to E}(\sum_{i=1}^{N-1} ||\varphi_i||_E^p \tau)^{\frac{1}{p}}.$$

By the definition of the grid function $\{\varphi_k\}_{k=1}^{N-1}$ we obtain that

$$\| \{p_k^1\}_{k=1}^{N-1} \|_{L_{p,\tau}(E)} \leq \sum_{j=1}^{N-1} M||(I - e^{-\tau B})e^{-t_{j-1}B}||_{E\to E}(\sum_{i=1}^{N-1} ||f_i||_E^p \tau)^{\frac{1}{p}}.$$

Using this estimate and estimate (5.33), we obtain that

$$\| \{p_k^1\}_{k=1}^{N-1} \|_{L_{p,\tau}(E)} \leq M_1 \min\left\{\ln \frac{1}{\tau}, 1 + |\ln \| B \|_{E\to E}|\right\} \| f^\tau \|_{L_{p,\tau}(E)}.$$

Let us estimate $\{p_k^2\}_{k=1}^{N-1}$. Using estimates (5.10), (5.11) and (5.36), we obtain

$$||p_k^2||_E \leq [||(I - e^{-2t_N B})^{-1}||_{E\to E}||I - e^{-\tau B}||_{E\to E}$$

$$\times||e^{-(t_N - t_{k+1})B} - e^{-(t_N + t_{k-1})B}||_{E\to E}$$

$$\times||e^{-(t_N - t_k)B} - e^{-(t_N + t_k)B}||_{E\to E}||(I + e^{-\tau B})^{-1}||_{E\to E} + 2||(I + e^{-\tau B})^{-1}||_{E\to E}$$

$$+||(I - e^{-\tau B})e^{-t_{2k-1}B}||_{E\to E}||(I + e^{-\tau B})^{-1}||_{E\to E}]||f_k||_E \leq M||f_k||_E.$$

From this it follows that

$$\| \{p_k^2\}_{k=1}^{N-1} \|_{L_{p,\tau}(E)} \leq M(\sum_{k=1}^{N-1} ||f_k||_E^p \tau)^{\frac{1}{p}}.$$

Let us estimate $\{p_k^3\}_{k=1}^{N-1}$. Using estimates (5.10), (5.11) and (5.36), we obtain that

$$||p_k^3||_E \leq ||(I - e^{-2t_N B})^{-1}||_{E\to E}||I - e^{-2t_k B}||_{E\to E}$$

$$\times \sum_{i=k+1}^{N-1} ||I - e^{-2t_{N-i}B}||_{E\to E}||(I - e^{-\tau B})e^{-(t_i - t_{k+1})B}||_{E\to E}$$

$$\times||(I + e^{-\tau B})^{-1}||_{E\to E}||f_i||_E \leq M \sum_{i=k+1}^{N-1} ||(I - e^{-\tau B})e^{-(t_i - t_{k+1})B}||_{E\to E}||f_i||_E$$

$$= M \sum_{j=1}^{N-k-1} ||(I - e^{-\tau B})e^{-t_{j-1}B}||_{E\to E}||f_{k+j}||_E, 1 \leq k \leq N-2.$$

Set $\varphi_i = f_i$ if $i = k+1, \dots, N-1$, and $\varphi_i = 0$ otherwise. We have that

$$||p_k^3||_E \le M \sum_{j=1}^{N-1} ||(I - e^{-\tau B})e^{-t_{j-1}B}||_{E\to E}||\varphi_{k+j}||_E, 1 \le k \le N-2.$$

Using this estimate and Minkowski sum inequality it follows that

$$\| \{p_k^3\}_{k=1}^{N-1} \|_{L_{p,\tau}(E)} = (\sum_{k=1}^{N-2} \| p_k^3 \|_E^p \tau)^{\frac{1}{p}}$$

$$\le \sum_{j=1}^{N-1} (\sum_{k=1}^{N-2} (M||(I - e^{-\tau B})e^{-t_{j-1}B}||_{E\to E}||\varphi_{k+j}||_E)^p \tau)^{\frac{1}{p}}$$

$$\le \sum_{j=1}^{N-1} M||(I - e^{-\tau B})e^{-t_{j-1}B}||_{E\to E}(\sum_{i=1}^{N-1} ||\varphi_i||_E^p \tau)^{\frac{1}{p}}.$$

By the definition of the grid function $\{\varphi_k\}_{k=1}^{N-1}$ we obtain that

$$\| \{p_k^3\}_{k=1}^{N-1} \|_{L_{p,\tau}(E)} \le \sum_{j=1}^{N-1} M||(I - e^{-\tau B})e^{-t_{j-1}B}||_{E\to E}(\sum_{i=1}^{N-1} ||f_i||_E^p \tau)^{\frac{1}{p}}.$$

Using this estimate and estimate (5.33), we obtain that

$$\| \{p_k^3\}_{k=1}^{N-1} \|_{L_{p,\tau}(E)} \le M_1 \min \left\{ \ln \frac{1}{\tau}, 1 + |\ln \| B \|_{E\to E}| \right\} \| f^\tau \|_{L_{p,\tau}(E)}.$$

Combining the estimates for $\{p_k^m\}_{k=1}^{N-1}$, $m = 1, 2, 3$ and using the triangle inequality we obtain estimate (5.39). Theorem 5.2.23 is proved.

Furthermore, the method of proof of Theorem 5.2.23 enables us to establish the following facts.

Theorem 5.2.24. *Suppose that the assumptions of Theorem* 5.2.6 *hold. Then the solutions of the difference problem* (5.29) *in* $L_{p,\tau}(E)$ *obey the almost coercive inequality*

$$\| \{\tau^{-2}(u_{k+1} - 2u_k + u_{k-1})\}_1^{N-1} \|_{L_{p,\tau}(E)}$$

$$+ \| \{\tau^{-2}[(I - \exp(-\tau B))(u_{k-1} + u_{k+1}) + (\exp(-2\tau B) - I)u_k]\}_1^{N-1} \|_{L_{p,\tau}(E)}$$

$$\le M_1[\| Au_0\|_E + ||Au_N||_E$$

$$+ \min \left\{ \ln \frac{1}{\tau}, 1 + |\ln \| B \|_{E\to E}| \right\} \| (I + \exp(-\tau B))^{-1} f^\tau \|_{L_{p,\tau}(E)}],$$

where M_1 *does not depend on* f^τ, u_0, u_N, p *and* τ.

Theorem 5.2.25. *Suppose that the assumptions of Theorem* 5.2.7 *hold. Then the solutions of the difference problem* (5.29) *in* $L_{p,\tau}(E)$ *obey the almost coercive inequality*

$$\| \{\tau^{-2}(u_{k+1} - 2u_k + u_{k-1})\}_1^{N-1} \|_{L_{p,\tau}(E)}$$
$$+ \| \{\tau^{-2}[(I - \exp(-\tau B))(u_{k-1} + u_{k+1}) + (\exp(-2\tau B) - I)u_k]\}_1^{N-1} \|_{L_{p,\tau}(E)}$$
$$\leq M_1[\| Au_0\|_E + \|Au_N\|_E$$
$$+ \min\left\{\ln\frac{1}{\tau}, 1 + |\ln \| B \|_{E\to E}|\right\} \| Bf^\tau \|_{L_{p,\tau}(E)}],$$

where M_1 *does not depend on* f^τ, u_0, u_N, p *and* τ.

Finally, let us give, without proof, the following results about well-posedness of the boundary-value difference problem (5.29) in the spaces

$$L_{p,\tau}(E) = L_p([0,T]_\tau, E), 1 \leq p < \infty.$$

Theorem 5.2.26. *Suppose that the assumptions of Theorem* 5.2.5 *hold. Suppose that the difference problem* (5.29) *is well posed in* $L_{p_0,\tau}(E)$ *for some* p_0, $1 < p_0 < \infty$. *Then it is well posed in* $L_{p,\tau}(E)$ *for any* p, $1 < p < \infty$ *and the following coercivity inequality holds:*

$$\| \{\tau^{-2}(u_{k+1} - 2u_k + u_{k-1})\}_1^{N-1} \|_{L_{p,\tau}(E)}$$
$$+ \| \{\tau^{-2}[(I - \exp(-\tau B))(u_{k-1} + u_{k+1}) + (\exp(-2\tau B) - I)u_k]\}_1^{N-1} \|_{L_{p,\tau}(E)}$$
$$+\|u^\tau\|_{C_c(E_{1-1/p,p})} \leq M(p_0)[\| u_0\|_{E_{1-1/p,p}} + \|u_N\|_{E_{1-1/p,p}} + \frac{p^2}{p-1} \| f^\tau \|_{L_{p,\tau}(E)}],$$

where $M(p_0)$ *does not depend on* f^τ, u_0, u_N, p *and* τ.

Theorem 5.2.27. *Suppose that the assumptions of Theorem* 5.2.6 *hold. Suppose that the difference problem* (5.29) *is well posed in* $L_{p_0,\tau}(E)$ *for some* p_0, $1 < p_0 < \infty$. *Then it is well posed in* $L_{p,\tau}(E)$ *for any* p, $1 < p < \infty$ *and the following coercivity inequality holds:*

$$\| \{\tau^{-2}(u_{k+1} - 2u_k + u_{k-1})\}_1^{N-1} \|_{L_{p,\tau}(E)}$$
$$+ \| \{\tau^{-2}[(I - \exp(-\tau B))(u_{k-1} + u_{k+1}) + (\exp(-2\tau B) - I)u_k]\}_1^{N-1} \|_{L_{p,\tau}(E)}$$
$$+\|u^\tau\|_{C_c(E_{1-1/p,p})} \leq M(p_0)[\| u_0\|_{E_{1-1/p,p}} + \|u_N\|_{E_{1-1/p,p}}$$
$$+ \frac{p^2}{p-1} \| (I + \exp(-\tau B))^{-1} f^\tau \|_{L_{p,\tau}(E)}],$$

where $M(p_0)$ *does not depend on* f^τ, u_0, u_N, p *and* τ.

Theorem 5.2.28. *Suppose that the assumptions of Theorem 5.2.7 hold. Suppose that the difference problem (5.29) is well posed in $L_{p_0,\tau}(E)$ for some p_0, $1 < p_0 < \infty$. Then it is well posed in $L_{p,\tau}(E)$ for any p, $1 < p < \infty$ and the following coercivity inequality holds:*

$$\| \{\tau^{-2}(u_{k+1} - 2u_k + u_{k-1})\}_1^{N-1} \|_{L_{p,\tau}(E)}$$
$$+ \| \{\tau^{-2}[(I - \exp(-\tau B))(u_{k-1} + u_{k+1}) + (\exp(-2\tau B) - I)u_k]\}_1^{N-1} \|_{L_{p,\tau}(E)}$$
$$+\|u^\tau\|_{C_c(E_{1-1/p,p})} \leq M(p_0)[\|u_0\|_{E_{1-1/p,p}} + \|u_N\|_{E_{1-1/p,p}} + \frac{p^2}{p-1}\|Bf^\tau\|_{L_{p,\tau}(E)}],$$

where $M(p_0)$ does not depend on f^τ, u_0, u_N, p and τ.

Theorem 5.2.29. *Let $1 \leq p \leq \infty$ and $0 < \alpha < 1$. Suppose that the assumptions of Theorem 5.2.5 hold. Then problem (5.29) is well posed in $L_{p,\tau}(E_{\alpha,p})$ and the following coercivity inequality holds:*

$$\| \{\tau^{-2}(u_{k+1} - 2u_k + u_{k-1})\}_1^{N-1} \|_{L_{p,\tau}(E_{\alpha,p})}$$
$$+ \| \{\tau^{-2}[(I - \exp(-\tau B))(u_{k-1} + u_{k+1}) + (\exp(-2\tau B) - I)u_k]\}_1^{N-1} \|_{L_{p,\tau}(E_{\alpha,p})}$$
$$\leq M[\| Au_0\|_{E_{\alpha,p}} + \|Au_N\|_{E_{\alpha,p}} + \frac{1}{\alpha(1-\alpha)}\| f^\tau \|_{L_{p,\tau}(E_{\alpha,p})}],$$

where M does not depend on f^τ, u_0, u_N, p, α and τ.

Theorem 5.2.30. *Let $1 \leq p \leq \infty$ and $0 < \alpha < 1$. Suppose that the assumptions of Theorem 5.2.6 hold. Then problem (5.29) is well posed in $L_{p,\tau}(E_{\alpha,p})$ and the following coercivity inequality holds:*

$$\| \{\tau^{-2}(u_{k+1} - 2u_k + u_{k-1})\}_1^{N-1} \|_{L_{p,\tau}(E_{\alpha,p})}$$
$$+ \| \{\tau^{-2}[(I - \exp(-\tau B))(u_{k-1} + u_{k+1}) + (\exp(-2\tau B) - I)u_k]\}_1^{N-1} \|_{L_{p,\tau}(E_{\alpha,p})}$$
$$\leq M[\| Au_0\|_{E_{\alpha,p}} + \|Au_N\|_{E_{\alpha,p}} + \frac{1}{\alpha(1-\alpha)}\| (I+\exp(-\tau B))^{-1} f^\tau \|_{L_{p,\tau}(E_{\alpha,p})}],$$

where M does not depend on f^τ, u_0, u_N, p, α and τ.

Theorem 5.2.31. *Let $1 \leq p \leq \infty$ and $0 < \alpha < 1$. Suppose that the assumptions of Theorem 5.2.7 hold. Then problem (5.29) is well posed in $L_{p,\tau}(E_{\alpha,p})$ and the following coercivity inequality holds:*

$$\| \{\tau^{-2}(u_{k+1} - 2u_k + u_{k-1})\}_1^{N-1} \|_{L_{p,\tau}(E_{\alpha,p})}$$
$$+ \| \{\tau^{-2}[(I - \exp(-\tau B))(u_{k-1} + u_{k+1}) + (\exp(-2\tau B) - I)u_k]\}_1^{N-1} \|_{L_{p,\tau}(E_{\alpha,p})}$$
$$\leq M[\| Au_0\|_{E_{\alpha,p}} + \|Au_N\|_{E_{\alpha,p}} + \frac{1}{\alpha(1-\alpha)}\| Bf^\tau \|_{L_{p,\tau}(E_{\alpha,p})}],$$

where M does not depend on f^τ, u_0, u_N, p, α and τ.

Theorem 5.2.32. *Let $1 < p, q < \infty$ and $0 < \alpha < 1$. Suppose that the assumptions of Theorem 5.2.5 hold. Then the difference problem (5.29) is well posed in $L_{p,\tau}(E_{\alpha,q})$ and the following coercivity inequality holds:*

$$\| \{\tau^{-2}(u_{k+1} - 2u_k + u_{k-1})\}_1^{N-1} \|_{L_{p,\tau}(E_{\alpha,q})}$$

$$+ \| \{\tau^{-2}[(I - \exp(-\tau B))(u_{k-1} + u_{k+1}) + (\exp(-2\tau B) - I)u_k]\}_1^{N-1} \|_{L_{p,\tau}(E_{\alpha,q})}$$

$$\leq M(q)[\| Au_0\|_{E_{\alpha,q}} + \|Au_N\|_{E_{\alpha,q}} + \frac{p^2}{p-1} \| f^\tau \|_{L_{p,\tau}(E_{\alpha,q})}],$$

where $M(q)$ does not depend on f^τ, u_0, u_N, p, q, α and τ.

Theorem 5.2.33. *Let $1 < p, q < \infty$ and $0 < \alpha < 1$. Suppose that the assumptions of Theorem 5.2.6 hold. Then the difference problem (5.29) is well posed in $L_{p,\tau}(E_{\alpha,q})$ and the following coercivity inequality holds:*

$$\| \{\tau^{-2}(u_{k+1} - 2u_k + u_{k-1})\}_1^{N-1} \|_{L_{p,\tau}(E_{\alpha,q})}$$

$$+ \| \{\tau^{-2}[(I - \exp(-\tau B))(u_{k-1} + u_{k+1}) + (\exp(-2\tau B) - I)u_k]\}_1^{N-1} \|_{L_{p,\tau}(E_{\alpha,q})}$$

$$\leq M(q)[\| Au_0\|_{E_{\alpha,q}} + \|Au_N\|_{E_{\alpha,q}} + \frac{p^2}{p-1} \| (I + \exp(-\tau B))^{-1} f^\tau \|_{L_{p,\tau}(E_{\alpha,q})}],$$

where $M(q)$ does not depend on f^τ, u_0, u_N, p, α and τ.

Theorem 5.2.34. *Let $1 < p, q < \infty$ and $0 < \alpha < 1$. Suppose that the assumptions of Theorem 5.2.7 hold. Then the difference problem (5.29) is well posed in $L_{p,\tau}(E_{\alpha,q})$ and the following coercivity inequality holds:*

$$\| \{\tau^{-2}(u_{k+1} - 2u_k + u_{k-1})\}_1^{N-1} \|_{L_{p,\tau}(E_{\alpha,q})}$$

$$+ \| \{\tau^{-2}[(I - \exp(-\tau B))(u_{k-1} + u_{k+1}) + (\exp(-2\tau B) - I)u_k]\}_1^{N-1} \|_{L_{p,\tau}(E_{\alpha,q})}$$

$$\leq M(q)[\| Au_0\|_{E_{\alpha,q}} + \|Au_N\|_{E_{\alpha,q}} + \frac{p^2}{p-1} \| Bf^\tau \|_{L_{p,\tau}(E_{\alpha,q})}],$$

where $M(q)$ does not depend on f^τ, u_0, u_N, p, α and τ.

Now, we will consider the applications of the exact difference scheme (5.29). From (5.29) it is clear that for the approximate solutions of the problem (5.1) it is necessary to approximate the expressions $\exp(-\tau B)$ and

$$\frac{1}{\tau}\int_{t_{k-1}}^{t_k} \exp(-(t_k - z)B))f(z)dz,$$

$$\frac{1}{\tau}\int_{t_{k-1}}^{t_k} \exp(-(z - t_{k-1})B)f(z)dz.$$

Let us remark that in constructing difference schemes it is important to know how to construct a right-hand side $f_k^{j,l}$ that satisfies

$$(2B\tau)^{-1}[\frac{1}{\tau}\int_{t_{k-1}}^{t_k}\exp(-(t_k-z)B))f(z)dz+\frac{1}{\tau}\int_{t_k}^{t_{k+1}}\exp(-(z-t_k)B)f(z)dz \quad (5.56)$$

$$-\exp(-\tau B)(\frac{1}{\tau}\int_{t_k}^{t_{k+1}}\exp(-(t_{k+1}-z)B))f(z)dz$$

$$+\frac{1}{\tau}\int_{t_{k-1}}^{t_k}\exp(-(z-t_{k-1})B)f(z)dz)]-f_k^{j,l}=o(\tau^{j+l})$$

and is a sufficiently simple. The choice formula $f_k^{j,l}$ is not unique. Using Taylor's formula, with respect to function $\exp(-(t_k-z)B))f(z)$ on variable z, we obtain

$$\frac{1}{\tau}\int_{t_{k-1}}^{t_k}\exp(-(t_k-z)B))f(z)dz$$

$$=f(t_k)+\sum_{m=1}^{l+j}\sum_{\lambda=0}^{m}\binom{m}{\lambda}B^{m-\lambda}f^{(\lambda)}(t_k)\frac{(-1)^m\tau^{m+1}}{(m+1)!}+o(\tau^{l+j+1}).$$

Using Taylor's formula, with respect to function $\exp(-(z-t_{k-1})B)f(z)$ on variable z, we obtain

$$\frac{1}{\tau}\int_{t_{k-1}}^{t_k}\exp(-(z-t_{k-1})B)f(z)dz$$

$$=f(t_{k-1})+\sum_{m=1}^{l+j}\sum_{\lambda=0}^{m}\binom{m}{\lambda}(-1)^{m-\lambda}B^{m-\lambda}f^{(\lambda)}(t_{k-1})\frac{\tau^{m+1}}{(m+1)!}+o(\tau^{j+l+1}).$$

Further, using the last formulas and Pade fractions for the function e^{-z}, we can write

$$f_k^{j,l}=(2B\tau)^{-1}[f(t_k)+\sum_{m=1}^{j+l}\sum_{\lambda=0}^{m}\binom{m}{\lambda}B^{m-\lambda}f^{(\lambda)}(t_k)\frac{(-1)^m\tau^{m+1}}{(m+1)!}$$

$$+f(t_k)+\sum_{m=1}^{j+l}\sum_{\lambda=0}^{m}\binom{m}{\lambda}(-1)^{m-\lambda}B^{m-\lambda}f^{(\lambda)}(t_k)\frac{\tau^{m+1}}{(m+1)!}$$

$$-R_{q,p+1}(\tau B)(f(t_{k+1})+\sum_{m=1}^{j+l}\sum_{\lambda=0}^{m}\binom{m}{\lambda}B^{m-\lambda}f^{(\lambda)}(t_{k+1})\frac{(-1)^m\tau^{m+1}}{(m+1)!}+f(t_{k-1})$$

$$+\sum_{m=1}^{j+l}\sum_{\lambda=0}^{m}\binom{m}{\lambda}(-1)^{m-\lambda}B^{m-\lambda}f^{(\lambda)}(t_{k-1})\frac{\tau^{m+1}}{(m+1)!})].$$

Now, using the last formula and Pade fractions for the function e^{-z}, we obtain the difference schemes of $(j+l)$-order of accuracy

$$-\frac{1}{\tau^2}(u_{k+1}-2u_k+u_{k-1}) \tag{5.57}$$

$$+\frac{1}{\tau^2}(I-R_{j,l+1}(\tau B))(u_{k-1}+u_{k+1})+\frac{1}{\tau^2}(R_{j,l+1}^2(\tau B)-I)u_k$$

$$=f_k^{j,l}, 1\le k\le N-1, u_0=v_0, u_N=v_T$$

for the approximate solutions of the boundary-value problem (5.1).

Note that the difference schemes (5.57) for $j=l$ and $j=l-1$ include difference schemes of arbitrary high order of approximation. Moreover, the corresponding functions $R_{j,l+1}(z)$ tend to 0 as $z\to\infty$ for $j=l-1, l$. Such difference schemes are simplest, in the sense that the degrees of the denominators of the corresponding Pade approximants of the function $\exp\{-z\}$ are minimal for a fixed order of approximation of the difference schemes.

It is clear that this problem is uniquely solvable, and the following formula holds:

$$u_k=(I-R_{j,l+1}^{2N}(\tau B))^{-1}\{(R_{j,l+1}^{k}(\tau B)-R_{j,l+1}^{2N-k}(\tau B))u_0 \tag{5.58}$$

$$+(R_{j,l+1}^{N-k}(\tau B)-R_{j,l+1}^{N+k}(\tau B))u_N$$

$$-(R_{j,l+1}^{N-k}(\tau B)-R_{j,l+1}^{N+k}(\tau B))\tau^2\sum_{i=1}^{N-1}(R_{j,l+1}^{N-i}(\tau B)$$

$$-R_{j,l+1}^{N+i}(\tau B))(I-R_{j,l+1}^2(\tau B))^{-1}f_i^{j,l}\}$$

$$+\tau^2\sum_{i=1}^{N-1}(R_{j,l+1}^{|k-i|}(\tau B)-R_{j,l+1}^{k+i}(\tau B))(I-R_{j,l+1}^2(\tau B))^{-1}f_i^{j,l}, 1\le k\le N-1.$$

Let us reduce such difference schemes to an operator problem in the space $F_\tau(E)$. In addition to the operator D_τ^2, acting from the space $E\times F_\tau(E)\times E$ of vectors $w^\tau=\{w_k\}_{k=0}^N$ into the space $F_\tau(E)$ of vectors $v^\tau=\{v_k\}_{k=1}^{N-1}$ by the rule

$$v^\tau=D_\tau^2u^\tau, v_k=\frac{1}{\tau^2}(w_{k+1}-2w_k+w_{k-1}), k=1,\dots,N-1,$$

define an operator $A_\tau^{j,l}$ from the space $E\times F_\tau(E)\times E$ of vectors $w^\tau=\{w_k\}_{k=0}^N$ into the space $F_\tau(E)$ of vectors $v^\tau=\{v_k\}_{k=1}^{N-1}$ by the rule

$$v^\tau=A_\tau^{j,l}u^\tau, v_k=\frac{1}{\tau^2}[(I-R_{j,l+1}(\tau B))(w_{k-1}+w_{k+1})$$

$$+(R_{j,l+1}^2(\tau B)-I)w_k], k=1,\dots,N-1.$$

Then the difference schemes (5.57) can obviously be rewritten as the equivalent operator equation

$$-D_\tau^2\Pi(u_0,u_N)u^\tau + A_\tau^{j,l}\Pi(u_0,u_N)u^\tau = f_{j,l}^\tau.$$

Here $f_{j,l}^\tau$ is defined by the formula

$$f_{j,l}^\tau = (f_1^{j,l},\dots,f_{N-1}^{j,l}).$$

The last operator problem will be considered in the space $F_\tau(E)$. From its unique solvability for any u_0, $u_N \in E$ and $f_{j,l}^\tau \in F_\tau(E)$ it follows that its solution u^τ defines a continuous additive and homogeneous operator $u^\tau(f_{j,l}^\tau,u_0,u_N)$.

The boundary-value problem (5.57) is said to be stable in $F_\tau(E)$ if we have the inequality

$$\| u^\tau(f_{j,l}^\tau,u_0,u_N) \|_{F_\tau(E)} \le M[\| f_{j,l}^\tau \|_{F_\tau(E)} + \| u_0\|_E + \|u_N\|_E],$$

where M is independent not only of $f_{j,l}^\tau$, u_0, u_N, but also of τ.

From formula (5.58) one derives the following result.

Theorem 5.2.35. *Let A be the positive operator in a Banach space E. Then the difference problem* (5.57) *is stable in $L_{p,\tau}(E)$, $1 \le p \le \infty$.(Here $C_\tau(E) = L_{\infty,\tau}(E)$.)*

The proof of this theorem is based on formula (5.58) and the estimates

$$||R_{j,l+1}^k(\tau B)||_{E\to E} \le M, 1 \le k \le N, \tag{5.59}$$

$$||(I - R_{j,l+1}^{2N}(\tau B))^{-1}||_{E\to E} \le M. \tag{5.60}$$

The proof of estimate (5.60) is based on the estimates

$$||(I - \exp(-2N\tau B))^{-1}||_{E\to E} \le M$$

and

$$||\exp(-k\tau B) - R_{j,l+1}^k(\tau B)||_{E\to E} \le Mk^{-j-l-1}, 1 \le k \le N. \tag{5.61}$$

Here M does not depend on k and τ.

Theorem 5.2.36. *Let A be the positive operator in a Banach space E. Then the difference problem* (5.57) *is stable in $C_\tau^{\alpha,a}(E)$, $0 < \alpha < 1$.*

The proof of this theorem is based on formula (5.58), estimates (5.59), (5.60) and

$$\left\|R_{j,l+1}^k(\tau B) - R_{j,l+1}^{k+r}(\tau B)\right\|_{E\to E} \le M\frac{r^\alpha}{(k+r)^\alpha}, \tag{5.62}$$

for all k, r, $1 \le k < k+r \le N-1$ and all $\tau > 0$,where M does not depend on k, r and τ. Estimate (5.62) follows from (5.59) and

$$k\tau||BR_{j,l+1}^k(\tau B)||_{E\to E} \le M, 1 \le k \le N. \tag{5.63}$$

The boundary-value difference problem (5.57) is said to be well posed (coercively stable) in $F_\tau(E)$ if we have the coercive inequality

$$\| \{\tau^{-2}(u_{k+1} - 2u_k + u_{k-1})\}_1^{N-1} \|_{F_\tau(E)}$$
$$+ \| \{\tau^{-2}[(I - R_{j,l+1}(\tau B))(u_{k-1} + u_{k+1}) + (R_{j,l+1}^2(\tau B) - I)u_k]\}_1^{N-1} \|_{F_\tau(E)}$$
$$\leq M[\| f_{j,l}^\tau \|_{F_\tau(E)} + \| Au_0 \|_E + \|Au_N\|_E],$$

where M is independent not only of $f_{j,l}^\tau$, u_0, u_N, but also of τ. Since the difference boundary-value problem (5.57) in $L_{p,\tau}(E)$, $1 \leq p \leq \infty$ is not well posed for the general positive operator A and space E, then the coercive norm

$$\| u^\tau \|_{K_{p,\tau}(E)} = \| \{\tau^{-2}(u_{k+1} - 2u_k + u_{k-1})\}_1^{N-1} \|_{L_{p,\tau}(E)}$$
$$+ \| \{\tau^{-2}[(I - R_{j,l+1}(\tau B))(u_{k-1} + u_{k+1}) + (R_{j,l+1}^2(\tau B) - I)u_k]\}_1^{N-1} \|_{L_{p,\tau}(E)}$$

tends to ∞ as $\tau \to +0$. We have

Theorem 5.2.37. *Let A be a strongly positive operator in a Banach space E with spectral angle $\phi(A, E) < \frac{\pi}{l}$ and u_0, $u_N \in D(A)$. Then the solutions of the difference problem (5.57) in $L_{p,\tau}(E)$ obey the almost coercive inequality*

$$\| u^\tau \|_{K_{p,\tau}(E)} \leq M_1[\| Au_0 \|_E + \|Au_N\|_E$$
$$+ \min\left\{\ln\frac{1}{\tau}, 1 + |\ln \| B \|_{E\to E}|\right\} \| f_{j,l}^\tau \|_{L_{p,\tau}(E)}],$$

where M_1 is independent not only of $f_{j,l}^\tau$, u_0, u_N, but also of τ.

The proof of Theorem 5.2.37 follows the scheme of the proofs of Theorem 5.2.5 and 5.2.23, and relies on the formula

$$\tau^{-2}(u_{k+1} - 2u_k + u_{k-1}) \tag{5.64}$$
$$= (I - R_{j,l+1}^{2N}(\tau B))^{-1}(I - R_{j,l+1}(\tau B))^2\tau^{-2}\{(R_{j,l+1}^{k-1}(\tau B) - R_{j,l+1}^{2N-k-1}(\tau B))u_0$$
$$+(R_{j,l+1}^{N-k-1}(\tau B) - R_{j,l+1}^{N+k-1}(\tau B))u_N\}$$
$$-(I - R_{j,l+1}^{2N}(\tau B))^{-1}(I - R_{j,l+1}(\tau B))(R_{j,l+1}^{N-k-1}(\tau B) - R_{j,l+1}^{N+k-1}(\tau B))$$
$$\times \sum_{i=1}^{N-1}(R_{j,l+1}^{N-i}(\tau B) - R_{j,l+1}^{N+i}(\tau B))(I + R_{j,l+1}(\tau B))^{-1}f_i^{j,l}$$
$$+(I - R_{j,l+1}(\tau B))\sum_{i=1}^{k-1} R_{j,l+1}^{k-1-i}(\tau B)(I + R_{j,l+1}(\tau B))^{-1}f_i^{j,l}$$
$$+(I - R_{j,l+1}(\tau B))\sum_{i=k+1}^{N-1} R_{j,l+1}^{i-k-1}(\tau B)(I + R_{j,l+1}(\tau B))^{-1}f_i^{j,l}$$
$$-2(I + R_{j,l+1}(\tau B))^{-1}f_k^{j,l},$$

and on estimates (5.59), (5.63) and

$$\sum_{j=1}^{N-1} \|(I - R_{j,l+1}(\tau B))R_{j,l+1}^{j-1}(\tau B)\|_{E\to E} \leq M \min\left(\ln(1/\tau), 1 + \left|\ln \|B\|\right|\right), \quad (5.65)$$

$$\|(I - R_{j,l+1}(\tau B))^2(\tau B)^{-2}\|_{E\to E} \leq M, \quad (5.66)$$

$$\|(I + R_{j,l+1}(\tau B))^{-1}\|_{E\to E} \leq M. \quad (5.67)$$

Estimate (5.66) is obvious. The proof of estimate (5.65) follows the scheme of proof of Lemma 5.2.4 and is based on estimates (5.59) and (5.63). The proof of estimate (5.67) is based on the following lemma.

Lemma 5.2.38. *Let B be a strongly positive operator in a Banach space E with spectral angle $\phi(B,E) < \frac{\pi}{2l}$. Then the operator $I + R_{j,l+1}(\tau B)$ is invertible and one has estimate* (5.67).

Proof. Since

$$(1 + R_{j,l+1}(z))^{-1} = \frac{Q_{j,l+1}(z)}{P_{j,l+1}(z) + Q_{j,l+1}(z)}$$

is a rational function, it suffices to show that the function $P_{j,l+1}(z) + Q_{j,l+1}(z)$ has no zeros in the sector $\{z \in C : z = \rho e^{\pm\phi}, 0 \leq \rho < \infty, 0 \leq \phi \leq \frac{\pi}{2l}\}$. Since

$$P_{j,l+1}(z) + Q_{j,l+1}(z) = \sum_{i=0}^{j} \frac{(j+l+1-i)!}{(j+l+1)!i!}\left[\frac{(l+1)!}{(l+1-i)!} + (-1)^{-i}\frac{j!}{(j-i)!}\right]z^i$$

$$+ \sum_{i=j+1}^{l+1} \frac{(j+l+1-i)!}{(j+l+1)!i!}\frac{(l+1)!}{(l+1-i)!}z^i$$

and $\frac{(l+1)!}{(l+1-i)!} + (-1)^{-i}\frac{j!}{(j-i)!} > 0$ for $j = l-1,\ l$, we obviously have $|P_{j,l+1}(z) + Q_{j,l+1}(z)| \geq 2$. Lemma 5.2.38 is proved.

Now let us study the well-posedness of the difference problem (5.57) in various Banach spaces.

Theorem 5.2.39. *Let A be a strongly positive operator in a Banach space E with spectral angle $\phi(A,E) < \frac{\pi}{l}$ and $u_0,\ u_N \in D(A)$. Then the solutions of the difference problem* (5.57) *in $C_\tau^{\alpha,\alpha}(E)$ obey the coercivity inequality*

$$\| \{\tau^{-2}(u_{k+1} - 2u_k + u_{k-1})\}_1^{N-1} \|_{C_\tau^{\alpha,\alpha}(E)}$$

$$+ \| \{\tau^{-2}[(I - R_{j,l+1}(\tau B))(u_{k-1} + u_{k+1}) + (R_{j,l+1}^2(\tau B) - I)u_k]\}_1^{N-1} \|_{C_\tau^{\alpha,\alpha}(E)}$$

$$\leq M_1\left[\frac{1}{\alpha(1-\alpha)}\| f_{j,l}^\tau \|_{C_\tau^{\alpha,\alpha}(E)} + \| Au_0\|_E + \|Au_N\|_E\right],$$

where M_1 is independent not only of $f_{j,l}^\tau$, u_0, u_N, α, but also of τ.

The proof of Theorem 5.2.39 follows the scheme of proof of Theorem 5.2.8, and relies on the formula

$$
\begin{aligned}
&\tau^{-2}(u_{k+1}-2u_k+u_{k-1}) \qquad (5.68)\\
&=(I-R_{j,l+1}^{2N}(\tau B))^{-1}(I-R_{j,l+1}(\tau B))^2\tau^{-2}(R_{j,l+1}^{k-1}(\tau B)-R_{j,l+1}^{2N-k-1}(\tau B))u_0\\
&+(I-R_{j,l+1}^{2N}(\tau B))^{-1}(I-R_{j,l+1}(\tau B))^2\tau^{-2}(R_{j,l+1}^{N-k-1}(\tau B)-R_{j,l+1}^{N+k-1}(\tau B))u_N\\
&+\{-I+(I-R_{j,l+1}^{2N}(\tau B))^{-1}(I-R_{j,l+1}^{N}(\tau B))[I-R_{j,l+1}^{k-1}(\tau B)-R_{j,l+1}^{N-k-1}(\tau B)\\
&+R_{j,l+1}^{N}(\tau B)]\}f_k^{j,l}+(I-R_{j,l+1}^{2N}(\tau B))^{-1}(I-R_{j,l+1}^{2N-2k}(\tau B))(I-R_{j,l+1}(\tau B))\\
&\times\sum_{i=1}^{k-1}R_{j,l+1}^{k-1-i}(\tau B)(I-R_{j,l+1}^{2i}(\tau B))(I+R_{j,l+1}(\tau B))^{-1}(f_i^{j,l}-f_k^{j,l})\\
&+(I-R_{j,l+1}^{2N}(\tau B))^{-1}(I-R_{j,l+1}^{2k}(\tau B))(I-R_{j,l+1}(\tau B))\\
&\times\sum_{i=k+1}^{N-1}R_{j,l+1}^{i-k-1}(\tau B)(I-R_{j,l+1}^{2N-2i}(\tau B))(I+R_{j,l+1}(\tau B))^{-1}(f_i^{j,l}-f_k^{j,l})
\end{aligned}
$$

and on estimates (5.59), (5.60), (5.62), (5.66) and (5.67). The coercivity inequality

$$
\begin{aligned}
&\| \{\tau^{-2}(u_{k+1}-2u_k+u_{k-1})\}_1^{N-1} \|_{C_\tau^\alpha(E)}\\
&+\| \{\tau^{-2}[(I-R_{j,l+1}(\tau B))(u_{k-1}+u_{k+1})+(R_{j,l+1}^2(\tau B)-I)u_k]\}_1^{N-1} \|_{C_\tau^\alpha(E)}\\
&\le M_1[\frac{1}{\alpha(1-\alpha)}\| f_{j,l}^\tau \|_{C_\tau^\alpha(E)}+\| Au_0\|_E+\|Au_N\|_E]
\end{aligned}
$$

fails. Nevertheless, we have the following result.

Theorem 5.2.40. *Let A be a strongly positive operator in a Banach space E with spectral angle $\phi(A,E)<\frac{\pi}{l}$ and $(I-R_{j,l+1}(\tau B))^2\tau^{-2}u_0-f_1^{j,l}$, $(I-R_{j,l+1}(\tau B))^2\tau^{-2}\times u_N-f_{N-1}^{j,l}\in E'_\alpha$. Then the solutions of the difference problem (5.57) in $C_\tau^\alpha(E)$ obey the coercivity inequality*

$$
\begin{aligned}
&\| \{\tau^{-2}(u_{k+1}-2u_k+u_{k-1})\}_1^{N-1} \|_{C_\tau^\alpha(E)}\\
&+\| \{\tau^{-2}[(I-R_{j,l+1}(\tau B))(u_{k-1}+u_{k+1})+(R_{j,l+1}^2(\tau B)-I)u_k]\}_1^{N-1} \|_{C_\tau^\alpha(E)}\\
&+\| \{\tau^{-2}(u_{k+1}-2u_k+u_{k-1})\}_1^{N-1} \|_{C_\tau(E'_\alpha)}\\
&\le\frac{M_1}{\alpha(1-\alpha)}[\|| f_{j,l}^\tau \|_{C_\tau^\alpha(E)}+\| (I-R_{j,l+1}(\tau B))^2\tau^{-2}u_0-f_1^{j,l} \|_{E'_\alpha}\\
&+\|(I-R_{j,l+1}(\tau B))^2\tau^{-2}u_N-f_{N-1}^{j,l}\|_{E'_\alpha}],
\end{aligned}
$$

where M_1 is independent not only of $f_{j,l}^{\tau}$, u_0, u_N, α, but also of τ. Here, the Banach space $E'_\alpha = E'_\alpha(E, B)$ $(0 < \alpha < 1)$ consists of those $v \in E$ for which the norm

$$\| v \|_{E'_\alpha} = \sup_{z>0} z^\alpha \| B(zI + B)^{-1} v \|_E + \| v \|_E$$

is finite.

Proof. To prove the theorem it suffices to establish the estimates for

$$\{\tau^{-2}(u_{k+1} - 2u_k + u_{k-1})\}_1^{N-1}$$

in $C_\tau^\alpha(E)$ and in $C_\tau(E'_\alpha)$. By formula (5.64), we have

$$\begin{aligned}
&\tau^{-2}(u_{k+1} - 2u_k + u_{k-1}) \\
&= (I - R_{j,l+1}^{2N}(\tau B))^{-1}(R_{j,l+1}^{k-1}(\tau B) - R_{j,l+1}^{2N-k-1}(\tau B)) \\
&\quad \times [(I - R_{j,l+1}(\tau B))^2 \tau^{-2} u_0 - f_1^{j,l}] \\
&\quad + (I - R_{j,l+1}^{2N}(\tau B))^{-1}(R_{j,l+1}^{N-k-1}(\tau B) - R_{j,l+1}^{N+k-1}(\tau B)) \\
&\quad \times [(I - R_{j,l+1}(\tau B))^2 \tau^{-2} u_N - f_{N-1}^{j,l}] \\
&\quad + (I - R_{j,l+1}^{2N}(\tau B))^{-1}[R_{j,l+1}^{k-1}(\tau B) - R_{j,l+1}^{2N-k-1}(\tau B)](f_1^{j,l} - f_k^{j,l}) \\
&\quad + (I - R_{j,l+1}^{2N}(\tau B))^{-1}[R_{j,l+1}^{N-k-1}(\tau B) - R_{j,l+1}^{N+k-1}(\tau B)](f_k^{j,l} - f_{N-1}^{j,l}) \\
&\quad + (I - R_{j,l+1}^{2N}(\tau B))^{-1}(I - R_{j,l+1}^{2N-2k}(\tau B))(I - R_{j,l+1}(\tau B)) \\
&\quad \times \sum_{i=1}^{k-1} R_{j,l+1}^{k-1-i}(\tau B)(I - R_{j,l+1}^{2i}(\tau B))(I + R_{j,l+1}(\tau B))^{-1}(f_i^{j,l} - f_k^{j,l}) \\
&\quad + (I - R_{j,l+1}^{2N}(\tau B))^{-1}(I - R_{j,l+1}^{2k}(\tau B))(I - R_{j,l+1}(\tau B)) \\
&\quad \times \sum_{i=k+1}^{N-1} R_{j,l+1}^{i-k-1}(\tau B)(I - R_{j,l+1}^{2N-2i}(\tau B))(I + R_{j,l+1}(\tau B))^{-1}(f_i^{j,l} - f_k^{j,l}) \\
&= \sum_{m=1}^{6} H_k^m,
\end{aligned}$$

where

$$\begin{aligned}
H_k^1 &= (I - R_{j,l+1}^{2N}(\tau B))^{-1}(R_{j,l+1}^{k-1}(\tau B) - R_{j,l+1}^{2N-k-1}(\tau B)) \\
&\quad \times [(I - R_{j,l+1}(\tau B))^2 \tau^{-2} u_0 - f_1^{j,l}], \\
H_k^2 &= (I - R_{j,l+1}^{2N}(\tau B))^{-1}(R_{j,l+1}^{N-k-1}(\tau B) - R_{j,l+1}^{N+k-1}(\tau B)) \\
&\quad \times [(I - R_{j,l+1}(\tau B))^2 \tau^{-2} u_N - f_{N-1}^{j,l}], \\
H_k^3 &= (I - R_{j,l+1}^{2N}(\tau B))^{-1}[R_{j,l+1}^{k-1}(\tau B) - R_{j,l+1}^{2N-k-1}(\tau B)](f_1^{j,l} - f_k^{j,l}), \\
H_k^4 &= (I - R_{j,l+1}^{2N}(\tau B))^{-1}[R_{j,l+1}^{N-k-1}(\tau B) - R_{j,l+1}^{N+k-1}(\tau B)](f_k^{j,l} - f_{N-1}^{j,l}),
\end{aligned}$$

$$H_k^5 = (I - R_{j,l+1}^{2N}(\tau B))^{-1}(I - R_{j,l+1}^{2N-2k}(\tau B))(I - R_{j,l+1}(\tau B))$$

$$\times \sum_{i=1}^{k-1} R_{j,l+1}^{k-1-i}(\tau B)(I - R_{j,l+1}^{2i}(\tau B))(I + R_{j,l+1}(\tau B))^{-1}(f_i^{j,l} - f_k^{j,l}),$$

$$H_k^6 = (I - R_{j,l+1}^{2N}(\tau B))^{-1}(I - R_{j,l+1}^{2k}(\tau B))(I - R_{j,l+1}(\tau B))$$

$$\times \sum_{i=k+1}^{N-1} R_{j,l+1}^{i-k-1}(\tau B)(I - R_{j,l+1}^{2N-2i}(\tau B))(I + R_{j,l+1}(\tau B))^{-1}(f_i^{j,l} - f_k^{j,l}).$$

The proof of estimates

$$||\{H_k^m\}_{k=1}^{N-1}||_{C_\tau^\alpha(E)} \leq \frac{M}{\alpha(1-\alpha)} \| f_{j,l}^\tau \|_{C_\tau^\alpha(E)}$$

for $m = 3, 4, 5, 6$ follows the scheme of proof of the estimates

$$||\{G_k^m\}_{k=1}^{N-1}||_{C_\tau^\alpha(E)} \leq \frac{M}{\alpha(1-\alpha)} \| f_{j,l}^\tau \|_{C_\tau^\alpha(E)}$$

for $m = 3, 4, 5, 6$ of Theorem 5.2.11, and relies on estimates (5.59), (5.60), (5.62), (5.66) and (5.67). Therefore to prove the theorem it suffices to establish the estimates for $\{H_k^m\}_1^{N-1}$, $m = 1, 2$ in $C_\tau^\alpha(E)$ and for $\{H_k^m\}_1^{N-1}$, $m = 1, 2, 3, 4, 5, 6$ in $C_\tau(E'_\alpha)$. Let us estimate $H_m^\tau = \{H_k^m\}_{k=1}^{N-1}$ for any $m = 1, 2$ in $C_\tau^\alpha(E)$. We start with H_1^τ.

Using estimates (5.59), (5.60) and the definition of the spaces E'_α, we show that, for $1 \leq k \leq N-1$,

$$||H_k^1||_E \leq ||(I - R_{j,l+1}^{2N}(\tau B))^{-1}||_{E\to E}||R_{j,l+1}^{k-1}(\tau B) - R_{j,l+1}^{2N-k-1}(\tau B)||_{E\to E}$$

$$\times||(I - R_{j,l+1}(\tau B))^2\tau^{-2}u_0 - f_1^{j,l}||_E \leq M||(I - R_{j,l+1}(\tau B))^2\tau^{-2}u_0 - f_1^{j,l}||_{E'_\alpha}.$$

Further, applying estimates (5.59) and (5.60), we show that, for $1 \leq k < k+r \leq N-1$,

$$||H_k^1 - H_{k+r}^1||_E \leq ||(I - R_{j,l+1}^{2N}(\tau B))^{-1}||_{E\to E}$$

$$\times[||(R_{j,l+1}^{k-1}(\tau B) - R_{j,l+1}^{k+r-1}(\tau B))[(I - R_{j,l+1}(\tau B))^2\tau^{-2}u_0 - f_1^{j,l}]||_E$$

$$+||R_{j,l+1}^{2N-2k-r}(\tau B)||_{E\to E}$$

$$\times||(R_{j,l+1}^{k-1}(\tau B) - R_{j,l+1}^{k+r-1}(\tau B))[(I - R_{j,l+1}(\tau B))^2\tau^{-2}u_0 - f_1^{j,l}]||_E]$$

$$\leq M[||(R_{j,l+1}^{k-1}(\tau B) - R_{j,l+1}^{k+r-1}(\tau B))[(I - R_{j,l+1}(\tau B))^2\tau^{-2}u_0 - f_1^{j,l}]||_E.$$

To estimate

$$(R_{j,l+1}^{k-1}(\tau B) - R_{j,l+1}^{k-1+r}(\tau B))[(I - R_{j,l+1}(\tau B))^2\tau^{-2}u_0 - f_1^{j,l}]$$

in the norm of E we use the following Cauchy–Riesz representation formula for the operator $(R_{j,l+1}^{k-1}(\tau B) - R_{j,l+1}^{k-1+r}(\tau B))(\tau B)^{-1}$ (see Chapter 4):

$$(R_{j,l+1}^{k-1}(\tau B) - R_{j,l+1}^{k-1+r}(\tau B))(\tau B)^{-1} y$$
$$= \frac{1}{2\pi i} \int\limits_{S_1 \cup S_2} (R_{j,l+1}^{k-1}(z) - R_{j,l+1}^{k-1+r}(z)) z^{-1} \tau B (z - \tau B)^{-1} y dz,$$

where $S_1 = \{\rho e^{i\psi}, 0 \leq \rho < \infty\}$ and $S_2 = \{\rho e^{i\psi}, 0 \leq \rho < \infty\}, 0 \leq \psi < \frac{\pi}{2l}$. We obtain

$$(R_{j,l+1}^{k-1}(\tau B) - R_{j,l+1}^{k-1+r}(\tau B))(\tau B)^{-1} y$$
$$= \frac{1}{2\pi i} \int\limits_{S_1 \cup S_2} \sum_{s=0}^{r-1} R_{j,l+1}^{k-1+s}(z)(I - R_{j,l+1}(z)) z^{-1} B(\frac{z}{\tau} - B)^{-1} y dz.$$

Since $z = \rho e^{\pm i\psi}$, with $|\psi| < \frac{\pi}{2l}$, from the strong positivity of B it follows that

$$|(\frac{z}{\tau})^{\alpha}| \parallel B(\frac{z}{\tau} - B)^{-1} y \parallel_E \leq M (\frac{\rho}{\tau})^{\alpha} \parallel B(\frac{\rho}{\tau} + B)^{-1} y \parallel_E \leq M \, ||y||_{E'_{\alpha}},$$

$$\sum_{s=0}^{r-1} |R_{j,l+1}^{k-1+s}(z)(I - R_{j,l+1}(z)) z^{-1}| \leq M \min\{\rho^{-1}, r\}.$$

Hence,

$$|| \, (R_{j,l+1}^{k}(\tau B) - R_{j,l+1}^{k+r}(\tau B))(\tau B)^{-1} y||_E \leq M_1 \int\limits_0^{\infty} \frac{\tau^{\alpha} \rho^{-\alpha}}{\rho + \frac{1}{r}} d\rho ||y||_{E'_{\alpha}}$$

$$= M_1 \int\limits_0^{\infty} \frac{r^{\alpha} \tau^{\alpha} s^{-\alpha}}{s + 1} ds ||y||_{E'_{\alpha}} = t_r^{\alpha} M_1 \int\limits_0^{\infty} \frac{s^{-\alpha}}{s+1} ds ||y||_{E'_{\alpha}} \leq t_r^{\alpha} \frac{M_2}{\alpha(1-\alpha)} ||y||_{E'_{\alpha}}.$$

This shows that

$$||H_k^1 - H_{k+r}^1||_E \leq ||(R_{j,l+1}^{k-1}(\tau B) - R_{j,l+1}^{k-1+r}(\tau B))[(I - R_{j,l+1}(\tau B))^2 \tau^{-2} u_0 - f_1^{j,l}]||_E$$

for any $1 \leq k < k + r \leq N - 1$. Thus, we have proved that

$$||H_1^{\tau}||_{C_{\tau}^{\alpha}(E)} \leq \frac{M_1}{\alpha(1-\alpha)} ||(I - R_{j,l+1}(\tau B))^2 \tau^{-2} u_0 - f_1^{j,l}||_{E'_{\alpha}}.$$

In a similar manner we can show that

$$||H_2^{\tau}||_{C_{\tau}^{\alpha}(E)} \leq \frac{M_1}{\alpha(1-\alpha)} ||(I - R_{j,l+1}(\tau B))^2 \tau^{-2} u_N - f_{N-1}^{j,l}||_{E'_{\alpha}}.$$

Now, let us estimate $H_m^\tau = \{H_k^m\}_{k=1}^{N-1}$ for any $m = 1,2,3,4,5,6$ in $C_\tau(E'_\alpha)$ separately. We start with H_1^τ. Using estimates (5.59), (5.60) and the definition of the spaces E'_α we show that, for $1 \leq k \leq N-1$,

$$||H_k^1||_{E'_\alpha} \leq ||(I - R_{j,l+1}^{2N}(\tau B))^{-1}||_{E\to E}||R_{j,l+1}^{k-1}(\tau B) - R_{j,l+1}^{2N-k-1}(\tau B)||_{E\to E}$$

$$\times||(I - R_{j,l+1}(\tau B))^2\tau^{-2}u_0 - f_1^{j,l}||_{E'_\alpha} \leq M||(I - R_{j,l+1}(\tau B))^2\tau^{-2}u_0 - f_1^{j,l}||_{E'_\alpha}.$$

Thus, we have proved that

$$||H_1^\tau||_{C_\tau(E'_\alpha)} \leq M||(I - R_{j,l+1}(\tau B))^2\tau^{-2}u_0 - f_1^{j,l}||_{E'_\alpha}.$$

In a similar manner we can show that

$$||H_2^\tau||_{C_\tau(E'_\alpha)} \leq M||(I - R_{j,l+1}(\tau B))^2\tau^{-2}u_N - f_{N-1}^{j,l}||_{E'_\alpha}.$$

By the definition of the spaces E'_α and using estimates (5.59), (5.60), we obtain that

$$\| \lambda^\alpha B(\lambda + B)^{-1}H_k^3 \|_E \leq \lambda^\alpha||(I - R_{j,l+1}^{2N}(\tau B))^{-1}||_{E\to E}||I - R_{j,l+1}^{2N-2k}(\tau B)||_{E\to E}$$

$$\times||B(\lambda + B)^{-1}R_{j,l+1}^{k-1}(\tau B)(f_1^{j,l} - f_k^{j,l})||_E \leq M\lambda^\alpha t_{k-1}^\alpha\|f_{j,l}^\tau\|_{C_\tau^\alpha(E)}$$

for all λ, $\lambda > 0$ and for $1 \leq k \leq N-1$. Using the estimate

$$||B(\lambda + B)^{-1}R_{j,l+1}^{k-1}(\tau B)||_{E\to E} \leq M\min\{\frac{1}{\lambda t_{k-1}}, 1\},$$

we obtain that

$$\| \lambda^\alpha B(\lambda + B)^{-1}H_k^3 \|_E \leq M_1\lambda^\alpha t_{k-1}^\alpha \min\{\frac{1}{\lambda t_{k-1}}, 1\}\|f_{j,l}^\tau\|_{C_\tau^\alpha(E)} \leq M_1\|f_{j,l}^\tau\|_{C_\tau^\alpha(E)}$$

for all λ, $\lambda > 0$ and for $2 \leq k \leq N-1$. So, we have established that

$$||H_3^\tau||_{C_\tau(E'_\alpha)} \leq M_1\|f_{j,l}^\tau\|_{C_\tau^\alpha(E)}.$$

In a similar manner we can show that

$$||H_4^\tau||_{C_\tau(E'_\alpha)} \leq M_1\|f_{j,l}^\tau\|_{C_\tau^\alpha(E)}.$$

By the definition of the spaces E'_α and using estimates (5.59), (5.60) and (5.67), we obtain that

$$\| \lambda^\alpha B(\lambda + B)^{-1}H_k^5 \|_E \leq \lambda^\alpha||(I - R_{j,l+1}^{2N}(\tau B))^{-1}||_{E\to E}$$

$$\times\sum_{i=1}^{k-1}||I - R_{j,l+1}^{2N-2k}(\tau B) \|_{E\to E} \, ||I - R_{j,l+1}^{2i}(\tau B) \|_{E\to E}$$

$$\times||(I+R_{j,l+1}(\tau B))^{-1}||_{E\to E}\|(I-R_{j,l+1}(\tau B))B(\lambda+B)^{-1}R_{j,l+1}^{k-1-i}(\tau B)(f_i^{j,l}-f_k^{j,l}) \|_E$$

$$\leq M\lambda^\alpha\sum_{i=1}^{k-1}(t_k-t_i)^\alpha\|(I-R_{j,l+1}(\tau B))B(\lambda+B)^{-1}R_{j,l+1}^{k-1-i}(\tau B) \|_{E\to E} \, \| f_{j,l}^\tau \|_{C_\tau^\alpha(E)}$$

for all λ, $\lambda > 0$ and for $2 \leq k \leq N-1$. Using the estimate

$$||(I - R_{j,l+1}(\tau B))B(\lambda + B)^{-1}R_{j,l+1}^{k-1-i}(\tau B)||_{E\to E} \leq M\tau \min\{\frac{1}{t_{k-i}}, \frac{1}{\lambda t_{k-i}^2}\},$$

we obtain that

$$\| \lambda^\alpha B(\lambda + B)^{-1}H_k^5 \|_E \leq M_1\lambda^\alpha \sum_{i=1}^{k-1}(t_k - t_i)^\alpha \tau \min\{\frac{1}{t_{k-i}}, \frac{1}{\lambda t_{k-i}^2}\}\| f_{j,l}^\tau \|_{C_\tau^\alpha(E)}$$

$$\leq M_2\lambda^\alpha[\sum_{i=1}^{k-1} \frac{\tau}{t_i^{1-\alpha}(1 + \lambda t_i)}]\| f_{j,l}^\tau \|_{C_\tau^\alpha(E)}.$$

The sum enclosed in the right-hand side square brackets is the lower Darboux integral sum for the integral

$$\int_0^{t_k} \frac{ds}{s^{1-\alpha}(1 + s\lambda)}.$$

Since

$$\lambda^\alpha \int_0^{t_k} \frac{ds}{s^{1-\alpha}(1 + s\lambda)} \leq \int_0^\infty \frac{dp}{p^{1-\alpha}(1 + p)},$$

it follows that

$$\| \lambda^\alpha B(\lambda + B)^{-1}H_k^5 \|_E \leq \frac{M_3}{\alpha(1-\alpha)}\| f_{j,l}^\tau \|_{C_\tau^\alpha(E)}$$

for all λ, $\lambda > 0$ and for $2 \leq k \leq N-1$. So, we have established that

$$||H_5^\tau||_{C_\tau(E_\alpha')} \leq \frac{M_3}{\alpha(1-\alpha)}\| f_{j,l}^\tau \|_{C_\tau^\alpha(E)}.$$

In a similar manner we can show that

$$||H_6^\tau||_{C_\tau(E_\alpha')} \leq \frac{M_3}{\alpha(1-\alpha)}\| f_{j,l}^\tau \|_{C_\tau^\alpha(E)}.$$

Theorem 5.2.40 is proved.

Let us give, without proof, the following result.

Theorem 5.2.41. *Let A be a strongly positive operator in a Banach space E with spectral angle $\phi(A, E) < \frac{\pi}{l}$ and $(I - R_{j,l+1}(\tau B))^2\tau^{-2}u_0 - f_1^{j,l}$, $(I - R_{j,l+1}(\tau B))^2 \times\tau^{-2}u_N - f_{N-1}^{j,l} \in E_{\beta-\gamma}'$. Then the solutions of the difference problem* (5.57) *in $C_\tau^{\beta,\gamma}(E)$ $(0 \leq \gamma \leq \beta, 0 < \beta < 1)$ obey the coercivity inequalities*

$$\| \{\tau^{-2}(u_{k+1} - 2u_k + u_{k-1})\}_1^{N-1} \|_{C_\tau(E_{\beta-\gamma}')} \leq \frac{M_1}{\beta(1-\beta)}\| f_{j,l}^\tau \|_{C_\tau^{\beta,\gamma}(E)}$$

$$+M_1[\| (I - R_{j,l+1}(\tau B))^2\tau^{-2}u_0 - f_1^{j,l} \|_{E_{\beta-\gamma}'}$$

$$+\|(I-R_{j,l+1}(\tau B))^2\tau^{-2}u_N - f^{j,l}_{N-1}\|_{E'_{\beta-\gamma}}],$$

$$\| \{\tau^{-2}(u_{k+1}-2u_k+u_{k-1})\}_1^{N-1} \|_{C^{\beta,\gamma}_\tau(E)}$$

$$+ \| \{\tau^{-2}[(I-R_{j,l+1}(\tau B))(u_{k-1}+u_{k+1}) + (R^2_{j,l+1}(\tau B)-I)u_k]\}_1^{N-1} \|_{C^{\beta,\gamma}_\tau(E)}$$

$$\leq \frac{M_1}{\beta(1-\beta)}\| f^\tau_{j,l} \|_{C^{\beta,\gamma}_\tau(E)} + M_1[|(I-R_{j,l+1}(\tau B))^2\tau^{-2}u_0 - f^{j,l}_1|^{\beta,\gamma}_0$$
$$+|(I-R_{j,l+1}(\tau B))^2\tau^{-2}u_N - f^{j,l}_{N-1}|^{\beta,\gamma}_0],$$

where M_1 is independent not only of $f^\tau_{j,l}$, u_0, u_N, β, γ but also of τ.

Here, $|w|^{\beta,\gamma}_0$ denotes that the norm of the Banach space $\widetilde{E}^{\beta,\gamma}_0$ consists of those $w \in E$ for which the norm

$$|w|^{\beta,\gamma}_0 = \max_{1\leq n\leq N-1} \|R^n_{j,l+1}(\tau B)w\|_E$$

$$+ \sup_{1\leq n<n+r\leq N-1} t_\tau^{-\beta}(t_n+t_r)^\gamma(t_N-t_n)^\gamma\|(R^{n+r}_{j,l+1}(\tau B)-R^n_{j,l+1}(\tau B))w\|_E$$

is finite.

Note that the parameter γ can be chosen freely in $[0,\beta)$, which increases the number of spaces of grid functions in which difference problem (5.57) is well posed. We have

Theorem 5.2.42. *Let A be a strongly positive operator in a Banach space E with spectral angle $\phi(A,E) < \frac{\pi}{l}$. Then the solutions of the difference problem* (5.57) *in $C_\tau(E'_\alpha)$ obey the coercivity inequality*

$$\| \{\tau^{-2}(u_{k+1}-2u_k+u_{k-1})\}_1^{N-1} \|_{C_\tau(E'_\alpha)}$$

$$+ \| \{\tau^{-2}[(I-\exp(-\tau B))(u_{k-1}+u_{k+1}) + (\exp(-2\tau B)-I)u_k]\}_1^{N-1} \|_{C_\tau(E'_\alpha)}$$

$$\leq M_1[\frac{1}{\alpha(1-\alpha)}\| f^\tau \|_{C_\tau(E'_\alpha)} + \| Au_0\|_{E'_\alpha} + \|Au_N\|_{E'_\alpha}],$$

where M_1 is independent not only of f^τ, u_0, u_N, α, but also of τ.

Proof. Using formula (5.64), we obtain

$$\tau^{-2}(u_{k+1}-2u_k+u_{k-1})$$

$$= (I-R^{2N}_{j,l+1}(\tau B))^{-1}(I-R_{j,l+1}(\tau B))^2\tau^{-2}\{(R^{k-1}_{j,l+1}(\tau B)-R^{2N-k-1}_{j,l+1}(\tau B))u_0$$
$$+(R^{N-k-1}_{j,l+1}(\tau B)-R^{N+k-1}_{j,l+1}(\tau B))u_N\}$$
$$-(I-R^{2N}_{j,l+1}(\tau B))^{-1}(I-R_{j,l+1}(\tau B))(R^{N-k-1}_{j,l+1}(\tau B)-R^{N+k-1}_{j,l+1}(\tau B))$$

$$\times \sum_{i=1}^{N-1} (R_{j,l+1}^{N-i}(\tau B) - R_{j,l+1}^{N+i}(\tau B))(I + R_{j,l+1}(\tau B))^{-1} f_i^{j,l}$$

$$+(I - R_{j,l+1}(\tau B)) \sum_{i=1}^{k-1} R_{j,l+1}^{k-1-i}(\tau B)(I + R_{j,l+1}(\tau B))^{-1} f_i^{j,l}$$

$$+(I - R_{j,l+1}(\tau B)) \sum_{i=k+1}^{N-1} R_{j,l+1}^{i-k-1}(\tau B)(I + R_{j,l+1}(\tau B))^{-1} f_i^{j,l}$$

$$-2(I + R_{j,l+1}(\tau B))^{-1} f_k^{j,l} = g_k + p_k,$$

where

$$g_k = (I - R_{j,l+1}^{2N}(\tau B))^{-1}(I - R_{j,l+1}(\tau B))^2 \tau^{-2}\{(R_{j,l+1}^{k-1}(\tau B) - R_{j,l+1}^{2N-k-1}(\tau B))u_0$$

$$+(R_{j,l+1}^{N-k-1}(\tau B) - R_{j,l+1}^{N+k-1}(\tau B))u_N\},$$

$$p_k = -(I - R_{j,l+1}^{2N}(\tau B))^{-1}(I - R_{j,l+1}(\tau B))(R_{j,l+1}^{N-k-1}(\tau B) - R_{j,l+1}^{N+k-1}(\tau B))$$

$$\times \sum_{i=1}^{N-1} (R_{j,l+1}^{N-i}(\tau B) - R_{j,l+1}^{N+i}(\tau B))(I + R_{j,l+1}(\tau B))^{-1} f_i^{j,l}$$

$$+(I - R_{j,l+1}(\tau B)) \sum_{i=1}^{k-1} R_{j,l+1}^{k-1-i}(\tau B)(I + R_{j,l+1}(\tau B))^{-1} f_i^{j,l}$$

$$+(I - R_{j,l+1}(\tau B)) \sum_{i=k+1}^{N-1} R_{j,l+1}^{i-k-1}(\tau B)(I + R_{j,l+1}(\tau B))^{-1} f_i^{j,l}$$

$$-2(I + R_{j,l+1}(\tau B))^{-1} f_k^{j,l}.$$

To this end it suffices to show that

$$\| g^\tau \|_{C_\tau(E'_\alpha)} \le M_1[\| Au_0\|_{E'_\alpha} + Au_N\|_{E'_\alpha}], \tag{5.69}$$

$$\| p^\tau \|_{C_\tau(E'_\alpha)} \le M_1 \alpha^{-1}(1-\alpha)^{-1} \| f_{j,l}^\tau \|_{C_\tau(E'_\alpha)}. \tag{5.70}$$

Estimate (5.69) follows from the estimate

$$||(I - R_{j,l+1}(\tau B))^2 (\tau B)^{-2}||_{E \to E} \le M$$

and the fact that the homogeneous difference equation ($f_{j,l}^\tau = 0$) is stable in $C_\tau(E'_\alpha)$ uniformly in τ. By the definition of the spaces E'_α and using estimates (5.59), (5.60) and (5.67), we obtain

$$\| \lambda^\alpha B(\lambda + B)^{-1} p_k \|_E \le \lambda^\alpha ||(I - R_{j,l+1}^{2N}(\tau B))^{-1}||_{E \to E}$$

$$\times ||R_{j,l+1}^{N-k-1}(\tau B) - R_{j,l+1}^{N+k-1}(\tau B)||_{E \to E}$$

$$\times \sum_{i=1}^{N-1} ||R_{j,l+1}(\tau B) - R_{j,l+1}^{N+1+2i}(\tau B)||_{E\to E}||(I + R_{j,l+1}(\tau B))^{-1}||_{E\to E}$$

$$\times||(I - R_{j,l+1}(\tau B))B(\lambda + B)^{-1}R_{j,l+1}^{N-1-i}(\tau B)f_i^{j,l}||_E$$

$$+\sum_{i=1}^{k-1}||(I+R_{j,l+1}(\tau B))^{-1}||_{E\to E}||\lambda^\alpha B(\lambda+B)^{-1}(I-R_{j,l+1}(\tau B))R_{j,l+1}^{k-1-i}(\tau B)f_i^{j,l}||_E$$

$$+ \sum_{i=k+1}^{N-1} ||(I + R_{j,l+1}(\tau B))^{-1}||_{E\to E}$$

$$\times||\lambda^\alpha B(\lambda + B)^{-1}(I - R_{j,l+1}(\tau B))R_{j,l+1}^{i-k-1}(\tau B)f_i^{j,l}||_E$$

$$+2||(I + R_{j,l+1}(\tau B))^{-1}||_{E\to E}||\lambda^\alpha B(\lambda + B)^{-1}f_k^{j,l}||_E$$

$$+ \sum_{i=1}^{N-1} ||R_{j,l+1}^k(\tau B)||_{E\to E}||(I + R_{j,l+1}(\tau B))^{-1}||_{E\to E}$$

$$\times||\lambda^\alpha B(\lambda + B)^{-1}(I - R_{j,l+1}(\tau B))R_{j,l+1}^{i-1}(\tau B)f_i^{j,l}||_E$$

$$\leq M(\| f_{j,l}^\tau \|_{C_\tau(E'_\alpha)} + \lambda^\alpha \sum_{j=1}^{N-1} ||B(\lambda + B)^{-1}(I - R_{j,l+1}(\tau B))R_{j,l+1}^{i-1}(\tau B)f_i^{j,l}||_E).$$

To estimate the last sum we use the following Cauchy–Riesz representation formula for the operator $B(\lambda + B)^{-1}(I - R_{j,l+1}(\tau B))R_{j,l+1}^{i-1}(\tau B)$ (see Chapter 4):

$$\lambda^\alpha B(\lambda + B)^{-1}(I - R_{j,l+1}(\tau B))R_{j,l+1}^{i-1}(\tau B)f_i^{j,l}$$
$$= \frac{1}{2\pi i}\int_{S_1\cup S_2} (\lambda + \frac{z}{\tau})^{-1}(I - R_{j,l+1}(z))R_{j,l+1}^{i-1}(z)B(z - \tau B)^{-1}f_i^{j,l}dz$$

where $S_1 = \{\rho e^{i\psi}, 0 \leq \rho < \infty\}$ and $S_2 = \{\rho e^{i\psi}, 0 \leq \rho < \infty\}, 0 \leq \psi < \frac{\pi}{2l}$. Since $z = \rho e^{\pm i\psi}$, with $|\psi| < \frac{\pi}{2l}$, from the strong positivity of A it follows that

$$|(\frac{z}{\tau})^\alpha| \parallel B(\frac{z}{\tau} - B)^{-1}f_i^{j,l} \parallel_E \leq M(\frac{\rho}{\tau})^\alpha \parallel B(\frac{\rho}{\tau} + B)^{-1}f_i^{j,l} \parallel_E .$$

From this estimate and the estimates

$$\frac{1}{|\lambda\tau + z|} \leq \frac{M}{\lambda\tau + \rho},$$

$$|(I - R_{j,l+1}(z))R_{j,l+1}^{i-1}(z)| \leq M\rho(1 + 2a\rho\cos\psi + a^2\rho^2)^{-\frac{i-1}{2}} \ (a > 0)$$

it follows that

$$\lambda^{\alpha}\sum_{j=1}^{N-1}\|B(\lambda+B)^{-1}(I-R_{j,l+1}(\tau B))R_{j,l+1}^{i-1}(\tau B)f_i^{j,l}\|_E$$

$$\leq M_1\int_0^{\infty}\sum_{j=1}^{N-1}\frac{\rho^{1-\alpha}}{(1+2a\rho\cos\psi+a^2\rho^2)^{\frac{i-1}{2}}}\frac{(\tau\lambda)^{\alpha}}{\lambda\tau+\rho}\left(\frac{\rho}{\tau}\right)^{\alpha}\| B\left(\frac{\rho}{\tau}+B\right)^{-1}f_i^{j,l}\|_E\,d\rho$$

$$\leq M_1\int_0^{\infty}\sum_{j=1}^{N-1}\frac{\rho^{1-\alpha}}{(1+2a\rho\cos\psi+a^2\rho^2)^{\frac{i-1}{2}}}\frac{(\tau\lambda)^{\alpha}}{\lambda\tau+\rho}d\rho\| f_{j,l}^{\tau}\|_{C_\tau(E'_\alpha)}.$$

Summing the geometric progression, we get

$$\lambda^{\alpha}\sum_{j=1}^{N-1}\|B(\lambda+B)^{-1}(I-R_{j,l+1}(\tau B))R_{j,l+1}^{i-1}(\tau B)f_i^{j,l}\|_E$$

$$\leq\quad M(a,\psi)\int_0^{\infty}\frac{\rho^{-\alpha}(\tau\lambda)^{\alpha}}{\lambda\tau+\rho}d\rho\| f_{j,l}^{\tau}\|_{C_\tau(E'_\alpha)}\leq\frac{M_2}{\alpha(1-\alpha)}\| f_{j,l}^{\tau}\|_{C_\tau(E'_\alpha)}.$$

So, it follows that

$$\|\lambda^{\alpha}B(\lambda+B)^{-1}p_k\|_E\leq\frac{M_3}{\alpha(1-\alpha)}\| f_{j,l}^{\tau}\|_{C_\tau(E'_\alpha)}$$

for all λ, $\lambda>0$ and for $1\leq k\leq N-1$. From the last estimate follows estimate (5.70). Theorem 5.2.42 is proved.

Let us give, without proof, the following result.

Theorem 5.2.43. *Let A be a strongly positive operator in a Banach space E with spectral angle $\phi(A,E)<\frac{\pi}{l}$ and*

$$(I-R_{j,l+1}(\tau B))^2\tau^{-2}u_0-f_1^{j,l},\quad (I-\ R_{j,l+1}(\tau B))^2\tau^{-2}u_N-f_{N-1}^{j,l}\in E'_{\alpha-\gamma}\,.$$

Then the solutions of the difference problem (5.57) in $C_\tau^{\beta,\gamma}(E'_{\alpha-\beta})$ $(0\leq\gamma\leq\beta\leq\alpha,0<\alpha<1)$ obey the coercivity inequalities

$$\|\{\tau^{-2}(u_{k+1}-2u_k+u_{k-1})\}_1^{N-1}\|_{C_\tau^{\beta,\gamma}(E'_{\alpha-\beta})}$$

$$+\|\{\tau^{-2}[(I-R_{j,l+1}(\tau B))(u_{k-1}+u_{k+1})+(R_{j,l+1}^2(\tau B)-I)u_k]\}_1^{N-1}\|_{C_\tau^{\beta,\gamma}(E'_{\alpha-\beta})}$$

$$+\|\{\tau^{-2}(u_{k+1}-2u_k+u_{k-1})\}_1^{N-1}\|_{C_\tau(\widetilde{E}_{\alpha-\beta}^{\beta,\gamma})}$$

$$\leq\frac{M_1}{\alpha(1-\alpha)}\| f_{j,l}^{\tau}\|_{C_\tau^{\beta,\gamma}(E'_{\alpha-\beta})}$$

$$+M_1[|(I-R_{j,l+1}(\tau B))^2\tau^{-2}u_0 - f_1^{j,l}|_{\alpha-\beta}^{\beta,\gamma} + |(I-R_{j,l+1}(\tau B))^2\tau^{-2}u_N - f_{N-1}^{j,l}|_{\alpha-\beta}^{\beta,\gamma}],$$

$$\| \{\tau^{-2}(u_{k+1}-2u_k+u_{k-1})\}_1^{N-1} \|_{C_\tau^{\beta,\gamma}(E'_{\alpha-\beta})}$$

$$+ \| \{\tau^{-2}[(I-R_{j,l+1}(\tau B))(u_{k-1}+u_{k+1}) + (R_{j,l+1}^2(\tau B)-I)u_k]\}_1^{N-1} \|_{C_\tau^{\beta,\gamma}(E'_{\alpha-\beta})}$$

$$+ \| \{\tau^{-2}(u_{k+1}-2u_k+u_{k-1})\}_1^{N-1} \|_{C_\tau(E'_{\alpha-\gamma})}$$

$$\leq \frac{M_1}{\alpha(1-\alpha)}[\| f_{j,l}^\tau \|_{C_\tau^{\beta,\gamma}(E'_{\alpha-\beta})}$$

$$+M_1[\||(I-R_{j,l+1}(\tau B))^2\tau^{-2}u_0 - f_1^{j,l}\||_{E'_{\alpha-\gamma}}$$

$$+\||(I-R_{j,l+1}(\tau B))^2\tau^{-2}u_N - f_{N-1}^{j,l}\||_{E'_{\alpha-\gamma}}]],$$

where M_1 is independent not only of $f_{j,l}^\tau$, u_0, u_N, α, β, γ but also of τ.

Here, $|w|_{\alpha-\beta}^{\beta,\gamma}$ denotes that the norm of the Banach space $\widetilde{E}_{\alpha-\beta}^{\beta,\gamma}$ consists of those $w \in E$ for which the norm

$$|w|_{\alpha-\beta}^{\beta,\gamma} = \max_{1\leq n\leq N-1} \|R_{j,l+1}^n(\tau B)w\|_{E_{\alpha-\beta}}$$

$$+ \sup_{1\leq n<n+r\leq N-1} t_r^{-\beta}(t_n+t_r)^\gamma(t_N-t_n)^\gamma\|(R_{j,l+1}^{n+r}(\tau B)-R_{j,l+1}^n(\tau B))w\|_{E_{\alpha-\beta}}$$

is finite.

Finally, let us give, without proof, the following results about well-posedness of the boundary-value difference problem (5.57) in the spaces

$$L_{p,\tau}(E) = L_p([0,T]_\tau, E), 1 \leq p < \infty.$$

Theorem 5.2.44. *Let A be a strongly positive operator in a Banach space E with spectral angle $\phi(A,E) < \frac{\pi}{l}$. Suppose that the difference problem (5.57) is well posed in $L_{p_0,\tau}(E)$ for some p_0, $1 < p_0 < \infty$. Then it is well posed in $L_{p,\tau}(E)$ for any p, $1 < p < \infty$ and the following coercivity inequality holds:*

$$\| \{\tau^{-2}(u_{k+1}-2u_k+u_{k-1})\}_1^{N-1} \|_{L_{p,\tau}(E)}$$

$$+ \| \{\tau^{-2}[(I-R_{j,l+1}(\tau B))(u_{k-1}+u_{k+1}) + (R_{j,l+1}^2(\tau B)-I)u_k]\}_1^{N-1} \|_{L_{p,\tau}(E)}$$

$$+\|u^\tau\|_{C_\tau(E_{1-1/p,p})} \leq M(p_0)[\| u_0\|_{E_{1-1/p,p}} + \|u_N\|_{E_{1-1/p,p}}$$

$$+\frac{p^2}{p-1}\| f_{j,l}^\tau \|_{L_{p,\tau}(E)}],$$

where $M(p_0)$ does not depend on $f_{j,l}^\tau$, u_0, u_N, p and τ.

Theorem 5.2.45. *Let $1 \le p \le \infty$ and $0 < \alpha < 1$. Suppose that A is a strongly positive operator in a Banach space E with spectral angle $\phi(A,E) < \frac{\pi}{l}$. Then the difference problem* (5.57) *is well posed in $L_{p,\tau}(E_{\alpha,p})$ and the following coercivity inequality holds:*

$$\| \{\tau^{-2}(u_{k+1} - 2u_k + u_{k-1})\}_1^{N-1} \|_{L_{p,\tau}(E_{\alpha,p})}$$

$$+ \| \{\tau^{-2}[(I - R_{j,l+1}(\tau B))(u_{k-1} + u_{k+1}) + (R_{j,l+1}^2(\tau B) - I)u_k]\}_1^{N-1} \|_{L_{p,\tau}(E_{\alpha,p})}$$

$$\le M[\| Au_0\|_{E_{\alpha,p}} + \|Au_N\|_{E_{\alpha,p}} + \frac{1}{\alpha(1-\alpha)} \| f_{j,l}^{\tau} \|_{L_{p,\tau}(E_{\alpha,p})}],$$

where M does not depend on $f_{j,l}^{\tau}$, u_0, u_N, p, α and τ.

Theorem 5.2.46. *Let $1 < p,q < \infty$ and $0 < \alpha < 1$. Suppose that A is a strongly positive operator in a Banach space E with spectral angle $\phi(A,E) < \frac{\pi}{l}$. Then the difference problem* (5.57) *is well posed in $L_{p,\tau}(E_{\alpha,q})$ and the following coercivity inequality holds:*

$$\| \{\tau^{-2}(u_{k+1} - 2u_k + u_{k-1})\}_1^{N-1} \|_{L_{p,\tau}(E_{\alpha,q})}$$

$$+ \| \{\tau^{-2}[(I - R_{j,l+1}(\tau B))(u_{k-1} + u_{k+1}) + (R_{j,l+1}^2(\tau B) - I)u_k]\}_1^{N-1} \|_{L_{p,\tau}(E_{\alpha,q})}$$

$$\le M(q)[\| Au_0\|_{E_{\alpha,q}} + \|Au_N\|_{E_{\alpha,q}} + \frac{p^2}{p-1} \| f_{j,l}^{\tau} \|_{L_{p,\tau}(E_{\alpha,q})}],$$

where $M(q)$ does not depend on $f_{j,l}^{\tau}$, u_0, u_N, p, α and τ.

Here, the Banach spaces $E'_{\alpha,p} = E'_{\alpha,p}(E,B)$ $(0 < \alpha < 1, 1 \le p \le \infty)$ consist of those $v \in E$ for which the norm

$$\| v \|_{E'_{\alpha,p}} = (\int_0^{\infty} \|z^{\alpha}B(zI+B)^{-1}v \|_E^p \frac{dz}{z})^{\frac{1}{p}}, 1 \le p < \infty,$$

are finite and $\| v \|_{E'_{\alpha,\infty}} = \| v \|_{E'_{\alpha}}$.

Now, the abstract theorems given above are applied in the investigation of difference schemes of higher order of accuracy with respect to the set all variables for approximate solution of the boundary-value problem (5.27). The discretization of problem (5.27) is carried out in two steps. In the first step let us give the difference operator A_h^x by the formula

$$A_h^x u_x^h = \sum_{2m \le |r| \le S} b_r^x D_h^r u_x^h + \delta u_x^h. \tag{5.71}$$

The coefficients are chosen in such a way that the operator A_h^x approximates in a specified way the operator

$$\sum_{|r|=2m} a_r(x) \frac{\partial^{|r|}}{\partial x_1^{r_1} \cdots \partial x_n^{r_n}} + \delta.$$

We shall assume that for $|\xi_k h| \leq \pi$ the symbol $A(\xi h, h)$ of the operator $A_h^x - \delta$ satisfies the inequalities

$$(-1)^m A^x(\xi h, h) \geq M_1|\xi|^{2m}, |\arg A^x(\xi h, h)| \leq \phi < \phi_0 \leq \frac{\pi}{l}. \tag{5.72}$$

With the help of A_h^x we arrive at the boundary-value problem

$$-\frac{d^2 v^h(y,x)}{dy^2} + A_h^x v^h(y,x) = f^h(y,x), 0 < y < T, \tag{5.73}$$

$$v^h(0,x) = \varphi^h(x), v^h(T,x) = \psi^h(x), x \in \mathbb{R}_h^n,$$

for an infinite system of ordinary differential equations.

In the second step we replace problem (5.73) by the difference scheme

$$-\frac{1}{\tau^2}(u_{k+1}^h(x) - 2u_k^h(x) + u_{k-1}^h(x)) \tag{5.74}$$

$$+\frac{1}{\tau^2}(I - R_{j,l+1}(\tau B_h^x))(u_{k-1}^h(x) + u_{k+1}^h(x)) + \frac{1}{\tau^2}(R_{j,l+1}^2(\tau B_h^x) - I)u_k^h(x)$$

$$= f_k^{j,l}(x), 1 \leq k \leq N-1, u_0^h(x) = v_0^h(x),\ u_N^h = v_T^h,\ x \in R_h^n, (B_h^x)^2 = A_h^x.$$

Let us give a number of corollaries of the abstract theorems given above.

Theorem 5.2.47. *The solutions of the difference schemes* (5.74) *satisfy the following stability estimates:*

$$\| u^{\tau,h} \|_{C_\tau^{\alpha,\alpha}(C_h^\beta)} \leq M[||u_0^h||_{C_h^\beta} + ||u_N^h||_{C_h^\beta} + \| f_{j,l}^{\tau,h} \|_{C_\tau^{\alpha,\alpha}(C_h^\beta)}],$$
$$0 < \alpha < 1, 0 \leq \beta < 1,$$

$$\| u^{\tau,h} \|_{L_{p,\tau}(C_h^\beta)} \leq M[||u_0^h||_{C_h^\beta} + ||u_N^h||_{C_h^\beta} + \| f_{j,l}^{\tau,h} \|_{L_{p,\tau}(C_h^\beta)}],$$
$$1 \leq p \leq \infty, 0 \leq \beta < 1,$$

where M does not depend on $f_{j,l}^\tau$, u_0^h, u_N^h, α, β, p, h and τ

The proof of Theorem 5.2.47 is based on the abstract Theorems 5.2.35 and 5.2.36, the positivity of the operator A_h^x in C_h^β and on the fact that for any $0 < \beta < \frac{1}{2m}$ the norms in the spaces $E'_\beta(C_h, A_h^x)$ and $C_h^{2m\beta}$ are equivalent uniformly in h and on the following theorem on the structure of the fractional spaces $E'_\alpha(C_h, (A_h^x)^{\frac{1}{2}})$.

Theorem 5.2.48. *Let A be a strongly positive operator in a Banach space E with spectral angle $\phi(A,E) < \frac{\pi}{2}$. Then for $0 < \alpha < \frac{1}{2}$ the norms of the spaces $E'_\alpha(E, A^{\frac{1}{2}})$ and $E'_{\frac{\alpha}{2}}(E, A)$ are equivalent.*

Proof. Using the following Cauchy–Riesz representation formula for the operator $A^{\frac{1}{2}}(\lambda + A^{\frac{1}{2}})^{-1}$ (see Chapter 4),

$$\lambda^{\alpha} A^{-\frac{1}{2}}(\lambda + A^{\frac{1}{2}})^{-1} y$$

$$= \frac{1}{2\pi i} \int\limits_{S_1 \cup S_2} \lambda^{\alpha} z^{-\frac{1}{2}}(\lambda + z^{\frac{1}{2}})^{-1} A^{\frac{1}{2}}(z - A)^{-1} y dz$$

where $S_1 = \{\rho e^{i\psi}, 0 \leq \rho < \infty\}$ and $S_2 = \{\rho e^{i\psi}, 0 \leq \rho < \infty\}, 0 \leq \psi < \frac{\pi}{2}$. Let $y \in E'_{\frac{\alpha}{2}}(A, E)$. Then from the strong positivity of A it follows that

$$|z^{\frac{\alpha}{2}}| \parallel A(z - A)^{-1} y \parallel_E \leq M \rho^{\frac{\alpha}{2}} \parallel A(\rho + A)^{-1} y \parallel_E \leq M \parallel y \parallel_{E'_{\frac{\alpha}{2}}(A,E)} .$$

From this estimate and the estimates

$$\frac{1}{|\lambda + z^{\frac{1}{2}}|} \leq \frac{M}{\lambda + \sqrt{\rho}},$$

it follows that

$$\lambda^{\alpha} ||A^{\frac{1}{2}}(\lambda + A^{\frac{1}{2}})^{-1} y||_E$$

$$\leq \int\limits_0^{\infty} \frac{M}{\lambda + \sqrt{\rho}} \frac{\lambda^{\alpha}}{\sqrt{\rho}\rho^{\frac{\alpha}{2}}} M \rho^{\frac{\alpha}{2}} \parallel A(\rho + A)^{-1} y \parallel_E d\rho$$

$$\leq M_1 \int\limits_0^{\infty} \frac{1}{\lambda + \sqrt{\rho}} \frac{\lambda^{\alpha}}{\sqrt{\rho}\rho^{\frac{\alpha}{2}}} d\rho \parallel y \parallel_{E'_{\frac{\alpha}{2}}(A,E)} .$$

Since

$$\int\limits_0^{\infty} \frac{1}{\lambda + \sqrt{\rho}} \frac{\lambda^{\alpha}}{\sqrt{\rho}\rho^{\frac{\alpha}{2}}} d\rho = 2 \int\limits_0^{\infty} \frac{dr}{(1 + r) r^{\alpha}} \leq \frac{M}{\alpha(1 - \alpha)},$$

we have

$$\lambda^{\alpha} ||A^{\frac{1}{2}}(\lambda + A^{\frac{1}{2}})^{-1} y||_E \leq \frac{M_2}{\alpha(1 - \alpha)} \parallel y \parallel_{E'_{\frac{\alpha}{2}}(A,E)}$$

for all $\lambda > 0$. This shows that $y \in E'_{\alpha}(A^{\frac{1}{2}}, E)$ and

$$\parallel y \parallel_{E'_{\alpha}(A^{\frac{1}{2}},E)} \leq \frac{M_2}{\alpha(1 - \alpha)} \parallel y \parallel_{E'_{\frac{\alpha}{2}}(A,E)} .$$

Next, using the the following Cauchy–Riesz representation formula for the operator $A^{\frac{1}{2}}(\lambda + A)^{-1}$ (see Chapter 4),

$$\lambda^{\frac{\alpha}{2}} A(\lambda + A)^{-1} y = \frac{1}{2\pi i} \int\limits_{S_1 \cup S_2} \lambda^{\frac{\alpha}{2}} z(\lambda + z^2)^{-1} A^{\frac{1}{2}}(z - A^{\frac{1}{2}})^{-1} y dz$$

where $S_1 = \{\rho e^{i\psi}, 0 \leq \rho < \infty\}$ and $S_2 = \{\rho e^{i\psi}, 0 \leq \rho < \infty\}, 0 \leq \psi < \frac{\pi}{2}$. Let $y \in E'_\alpha(A^{\frac{1}{2}}, E)$. Then from the strong positivity of A it follows that

$$|z^\alpha| \parallel A^{\frac{1}{2}}(z - A^{\frac{1}{2}})^{-1} y \parallel_E \leq M\rho^\alpha \parallel A^{\frac{1}{2}}(\rho + A^{\frac{1}{2}})^{-1} y \parallel_E \leq M \parallel y \parallel_{E'_\alpha(A^{\frac{1}{2}},E)} .$$

From this estimate and the estimates

$$\frac{1}{|\lambda + z^2|} \leq \frac{M}{\lambda + \rho^2},$$

it follows that

$$\lambda^{\frac{\alpha}{2}} ||A(\lambda + A)^{-1} y||_E$$

$$\leq \int_0^\infty \frac{\rho \lambda^{\frac{\alpha}{2}} M}{\lambda + \rho^2} \rho^{-\alpha} M \rho^\alpha \parallel A^{\frac{1}{2}}(\rho + A^{\frac{1}{2}})^{-1} y \parallel_E d\rho$$

$$\leq M_1 \int_0^\infty \frac{\lambda^{\frac{\alpha}{2}}}{\lambda + \rho^2} \rho^{1-\alpha} d\rho \parallel y \parallel_{E'_\alpha(A^{\frac{1}{2}},E)} .$$

Since

$$\int_0^\infty \frac{\lambda^{\frac{\alpha}{2}}}{\lambda + \rho^2} \rho^{1-\alpha} d\rho = \frac{1}{2} \int_0^\infty \frac{dr}{(1+r) r^\alpha} \leq \frac{M}{\alpha(1-\alpha)},$$

we have

$$\lambda^{\frac{\alpha}{2}} ||A(\lambda + A)^{-1} y||_E \leq \frac{M_2}{\alpha(1-\alpha)} \parallel y \parallel_{E'_\alpha(A^{\frac{1}{2}},E)}$$

for all $\lambda > 0$. This shows that $y \in E'_{\frac{\alpha}{2}}(A, E)$ and

$$\parallel y \parallel_{E'_{\frac{\alpha}{2}}(A,E)} \leq \frac{M_2}{\alpha(1-\alpha)} \parallel y \parallel_{E'_\alpha(A^{\frac{1}{2}},E)} .$$

Theorem 5.2.48 is proved.

Theorem 5.2.49. *The solutions of the difference schemes* (5.74) *satisfy the following almost coercive stability estimates:*

$$\parallel \{\tau^{-2}(u^h_{k+1} - 2u^h_k + u^h_{k-1})\}_1^{N-1} \parallel_{Lp,\tau(C_h)}$$

$$\leq M[\ln \frac{1}{h}[\sum_{|r|=2m} ||D^r_h u^h_0||_{C_h} + \sum_{|r|=2m} ||D^r_h u^h_N||_{C_h}]$$

$$+ \ln \frac{1}{\tau + h} \parallel f^{\tau,h}_{j,l} \parallel_{L_{p,\tau}(C_h)}], 1 \leq p \leq \infty,$$

where M does not depend on $f^\tau_{j,l}$, u^h_0, u^h_N, p, h and τ.

The proof of Theorem 5.2.49 is based on the abstract Theorems 5.2.37 and 5.2.48, the positivity of the operator A_h^x in C_h^β and on the almost coercivity inequality for an elliptic operator A_h^x in C_h and on the estimate

$$\min\left\{\ln\frac{1}{\tau}, 1+\left|\ln \| B_h^x \|_{C_h^\beta \to C_h^\beta}\right|\right\} \leq M \ln\frac{1}{\tau+h}.$$

Theorem 5.2.50. *The solutions of the difference schemes* (5.74) *satisfy the following coercivity estimates:*

$$\| \{\tau^{-2}(u_{k+1}^h - 2u_k^h + u_{k-1}^h)\}_1^{N-1} \|_{C_\tau^{\alpha,\alpha}(C_h^\beta)}$$

$$\leq M(\alpha,\beta)[\sum_{|r|=2m} ||D_h^r u_0^h||_{C_h^\beta} + \sum_{|r|=2m} ||D_h^r u_N^h||_{C_h^\beta} + \| f_{j,l}^{\tau,h} \|_{C_\tau^{\alpha,\alpha}(C_h^\beta)}],$$

$$0 \leq \alpha < 1, 0 < \beta < 1,$$

$$\| \{\tau^{-2}(u_{k+1}^h - 2u_k^h + u_{k-1}^h)\}_1^{N-1} \|_{L_{p,\tau}(W_{q,h}^{m\alpha})}$$

$$\leq M(\alpha,p,q)[||u_0^h||_{E'_{1-\frac{1}{p},p}(W_{q,h}^{m\alpha})} + ||u_N^h||_{E'_{1-\frac{1}{p},p}(W_{q,h}^{m\alpha})} + \| f_{j,l}^{\tau,h} \|_{L_{p,\tau}(W_{q,h}^{m\alpha})}],$$

$$1 < p, q < \infty, 0 < \alpha < \frac{1}{m},$$

where $M(\alpha,\beta)$ *and* $M(\alpha,p,q)$ *do not depend on* $f_{j,l}^\tau$, u_0^h, u_N^h, h *and* τ

The proof of Theorem 5.2.47 is based on the abstract Theorems 5.2.39 and 5.2.45, the positivity of the operator A_h^x in $W_{q,h}^\beta$, on the almost coercivity inequality for an elliptic operator A_h^x in C_h^β, $W_{q,h}^\beta$, $1 < q < \infty$, $0 < \beta < 1$, on the fact that for any $0 < \beta < \frac{1}{2m}$ and $1 \leq q \leq \infty$ the norms in the spaces $E'_{\beta,q}(L_{q,h}, A_h^x)$ and $W_{q,h}^{2m\beta}$ are equivalent uniformly in h and on the following theorem on the structure of the fractional spaces $E'_{\alpha,q}(L_{q,h}, (A_h^x)^{\frac{1}{2}})$.

Theorem 5.2.51. *Let A be a strongly positive operator in a Banach space E with spectral angle $\phi(A,E) < \frac{\pi}{2}$. Then for $0 < \alpha < \frac{1}{2}, 1 \leq p \leq \infty$ the norms of the spaces $E'_{\alpha,p}(E, A^{\frac{1}{2}})$ and $E'_{\frac{\alpha}{2},p}(E, A)$ are equivalent.*

This result was proved in the case $p = \infty$ in Theorem 5.2.48. In the case $p = 1$ the proof follows the same scheme. In the case $1 < p < \infty$ the proof is based on Minkowski's inequality and follows the same scheme.

Note that in a similar manner we can construct the difference schemes of a high order of accuracy with respect to one variable for approximate solutions of the boundary-value problems (5.25) and (5.26). Abstract theorems given above permit us to obtain the stability, the almost stability and the coercive stability estimates for the solutions of these difference schemes.

5.3 Two-Step Difference Schemes Generated by Taylor's Decomposition

We consider again the boundary-value problem (5.1). Let the function $v(t)(0 \leq t \leq T)$ have a $(2l+2j+2)$-th continuous derivative and $t_{k-1}, t_k, t_{k+1} \in [0,T]_\tau$. Then by Theorem 3.2.1 we have the following Taylor decomposition on three points:

$$v(t_{k+1}) - 2v(t_k) + v(t_{k-1}) - \sum_{i=1}^{l} \alpha_i v^{(2i)}(t_k)\tau^{2i} \tag{5.75}$$

$$-\sum_{i=1}^{j} \eta_i (v^{(2i)}(t_{k-1}) + v^{(2i)}(t_{k+1}))\tau^{2i} = o(\tau^{2l+2j+2}),$$

where $\eta_m, m = 1, \dots, j$ is a solution of system

$$\sum_{m=1}^{j} \frac{(2i)!\eta_m}{(2i-2m)!} = 1, l+1 \leq i \leq l+j, \tag{5.76}$$

and

$$\begin{cases} \alpha_1 = 1 - 2\eta_1, \alpha_i = (\frac{1}{(2i)!} - \eta_i - \sum\limits_{m=1}^{i-1} \eta_m \frac{1}{(2(i-m))!})2 \quad \text{for any } i, 2 \leq i \leq j, \\ \alpha_i = (\frac{1}{(2i)!} - \sum\limits_{m=1}^{j} \eta_m \frac{1}{(2(i-m))!})2 \text{ for any } i, \; j+1 \leq \; i \leq l. \end{cases}$$

Now, we will consider the applications of Taylor's decomposition (5.75) of function $v(t)$ on three points to approximate solutions of the boundary-value problem (5.1). From (5.75) it is clear that for the approximate solution of problem (5.1) it is necessary to find $v^{(2i)}_{(t_k)}$ for any $i, 1 \leq i \leq l$ and $v^{(2i)}_{(t_{k-1})}$, $v^{(2i)}_{(t_{k+1})}$ for any $i, 1 \leq i \leq j$. Using the equation

$$v''(t) = Av(t) - f(t),$$

we obtain

$$v^{(2n)}(t) = A^n v(t) - \sum_{\lambda=1}^{n} A^{n-\lambda} f^{(2\lambda-2)}(t), n = 2, \dots. \tag{5.77}$$

Now, using formulas (5.75) and (5.77), we obtain the difference schemes

$$-\frac{u_{k+1} - 2u_k + u_{k-1}}{\tau^2} + \sum_{i=1}^{l} \alpha_i A^i \tau^{2i-2} u_k \tag{5.78}$$

$$+\sum_{i=1}^{j} \eta_i A^i \tau^{2i-2}(u_{k-1} + u_{k+1}) = f_k^{j,l},$$

$$f_k^{j,l} = \sum_{i=1}^{j} \eta_i \sum_{\lambda=1}^{i} A^{i-\lambda}(f^{(2\lambda-2)}(t_{k+1}) + f^{(2\lambda-2)}(t_{k-1}))\tau^{2i-2}$$

$$+\sum_{i=1}^{l} \alpha_i \sum_{\lambda=1}^{i} A^{i-\lambda} f^{(2\lambda-2)}(t_k)\tau^{2i-2}, 1 \leq k \leq N-1, u_0 = v_0, u_N = v_T.$$

Thus, we have constructed the difference schemes of $(2l+2j)$-order of accuracy for the approximate solution of the boundary-value problem (5.1).

This problem is uniquely solvable, and the following formula holds:

$$u_k = (I - R^{2N}(\tau B))^{-1}\{(R^k(\tau B) - R^{2N-k}(\tau B))u_0 \tag{5.79}$$

$$+(R^{N-k}(\tau B) - R^{N+k}(\tau B))u_N - (R^{N-k}(\tau B) - R^{N+k}(\tau B))(I + \tau B)$$

$$\times(2I + \tau B)^{-1}B^{-1} \sum_{i=1}^{N-1} (R^{N-i}(\tau B) - R^{N+i}(\tau B))Cf_i^{j,l}\tau$$

$$+(I + \tau B)(2I + \tau B)^{-1}B^{-1} \sum_{i=1}^{N-1} (R^{|k-i|}(\tau B) - R^{k+i}(\tau B))Cf_i^{j,l}\tau, 1 \leq k \leq N-1,$$

where

$$B = B(\tau, A) = \frac{\tau^2 A_l}{2} + \sqrt{(\frac{\tau^2 A_l}{2})^2 + A_l},$$

$$A_l = (\sum_{i=1}^{j}(\alpha_i + 2\eta_i)A^i\tau^{2i-2} + \sum_{i=j+1}^{l} \alpha_i A^i \tau^{2i-2})C,$$

$$C = (I - \sum_{i=1}^{j} \eta_i A^i \tau^{2i})^{-1}.$$

Actually, problem (5.78) can obviously be rewritten as the equivalent boundary-value problem

$$-\frac{u_{k+1} - 2u_k + u_{k-1}}{\tau^2} + A_l u_k = Cf_k^{j,l}, 1 \leq k \leq N-1, u_0 = v_0, u_N = v_T.$$

In a similar manner with the boundary-value problem (5.1) it can be rewritten as the equivalent boundary-value problem for first-order linear difference equations

$$\begin{cases} \tau^{-1}(u_k - u_{k-1}) + Bu_k = w_k, 1 \leq k \leq N, \ u_0 = v(0), u_N = v(T), \\ -\tau^{-1}(w_{k+1} - w_k) + Bw_k = (I + \tau B)Cf_k^{j,l}, 1 \leq k \leq N-1, \end{cases}$$

with operator $B = B(\tau, A)$. Note that $B(\tau, A) \neq A^{\frac{1}{2}}$, but then $B(\tau, A) \to A^{\frac{1}{2}}$ as $\tau \to 0$ and it has the same spectral properties of $A^{\frac{1}{2}}$ under some assumption for A

and for some cases of j and l. From the last system follows the system of recursion formulas

$$\begin{cases} u_k = R(\tau B)u_{k-1} + \tau R(\tau B)w_k, 1 \le k \le N, \\ w_k = R(\tau B)w_{k+1} + \tau C f_k^{j,l}, 1 \le k \le N-1, \end{cases}$$

where $R(\tau B) = (I + \tau B)^{-1}$. Hence

$$\begin{cases} u_k = R^k(\tau B)u_0 + \sum\limits_{i=1}^{k} R^{k-i+1}(\tau B)\tau w_i, 1 \le k \le N, \\ w_k = R^{N-k}(\tau B)w_N + \sum\limits_{s=k}^{N-1} R^{s-k}(\tau B)C f_s^{j,l}\tau, 1 \le k \le N-1. \end{cases}$$

From this it follows that

$$u_N = R^N(\tau B)u_0 + \sum_{i=1}^{N-1} R^{N+1-i}(\tau B)\tau w_i + \tau R(\tau B)w_N.$$

Since

$$\sum_{i=1}^{N-1} R^{N+1-i}(\tau B)\tau R^{N-i}(\tau B) + \tau R(\tau B) = \sum_{i=1}^{N} R^{2N-2i+1}(\tau B)\tau$$

$$= \tau(I - R^2(\tau B))^{-1}(I - R^{2N}(\tau B))R(\tau B),$$

$$\sum_{i=1}^{N-1} R^{N-i+1}(\tau B)\tau \sum_{s=i}^{N-1} R^{s-i}(\tau B)C f_s^{j,l}\tau = \sum_{s=1}^{N-1} \tau \sum_{i=1}^{s} R^{N+1+s-2i}(\tau B)C f_s^{j,l}\tau$$

$$= \sum_{s=1}^{N-1} \tau^2(I - R^2(\tau B))^{-1}R(\tau B)(R^{N-s}(\tau B) - R^{N+s}(\tau B))C f_s^{j,l},$$

we have that

$$w_N = \tau^{-1}(I - R^2(\tau B))(I - R^{2N}(\tau B))^{-1}R^{-1}(\tau B)\{u_N - R^N(\tau B)u_0$$

$$- \sum_{s=1}^{N-1} \tau^2(I - R^2(\tau B))^{-1}R(\tau B)(R^{N-s}(\tau B) - R^{N+s}(\tau B))C f_s^{j,l}\}$$

$$= \tau^{-1}(I - R^2(\tau B))(I - R^{2N}(\tau B))^{-1}R^{-1}(\tau B)\{u_N - R^N(\tau B)u_0\}$$

$$-(I - R^{2N}(\tau B))^{-1} \sum_{s=1}^{N-1} \tau(R^{N-s}(\tau B) - R^{N+s}(\tau B))C f_s^{j,l}.$$

Therefore

$$u_k = R^k(\tau B)u_0 + \sum_{i=1}^{k} R^{k-i+1}(\tau B)\tau(R^{N-i}(\tau B)w_N$$

$$+\sum_{s=i}^{N-1} R^{s-i}(\tau B)Cf_s^{j,l}\tau) = R^k(\tau B)u_0 + \sum_{i=1}^{k} R^{k-i+1}(\tau B)\tau R^{N-i}(\tau B)w_N$$

$$+\sum_{i=1}^{k} R^{k-i+1}(\tau B)\tau \sum_{s=i}^{N-1} R^{s-i}(\tau B)Cf_s^{j,l}\tau.$$

Since

$$\sum_{i=1}^{k} R^{k-i+1}(\tau B)\tau R^{N-i}(\tau B) = \sum_{i=1}^{k} R^{N+k-2i+1}(\tau B)\tau$$

$$= \tau(I - R^2(\tau B))^{-1}(R^{N-k}(\tau B) - R^{N+k}(\tau B))R(\tau B),$$

$$\sum_{i=1}^{k} R^{k-i+1}(\tau B)\tau \sum_{s=i}^{N-1} R^{s-i}(\tau B)Cf_s^{j,l}\tau$$

$$= \sum_{s=1}^{k} \tau \sum_{i=1}^{s} R^{k+s+1-i}(\tau B)Cf_s^{j,l}\tau + \sum_{s=k+1}^{N-1} \tau \sum_{i=1}^{k} R^{k+s-2i+1}(\tau B)Cf_s^{j,l}\tau$$

$$= (I - R^2(\tau B))^{-1} \sum_{s=1}^{k} \tau R(\tau B)(R^{k-s}(\tau B) - R^{k+s}(\tau B))Cf_s^{j,l}\tau$$

$$+(I - R^2(\tau B))^{-1} \sum_{s=k+1}^{N-1} \tau R(\tau B)(R^{s-k}(\tau B) - R^{k+s}(\tau B))Cf_s^{j,l}\tau$$

$$= (I - R^2(\tau B))^{-1} \sum_{s=1}^{N-1} \tau R(\tau B)(R^{|k-s|}(\tau B) - R^{k+s}(\tau B))Cf_s^{j,l}\tau,$$

we have that

$$u_k = R^k(\tau B)u_0 + \tau(I - R^2(\tau B))^{-1}(R^{N-k}(\tau B) - R^{N+k}(\tau B))R(\tau B)w_N$$

$$+(I - R^2(\tau B))^{-1} \sum_{s=1}^{N-1} \tau R(\tau B)(R^{|k-s|}(\tau B) - R^{k+s}(\tau B))Cf_s^{j,l}\tau.$$

From this formula and using the formula for w_N we can obtain formula (5.79).

Let us reduce the difference schemes (5.78) to an operator problem in the space $F_\tau(E)$. In addition to the operator D_τ^2, acting from the space $E \times F_\tau(E) \times E$ of vectors $w^\tau = \{w_k\}_{k=0}^N$ into the space $F_\tau(E)$ of vectors $v^\tau = \{v_k\}_{k=1}^{N-1}$ by the rule

$$v^\tau = D_\tau^2 u^\tau, v_k = \frac{1}{\tau^2}(w_{k+1} - 2w_k + w_{k-1}), k = 1, \ldots, N-1,$$

define an operator $A_\tau^{j,l}$ from the space $E \times F_\tau(E) \times E$ of vectors $w^\tau = \{w_k\}_{k=0}^N$ into the space $F_\tau(E)$ of vectors $v^\tau = \{v_k\}_{k=1}^{N-1}$ by the rule

$$v^\tau = A_\tau^{j,l} u^\tau, \; v_k = \sum_{i=1}^{j} \eta_i A^i \tau^{2i-2} (w_{k-1} + w_{k+1})$$

$$+ \sum_{i=1}^{l} \alpha_i A^i \tau^{2i-2} w_k, \; k = 1, \dots, N-1.$$

Then the difference schemes (5.78) can obviously be rewritten as the equivalent operator equation

$$-D_\tau^2 \Pi(u_0, u_N) u^\tau + A_\tau^{j,l} \Pi(u_0, u_N) u^\tau = f_{j,l}^\tau.$$

Here $f_{j,l}^\tau$ is defined by the formula

$$f_{j,l}^\tau = (f_1^{j,l}, \dots, f_{N-1}^{j,l}).$$

The last operator problem will be considered in the space $F_\tau(E)$. From its unique solvability for any u_0, $u_N \in E$ and $f_{j,l}^\tau \in F_\tau(E)$, it follows that its solution u^τ defines a continuous additive and homogeneous operator $u^\tau(f_{j,l}^\tau, u_0, u_N)$.

The boundary-value problem (5.78) is said to be stable in $F_\tau(E)$ if we have the inequality

$$\| u^\tau(f_{j,l}^\tau, u_0, u_N) \|_{F_\tau(E)} \leq M[\| f_{j,l}^\tau \|_{F_\tau(E)} + \| u_0 \|_E + \|u_N\|_E],$$

where M is independent not only of $f_{j,l}^\tau$, u_0, u_N, but also of τ.

The boundary-value difference problem (5.78) is said to be well posed (coercively stable) in $F_\tau(E)$ if we have the coercive inequality

$$\| \{\tau^{-2}(u_{k+1} - 2u_k + u_{k-1})\}_1^{N-1} \|_{F_\tau(E)}$$

$$+ \| \{\sum_{i=1}^{j} \eta_i A^i \tau^{2i-2} (u_{k-1} + u_{k+1}) + \sum_{i=1}^{l} \alpha_i A^i \tau^{2i-2} u_k\}_1^{N-1} \|_{F_\tau(E)}$$

$$\leq M[\| f_{j,l}^\tau \|_{F_\tau(E)} + \| Au_0 \|_E + \|Au_N\|_E],$$

where M is independent not only of $f_{j,l}^\tau$, u_0, u_N, but also of τ. From formula (5.58) it follows that the investigation of the stability and well-posedness of difference schemes (5.78) relies in an essential manner on a number of properties of the powers of the operator $(I + \tau B)^{-1}$. We were not able to obtain estimates for powers of the operator $(I + \tau B)^{-1}$ in the general cases of operator A and numbers j and l. We begin by deriving some estimates for powers of the operator $(I + \tau B)^{-1}$ in the cases $j = 0$ and all l, $l \geq 1$ and a strongly positive operator A in a Banach space E with spectral angle $\phi(A, E) < \frac{\pi}{2l}$.

Lemma 5.3.1. *Let A be a strongly positive operator in a Banach space E with spectral angle $\phi(A,E) < \frac{\pi}{2l}$. Then $-A_l$ is a generator of the analytic semigroup $\exp\{-tA_l\}$ $(t \geq 0)$ with exponentially decreasing norm, when $t \longrightarrow +\infty$, i.e., we have the estimates*

$$\|\exp\{-tA_l\}\|_{E\longrightarrow E} \leq M_l\ e^{-t\delta_l}\ (t > 0), \tag{5.80}$$

$$\|tA_l \exp\{-tA_l\}\|_{E\longrightarrow E} \leq M_l\ e^{-t\delta_l}(t > 0) \tag{5.81}$$

for some $1 \leq M_l < +\infty$, $0 < \delta_l < +\infty$. Here M_l does not depend on τ.

Proof. Let us show by induction that for any integer m, $2 \leq m \leq l$, A^m is a strongly positive operator in a Banach space E. Let $m = 2$ and $\lambda = |\lambda|e^{i\varphi}, \pi > |\varphi| \geq \frac{\pi}{2}$. We consider the identity

$$\lambda I - A^2 = (\lambda^{\frac{1}{2}} I - A)(\lambda^{\frac{1}{2}} I + A).$$

Since $\lambda^{\frac{1}{2}} = |\lambda|^{\frac{1}{2}} e^{i\frac{\varphi}{2}}, \frac{\pi}{2} > |\frac{\varphi}{2}| \geq \frac{\pi}{4}$, by the definition of a strongly positive operator in a Banach space E we have that the operator $\lambda^{\frac{1}{2}} I - A$ has a bounded inverse, and the estimate

$$\left\|(\lambda^{\frac{1}{2}} I - A)^{-1}\right\|_{E\to E} \leq M_A(1 + \sqrt{|\lambda|})^{-1}$$

holds for some $M_A \in [1, +\infty)$ that does not depend on λ. It is readily proved that $-(\lambda^{\frac{1}{2}} I + A)$ is a generator of the analytic semigroup $\exp\left\{-t(\lambda^{\frac{1}{2}} I + A)\right\}$ $(t \geq 0)$ with exponentially decreasing norm, when $t \longrightarrow +\infty$, i.e., we have the following estimates:

$$\left\|\exp\{-t(\lambda^{\frac{1}{2}} I + A)\}\right\|_{E\longrightarrow E} \leq M\ e^{-t(|\lambda|^{\frac{1}{2}} \cos\frac{\varphi}{2} + \delta)}\ (t \geq 0)$$

for some $1 \leq M < +\infty$, $0 < \delta < +\infty$. From this it follows that the operator $\lambda^{\frac{1}{2}} I + A$ has a bounded inverse, and the estimate

$$\left\|(\lambda^{\frac{1}{2}} I + A)^{-1}\right\|_{E\to E} \leq M(1 + |\lambda|^{\frac{1}{2}} \cos\frac{\varphi}{2})^{-1}$$

holds for some $M \in [1, +\infty)$ that does not depend on λ. Therefore, the operator $\lambda I - A^2$ has a bounded inverse, and the estimate

$$\|(\lambda I - A^2)^{-1}\|_{E\to E} \leq M_{A^2}(1 + |\lambda|)^{-1}$$

holds for some $M_{A^2} \in [1, +\infty)$ and for all λ, $\lambda = |\lambda|e^{i\varphi}, \pi > |\varphi| \geq \frac{\pi}{2}$. This means that A^2 is a strongly positive operator in a Banach space E. Suppose that for any integer k, $2 \leq k \leq m-1$, A^k is a strongly positive operator in a Banach space E. Then, we will prove that A^m is a strongly positive operator in a Banach space E. We consider the identity

$$\lambda I - A^m = (\lambda^{\frac{1}{m}} I - A)(A^{m-1} + A^{m-2}\lambda^{\frac{1}{m}} + \cdots + A\lambda^{\frac{m-2}{m}} + \lambda^{\frac{m-1}{m}} I).$$

We have that for any integer k, $1 \leq k \leq m$, A^{m-k} is a strongly positive operator in a Banach space E. Therefore, the operator $\lambda I - A^{m-k}$ has a bounded inverse, and the estimate

$$\left\|(zI - A^{m-k})^{-1}\right\|_{E \to E} \leq M_{A^{m-k}}(1+|z|)^{-1}$$

holds for some $M_{A^{m-k}} \in [1,+\infty)$ and for all z, $z = |z|e^{i\psi}, \pi > |\psi| \geq \frac{\pi}{2}$. It is readily proved that the operator $\lambda^{\frac{k-1}{m}} A^{m-k}$ has a bounded inverse, and the estimate

$$\left\|(zI - \lambda^{\frac{k-1}{m}} A^{m-k})^{-1}\right\|_{E \to E} \leq M_{A^{m-k}}(\delta+|z|)^{-1}$$

holds for some $M_{A^{m-k}} \in [1,+\infty)$, for all z, $z = |z|e^{i\psi}, \pi > |\psi| \geq \frac{\pi}{2}$ and $|\lambda| \geq \delta > 0$. Indeed,

$$\begin{aligned}\left\|(zI - \lambda^{\frac{k-1}{m}} A^{m-k})^{-1}\right\|_{E \to E} &= \left\|\lambda^{\frac{k-1}{m}}(z\lambda^{-\frac{k-1}{m}} I - A^{m-k})^{-1}\right\|_{E \to E} \\ &\leq M_{A^{m-k}}|\lambda|^{\frac{k-1}{m}}(1+|\lambda|^{-\frac{k-1}{m}}|z|)^{-1}\end{aligned}$$

$$= M_{A}{}^{m-k}|(|\lambda|^{\frac{k-1}{m}} + |z|)^{-1} \leq M_{A^{m-k}}(\delta+|z|)^{-1}.$$

From this it follows that $-\lambda^{\frac{k-1}{m}} A^{m-k}$ is a generator of the analytic semigroup $\exp\left\{-t\lambda^{\frac{k-1}{m}} A^{m-k}\right\}$ $(t \geq 0)$ with exponentially decreasing norm, when $t \longrightarrow +\infty$, i.e., we have the estimates

$$\left\|\exp\{-t\lambda^{\frac{k-1}{m}} A^{m-k}\}\right\|_{E \longrightarrow E} \leq \widetilde{M}\, e^{-t\tilde{\delta}} \ (t \geq 0)$$

for some $1 \leq \tilde{M} < +\infty$, $0 < \tilde{\delta} < +\infty$.

Since the operators $\lambda^{\frac{k-1}{m}} A^{m-k}$ commute, we have that $-\sum\limits_{k=1}^{m} \lambda^{\frac{k-1}{m}} A^{m-k}$ is a generator of the analytic semigroup $\exp\left\{-t\sum\limits_{k=1}^{m} \lambda^{\frac{k-1}{m}} A^{m-k}\right\}$ $(t \geq 0)$ with exponentially decreasing norm, when $t \longrightarrow +\infty$. Therefore, $\sum\limits_{k=1}^{m} \lambda^{\frac{k-1}{m}} A^{m-k}$ has a bounded inverse, and the estimate

$$||(\sum_{k=1}^{m} \lambda^{\frac{k-1}{m}} A^{m-k})^{-1}||_{E \to E} \leq \int_{o}^{\infty} ||e^{-t\sum\limits_{k=1}^{m} \lambda^{\frac{k-1}{m}} A^{m-k}}||_{E \to E} dt$$

$$\leq \int_{o}^{\infty} ||e^{-t\sum\limits_{k=1}^{m-1} \lambda^{\frac{k-1}{m}} A^{m-k}}||_{E \to E}|e^{-t\lambda^{\frac{m-1}{m}}}|dt \leq (\widetilde{M})^{m-1} e^{-(m-1)\tilde{\delta}}|\lambda|^{-\frac{m-1}{m}}.$$

holds. From this it follows that the operator $\lambda I - A^m$ has a bounded inverse, and the estimate

$$\|(\lambda I - A^m)^{-1}\|_{E\to E} \leq \left\|(\lambda^{\frac{1}{m}} I - A)^{-1}\right\|_{E\to E} \|(\sum_{k=1}^{m} \lambda^{\frac{k-1}{m}} A^{m-k})^{-1}\|_{E\to E}$$

$$\leq M_A(1+|\lambda|^{\frac{1}{m}})^{-1}(\widetilde{M})^{m-1}e^{-(m-1)\widetilde{\delta}}|\lambda|^{-\frac{m-1}{m}} = M_{A^m}(1+|\lambda|)^{-1}$$

holds for some $M_{A^m} \in [1,+\infty)$ and for all λ, $\lambda = |\lambda|e^{i\psi}, \pi > |\psi| \geq \frac{\pi}{2}$. This means that A^m is a strongly positive operator in a Banach space E. Finally, we will prove that $-A_l$ is a generator of the analytic semigroup $\exp\{-tA_l\}$ $(t \geq 0)$ with exponentially decreasing norm, when $t \longrightarrow +\infty$. The operators $\alpha_i A^i \tau^{2i-2}$ commute. Therefore, $-\sum_{i=1}^{l} \alpha_i A^i \tau^{2i-2}$ is a generator of the analytic semigroup $\exp\left\{-t\sum_{i=1}^{l} \alpha_i A^i \tau^{2i-2}\right\}$ $(t \geq 0)$ with exponentially decreasing norm, when $t \longrightarrow +\infty$. Indeed,

$$\|\exp\left\{-t\sum_{i=1}^{l}\alpha_i A^i \tau^{2i-2}\right\}\|_{E\to E} \leq \|exp\{-tA\}\|_{E\to E}$$

$$\leq \|exp\{-tA\}\|_{E\to E}\prod_{i=2}^{l}\|exp\{-t\alpha_i A^i\tau^{2i-2}\}\|_{E\to E}$$

$$\leq M_A exp\{-t\delta\}\prod_{i=2}^{l} M_{A^i} exp\left\{-t\alpha_i\widetilde{\delta}_i\tau^{2i-2}\right\} \leq M_{A_l} exp\{-t\delta_l\}.$$

Estimate (5.80) is established. In a similar manner one can obtain estimate (5.81). Lemma 5.3.1 is established.

Lemma 5.3.2. *Let $-A$ be a generator of the analytic semigroup $\exp\{-tA\}$ $(t \geq 0)$ with exponentially decreasing norm, when $t \longrightarrow +\infty$. Then the following estimates hold for any $k \geq 1$:*

$$\left\|(\lambda I + \tau B)^{-k}\right\|_{E\to E} \leq M[\lambda + \tau^2 a(A)]^{-k}, \tag{5.82}$$

$$\|k\tau B(I+\tau B)^{-k}\|_{E\to E} \leq M, \tag{5.83}$$

where M does not depend on τ. Here

$$B = B(\tau A) = \tau A/2 + \sqrt{(\frac{\tau A}{2})^2 + A}.$$

Proof. We consider the operator

$$B_1 = B_1(A_1) = \tau B(\tau A) = \tau^2 A/2 + \sqrt{(\frac{\tau^2 A}{2})^2 + \tau^2 A} = A_1/2 + \sqrt{\frac{A_1^2}{4} + A_1},$$

where $A_1 = \tau^2 A$. Using the following Cauchy–Riesz representation formula for the operator $(\lambda + B_1)^{-1}$ (see Chapter 4), we can write

$$(\lambda I + B_1)^{-1} = \frac{1}{2\pi i} \int\limits_{\Gamma} (\lambda + z/2 + \sqrt{\frac{z^2}{4} + z})^{-1} (zI - A_1)^{-1} dz.$$

Here the contour Γ is chosen with $\operatorname{Re}\{z\} \le 0$. If $z = \rho e^{\pm i\pi}$ and $\rho > 4$, then

$$(\lambda + z/2 + \sqrt{\frac{z^2}{4} + z})^{-1} = (\lambda - \rho/2 + \sqrt{\frac{\rho^2}{4} - \rho})^{-1}.$$

Therefore

$$\lim_{\in \to 0} [\frac{1}{2\pi i} \int\limits_{-\infty + i\in}^{-4 + i\in} (\lambda + z/2 + \sqrt{\frac{z^2}{4} + z})^{-1} (zI - A_1)^{-1} dz$$
$$+ \frac{1}{2\pi i} \int\limits_{-4 - i\in}^{-\infty - i\in} (\lambda + z/2 + \sqrt{\frac{z^2}{4} + z})^{-1} (zI - A_1)^{-1} dz\ ] = 0.$$

If $z = \rho e^{\pm i\pi}$ and $0 \le \rho \le 4$, then

$$(\lambda + z/2 + \sqrt{\frac{z^2}{4} + z})^{-1} = (\lambda - \rho/2 + \frac{1}{2}\sqrt{e^{\pm i\pi} \rho (4 - \rho)})^{-1}$$
$$= (\lambda - \rho/2 + \frac{1}{2} e^{\pm \frac{1}{2} i\pi} \sqrt{\rho(4 - \rho)})^{-1} = (\lambda - \rho/2 \pm \frac{1}{2} i \sqrt{\rho(4 - \rho)})^{-1}.$$

Therefore

$$\lim_{\in \to 0} [\frac{1}{2\pi i} \int\limits_{-4 + i\in}^{-0 + i\in} (\lambda + z/2 + \sqrt{\frac{z^2}{4} + z})^{-1} (zI - A_1)^{-1} dz$$
$$+ \frac{1}{2\pi i} \int\limits_{-0 - i\in}^{-4 - i\in} (\lambda + z/2 + \sqrt{\frac{z^2}{4} + z})^{-1} (zI - A_1)^{-1} dz\ ]$$
$$= -\frac{1}{2\pi i} \int\limits_{0}^{4} [(\lambda - \rho/2 - \frac{1}{2} i \sqrt{\rho(4 - \rho)})^{-1}$$
$$- (\lambda - \rho/2 + \frac{1}{2} i \sqrt{\rho(4 - \rho)})^{-1}] (-\rho I - A_1)^{-1} d\rho$$

$$= \frac{1}{2\pi i} \int_0^4 \frac{i\sqrt{\rho(4-\rho)}}{(\lambda-\rho/2)^2 + \frac{1}{4}\rho(4-\rho)} (\rho I + A_1)^{-1} d\rho$$

$$= \frac{1}{2\pi} \int_0^4 \frac{\sqrt{\rho(4-\rho)}}{\lambda^2 - \lambda\rho + \rho} (\rho I + A_1)^{-1} d\rho.$$

Then the deformation of integration contour Γ leads us to obtain the formula

$$(\lambda I + B_1)^{-1} = \int_0^4 M_1(\lambda, \rho)(\rho I + A_1)^{-1} d\rho, \tag{5.84}$$

where

$$M_1(\lambda, \rho) = \frac{1}{2\pi} \frac{\sqrt{\rho(4-\rho)}}{\lambda^2 - \lambda\rho + \rho}. \tag{5.85}$$

In a similar manner one can obtain

$$(\lambda I + B_1)^{-k} = \int_0^4 M_k(\lambda, \rho)(\rho I + A_1)^{-1} d\rho, k = 2, 3, \ldots, \tag{5.86}$$

where

$$M_k(\lambda, \rho) = \frac{1}{2\pi i}[(\lambda - \rho/2 - \frac{1}{2}i\sqrt{\rho(4-\rho)})^{-k} \tag{5.87}$$

$$-(\lambda - \rho/2 + \frac{1}{2}i\sqrt{\rho(4-\rho)})^{-k}]$$

$$= \frac{1}{2\pi} \frac{2\sum\limits_{m=1}^{[\frac{k+1}{2}]} \binom{k}{2m-1} (-1)^m (\lambda-\rho/2)^{k-2m+1} (\frac{1}{2}\sqrt{\rho(4-\rho)})^{2m-1}}{(\lambda^2 - \lambda\rho + \rho)^k}.$$

Clearly, $M_1(\lambda, \rho) \geq 0$ for $0 \leq \rho \leq 4$ and $\lambda \geq 0$. Therefore, using the estimate

$$||(\lambda I + A_1)^{-1}||_{E \to E} \leq M[\lambda + \tau^2 a(A)]^{-1}, \ \lambda \geq 0, \ 1 \leq M < +\infty, \ a(A) > 0,$$

we obtain that

$$||(\lambda I + B_1)^{-1}||_{E \to E} \leq \int_0^4 M_1(\lambda, \rho)||(\rho I + A_1)^{-1}||_{E \to E} d\rho$$

$$\leq M \int_0^4 M_1(\lambda, \rho)(\rho + a_1)^{-1} d\rho,$$

where $a_1 = \tau^2 a(A)$. Now, using formula (5.84) for $A_1 = a_1 I$, we obtain

$$\int_0^4 M_1(\lambda,\rho)(\rho + a(A_1))^{-1} d\rho = (\lambda + a_1/2 + \sqrt{\frac{(a_1)^2}{4} + a_1})^{-1}.$$

Therefore,

$$\|(\lambda I + B_1(A_1))^{-1}\|_{E\to E} \le M(\lambda + B_1(a_1))^{-1}, \tag{5.88}$$

where

$$B_1(a_1) = a_1/2 + \sqrt{\frac{(a_1)^2}{4} + a_1} > 0.$$

So, from estimate (5.88) it follows that $B_1(A_1)$ is a positive operator in E. Now, we will estimate $(\lambda I + B_1)^{-k}$ for $k \ge 2$. Clearly, the function $M_k(\lambda,\rho)$ for $k \ge 2$ changes sign on the segment $0 \le \rho \le 4$. Therefore, the last method of estimate of $(\lambda I + B_1(A_1))^{-1}$ does not work in the general cases $k \ge 2$. Now, we will use another formula for $(\lambda I + B_1)^{-k}$. Since $-A$ is a generator of the strongly continuous semigroup $\exp\{-tA\}$ $(t \ge 0)$ with exponentially decreasing norm, when $t \longrightarrow +\infty$, we have the estimate

$$\|\exp\{-tA\}\|_{E\longrightarrow E} \le M\, e^{-ta(A)}\ (t \ge 0)$$

for some $1 \le M < +\infty$, $0 < a(A) < +\infty$. Then, there exists a bounded inverse $(\lambda I + A_1)^{-1}$ for any complex number λ with $\mathrm{Re}\lambda > -a_1(A_1)$, and the following formula holds:

$$(\lambda I + A_1)^{-1} = \int_0^{+\infty} e^{-\lambda t} \exp\{-tA_1\} dt. \tag{5.89}$$

Then, from formulas (5.86) and (5.89) it follows that

$$\begin{aligned}(\lambda I + B_1)^{-k} &= \int_0^4 M_k(\lambda,\rho) \int_0^{+\infty} e^{-\rho t} \exp\{-tA_1\} dt d\rho \\ &= \int_0^{+\infty}\int_0^4 M_k(\lambda,\rho) e^{-\rho t} d\rho \exp\{-tA_1\} dt \\ &= \int_0^{+\infty} L_k(\lambda,t)\exp\{-tA_1\} dt,\end{aligned} \tag{5.90}$$

where

$$L_k(\lambda,t) = \int_0^4 M_k(\lambda,\rho) e^{-\rho t} d\rho. \tag{5.91}$$

Note if A_1 is a complex number with $Re\ A_1 > 0$, then $(\lambda I + B_1)^{-k}$ is the Laplace transform of the function $L_k(\lambda, t)$. Since $(\lambda I + B_1)^{-k-1} = (\lambda I + B_1)^{-k}(\lambda I + B_1)^{-1} = \prod_{m=1}^{k+1}(\lambda I + B_1)^{-1}$, from properties of the Laplace transform it follows that $L_k(\lambda, t)$ is a convolution of the k functions $L_1(\lambda, t)$ defined by the recurrence relation

$$L_{k+1}(\lambda, t) = \int\limits_0^t L_k(\lambda, s)L_1(\lambda, t-s)ds\, k = 1, 2, \dots. \tag{5.92}$$

Since $M_1(\lambda, \rho) \geq 0$, then from formula (5.91) it follows that $L_1(\lambda, \rho) \geq 0$. Therefore, from formula (5.92) it follows that

$$L_k(\lambda, \rho) \geq 0 \text{ for all } k, k = 1, 2, \dots. \tag{5.93}$$

Formula (5.90) and inequality (5.93) permit us to apply the same approach of estimate (5.88) for the estimate of $(\lambda I + B_1(A_1))^{-k}$ in the general cases $k \geq 2$. Namely, we have that

$$\|(\lambda I + B_1(A_1))^{-k}\|_{E\to E} \leq \int\limits_0^{+\infty} L_k(\lambda, t)\|\exp\{-tA_1\}\|_{E\to E}dt$$

$$\leq M \int\limits_0^{+\infty} L_k(\lambda, t)\exp\{-ta_1\}dt = M[\lambda + a_1]^{-k}$$

for $\lambda \geq 0$. Therefore, $-B_1(A_1)$ is a generator of the strongly continuous semigroup $\exp\{-tB_1(A_1)\}$ $(t \geq 0)$ with exponentially decreasing norm, when $t \longrightarrow +\infty$, i.e., we have the estimate

$$\|\exp\{-tB_1(A_1)\}\|_{E\longrightarrow E} \leq M\ e^{-tB_1(a_1)}\ (t \geq 0)$$

for some $1 \leq M < +\infty$. Now we consider $A_1^{\alpha}(\lambda I + B_1)^{-k}$ for $0 < \alpha < 1$. Applying the formula

$$A_1^{-\beta} = \frac{1}{\Gamma(\beta)} \int\limits_0^{+\infty} z^{\beta-1}\exp\{-zA_1\}dz, \beta > 0$$

and (5.90), we obtain

$$A_1^{\alpha}(\lambda I + B_1)^{-k}u = A_1^{-(1-\alpha)}(\lambda I + B_1)^{-k}A_1 u$$

$$= A_1^{-(1-\alpha)} \int\limits_0^{+\infty} L_k(\lambda, t)\exp\{-tA_1\}dtA_1 u$$

$$= \frac{1}{\Gamma(1-\alpha)} \int\limits_0^{+\infty} z^{-\alpha}\exp\{-zA_1\}dz \int\limits_0^{+\infty} L_k(\lambda, t)\exp\{-tA_1\}dtA_1 u$$

$$= \frac{1}{\Gamma(1-\alpha)} \int_0^{+\infty} \int_0^{+\infty} z^{-\alpha} L_k(\lambda, t) \exp\{-(t+z)A_1\} dz dt A_1 u$$

for $u \in D(A_1)$. Using the last formula and the estimate

$$(t+z)||A_1 \exp\{-(t+z)A_1\}||_{E \longrightarrow E} \leq M\ e^{-(t+z)a_1},$$

we obtain that

$$\begin{aligned} ||A_1^\alpha (\lambda I + B_1)^{-k}||_{E \longrightarrow E} &\leq \frac{M}{\Gamma(1-\alpha)} \int_0^{+\infty} \int_0^{+\infty} z^{-\alpha} L_k(\lambda, t) \frac{1}{t+z} \\ &\quad \times \exp\{-(t+z)a_1\} dz dt \\ &= M\Gamma(\alpha) \int_0^{+\infty} L_k(\lambda, t) \exp\{-t a_1\} dt \\ &\quad \times \frac{1}{\Gamma(\alpha)\Gamma(1-\alpha)} \int_0^{+\infty} z^{-\alpha} (t+z)^{-1} dz. \end{aligned}$$

Since

$$t^{-\alpha} = \frac{1}{\Gamma(\alpha)\Gamma(1-\alpha)} \int_0^{+\infty} z^{-\alpha} (t+z)^{-1} dz, \tag{5.94}$$

we have that

$$||A_1^\alpha (\lambda I + B_1)^{-k}||_{E \longrightarrow E} \leq M\Gamma(\alpha) \int_0^{+\infty} L_k(\lambda, t) t^{-\alpha} \exp\{-t a_1\} dt.$$

Using formula (5.91), we obtain

$$\begin{aligned} &||A_1^\alpha (\lambda I + B_1)^{-k}||_{E \longrightarrow E} \\ &\leq M\Gamma(\alpha) \int_0^{+\infty} \int_0^{4} M_k(\lambda, \rho) e^{-\rho t} d\rho\, t^{-\alpha} \exp\{-t a_1\} dt \\ &= M\Gamma(\alpha)\Gamma(1-\alpha) \int_0^{4} M_k(\lambda, \rho) [\frac{1}{\Gamma(1-\alpha)} \int_0^{+\infty} e^{-\rho t} t^{-\alpha} \\ &\quad \times \exp\{-t a_1\} dt] d\rho \\ &= M\Gamma(\alpha)\Gamma(1-\alpha) \int_0^{4} M_k(\lambda, \rho) (\rho + a_1)^{\alpha - 1} d\rho. \end{aligned}$$

Using formula (5.94), we obtain

$$||A_1^\alpha(\lambda I+B_1)^{-k}||_{E\longrightarrow E}\le M\int_0^4 M_k(\lambda,\rho)\int_0^{+\infty} z^{-(1-\alpha)}(z+\rho+a_1)^{-1}dzd\rho$$

$$=M\int_0^{+\infty} z^{-(1-\alpha)}\int_0^4 M_k(\lambda,\rho)(z+\rho+a_1)^{-1}d\rho dz$$

$$=M\int_0^{+\infty} z^{-(1-\alpha)}(\lambda+B_1(z+a_1))^{-k}dz.$$

Finally, by passing to the limit $\alpha\to 1$ in the last inequality one can obtain the estimate

$$||A_1(\lambda I+B_1)^{-k}||_{E\longrightarrow E}\le M\int_0^{+\infty}(\lambda+B_1(z+a_1))^{-k}dz.$$

Since

$$B_1(z+a_1)=(z+a_1)/2+\sqrt{\frac{(z+a_1)^2}{4}+z+a_1}$$

and $a_1>0$, it follows that

$$||A_1(I+B_1)^{-k}||_{E\longrightarrow E}\le M\int_0^{+\infty}(1+z/2+\sqrt{\frac{z^2}{4}+z})^{-k}dz.$$

Let us make the substitution

$$x=z/2+\sqrt{\frac{z^2}{4}+z}.$$

Clearly,

$$x^2-zx-z=0.$$

Therefore, $z=\frac{x^2}{x+1}$ and $dz=\frac{x^2+2x}{(x+1)^2}dx$. From this it follows that

$$||A_1(I+B_1)^{-k}||_{E\longrightarrow E}\le M\int_0^{+\infty}(1+x)^{-(k+2)}(x^2+2x)dx.$$

Integrating the last integral by parts, we obtain

$$\int_0^{+\infty}(1+x)^{-(k+2)}(x^2+2x)dx=-\frac{(1+x)^{-(k+1)}}{k+1}(x^2+2x)|_0^\infty$$

$$+\frac{1}{k+1}\int_0^{+\infty}(1+x)^{-(k+1)}(2x+2)dx$$

$$= -\frac{(1+x)^{-k}}{(k+1)k}(2x+2)|_0^\infty + \frac{1}{(k+1)k}\int\limits_0^{+\infty}(1+x)^{-k}2dx$$

$$= \frac{2}{(k+1)k} - \frac{(1+x)^{-k+1}}{(k+1)k(k-1)}2|_0^\infty$$

$$= \frac{2}{(k+1)k} + \frac{2}{(k+1)k(k-1)} = \frac{2[k-1+1]}{(k^2-1)k} = \frac{2}{(k^2-1)}$$

for $k \geq 2$. Therefore,

$$||A_1(I+B_1)^{-k}||_{E\longrightarrow E} \leq \frac{2M}{k^2-1}$$

for $k \geq 2$. Using the identity

$$A_1 = B_1^2(I+B_1)^{-1},$$

we obtain

$$||B_1^2(I+B_1)^{-(k+1)}||_{E\longrightarrow E} \leq \frac{2M}{k^2-1}$$

for $k \geq 2$. The estimate

$$||(I+B_1)^{-(k+1)}||_{E\longrightarrow E} \leq M$$

for all k, $k \geq 1$ follows from estimate (5.82). Since $B_1(A_1)$ is a positive operator in E, we have the moment inequality

$$||B_1^2 u||_E \leq M||B_1^2 u||_E^{\frac{1}{2}}||u||_E^{\frac{1}{2}}(u \ \in D(B_1^2)).$$

From this it follows that

$$||kB_1(I+B_1)^{-k}||_{E\longrightarrow E} \leq M(k \geq 1).$$

Lemma 5.3.2 is established.

We have the following results.

Theorem 5.3.3. *Let A be a strongly positive operator in a Banach space E with spectral angle $\phi(A,E) < \frac{\pi}{2l}$. Then difference problem (5.78) in the case $j=0$ is stable in $C_\tau^{\alpha,\alpha}(E), (0 \leq \alpha \leq 1)$ and in $L_{p,\tau}(E)$, $1 \leq p \leq \infty$ (here $L_{\infty,\tau}(E) = C_\tau(E)$), since the solutions of the difference problem (5.78) satisfy the stability inequalities*

$$\| u^\tau \|_{C_\tau^{\alpha,\alpha}(E)} \leq M[\| f_{j,l}^\tau \|_{C_\tau^{\alpha,\alpha}(E)} + \| u_0\|_E + ||u_N||_E],$$

$$\| u^\tau \|_{L_{p,\tau}(E)} \leq M[\| f_{j,l}^\tau \|_{L_{p,\tau}(E)} + \| u_0\|_E + ||u_N||_E],$$

where M does not depend on $f_{j,l}^\tau$, u_0, u_N, p, α and τ.

Theorem 5.3.4. *Let A be a strongly positive operator in a Banach space E with spectral angle $\phi(A,E) < \frac{\pi}{2l}$ and u_0, $u_N \in D(A)$. Then the solutions of the difference problem (5.78) for $j=0$ in $L_{p,\tau}(E)$ obey the almost coercive inequality*

$$\| \{\tau^{-2}(u_{k+1}-2u_k+u_{k-1})\}_1^{N-1} \|_{L_{p,\tau}(E)} + \| \{\sum_{i=1}^{l} \alpha_i A^i \tau^{2i-2} u_k\}_1^{N-1} \|_{L_{p,\tau}(E)}$$

$$\leq M_1[\| Au_0\|_E + \|Au_N\|_E$$

$$+ \min\left\{\ln\frac{1}{\tau}, 1+|\ln \| B \|_{E\to E}|\right\} \| f_{j,l}^{\tau} \|_{L_{p,\tau}(E)}],$$

where M_1 is independent not only of $f_{j,l}^{\tau}$, u_0, u_N, but also of τ.

In the case $j=0$ we have the estimate

$$\|AA_l^{-1}\|_{E\to E} \leq M.$$

Therefore, we have the following results.

Theorem 5.3.5. *Let A be a strongly positive operator in a Banach space E with spectral angle $\phi(A,E) < \frac{\pi}{2l}$ and u_0, $u_N \in D(A)$. Then the solutions of the difference problem (5.78) for $j=0$ in $L_{p,\tau}(E)$ obey the almost coercive inequality*

$$\| \{\tau^{-2}(u_{k+1}-2u_k+u_{k-1})\}_1^{N-1} \|_{L_{p,\tau}(E)} + \| \{Au_k\}_1^{N-1} \|_{L_{p,\tau}(E)}$$

$$\leq M_1[\| Au_0\|_E + \|Au_N\|_E$$

$$+ \min\left\{\ln\frac{1}{\tau}, 1+|\ln \| B \|_{E\to E}|\right\} \| f_{j,l}^{\tau} \|_{L_{p,\tau}(E)}],$$

where M_1 is independent not only of $f_{j,l}^{\tau}$, u_0, u_N, but also of τ.

Theorem 5.3.6. *Let A be a strongly positive operator in a Banach space E with spectral angle $\phi(A,E) < \frac{\pi}{2l}$ and u_0, $u_N \in D(A)$. Then the solutions of the difference problem (5.78) for $j=0$ in $C_{\tau}^{\alpha,\alpha}(E)$ obey the coercivity inequality*

$$\| \{\tau^{-2}(u_{k+1}-2u_k+u_{k-1})\}_1^{N-1} \|_{C_{\tau}^{\alpha,\alpha}(E)} + \| \{Au_k\}_1^{N-1} \|_{C_{\tau}^{\alpha,\alpha}(E)}$$

$$\leq M_1[\frac{1}{\alpha(1-\alpha)}\| f_{j,l}^{\tau} \|_{C_{\tau}^{\alpha,\alpha}(E)} + \| Au_0\|_E + \|Au_N\|_E],$$

where M_1 is independent not only of $f_{j,l}^{\tau}$, u_0, u_N, α, but also of τ.

Theorem 5.3.7. *Let A be a strongly positive operator in a Banach space E with spectral angle $\phi(A,E) < \frac{\pi}{2l}$ and $(I-R(\tau B))^2\tau^{-2}u_0 - f_1^{j,l}$, $(I-R(\tau B))^2\tau^{-2}u_N - f_{N-1}^{j,l} \in E'_{\alpha}$. Then the solutions of the difference problem (5.78) for $j=0$ in $C_{\tau}^{\alpha}(E)$ obey the coercivity inequality*

$$\| \{\tau^{-2}(u_{k+1}-2u_k+u_{k-1})\}_1^{N-1} \|_{C_{\tau}^{\alpha}(E)} + \| \{Au_k\}_1^{N-1} \|_{C_{\tau}^{\alpha}(E)}$$

$$+ \| \{\tau^{-2}(u_{k+1}-2u_k+u_{k-1})\}_1^{N-1} \|_{C_{\tau}(E'_{\alpha})}$$

$$\leq \frac{M_1}{\alpha(1-\alpha)}[\| f_{j,l}^{\tau} \|_{C_\tau^\alpha(E)} + \| (I - R(\tau B))^2\tau^{-2}u_0 - f_1^{j,l} \|_{E'_\alpha}$$
$$+\|(I - R(\tau B))^2\tau^{-2}u_N - f_{N-1}^{j,l}\|_{E'_\alpha}],$$

where M_1 is independent not only of $f_{j,l}^{\tau}$, u_0, u_N, α, but also of τ. Here, the Banach space $E'_\alpha = E'_\alpha(E,B)$ $(0 < \alpha < 1)$ consists of those $v \in E$ for which the norm

$$\| v \|_{E'_\alpha} = \sup_{z>0} z^\alpha \| B(zI + B)^{-1}v \|_E + \| v \|_E$$

is finite.

Theorem 5.3.8. *Let A be a strongly positive operator in a Banach space E with spectral angle $\phi(A,E) < \frac{\pi}{2l}$ and $(I - R(\tau B))^2\tau^{-2}u_0 - f_1^{j,l}$, $(I - R(\tau B))^2\tau^{-2}u_N - f_{N-1}^{j,l} \in E'_{\beta-\gamma}$. Then the solutions of the difference problem (5.78) for $j = 0$ in $C_\tau^{\beta,\gamma}(E)$ $(0 \leq \gamma \leq \beta, 0 < \beta < 1)$ obey the coercivity inequalities*

$$\| \{\tau^{-2}(u_{k+1} - 2u_k + u_{k-1})\}_1^{N-1} \|_{C_\tau(E'_{\beta-\gamma})} \leq \frac{M_1}{\beta(1-\beta)}\| f_{j,l}^{\tau} \|_{C_\tau^{\beta,\gamma}(E)}$$
$$+M_1[\| (I - R(\tau B))^2\tau^{-2}u_0 - f_1^{j,l} \|_{E'_{\beta-\gamma}} + \|(I - R(\tau B))^2\tau^{-2}u_N - f_{N-1}^{j,l}\|_{E'_{\beta-\gamma}}],$$

$$\| \{\tau^{-2}(u_{k+1} - 2u_k + u_{k-1})\}_1^{N-1} \|_{C_\tau^{\beta,\gamma}(E)} + \| \{Au_k\}_1^{N-1} \|_{C_\tau^{\beta,\gamma}(E)}$$
$$\leq \frac{M_1}{\beta(1-\beta)}\| f_{j,l}^{\tau} \|_{C_\tau^{\beta,\gamma}(E)} + M_1[|(I - R(\tau B))^2\tau^{-2}u_0 - f_1^{j,l}|_0^{\beta,\gamma}$$
$$+|(I - R(\tau B))^2\tau^{-2}u_N - f_{N-1}^{j,l}|_0^{\beta,\gamma}],$$

where M_1 is independent not only of $f_{j,l}^{\tau}$, u_0, u_N, β, γ, but also of τ. Here, $|w|_0^{\beta,\gamma}$ denotes that the norm of the Banach space $\widetilde{E}_0^{\beta,\gamma}$ consists of those $w \in E$ for which the norm

$$|w|_0^{\beta,\gamma} = \max_{1\leq n\leq N-1} \|R^n(\tau B)w\|_E$$
$$+ \sup_{1\leq n<n+r\leq N-1} t_r^{-\beta}(t_n + t_r)^\gamma(t_N - t_n)^\gamma\|(R^{n+r}(\tau B) - R^n(\tau B))w\|_E$$

is finite.

Theorem 5.3.9. *Let A be a strongly positive operator in a Banach space E with spectral angle $\phi(A,E) < \frac{\pi}{2l}$. Then the solutions of the difference problem (5.78) for $j = 0$ in $C_\tau(E'_\alpha)$ obey the coercivity inequality*

$$\| \{\tau^{-2}(u_{k+1} - 2u_k + u_{k-1})\}_1^{N-1} \|_{C_\tau(E'_\alpha)} + \| \{Au_k\}_1^{N-1} \|_{C_\tau(E'_\alpha)}$$
$$\leq M_1[\frac{1}{\alpha(1-\alpha)}\| f^\tau \|_{C_\tau(E'_\alpha)} + \| Au_0\|_{E'_\alpha} + \|Au_N\|_{E'_\alpha}],$$

where M_1 is independent not only of f^τ, u_0, u_N, α, but also of τ.

Theorem 5.3.10. *Let A be a strongly positive operator in a Banach space E with spectral angle $\phi(A,E) < \frac{\pi}{2l}$ and $(I-R(\tau B))^2\tau^{-2}u_0 - f_1^{j,l}$, $(I-R(\tau B))^2\tau^{-2}u_N - f_{N-1}^{j,l} \in E'_{\alpha-\gamma}$. Then the solutions of the difference problem (5.78) for $j=0$ in $C_\tau^{\beta,\gamma}(E'_{\alpha-\beta})$ $(0 \le \gamma \le \beta \le \alpha, 0<\alpha<1)$ obey the coercivity inequalities*

$$\| \{\tau^{-2}(u_{k+1}-2u_k+u_{k-1})\}_1^{N-1} \|_{C_\tau^{\beta,\gamma}(E'_{\alpha-\beta})} + \| \{Au_k\}_1^{N-1} \|_{C_\tau^{\beta,\gamma}(E'_{\alpha-\beta})}$$

$$+ \| \{\tau^{-2}(u_{k+1}-2u_k+u_{k-1})\}_1^{N-1} \|_{C_\tau(\widetilde{E}^{\beta,\gamma}_{\alpha-\beta})} \le \frac{M_1}{\alpha(1-\alpha)} \| f^\tau \|_{C_\tau^{\beta,\gamma}(E'_{\alpha-\beta})}$$

$$+M_1[|(I-R(\tau B))^2\tau^{-2}u_0 - f_1^{j,l}|^{\beta,\gamma}_{\alpha-\beta} + |(I-R(\tau B))^2\tau^{-2}u_N - f_{N-1}^{j,l}|^{\beta,\gamma}_{\alpha-\beta}],$$

$$\| \{\tau^{-2}(u_{k+1}-2u_k+u_{k-1})\}_1^{N-1} \|_{C_\tau^{\beta,\gamma}(E'_{\alpha-\beta})} + \| \{Au_k\}_1^{N-1} \|_{C_\tau^{\beta,\gamma}(E'_{\alpha-\beta})}$$

$$+ \| \{\tau^{-2}(u_{k+1}-2u_k+u_{k-1})\}_1^{N-1} \|_{C_\tau(E'_{\alpha-\gamma})}$$

$$\le \frac{M_1}{\alpha(1-\alpha)}[\| (I+\tau B)^{-1} C f_{j,l}^\tau \|_{C_\tau^{\beta,\gamma}(E'_{\alpha-\beta})}$$

$$+M_1[\|(I-R(\tau B))^2\tau^{-2}u_0 - f_1^{j,l}\|_{E'_{\alpha-\gamma}} + \|(I-R(\tau B))^2\tau^{-2}u_N - f_{N-1}^{j,l}\|_{E'_{\alpha-\gamma}}],$$

where M_1 is independent not only of $f_{j,l}^\tau$, u_0, u_N, α, β, γ, but also of τ.

Here, $|w|^{\beta,\gamma}_{\alpha-\beta}$ denotes that the norm of the Banach space $\widetilde{E}^{\beta,\gamma}_{\alpha-\beta}$ consists of those $v \in E$ for which the norm

$$|w|^{\beta,\gamma}_{\alpha-\beta} = \max_{1\le n\le N-1} \|R^n(\tau B)w\|_{E_{\alpha-\beta}}$$

$$+ \sup_{1\le n<n+r\le N-1} t_r^{-\beta}(t_n+t_r)^\gamma(t_N-t_n)^\gamma \|(R^{n+r}(\tau B)-R^n(\tau B))w\|_{E_{\alpha-\beta}}$$

is finite.

Finally, let us give the following results about well-posedness of the boundary-value difference problem (5.78) in the spaces $L_{p,\tau}(E) = L_p([0,T]_\tau, E), 1 \le p < \infty$.

Theorem 5.3.11. *Let A be a strongly positive operator in a Banach space E with spectral angle $\phi(A,E) < \frac{\pi}{2l}$. Suppose that the difference problem (5.78) for $j=0$ is well posed in $L_{p_0,\tau}(E)$ for some p_0, $1<p_0<\infty$. Then it is well posed in $L_{p,\tau}(E)$ for any p, $1<p<\infty$ and the following coercivity inequality holds:*

$$\| \{\tau^{-2}(u_{k+1}-2u_k+u_{k-1})\}_1^{N-1} \|_{L_{p,\tau}(E)} + \| \{Au_k\}_1^{N-1} \|_{L_{p,\tau}(E)}$$

$$+\|u^\tau\|_{C_c(E_{1-1/p,p})} \le M(p_0)[\| u_0\|_{E_{1-1/p,p}} + \|u_N\|_{E_{1-1/p,p}}$$

$$+\frac{p^2}{p-1} \| f_{j,l}^\tau \|_{L_{p,\tau}(E)}],$$

where $M(p_0)$ does not depend on $f_{j,l}^\tau$, u_0, u_N, p and τ.

Theorem 5.3.12. *Let $1 \leq p \leq \infty$ and $0 < \alpha < 1$. Suppose that A is a strongly positive operator in a Banach space E with spectral angle $\phi(A, E) < \frac{\pi}{2l}$. Then the difference problem (5.78) for $j = 0$ is well posed in $L_{p,\tau}(E_{\alpha,p})$ and the following coercivity inequality holds:*

$$\| \{\tau^{-2}(u_{k+1} - 2u_k + u_{k-1})\}_1^{N-1} \|_{L_{p,\tau}(E_{\alpha,p})} + \| \{Au_k\}_1^{N-1} \|_{L_{p,\tau}(E_{\alpha,p})}$$
$$\leq M[\| Au_0\|_{E_{\alpha,p}} + \|Au_N\|_{E_{\alpha,p}} + \frac{1}{\alpha(1-\alpha)} \| f_{j,l}^{\tau} \|_{L_{p,\tau}(E_{\alpha,p})}],$$

where M does not depend on $f_{j,l}^{\tau}$, u_0, u_N, p, α and τ.

Theorem 5.3.13. *Let $1 < p, q < \infty$ and $0 < \alpha < 1$. Suppose that A is a strongly positive operator in a Banach space E with spectral angle $\phi(A, E) < \frac{\pi}{2l}$. Then the difference problem (5.78) for $j = 0$ is well posed in $L_{p,\tau}(E_{\alpha,q})$ and the following coercivity inequality holds:*

$$\| \{\tau^{-2}(u_{k+1} - 2u_k + u_{k-1})\}_1^{N-1} \|_{L_{p,\tau}(E_{\alpha,q})} + \| \{Au_k\}_1^{N-1} \|_{L_{p,\tau}(E_{\alpha,q})}$$
$$\leq M(q)[\| Au_0\|_{E_{\alpha,q}} + \|Au_N\|_{E_{\alpha,q}} + \frac{p^2}{p-1} \| f_{j,l}^{\tau} \|_{L_{p,\tau}(E_{\alpha,q})}],$$

where $M(q)$ does not depend on $f_{j,l}^{\tau}$, u_0, u_N, p, α and τ.

Now, the abstract theorems given above are applied in the investigation of difference schemes of higher order of accuracy with respect to the set of all variables for approximate solution of the boundary-value problem (5.78). The discretization of problem (5.27) is carried out in two steps. In the first step let us give the difference operator A_h^x by the formula

$$A_h^x u_x^h = \sum_{2m \leq |r| \leq S} b_r^x D_h^r u_x^h + \delta u_x^h. \tag{5.95}$$

The coefficients are chosen in such a way that the operator A_h^x approximates in a specified way the operator

$$\sum_{|r|=2m} a_r(x) \frac{\partial^{|r|}}{\partial x_1^{r_1} \cdots \partial x_n^{r_n}} + \delta.$$

We shall assume that for $|\xi_k h| \leq \pi$ the symbol $A(\xi h, h)$ of the operator $A_h^x - \delta$ satisfies the inequalities

$$(-1)^m A^x(\xi h, h) \geq M_1 |\xi|^{2m}, |\arg A^x(\xi h, h)| \leq \phi < \phi_0 \leq \frac{\pi}{2l}. \tag{5.96}$$

With the help of A_h^x we arrive at the boundary-value problem

$$-\frac{d^2 v^h(y,x)}{dy^2} + A_h^x v^h(y,x) = f^h(y,x), 0 < y < T, \tag{5.97}$$
$$v^h(0,x) = \varphi^h(x), v^h(T,x) = \psi^h(x), x \in \mathbb{R}_h^n,$$

for an infinite system of ordinary differential equations.

In the second step we replace problem (5.78) by the difference scheme

$$-\frac{1}{\tau^2}(u_{k+1}^h(x) - 2u_k^h(x) + u_{k-1}^h(x)) + \sum_{i=1}^{l} \alpha_i (A_h^x)^i \tau^{2i-2} u_k^h(x) \tag{5.98}$$

$$= f_k^{j,l}(x), 1 \le k \le N-1, u_0^h(x) = v_0^h(x),\ u_N^h = v_T^h,\ x \in R_h^n.$$

Let us give a number of corollaries of the abstract theorems given above.

Theorem 5.3.14. *The solutions of the difference schemes* (5.98) *satisfy the following stability estimates:*

$$\| u^{\tau,h} \|_{C_\tau^{\alpha,\alpha}(C_h^\beta)} \le M[||u_0^h||_{C_h^\beta} + ||u_N^h||_{C_h^\beta} + \| f_{j,l}^{\tau,h} \|_{C_\tau^{\alpha,\alpha}(C_h^\beta)}],$$
$$0 < \alpha < 1, 0 \le \beta < 1,$$

$$\| u^{\tau,h} \|_{L_{p,\tau}(C_h^\beta)} \le M[||u_0^h||_{C_h^\beta} + ||u_N^h||_{C_h^\beta} + \| f_{j,l}^{\tau,h} \|_{L_{p,\tau}(C_h^\beta)}],$$
$$1 \le p \le \infty, 0 \le \beta < 1,$$

where M does not depend on $f_{j,l}^\tau$, u_0^h, u_N^h, α, β, p, h and τ.

The proof of Theorem 5.3.14 is based on the abstract Theorem 5.3.3 and the positivity of the operator A_h^x in C_h^β.

Theorem 5.3.15. *The solutions of the difference schemes* (5.98) *satisfy the following almost coercive stability estimates:*

$$\| \{\tau^{-2}(u_{k+1}^h - 2u_k^h + u_{k-1}^h)\}_1^{N-1} \|_{Lp,\tau(C_h)} + \sum_{|r|=S} ||\{D_h^r u_k^h\}_1^{N-1}||_{Lp,\tau(C_h)}$$

$$\le M[\ln \frac{1}{h}[\sum_{|r|=2m} ||D_h^r u_0^h||_{C_h} + \sum_{|r|=2m} ||D_h^r u_N^h||_{C_h}]$$

$$+ \ln \frac{1}{\tau + h} \| f_{j,l}^{\tau,h} \|_{L_{p,\tau}(C_h)}], 1 \le p \le \infty,$$

where M does not depend on $f_{j,l}^\tau$, u_0^h, u_N^h, p, h and τ.

The proof of Theorem 5.3.15 is based on the abstract Theorem 5.3.5, the positivity of the operator A_h^x in C_h and on the almost coercivity inequality for an elliptic operator A_h^x in C_h^β.

Theorem 5.3.16. *The solutions of the difference schemes* (5.98) *satisfy the coercivity estimates*

$$\| \{\tau^{-2}(u_{k+1}^h - 2u_k^h + u_{k-1}^h)\}_1^{N-1} \|_{C_\tau^{\alpha,\alpha}(C_h^\beta)} + \sum_{|r|=2m} ||\{D_h^r u_k^h\}_1^{N-1}||_{C_\tau^{\alpha,\alpha}(C_h^\beta)}$$

$$\leq M(\alpha,\beta)[\sum_{|r|=2m} ||D_h^r u_0^h||_{C_h^\beta} + \sum_{|r|=2m} ||D_h^r u_N^h||_{C_h^\beta} + \| f_{j,l}^{\tau,h} \|_{C_\tau^{\alpha,\alpha}(C_h^\beta)}],$$

$$0 \leq \alpha < 1, 0 < \beta < 1,$$

where $M(\alpha,\beta)$ *does not depend on* $f_{j,l}^{\tau}$, u_0^h, u_N^h, h *and* τ

The proof of Theorem 5.3.16 is based on the abstract Theorem 5.3.6, the positivity of the operator A_h^x in C_h^β and on the coercivity inequality for an elliptic operator A_h^x in C_h^β. Note that in a similar manner we can construct the difference schemes of a high order of accuracy generated by Taylor's decomposition of the function on three points with respect to one variable for approximate solutions of the boundary-value problems (5.25) and (5.26). Abstract theorems given above permit us to obtain the stability, the almost stability and the coercive stability estimates for the solutions of these difference schemes.

Chapter 6

Partial Differential Equations of Hyperbolic Type

In the present chapter we consider the abstract Cauchy problem for differential equations of the hyperbolic type

$$v''(t) + Av(t) = f(t) \quad (0 \le t \le T), v(0) = v_0, v'(0) = v_0'$$

in an arbitrary Hilbert space H with the self-adjoint positive definite operator A. The high order of accuracy two-step difference schemes generated by an exact difference scheme or by Taylor's decomposition on three points for the numerical solutions of this problem are presented. Stability estimates for the solutions of these difference schemes are established. In applications, the stability estimates for solutions of a high order of accuracy difference schemes of the mixed type boundary-value problems for hyperbolic equations are obtained.

6.1 A Cauchy Problem

We consider the abstract Cauchy problem for hyperbolic equations

$$v''(t) + Av(t) = f(t) \quad (0 \le t \le T), v(0) = v_0, v'(0) = v_0' \tag{6.1}$$

in a Hilbert space H with the self-adjoint positive definite operator A.

A function $v(t)$ is called a solution of problem (6.1) if the following conditions are satisfied:

i. $v(t)$ is twice continuously differentiable on the segment $[0, T]$. The derivatives at the endpoints of the segment are understood as the appropriate unilateral derivatives.
ii. The element $v(t)$ belongs to $D(A)$ for all $t \in [0, T]$ and the function $Av(t)$ is continuous on the segment $[0, T]$.
iii. $v(t)$ satisfies the equations and initial conditions (6.1). If the function $f(t)$ is not only continuous, but also continuously differentiable on $[0, T], v_0 \in D(A)$

and $v_0' \in D(A^{\frac{1}{2}})$. It is easy to show that the formula

$$v(t) = c(t)v_0 + s(t)v_0' + \int_0^t s(t-\lambda)f(\lambda)d\lambda \tag{6.2}$$

gives a solution of problem (6.1). Here

$$c(t) = \frac{e^{itA^{1/2}} + e^{-itA^{1/2}}}{2}, s(t) = A^{-1/2}\frac{e^{itA^{1/2}} - e^{-itA^{1/2}}}{2i}.$$

Actually, obviously (6.1) can be rewritten as the equivalent initial-value problem for a system of first-order linear differential equations

$$\begin{cases} v'(t) + iBv(t) = z(t) \ (0 \le t \le T), v(0) = v_0, v'(0) = v_0', \\ z'(t) - iBz(t) = f(t), \end{cases}$$

where $B = A^{\frac{1}{2}}$. Integrating these, we at once obtain

$$\begin{cases} v(t) = e^{-iBt}v_0 + \int_0^t e^{-iB(t-s)}z(s)ds, \\ z(t) = e^{iBt}z_0 + \int_0^t e^{iB(t-p)}f(p)dp. \end{cases}$$

That is,

$$v(t) = e^{-iBt}v_0 + \int_0^t e^{-iB(t-s)}e^{iBs}z_0ds$$
$$+ \int_0^t e^{-iB(t-s)} \int_0^s e^{iB(s-p)}f(p)dpds,$$

so, using the condition $v'(0) + iBv(0) = z(0)$, we obtain

$$v(t) = (e^{-iBt} + iB\int_0^t e^{-iB(t-s)}e^{iBs}ds)v_0$$
$$+ \int_0^t e^{-iB(t-s)}e^{iBs}dsv'(0) + \int_0^t e^{-iB(t-s)} \int_0^s e^{iB(s-p)}f(p)dpds.$$

By an interchange of the order of integration, we obtain

$$v(t) = (e^{-iBt} + \frac{e^{iBt} - e^{-iBt}}{2})v_0$$
$$+B^{-1}\frac{e^{iBt} - e^{-iBt}}{2i}dsv'(0) + \int_0^t B^{-1}\frac{e^{iB(t-s)} - e^{-iB(t-s)}}{2i}f(s)ds.$$

Thus, by the definitions of B, $c(t)$ and $s(t)$ we have (6.2).

Theorem 6.1.1. *Suppose that $v_0 \in D(A)$, $v_0' \in D(A^{\frac{1}{2}})$ and $f(t)$ is a continuously differentiable function on $[0,T]$. Then there is a unique solution of problem (6.1) and the stability inequalities*

$$\max_{0\le t\le T} \| v(t) \|_H \le M \left[\| v_0 \|_H + \| A^{-1/2} v_0' \|_H + \max_{0\le t\le T} \| A^{-1/2} f(t) \|_H \right],$$

$$\max_{0\le t\le T} \| A^{1/2} v(t) \|_H \le M \left[\| A^{1/2} v_0 \|_H + \| v_0' \|_H + \max_{0\le t\le T} \| f(t) \|_H \right],$$

$$\max_{0\le t\le T} \| \frac{d^2 v(t)}{dt^2} \|_H + \max_{0\le t\le T} \| Av(t) \|_H \le M[\| Av_0 \|_H + \| A^{1/2} v_0' \|_H + \| f(0) \|_H + \int_0^T \| f'(t) \|_H \, dt]$$

hold, where M does not depend on $f(t)$, $t \in [0,T]$ and v_0, v_0'.

Proof. Using formula (6.2) and estimates

$$\|c(t)\|_{H\to H} \le 1, \|A^{\frac{1}{2}} s(t)\|_{H\to H} \le 1, \tag{6.3}$$

we obtain

$$\|v(t)\|_H \le \|c(t)\|_{H\to H} \| v_0 \|_H + \|A^{\frac{1}{2}} s(t)\|_{H\to H} \| A^{-\frac{1}{2}} v_0' \|_H + \int_0^t \|A^{\frac{1}{2}} s(t-\lambda)\|_{H\to H} \|A^{-\frac{1}{2}} f(\lambda)\|_H d\lambda$$
$$\le \| v_0 \|_H + \| A^{-\frac{1}{2}} v_0' \|_H + T \max_{0\le t\le T} \| A^{-1/2} f(t) \|_H .$$

Applying $A^{\frac{1}{2}}$ to the formula (6.2) and using estimates (6.3), in a similar manner, we obtain

$$\|A^{\frac{1}{2}} v(t)\|_H \le \|c(t)\|_{H\to H} \| A^{\frac{1}{2}} v_0 \|_H + \|A^{\frac{1}{2}} s(t)\|_{H\to H} \| v_0' \|_H + \int_0^t \|A^{\frac{1}{2}} s(t-\lambda)\|_{H\to H} \|f(\lambda)\|_H d\lambda$$
$$\le \| A^{\frac{1}{2}} v_0 \|_H + \| v_0' \|_H + T \max_{0\le t\le T} \| f(t) \|_H .$$

Now, we obtain an estimate for $\| Av(t) \|_H$. Applying A to formula (6.2) and using an integration by parts, we can write the formula

$$Av(t) = c(t)Av_0 + A^{\frac{1}{2}} s(t) A^{\frac{1}{2}} v_0' + f(T) - c(t)f(0) - \int_0^t c(t-\lambda) f'(\lambda) d\lambda.$$

Using the last formula and estimates (6.3), we obtain

$$\|Av(t)\|_H \leq \|c(t)\|_{H\to H} \parallel Av_0 \parallel_H + \|A^{\frac{1}{2}}s(t)\|_{H\to H} \parallel A^{\frac{1}{2}}v_0' \parallel_H$$

$$+ \parallel f(T) \parallel_H + \parallel c(t)\|_{H\to H}\|f(0) \parallel_H + \int_0^t \|c(t-\lambda)\|_{H\to H}\|f'(\lambda)\|_H d\lambda$$

$$\leq \parallel Av_0 \parallel_H + \parallel A^{\frac{1}{2}}v_0' \parallel_H + 2\|f(0) \parallel_H + 2T \max_{0\leq t\leq T} \parallel f'(t) \parallel_H .$$

Then from the last estimate, it follows that

$$\max_{0\leq t\leq T} \parallel Au(t) \parallel_H \leq \parallel Av_0 \parallel_H + \parallel A^{\frac{1}{2}}v_0' \parallel_H + 2\|f(0) \parallel_H + 2T \max_{0\leq t\leq T} \parallel f'(t) \parallel_H .$$

The estimate for $\max_{0\leq t\leq T} \parallel \frac{d^2v(t)}{dt^2} \parallel_H$ follows from the last estimate and the triangle inequality. Theorem 6.1.1 is proved.

Remark 6.1.2. Theorem 6.1.1 holds in an arbitrary Banach space E under the assumptions

$$\|c(t)\|_{E\to E} \leq M, \ \|A^{\frac{1}{2}}s(t)\|_{E\to E} \leq M, \ 0 \leq t \leq T. \tag{6.4}$$

Now, we consider the application of this abstract Theorem 6.1.1. First, we consider the mixed problem for hyperbolic equations

$$\begin{cases} u_{tt}(t,x) - (a(x)u_x)_x + u(t,x) = f(t,x), 0 \leq t \leq T, \ 0 \leq x \leq L, \\ u(0,x) = \varphi(x), u_t(0,x) = \psi(x), 0 \leq x \leq L, \\ u(t,0) = u(t,L), u_x(t,0) = u_x(t,L), 0 \leq t \leq T. \end{cases} \tag{6.5}$$

Problem (6.5) has a unique smooth solution $u(t,x)$ for smooth $a(x) > 0$ ($x \in [0,L]$), $\varphi(x)$, $\psi(x)$ ($x \in [0,L]$) and $f(t,x)$ ($t \in [0,T]$, $x \in [0,L]$) functions. This allows us to reduce the mixed problem (6.5) to the initial-value problem (6.1) in a Hilbert space $H = L_2[0,L]$ with a self-adjoint positive definite operator A defined by (6.5). Let us give a number of corollaries of the abstract Theorem 6.1.1.

Theorem 6.1.3. *For solutions of the mixed problem* (6.5) *the stability inequalities*

$$\max_{0\leq t\leq T} \parallel u(t,\cdot) \parallel_{W_2^1[0,L]} \leq M[\max_{0\leq t\leq T} \parallel f(t,\cdot) \parallel_{L_2[0,L]} + \parallel \varphi \parallel_{W_2^1[0,L]}$$

$$+ \parallel \psi \parallel_{L_2[0,L]}],$$

$$\max_{0\leq t\leq T} \parallel u(t,\cdot) \parallel_{W_2^2[0,L]} + \max_{0\leq t\leq T} \parallel u_{tt}(t,\cdot) \parallel_{L_2[0,L]}$$

$$\leq M\left[\max_{0\leq t\leq T} \parallel f_t(t,\cdot) \parallel_{L_2[0,L]} + \parallel f(0,\cdot) \parallel_{L_2[0,L]} + \parallel \varphi \parallel_{W_2^2[0,L]} + \parallel \psi \parallel_{W_2^1[0,L]}\right]$$

hold, where M does not depend on $f(t,x)$ and $\varphi(x)$, $\psi(x)$.

The proof of this theorem is based on the abstract Theorem 6.1.1 and the symmetry properties of the space operator generated by problem (6.5).

Second, let $\Omega \subset \mathbb{R}^n$ be a bounded open domain with smooth boundary S, $\overline{\Omega} = \Omega \cup S$. In $[0,T] \times \Omega$ we consider the mixed boundary-value problem for hyperbolic equations

$$u_{tt}(t,x) - \sum_{r=1}^{n}(a_r(x)u_{x_r})_{x_r} = f(t,x), x = (x_1,\dots,x_n) \in \overline{\Omega}, 0 \le t \le T, \qquad (6.6)$$

$$u(0,x) = \varphi(x), \frac{\partial u(0,x)}{\partial t} = \psi(x), x \in \overline{\Omega},$$

$$u(t,x) = 0, x \in S, 0 \le t \le T,$$

where $a_r(x), (x \in \overline{\Omega})$, $\varphi(x)$, $\psi(x)$ $(x \in \overline{\Omega})$ and $f(t,x)$ $(t \in [0,T],\ x \in \overline{\Omega})$ are given smooth functions and $a_r(x) > 0$. We introduce the Hilbert space $L_2(\overline{\Omega})$, the space of all integrable functions defined on $\overline{\Omega}$, equipped with the norm

$$\| f \|_{L_2(\overline{\Omega})} = \{ \int \cdots \int_{x\in\overline{\Omega}} |f(x)|^2 dx_1 \cdots dx_n \}^{\frac{1}{2}}.$$

Problem (6.6) has a unique smooth solution $u(t,x)$ for smooth $a_r(x) > 0$ and $f(t,x)$ functions. This allows us to reduce the mixed problem (6.6) to the initial-value problem (6.1) in Hilbert space $H = L_2(\overline{\Omega})$ with a self-adjoint positive definite operator A defined by (6.6). Let us give a number of corollaries of the abstract Theorem 6.1.1.

Theorem 6.1.4. *For solutions of the mixed problem* (6.6) *the stability inequalities*

$$\max_{0\le t\le T} \| u(t,\cdot) \|_{W_2^1(\overline{\Omega})} \le M \left[\max_{0\le t\le T} \| f(t,\cdot) \|_{L_2(\overline{\Omega})} + \| \varphi \|_{W_2^1(\overline{\Omega})} + \| \psi \|_{L_2(\overline{\Omega})} \right],$$

$$\max_{0\le t\le T} \| u(t,\cdot) \|_{W_2^2(\overline{\Omega})} + \max_{0\le t\le T} \| u_{tt}(t,\cdot) \|_{L_2(\overline{\Omega})}$$

$$\le M \left[\max_{0\le t\le T} \| f_t(t,\cdot) \|_{L_2(\overline{\Omega})} + \| f(0,\cdot) \|_{L_2(\overline{\Omega})} + \| \varphi \|_{W_2^2(\overline{\Omega})} + \| \psi \|_{W_2^1(\overline{\Omega})} \right]$$

hold, where M does not depend on $f(t,x)$ and $\varphi(x)$, $\psi(x)$.

The proof of this theorem is based on the abstract Theorem 6.1.1 and the symmetry properties of the space operator generated by problem (6.6).

6.2 Difference Schemes Generated by an Exact Difference Scheme

We consider the initial-value problem (6.1). On the segment $[0,T]$ we consider a uniform grid

$$[0,T]_\tau = \{t_k = k\tau, k = 0,1,\dots,N, N\tau = T\}$$

with step τ. By Theorem 3.1.1 we have the following two-step difference scheme for the solution of the initial-value problem (6.1):

$$\frac{1}{\tau}s(\tau)v(t_{k+1})-\frac{2}{\tau}s(\tau)c(\tau)v(t_k)+\frac{1}{\tau}s(\tau)v(t_{k-1})$$

$$=\frac{1}{\tau}s(\tau)\{\int_{t_k}^{t_{k+1}}s(t_{k+1}-z)f(z)dz+\int_{t_{k-1}}^{t_k}s(z-t_{k-1})f(z)dz\},$$

$$1\leq k\leq N-1, v(0)=v_0,$$

$$\tau^{-1}(v(\tau)-v(0))=\tau^{-1}(c(\tau)-I)v(0)+\tau^{-1}s(\tau)v'(0)+\tau^{-1}\int_0^\tau s(\tau-z)f(z)dz$$

or

$$\frac{1}{\tau^2}(u_{k+1}-2u_k+u_{k-1})=\frac{2}{\tau^2}(c(\tau)-I)u_k+f_k, \tag{6.7}$$

$$f_k=\tau^{-1}\{f_{1,k+1}+s(\tau)f_{2,k}-c(\tau)f_{1,k}\},$$

$$f_{1,k}=\tau^{-1}\int_{t_{k-1}}^{t_k}s(t_k-z)f(z)dz, f_{2,k}=\tau^{-1}\int_{t_{k-1}}^{t_k}c(t_k-z)f(z)dz, 1\leq k\leq N-1,$$

$$u_0=v_0, u_1=c(\tau)v(0)+s(\tau)v'(0)+\tau f_{1,1}.$$

The latter will be referred to as the exact two-step difference scheme for the initial-value problem (6.1).

Suppose the operators $(I-e^{2\tau iB})$ and $(I+e^{\tau iB})$ have the bounded inverses $(I-e^{2\tau iB})^{-1}$ and $(I+e^{\tau iB})^{-1}$. Then this problem is uniquely solvable and the following formula holds:

$$u_k=[e^{-k\tau iB}-(I-e^{2\tau iB})^{-1}(e^{-k\tau iB}-e^{k\tau iB})]u_0 \tag{6.8}$$

$$+e^{\tau iB}(I-e^{2\tau iB})^{-1}(e^{-k\tau iB}-e^{k\tau iB})u_1$$

$$+\tau^2\sum_{m=1}^{k-1}e^{\tau iB}(I-e^{2\tau iB})^{-1}(e^{-(k-m)\tau iB}-e^{(k-m)\tau iB})f_m, 2\leq k\leq N.$$

Actually, (6.7) can be rewritten as the equivalent initial-value problem for a system of first-order difference equations

$$\begin{cases}\tau^{-1}(u_k-u_{k-1})+\tau^{-1}(I-e^{-\tau iB})u_{k-1}=z_k, 1\leq k\leq N,\ u_0,\ u_1 \text{ are given,}\\ \tau^{-1}(z_{k+1}-z_k)+\tau^{-1}(I-e^{\tau iB})z_k=f_k, 1\leq k\leq N-1.\end{cases}$$

From this follows the system of recursion formulas

$$\begin{cases}u_k=e^{-\tau iB}u_{k-1}+\tau z_k, 1\leq k\leq N,\\ z_{k+1}=e^{\tau iB}z_k+\tau f_k, 1\leq k\leq N-1.\end{cases}$$

Hence

$$\begin{cases} u_k = e^{-k\tau iB}u_0 + \sum\limits_{s=1}^{k} e^{-(k-s)\tau iB}\tau z_s, 1 \le k \le N, \\ z_k = e^{(k-1)i\tau B}z_1 + \sum\limits_{m=1}^{k-1} e^{(k-1-m)\tau iB}\tau f_m, 2 \le k \le N. \end{cases}$$

From this it follows that

$$u_1 = e^{-\tau iB}u_0 + \tau z_1,$$

$$u_k = e^{-k\tau iB}u_0 + e^{-(k-1)\tau iB}\tau z_1$$

$$+\sum_{s=2}^{k} e^{-(k-s)\tau iB}\tau[e^{(s-1)i\tau B}z_1 + \sum_{m=1}^{s-1} e^{(s-1-m)\tau iB}\tau f_m]$$

$$= e^{-k\tau iB}u_0 + [e^{-(k-1)\tau iB} + \sum_{s=2}^{k} e^{-(k-s)\tau iB}e^{(s-1)i\tau B}]\tau z_1$$

$$+\sum_{s=2}^{k} e^{-(k-s)\tau iB}\sum_{m=1}^{s-1} e^{(s-1-m)\tau iB}\tau^2 f_m, 2 \le k \le N.$$

Since

$$e^{-(k-1)\tau iB} + \sum_{s=2}^{k} e^{-(k-s)\tau iB}e^{(s-1)i\tau B} = (I - e^{2\tau iB})^{-1}(e^{-(k-1)\tau iB} - e^{(k+1)\tau iB})$$

$$= (I - e^{2\tau iB})^{-1}e^{\tau iB}(e^{-k\tau iB} - e^{k\tau iB})$$

and

$$\sum_{s=2}^{k} e^{-(k-s)\tau iB}\sum_{m=1}^{s-1} e^{(s-1-m)\tau iB}\tau^2 f_m = \sum_{m=1}^{k-1}\sum_{s=m+1}^{k} e^{-(k-s)\tau iB}e^{(s-1-m)\tau iB}\tau^2 f_m$$

$$= \sum_{m=1}^{k-1}(I - e^{2\tau iB})^{-1}e^{\tau iB}(e^{-(k-m)\tau iB} - e^{(k-m)\tau iB})\tau^2 f_m,$$

we have that

$$u_k = e^{-k\tau iB}u_0 + (I - e^{2\tau iB})^{-1}e^{\tau iB}(e^{-k\tau iB} - e^{k\tau iB})\tau z_1$$

$$+\sum_{m=1}^{k-1}(I - e^{2\tau iB})^{-1}e^{\tau iB}(e^{-(k-m)\tau iB} - e^{(k-m)\tau iB})\tau^2 f_m, 2 \le k \le N.$$

Using this formula and the formula $\tau z_1 = u_1 - e^{-\tau iB}u_0$, we can obtain formula (6.8).

Let us investigate the stability of the exact two-step difference scheme (6.7).

Theorem 6.2.1. *Suppose the operators* $(I - e^{2\tau iB})$ *and* $(I + e^{\tau iB})$ *have the bounded inverses* $(I - e^{2\tau iB})^{-1}$ *and* $(I + e^{\tau iB})^{-1}$. *Suppose further that* $u_0 \in D(B^2(I + e^{\tau iB})^{-1})$, $u_1 - u_0 \in D(B^2(I - e^{2\tau iB})^{-1})$ *and* $f_k \in D(B(I - e^{2\tau iB})^{-1}), 1 \leq k \leq N-1$. *Then for the solution of the exact two-step difference scheme* (6.7) *the following stability inequalities hold:*

$$\max_{0\leq k\leq N}\|u_k\|_H \leq M[\max_{1\leq k\leq N-1}\|\tau(I - e^{2\tau iB})^{-1}f_k\|_H \tag{6.9}$$

$$+ \| (I - e^{2\tau iB})^{-1}(u_1 - u_0) \|_H + \| (I + e^{\tau iB})^{-1}u_0 \|_H],$$

$$\max_{1\leq k\leq N}\|\tau^{-1}(u_k - u_{k-1})\|_H + \max_{1\leq k\leq N}\|\tau^{-1}(I - e^{-\tau iB})u_{k-1}\|_H \tag{6.10}$$

$$\leq M[\max_{1\leq k\leq N-1}\|\tau B(I - e^{2\tau iB})^{-1}f_k\|_H$$

$$+ \| B(I - e^{2\tau iB})^{-1}(u_1 - u_0) \|_H + \| B(I + e^{\tau iB})^{-1}u_0 \|_H],$$

$$\max_{1\leq k\leq N-1}\|\tau^{-2}(u_{k+1} - 2u_k + u_{k-1})\|_H + \max_{1\leq k\leq N-1}\|\frac{2}{\tau^2}(c(\tau) - I)u_k\|_H \tag{6.11}$$

$$\leq M[\max_{2\leq k\leq N-1}\| B(I - e^{2\tau iB})^{-1}(f_k - f_{k-1}) \|_H + \| \tau B(I - e^{2\tau iB})^{-1}f_1 \|_H$$

$$+ \| B^2(I - e^{2\tau iB})^{-1}(u_1 - u_0) \|_H + \| B^2(I + e^{\tau iB})^{-1}u_0 \|_H],$$

where M *does not depend on* τ, f_k, $1 \leq k \leq N-1$ *and* u_0, u_1.

Proof. Using formula (6.8) and the estimates

$$\|e^{\pm k\tau iB}\|_{H\to H} \leq 1,\ 1 \leq k \leq N, \tag{6.12}$$

we obtain

$$\|u_k\|_H \leq \|(I + e^{\tau iB})e^{-k\tau iB} - (e^{-k\tau iB} - e^{k\tau iB})\|_{H\to H} \| (I + e^{\tau iB})^{-1}u_0 \|_H$$

$$+\|e^{-k\tau iB} - e^{k\tau iB}\|_{H\to H} \| (I - e^{2\tau iB})^{-1}(u_1 - u_0) \|_H$$

$$+\tau^2\sum_{m=1}^{k-1}\|e^{\tau iB}(e^{-(k-m)\tau iB} - e^{(k-m)\tau iB})\|_{H\to H}\|(I - e^{2\tau iB})^{-1}f_m\|_H$$

$$\leq M[\max_{1\leq m\leq N-1}\|\tau(I - e^{2\tau iB})^{-1}f_m\|_H$$

$$+ \| (I - e^{2\tau iB})^{-1}(u_1 - u_0) \|_H + \| (I + e^{\tau iB})^{-1}u_0 \|_H]$$

for $k \geq 2$. It obviously holds also for $k = 0, 1$. The estimate (6.9) is established. Applying $\tau^{-1}(I - e^{-\tau iB})$ to formula (6.8) and using estimates (6.12) and

$$\|\tau^{-1}(I - e^{-\tau iB})B^{-1}\|_{H\to H} \leq 1, \tag{6.13}$$

in a similar manner as (6.9), we obtain

$$\|\tau^{-1}(I-e^{-\tau iB})u_{k-1}\|_H \le \|\tau^{-1}(I-e^{-\tau iB})B^{-1}\|_{H\to H}$$
$$\times\{\|(I+e^{\tau iB})e^{-(k-1)\tau iB}-(e^{-(k-1)\tau iB}-e^{(k-1)\tau iB})\|_{H\to H}\parallel B(I+e^{\tau iB})^{-1}u_0\parallel_H$$
$$+\|e^{-(k-1)\tau iB}-e^{(k-1)\tau iB}\|_{H\to H}\parallel B(I-e^{2\tau iB})^{-1}(u_1-u_0)\parallel_H$$
$$+\tau\sum_{m=1}^{k-2}\|e^{\tau iB}(e^{-(k-1-m)\tau iB}-e^{(k-1-m)\tau iB})\|_{H\to H}\|\tau B(I-e^{2\tau iB})^{-1}f_m\|_H\}$$
$$\le M[\max_{1\le m\le N-1}\|B\tau(I-e^{2\tau iB})^{-1}f_m\|_H$$
$$+\parallel B(I-e^{2\tau iB})^{-1}(u_1-u_0)\parallel_H+\parallel B(I+e^{\tau iB})^{-1}u_0\parallel_H]$$

for $k\ge 3$. It obviously holds also for $k=1,2$. So, we have that

$$\max_{1\le k\le N}\|\tau^{-1}(I-e^{-\tau iB})u_{k-1}\|_H$$
$$\le M[\max_{1\le k\le N-1}\|B\tau(I-e^{2\tau iB})^{-1}f_k\|_H$$
$$+\parallel B(I-e^{2\tau iB})^{-1}(u_1-u_0)\parallel_H+\parallel B(I+e^{\tau iB})^{-1}u_0\parallel_H].$$

In a similar manner one can show that

$$\max_{1\le k\le N}\|\tau^{-1}(u_k-u_{k-1})\|_H$$
$$\le M[\max_{1\le k\le N-1}\|B\tau(I-e^{2\tau iB})^{-1}f_k\|_H$$
$$+\parallel B(I-e^{2\tau iB})^{-1}(u_1-u_0)\parallel_H+\parallel B(I+e^{\tau iB})^{-1}u_0\parallel_H].$$

Estimate (6.10) is established. Now, we obtain the estimate for $\parallel \frac{2}{\tau^2}(c(\tau)-I)u_k\parallel_H$. Applying $\frac{2}{\tau^2}(c(\tau)-I)=\frac{1}{\tau^2}e^{-\tau iB}(I-e^{\tau iB})^2$ to formula (6.8) and using Abel's formula, we can write the formula

$$\frac{2}{\tau^2}(c(\tau)-I)u_k=\frac{1}{\tau^2}e^{-\tau iB}(I-e^{\tau iB})^2[e^{-k\tau iB} \tag{6.14}$$
$$-(I+e^{\tau iB})^{-1}(e^{-k\tau iB}-e^{k\tau iB})]u_0$$
$$+\frac{1}{\tau^2}(I-e^{\tau iB})^2(I-e^{2\tau iB})^{-1}(e^{-k\tau iB}-e^{k\tau iB})(u_1-u_0)$$
$$+\frac{1}{\tau}(I-e^{\tau iB})\sum_{m=1}^{k-1}\tau(I-e^{\tau iB})(I-e^{2\tau iB})^{-1}(e^{-(k-m)\tau iB}-e^{(k-m)\tau iB})f_m$$
$$=\frac{1}{\tau^2}e^{-\tau iB}(I-e^{\tau iB})^2[e^{-k\tau iB}-(I+e^{\tau iB})^{-1}(e^{-k\tau iB}-e^{k\tau iB})]u_0$$

$$+\frac{1}{\tau^2}(I-e^{\tau iB})^2(I-e^{2\tau iB})^{-1}(e^{-k\tau iB}-e^{k\tau iB})(u_1-u_0)$$

$$+\frac{1}{\tau}(I-e^{\tau iB})\sum_{m=1}^{k-1}\tau(I-e^{2\tau iB})^{-1}(e^{-(k-m)\tau iB}+e^{(k+1-m)\tau iB})(f_m-f_{m-1})$$

$$+\frac{1}{\tau}(I-e^{\tau iB})(e^{-(k-1)\tau iB}+e^{k\tau iB})(I-e^{2\tau iB})^{-1}\tau f_1$$

$$-\frac{1}{\tau}(I-e^{\tau iB})(I+e^{\tau iB})(I-e^{2\tau iB})^{-1}\tau f_{k-1}.$$

Using the last formula and estimates (6.12) and (6.13), we obtain

$$\frac{2}{\tau^2}||(c(\tau)-I)u_k||_H\le||\frac{1}{\tau}(I-e^{\tau iB})B^{-1}||^2_{H\to H}$$

$$\times||e^{-\tau iB}||_{H\to H}||(I+e^{\tau iB})e^{-k\tau iB}-(e^{-k\tau iB}-e^{k\tau iB})||_{H\to H}$$

$$\times||B^2(I+e^{\tau iB})^{-1}u_0||_H+||\frac{1}{\tau}(I-e^{\tau iB})B^{-1}||^2_{H\to H}||e^{-k\tau iB}-e^{k\tau iB}||_{H\to H}$$

$$\times||B^2(I-e^{2\tau iB})^{-1}(u_1-u_0)||_H+||\frac{1}{\tau}(I-e^{\tau iB})B^{-1}||_{H\to H}$$

$$\times\sum_{m=1}^{k-1}\tau||(e^{-(k-m)\tau iB}+e^{(k+1-m)\tau iB})||_{H\to H}||B(I-e^{2\tau iB})^{-1}(f_m-f_{m-1})||_H$$

$$+||\frac{1}{\tau}(I-e^{\tau iB})B^{-1}||_{H\to H}||e^{-(k-1)\tau iB}+e^{k\tau iB}||_{H\to H}||\tau B(I-e^{2\tau iB})^{-1}f_1||_H$$

$$+||\frac{1}{\tau}(I-e^{\tau iB})B^{-1}||_{H\to H}||I+e^{\tau iB}||_{H\to H}||\tau B(I-e^{2\tau iB})^{-1}f_{k-1}||_H$$

$$\le M[\max_{2\le m\le N-1}\|B(I-e^{2\tau iB})^{-1}(f_m-f_{m-1})\|_H+\|\tau B(I-e^{2\tau iB})^{-1}f_1\|_H$$

$$+\|B^2(I-e^{2\tau iB})^{-1}(u_1-u_0)\|_H+\|B^2(I+e^{\tau iB})^{-1}u_0\|_H]$$

for $k\ge 2$. It obviously holds also for $k=1$. So, we have that

$$\max_{1\le k\le N-1}\|\frac{2}{\tau^2}(c(\tau)-I)u_k\|_H$$

$$\le M[\max_{2\le k\le N-1}\|\tau B(I-e^{2\tau iB})^{-1}(f_k-f_{k-1})\|_H+\|\tau B(I-e^{2\tau iB})^{-1}f_1\|_H$$

$$+\|B^2(I-e^{2\tau iB})^{-1}(u_1-u_0)\|_H+\|B^2(I+e^{\tau iB})^{-1}u_0\|_H].$$

The estimate for $\max\limits_{1\le k\le N-1}\|\tau^{-2}(u_{k+1}-2u_k+u_{k-1})\|_H$ follows from the last estimate and the triangle inequality. Theorem 6.2.1 is proved.

Note that we have not been able to obtain the stability inequalities

$$\max_{0\le k\le N}\|u_k\|_H\le M[\max_{1\le k\le N-1}\|B^{-1}f_k\|_H+\|B^{-1}\tau^{-1}(u_1-u_0)\|_H+\|u_0\|_H],$$

$$\max_{1\le k\le N}\|\tau^{-1}(u_k-u_{k-1})\|_H+\max_{1\le k\le N}\|\tau^{-1}(I-e^{-\tau iB})u_{k-1}\|_H$$

$$\leq M[\max_{1\leq k\leq N-1} \| f_m \|_H + \| Bu_0 \|_H + \| \tau^{-1}(u_1 - u_0) \|_H],$$

$$\max_{1\leq k\leq N-1} \|\tau^{-2}(u_{k+1} - 2u_k + u_{k-1})\|_H + \max_{1\leq k\leq N-1} \|\frac{2}{\tau^2}(c(\tau) - I)u_k\|_H$$

$$\leq M[\max_{2\leq k\leq N-1} \| \tau^{-1}(f_k - f_{k-1}) \|_H + \| f_1 \|_H$$

$$+ \| B\tau^{-1}(u_1 - u_0) \|_H + \| B^2 u_0 \|_H]$$

for the solution of the exact two-step difference scheme (6.7). Nevertheless, the following result holds.

Theorem 6.2.2. *Suppose the operators $(I - e^{2\tau iB})$ and $(I + e^{\tau iB})$ have the bounded inverses $(I - e^{2\tau iB})^{-1}$ and $(I + e^{\tau iB})^{-1}$. Suppose that $u_0 \in D(B^3)$, $u_1 - u_0 \in D(B^2)$ and $f_k \in D(B)$, $1 \leq k \leq N-1$. Then for the solution of the exact two-step difference scheme (6.7) the following stability inequalities hold:*

$$\max_{0\leq k\leq N} \|u_k\|_H \leq M[\max_{1\leq k\leq N-1} \|f_k\|_H + \| \tau^{-1}(u_1 - u_0) \|_H + \| Bu_0 \|_H], \tag{6.15}$$

$$\max_{1\leq k\leq N} \|\tau^{-1}(u_k - u_{k-1})\|_H + \max_{1\leq k\leq N} \|\tau^{-1}(I - e^{-\tau iB})u_{k-1}\|_H \tag{6.16}$$

$$\leq M[\max_{1\leq k\leq N-1} \| Bf_k \|_H + \| B^2 u_0 \|_H + \| B\tau^{-1}(u_1 - u_0) \|_H],$$

$$\max_{1\leq k\leq N-1} \|\tau^{-2}(u_{k+1} - 2u_k + u_{k-1})\|_H + \max_{1\leq k\leq N-1} \|\frac{2}{\tau^2}(c(\tau) - I)u_k\|_H \tag{6.17}$$

$$\leq M[\max_{2\leq k\leq N-1} \| B\tau^{-1}(f_k - f_{k-1}) \|_H + \| Bf_1 \|_H$$

$$+ \| B^2\tau^{-1}(u_1 - u_0) \|_H + \| B^3 u_0 \|_H]$$

where M does not depend on τ, f_k, $1 \leq k \leq N-1$ and u_0, u_1.

Proof. Using formula (6.8) and estimates (6.12), (6.13) and

$$\|(I - e^{2\tau iB})^{-1}(e^{-k\tau iB} - e^{k\tau iB})\|_{H\to H} \leq k, \tag{6.18}$$

we obtain

$$\|u_k\|_H \leq \|e^{-k\tau iB}\|_{H\to H} \| u_0 \|_H + \|\tau^{-1}(I - e^{-\tau iB})B^{-1}\|_{H\to H}$$

$$\times \|(I - e^{2\tau iB})^{-1}(e^{-k\tau iB} - e^{k\tau iB})\|_{H\to H}\tau \| Bu_0 \|_H$$

$$+ \|(I - e^{2\tau iB})^{-1}(e^{-k\tau iB} - e^{k\tau iB})\|_{H\to H}\tau \| \tau^{-1}(u_1 - u_0) \|_H$$

$$+\tau^2 \sum_{m=1}^{k-1} \|e^{\tau iB}\|_{H\to H} \|(I - e^{2\tau iB})^{-1}(e^{-(k-m)\tau iB} - e^{(k-m)\tau iB})\|_{H\to H} \|f_m\|_H$$

$$\leq M[\max_{1<m<N-1} \|f_m\|_H + \| \tau^{-1}(u_1 - u_0) \|_H + \| Bu_0 \|_H]$$

for $k \geq 2$. It obviously holds also for $k = 0, 1$. Estimate (6.15) is established. Applying $\tau^{-1}(I - e^{-\tau iB})$ to formula (6.8) and using estimates (6.12), (6.13) and (6.18) in a similar manner as (6.15), we obtain

$$\|\tau^{-1}(I - e^{-\tau iB})u_{k-1}\|_H \leq \|\tau^{-1}(I - e^{-\tau iB})B^{-1}\|_{H\to H}$$
$$\times\{\|e^{-(k-1)\tau iB}\|_{H\to H} \| Bu_0 \|_H + \|\tau^{-1}(I - e^{-\tau iB})B^{-1}\|_{H\to H}$$
$$\times\|(I - e^{2\tau iB})^{-1}(e^{-(k-1)\tau iB} - e^{(k-1)\tau iB})\|_{H\to H} \| B^2u_0 \|_H$$
$$+\|(I - e^{2\tau iB})^{-1}(e^{-(k-1)\tau iB} - e^{(k-1)\tau iB})\|_{H\to H}\tau \| B\tau^{-1}(u_1 - u_0) \|_H$$
$$+\tau\sum_{m=1}^{k-2} \|e^{\tau iB}\|_{H\to H}\|(I - e^{2\tau iB})^{-1}(e^{-(k-1-m)\tau iB} - e^{(k-1-m)\tau iB})\|_{H\to H}\|\tau Bf_m\|_H\}$$
$$\leq M[\max_{1\leq m\leq N-1}\|Bf_m\|_H + \| B\tau^{-1}(u_1 - u_0) \|_H + \| B^2u_0 \|_H]$$

for $k \geq 3$. It obviously holds also for $k = 1, 2$. So, we have that

$$\max_{1\leq k\leq N}\|\tau^{-1}(I - e^{-\tau iB})u_{k-1}\|_H$$
$$\leq M[\max_{1\leq k\leq N-1}\|Bf_k\|_H + \| B\tau^{-1}(u_1 - u_0) \|_H + \| B^2u_0 \|_H].$$

In a similar manner one can show that

$$\max_{1\leq k\leq N}\|\tau^{-1}(u_k - u_{k-1})\|_H$$
$$\leq M[\max_{1\leq k\leq N-1}\|Bf_k\|_H + \| B\tau^{-1}(u_1 - u_0) \|_H + \| B^2u_0 \|_H].$$

Estimate (6.16) is established. Now, we obtain the estimate for $\| \frac{2}{\tau^2}(c(\tau)-I)u_k \|_H$. Using formula (6.14) and estimates (6.12), (6.13) and (6.18), we obtain

$$\frac{2}{\tau^2}\|(c(\tau) - I)u_k\|_H \leq \|\frac{1}{\tau}(I - e^{\tau iB})B^{-1}\|^2_{H\to H}\|e^{-\tau iB}\|_{H\to H}[\|e^{-k\tau iB}\|_{H\to H}$$
$$\times\|B^2u_0\|_H + \|\frac{1}{\tau}(I - e^{\tau iB})B^{-1}\|_{H\to H}$$
$$\times\|(I - e^{2\tau iB})^{-1}(e^{-k\tau iB} - e^{k\tau iB})\|_{H\to H}\|B^3u_0\|_H]$$
$$+\|\frac{1}{\tau}(I - e^{\tau iB})B^{-1}\|^2_{H\to H}\|(I - e^{2\tau iB})^{-1}(e^{-k\tau iB} - e^{k\tau iB})\|_{H\to H}\|B^2(u_1 - u_0)\|_H$$
$$+\|\frac{1}{\tau}(I - e^{\tau iB})B^{-1}\|_{H\to H}\sum_{m=1}^{k-1}\tau\|(I - e^{2\tau iB})^{-1}(e^{-(k-m)\tau iB} - e^{(k-m)\tau iB})\|_{H\to H}$$
$$\times\|B(f_m - f_{m-1})\|_H + \sum_{m=1}^{k-1}\|(I - e^{\tau iB})(I - e^{2\tau iB})^{-1}(e^{(k-m)\tau iB} + e^{(k+1-m)\tau iB})\|_{H\to H}$$

$$\times \|(f_m - f_{m-1})\|_H + \|\frac{1}{\tau}(I - e^{\tau iB})B^{-1}\|_{H\to H}$$
$$\times \|(I - e^{2\tau iB})^{-1}[e^{-(k-1)\tau iB} - e^{(k-1)\tau iB}]\|_{H\to H}$$
$$\times \|\tau B f_1\|_H + \|(I - e^{\tau iB})(I - e^{2\tau iB})^{-1}[e^{(k-1)\tau iB} + e^{k\tau iB}]\|_{H\to H}\|f_1\|_H$$
$$+\|(I - e^{\tau iB})(I - e^{2\tau iB})^{-1}[I + e^{\tau iB}]\|_{H\to H}\|f_{k-1}\|_H$$
$$\leq M[\max_{2\leq m\leq N-1} \| B\tau^{-1}(f_m - f_{m-1}) \|_H + \max_{2\leq m\leq N-1} \| \tau^{-1}(f_m - f_{m-1}) \|_H$$
$$+ \| Bf_1 \|_H + \| f_1 \|_H + \| B^2\tau^{-1}(u_1 - u_0)\|_H + \| B^3u_0 \|_H + \|B^2u_0\|_H]$$
$$\leq M_1[\max_{2\leq m\leq N-1} \| B\tau^{-1}(f_m - f_{m-1}) \|_H + \| Bf_1 \|_H$$
$$+ \| B^2\tau^{-1}(u_1 - u_0)\|_H + \| B^3u_0 \|_H]$$

for $k \geq 2$. It obviously holds also for $k = 1$. So, we have that

$$\max_{1\leq k\leq N-1} \|\frac{2}{\tau^2}(c(\tau) - I)u_k\|_H$$
$$\leq M_1[\max_{2\leq k\leq N-1} \| B\tau^{-1}(f_k - f_{k-1}) \|_H + \| Bf_1 \|_H$$
$$+ \| B^2\tau^{-1}(u_1 - u_0)\|_H + \| B^3u_0 \|_H].$$

The estimate for $\max\limits_{1\leq k\leq N-1} \|\tau^{-2}(u_{k+1}-2u_k+u_{k-1})\|_H$ follows from the last estimate and the triangle inequality. Theorem 6.2.2 is proved.

Remark 6.2.3. Theorem 6.2.2 actually holds in an arbitrary Banach space E under the assumptions

$$\|e^{\pm k\tau iB}\|_{E\to E} \leq M, 1 \leq k \leq N, \tag{6.19}$$

where M does not depend on τ.

Now, we will consider the applications of the exact difference scheme (6.7). From (6.7) it is clear that for the approximate solutions of problem (6.1) it is necessary to approximate the expressions

$$s(\tau), c(\tau) \qquad \text{and} \qquad \int_{t_{k-1}}^{t_k} s(t_k - z)f(z)dz, \int_{t_{k-1}}^{t_k} c(t_k - z)f(z)dz.$$

Let us remark that in constructing difference schemes it is important to know how to construct $f_k^{j,l}$ and $f_{1,1}^{j,l}$ such that

$$\tau^{-2}\int_{t_k}^{t_{k+1}} s(t_{k+1} - z)f(z)dz - c(\tau)\tau^{-2}\int_{t_{k-1}}^{t_k} s(t_k - z)f(z)dz \tag{6.20}$$
$$+ s(\tau)\tau^{-2}\int_{t_{k-1}}^{t_k} c(t_k - z)f(z)dz - f_k^{j,l} = o(\tau^{j+l}),$$

$$\tau^{-1}\int_0^{\tau} s(\tau - z)f(z)dz - f_{1,1}^{j,l} = o(\tau^{j+l}) \tag{6.21}$$

and the formulas of $f_k^{j,l}$ and $f_{1,1}^{j,l}$ are sufficiently simple. The choice of $f_k^{j,l}$ and $f_{1,1}^{j,l}$ is not unique. Using Taylor's formula and integration by parts, we obtain the representation

$$\tau^{-2}\int_{t_{k-1}}^{t_k} s(t_k - z)f(z)dz \tag{6.22}$$

$$= \sum_{m=0}^{j+l-1} \beta_m f^{(m)}(t_{k-1}) + \tau^{-2}\int_{t_{k-1}}^{t_k} s(t_k - z)\int_{t_{k-1}}^{s} \frac{(s-z)^{j+l-1}}{(j+l-1)!} f^{(j+l)}(z)dzds,$$

in which

$$\beta_m = -A^{-1}\frac{\tau^{m-2}}{m!} + A^{-1}\beta_{m-2}, 2 \le m \le j+l-1,$$

$$\beta_0 = \tau^{-2}A^{-1}(c(\tau) - I), \beta_1 = \tau^{-2}A^{-1}(s(\tau) - \tau I).$$

In a similar manner one can obtain that

$$\tau^{-2}\int_{t_{k-1}}^{t_k} c(t_k - z)f(z)dz \tag{6.23}$$

$$= \sum_{m=0}^{j+l-1} \gamma_m f^{(m)}(t_{k-1}) + \tau^{-2}\int_{t_{k-1}}^{t_k} c(t_k - z)\int_{t_{k-1}}^{s} \frac{(s-z)^{j+l-1}}{(j+l-1)!} f^{(j+l)}(z)dzds,$$

in which

$$\gamma_m = \beta_{m-1}, 1 \le m \le j+l-1, \gamma_0 = \tau^{-2}s(\tau).$$

Using the definitions of $s(\tau)$ and $c(\tau)$ and Pade fractions for the function e^{-z} (see Chapter 2), we can write

$$\begin{cases} c(\tau) = \frac{R_{j,l+1}(i\tau B)+R_{j,l+1}(-i\tau B)}{2} + o(\tau^{j+l+2}), \\ s(\tau) = B^{-1}\frac{R_{j,l+1}(i\tau B)-R_{j,l+1}(-i\tau B)}{2i} + o(\tau^{j+l+2}). \end{cases} \tag{6.24}$$

Now, using formulas (6.20), (6.21), (6.22), (6.23) and (6.24) $f_k^{j,l}$ and $f_{1,1}^{j,l}$ can be defined by the following formulas:

$$f_k^{j,l} = \sum_{m=0}^{j+l-1} B_m f^{(m)}(t_k) + B^{-1}\frac{R_{j,l+1}(i\tau B) - R_{j,l+1}(-i\tau B)}{2i} \tag{6.25}$$

$$\times \sum_{m=0}^{j+l-1} \widetilde{B}_m f^{(m)}(t_{k-1}) - \frac{R_{j,l+1}(i\tau B) + R_{j,l+1}(-i\tau B)}{2}\sum_{m=0}^{j+l-1} B_m f^{(m)}(t_{k-1}),$$

$$f_{1,1}^{j,l} = \sum_{m=0}^{j+l-2} \widetilde{B}_m f^{(m)}(0),\ f_0^{j,l} = \tau f_{1,1}^{j,l}, \tag{6.26}$$

where

$$B_m = -A^{-1}\frac{\tau^{m-2}}{m!} + A^{-1}B_{m-2}, 2 \le m \le j+l-1,$$

$$B_0 = \tau^{-2}A^{-1}\left(\frac{R_{j,l+1}(i\tau B) + R_{j,l+1}(-i\tau B)}{2} - I\right),$$

$$B_1 = \tau^{-2}A^{-1}\left(B^{-1}\frac{R_{j,l+1}(i\tau B) - R_{j,l+1}(-i\tau B)}{2i} - \tau I\right),$$

$$\widetilde{B}_m = B_{m-1}, 1 \le m \le j+l-1,$$

$$\widetilde{B}_0 = \tau^{-2}B^{-1}\frac{R_{j,l+1}(i\tau B) - R_{j,l+1}(-i\tau B)}{2i}.$$

Now, using formulas (6.24), (6.25) and (6.26), we obtain the difference schemes to a $(j+l)$-th order of accuracy

$$\tau^{-2}(u_{k+1} - 2u_k + u_{k-1}) + \widetilde{A}u_k = f_k^{j,l}, 1 \le k \le N-1, u_0 = v_0, \tag{6.27}$$

$$\widetilde{A} = 2\tau^{-2}\left(I - \frac{R_{j,l+1}(i\tau B) + R_{j,l+1}(-i\tau B)}{2}\right),$$

$$\tau^{-1}(u_1 - u_0) = \tau^{-1}\left(\frac{R_{j,l}(i\tau B) + R_{j,l}(-i\tau B)}{2} - I\right)v_0$$

$$+\tau^{-1}B^{-1}\frac{R_{j,l}(i\tau B) - R_{j,l}(-i\tau B)}{2i}v'(0) + f_0^{j,l}$$

for the approximate solution of the initial-value problem (6.1).

Note that the difference schemes (6.27) for $j = l$, $j = l-1$ and $j = l+1$ include difference schemes of arbitrary high order of approximation. Moreover, the corresponding functions $|R_{j,l+1}(z)|$ tend to 0 as $z \to \infty$ for $j = l-1, l$ and $|R_{j,l+1}(z)| = 1$ for $j = l+1$. Such difference schemes are simplest, in the sense that the degrees of the denominators of the corresponding Pade approximants of the function $\exp\{-z\}$ are minimal for a fixed order of approximation of the difference schemes.

Suppose the operators $(I - R_{j,l}^2(i\tau B))$ and $(I + R_{j,l}(i\tau B))$ have the bounded inverses $(I - R_{j,l}^2(i\tau B))^{-1}$ and $(I + R_{j,l}(i\tau B))^{-1}$. It is clear that this problem is uniquely solvable and the following formula holds:

$$u_k = [R_{j,l}^k(-i\tau B) - (I - R_{j,l}^2(i\tau B))^{-1}(R_{j,l}^k(-i\tau B) - R_{j,l}^k(i\tau B))]u_0 \tag{6.28}$$

$$+R_{j,l}(i\tau B)(I - R_{j,l}^2(i\tau B))^{-1}(R_{j,l}^k(-i\tau B) - R_{j,l}^k(i\tau B))u_1$$

$$+\tau^2\sum_{m=1}^{k-1} R_{j,l}(i\tau B)(I - R_{j,l}^2(i\tau B))^{-1}(R_{j,l}^{k-m}(-i\tau B) - R_{j,l}^{k-m}(i\tau B))f_m^{j,l}, 2 \le k \le N,$$

$$u_1 = \frac{R_{j,l}(i\tau B) + R_{j,l}(-i\tau B)}{2}v_0 + B^{-1}\frac{R_{j,l}(i\tau B) - R_{j,l}(-i\tau B)}{2i}v'(0) + \tau f_0^{j,l},$$

$$u_0 = v_0.$$

Let us investigate the stability of the exact two-step difference scheme (6.27).

Theorem 6.2.4. *Suppose the operators* $(I - R_{j,l}^2(i\tau B))$ *and* $(I + R_{j,l}(i\tau B))$ *have the bounded inverses* $(I - R_{j,l}^2(i\tau B))^{-1}$ *and* $(I + R_{j,l}(i\tau B))^{-1}$. *Suppose that* $u_0 \in D(B^2)$, $u_1 - u_0 \in D(B)$. *Then for the solution of the two-step difference schemes* (6.27) *for* $j = l$ *and* $j = l - 1$ *the following stability inequalities hold:*

$$\max_{0\le k\le N}\|u_k\|_H \le M[\max_{1\le k\le N-1}\|B^{-1}f_k^{j,l}\|_H + \| B^{-1}\tau^{-1}(u_1 - u_0) \|_H + \| u_0 \|_H],$$

$$\max_{1\le k\le N}\|\tau^{-1}(u_k - u_{k-1})\|_H + \max_{1\le k\le N}\|\tau^{-1}(I - e^{-\tau iB})u_{k-1}\|_H$$
$$\le M[\max_{1\le k\le N-1}\|f_k^{j,l}\|_H + \| \tau^{-1}(u_1 - u_0) \|_H + \| Bu_0 \|_H],$$

$$\max_{1\le k\le N-1}||\tau^{-2}(u_{k+1} - 2u_k + u_{k-1})\|_H + \max_{1\le k\le N-1}\|\widetilde{A}u_k\|_H$$
$$\le M[\max_{2\le k\le N-1}\| \tau^{-1}(f_k^{j,l} - f_{k-1}^{j,l}) \|_H + \| f_1^{j,l} \|_H$$
$$+ \| B\tau^{-1}(u_1 - u_0) \|_H + \| B^2u_0 \|_H],$$

where M *does not depend on* τ, $f_k^{j,l}$, $1 \le k \le N - 1$ *and* u_0, u_1.

The proof of Theorem 6.2.4 follows the scheme of the proof of Theorem 6.2.1 and relies on formula (6.28) and on the estimates

$$||R_{j,l+1}(\pm i\tau B)||_{H\to H} \le 1, \tag{6.29}$$

$$||\tau B(I - R_{j,l+1}^2(i\tau B))^{-1}R_{j,l+1}(i\tau B)||_{H\to H} \le M. \tag{6.30}$$

Note that estimate (6.30) is not satisfied for $j = l+1$. Therefore we have not been able to obtain the same stability inequalities for the solution of the exact two-step difference scheme (6.27) for $j = l + 1$. Nevertheless, the following result holds.

Theorem 6.2.5. *Suppose the operators* $(I - R_{j,l}^2(i\tau B))$ *and* $(I + R_{j,l}(i\tau B))$ *have the bounded inverses* $(I - R_{j,l}^2(i\tau B))^{-1}$ *and* $(I + R_{j,l}(i\tau B))^{-1}$. *Suppose that* $u_0 \in D(B^3)$, $u_1 - u_0 \in D(B^2)$ *and* $f_k^{j,l} \in D(B)$, $1 \le k \le N - 1$. *Then for the solution of the two-step difference schemes* (6.27) *for* $j = l + 1$ *the following stability inequalities hold:*

$$\max_{0\le k\le N}\|u_k\|_H \le M[\max_{1\le k\le N-1}\|f_k^{j,l}\|_H + \| \tau^{-1}(u_1 - u_0) \|_H + \| Bu_0 \|_H],$$

$$\max_{1\le k\le N}\|\tau^{-1}(u_k - u_{k-1})\|_H + \max_{1\le k\le N}\|\tau^{-1}(I - e^{-\tau iB})u_{k-1}\|_H$$
$$\le M[\max_{1\le k\le N-1}\|Bf_k^{j,l}\|_H + \| B\tau^{-1}(u_1 - u_0) \|_H + \| B^2u_0 \|_H],$$

$$\max_{1\le k\le N-1}||\tau^{-2}(u_{k+1} - 2u_k + u_{k-1})\|_H + \max_{1\le k\le N-1}\|\widetilde{A}u_k\|_H$$
$$\le M[\max_{2\le k\le N-1}\| \tau^{-1}B(f_k^{j,l} - f_{k-1}^{j,l}) \|_H + \| Bf_1^{j,l} \|_H$$
$$+ \| B^2\tau^{-1}(u_1 - u_0) \|_H + \| B^3u_0 \|_H],$$

where M *does not depend on* τ, $f_k^{j,l}$, $1 \le k \le N - 1$ *and* u_0, u_1.

The proof of Theorem 6.2.5 follows the scheme of proof of Theorem 6.2.2 and relies on formula (6.28) and on estimates (6.29) and

$$\|(I - R_{j,l+1}(2i\tau B))^{-1}(R^k_{j,l+1}(-i\tau B) - R^k_{j,l+1}(i\tau B))\|_{H\to H} \le k. \tag{6.31}$$

Remark 6.2.6. Theorem 6.2.5 actually holds in an arbitrary Banach space E under the assumptions

$$\|R^k_{j,l+1}(\pm i\tau B)\|_{E\to E} \le M, 1 \le k \le N, \tag{6.32}$$

where M does not depend on τ.

Now, the abstract Theorems 6.2.4 and 6.2.5 are applied in the investigation of difference schemes of higher order of accuracy with respect to one variable for approximate solutions of the mixed boundary-value problem (6.6). The discretization of problem (6.6) is carried out in two steps. In the first step let us define the grid sets

$$\widetilde{\Omega}_h = \{x = x_m = (h_1 m_1, \dots, h_n m_n), m = (m_1, \dots, m_n),$$
$$0 \le m_r \le N_r, h_r N_r = L, r = 1, \dots, n\},$$
$$\Omega_h = \widetilde{\Omega}_h \cap \Omega, S_h = \widetilde{\Omega}_h \cap S.$$

We introduce the Banach space $L_2(\widetilde{\Omega}_h)$ of the grid functions $\varphi^h(x) = \{\varphi(h_1 m_1, \dots, h_n m_n)\}$ defined on $\widetilde{\Omega}_h$, equipped with the norm

$$\| \varphi^h \|_{L_2(\widetilde{\Omega}_h)} = (\sum_{x\in\overline{\Omega_h}} |\varphi^h(x)|^2 h_1 \cdots h_n)^{\frac{1}{2}}.$$

To the differential operator A^x generated by problem (6.6) we assign the difference operator A^x_h by the formula

$$A^x_h u^h_x = -\sum_{r=1}^{n} (a_r(x) u^h_{\bar{x}_r})_{x_r, j_r} \tag{6.33}$$

acting in the space of grid functions $u^h(x)$, satisfying the conditions $u^h(x) = 0$ for all $x \in S_h$. It is known that A^x_h is a self-adjoint positive definite operator in $L_2(\widetilde{\Omega}_h)$. With the help of A^x_h we arrive at the initial-value problem

$$\frac{d^2 v^h(t,x)}{dt^2} + A^x_h v^h(t,x) = f^h(t,x),\ 0 \le t \le T,\ x \in \widetilde{\Omega}_h, \tag{6.34}$$

$$v^h(0,x) = \varphi^h(x),\ \frac{dv^h(0,x)}{dt} = \psi^h(x),\ x \in \widetilde{\Omega}_h$$

for an infinite system of ordinary differential equations.

In the second step we replace problem (6.34) by the difference schemes (6.27)

$$\tau^{-2}(u^h_{k+1}(x) - 2u^h_k(x) + u^h_{k-1}(x)) + \widetilde{A^x_h} u^h_k(x) = f^h_k(x), \tag{6.35}$$

$$t_k = k\tau, 1 \le k \le N-1,\ x \in \widetilde{\Omega}_h,$$

$$u_0^h(x) = \varphi^h(x),\ \tau^{-1}(u_1^h(x) - u_0^h(x)) = \tau^{-1}(\frac{R_{j,l}(i\tau B_h^x) + R_{j,l}(-i\tau B_h^x)}{2} - I)\varphi^h(x)$$

$$+\tau^{-1}(B_h^x)^{-1}\frac{R_{j,l}(i\tau B_h^x) - R_{j,l}(-i\tau B_h^x)}{2i}\psi^h(x) + \tau f_0^h(x), x \in \widetilde{\Omega}_h,$$

$$\widetilde{A_h^x} = 2\tau^{-2}(I - \frac{R_{j,l+1}(i\tau B_h^x) + R_{j,l+1}(-i\tau B_h^x)}{2}),\ (B_h^x)^2 = A_h^x,$$

$$f_k^h(x) = \sum_{m=0}^{j+l-1} B_m f^{(m)}(t_k, x) + B^{-1}\frac{R_{j,l+1}(i\tau B_h^x) - R_{j,l+1}(-i\tau B_h^x)}{2i}$$

$$\times \sum_{m=0}^{j+l-1} \widetilde{B}_m f^{(m)}(t_{k-1}, x)$$

$$-\frac{R_{j,l+1}(i\tau B_h^x) + R_{j,l+1}(-i\tau B_h^x)}{2}\sum_{m=0}^{j+l-1} B_m f^{(m)}(t_{k-1}, x),$$

$$f_0^h(x) = \tau \sum_{m=0}^{j+l-2} \widetilde{B}_m f^{(m)}(0, x),$$

where

$$B_m = -(A_h^x)^{-1}\frac{\tau^{m-2}}{m!} + (A_h^x)^{-1}B_{m-2},\ 2 \le m \le j+l-1,$$

$$B_0 = \tau^{-2}(A_h^x)^{-1}(\frac{R_{j,l+1}(i\tau B_h^x) + R_{j,l+1}(-i\tau B_h^x)}{2} - I),$$

$$B_1 = \tau^{-2}(A_h^x)^{-1}((B_h^x)^{-1}\frac{R_{j,l+1}(i\tau B_h^x) - R_{j,l+1}(-i\tau B_h^x)}{2i} - \tau I),$$

$$\widetilde{B}_m = B_{m-1},\ 1 \le m \le j+l-1,$$

$$\widetilde{B}_0 = \tau^{-2}(B_h^x)^{-1}\frac{R_{j,l+1}(i\tau B_h^x) - R_{j,l+1}(-i\tau B_h^x)}{2i}.$$

Theorem 6.2.7. *Let τ and $|h|$ be sufficiently small numbers. Then the solutions of the difference schemes (6.35) for $j = l$ and $j = l - 1$ satisfy the following stability estimates:*

$$\max_{0\le k\le N} \|u_k^h\|_{L_2(\widetilde{\Omega}_h)} \le M_1[\max_{0\le k\le N-1} \|(B_h^x)^{-1}f_k^h\|_{L_2(\widetilde{\Omega}_h)}$$

$$+ \|(B_h^x)^{-1}\psi^h\|_{L_2(\widetilde{\Omega}_h)} + \|\varphi^h\|_{L_2(\widetilde{\Omega}_h)}],$$

$$\max_{1\le k\le N} \|\tau^{-1}(u_k^h - u_{k-1}^h)\|_{L_2(\widetilde{\Omega}_h)} \le M_1[\max_{0\le k\le N-1} \|f_k^h\|_{L_2(\widetilde{\Omega}_h)}$$

$$+ \|\psi^h\|_{L_2(\widetilde{\Omega}_h)} + \sum_{r=1}^{n} \|\varphi^h_{x_r,\bar{m}_r}\|_{L_2(\widetilde{\Omega}_h)}],$$

$$\max_{1\le k\le N-1}\|\tau^{-2}(u_{k+1}^h-2u_k^h+u_{k-1}^h)\|_{L_2(\tilde{\Omega}_h)}\le M_1\Big[\max_{1\le k\le N-1}\|\tau^{-1}(f_k^h-f_{k-1}^h)\|_{L_2(\tilde{\Omega}_h)}$$

$$+\|f_0^h\|_{L_2(\tilde{\Omega}_h)}+\sum_{r=1}^{n}\tau^{-1}\|\psi^h_{\bar{x}_r,m_r}\|_{L_2(\tilde{\Omega}_h)}+\sum_{r=1}^{n}\|(\varphi^h_{\bar{x}_r})_{x_r,m_r}\}\|_{L_2(\tilde{\Omega}_h)}\Big].$$

Here M_1 does not depend on τ, h, $\varphi^h(x)$, $\psi^h(x)$ and $f_k^h(x), 0\le k\le N-1$.

The proof of Theorem 6.2.7 is based on the abstract Theorems 6.2.2 and the symmetry properties of the difference operator A_h^x defined by formula (6.33).

Theorem 6.2.8. *Let τ and $|h|$ be sufficiently small numbers. Then the solutions of the difference schemes (6.35) for $j=l+1$ satisfy the following stability estimates:*

$$\max_{0\le k\le N}\|u_k^h\|_{L_2(\tilde{\Omega}_h)}\le M_1\Big[\max_{0\le k\le N-1}\|f_k^h\|_{L_2(\tilde{\Omega}_h)}$$

$$+\|\psi^h\|_{L_2(\tilde{\Omega}_h)}+\sum_{r=1}^{n}\|\varphi^h_{\bar{x}_r,m_r}\|_{L_2(\tilde{\Omega}_h)}\Big],$$

$$\max_{1\le k\le N}\|\tau^{-1}(u_k^h-u_{k-1}^h)\|_{L_2(\tilde{\Omega}_h)}\le M_1\Big[\max_{0\le k\le N-1}\sum_{r=1}^{n}\|(f_k^h)_{\bar{x}_r,m_r}\|_{L_2(\tilde{\Omega}_h)}$$

$$+\sum_{r=1}^{n}\|\psi^h_{\bar{x}_r,m_r}\|_{L_2(\tilde{\Omega}_h)}+\sum_{r=1}^{n}\|(\varphi^h_{\bar{x}_r})_{x_r,m_r}\|_{L_2(\tilde{\Omega}_h)}\Big],$$

$$\max_{1\le k\le N-1}\|\tau^{-2}(u_{k+1}^h-2u_k^h+u_{k-1}^h)\|_{L_2(\tilde{\Omega}_h)}$$

$$\le M_1\Big[\max_{1\le k\le N-1}\|\tau^{-1}\sum_{r=1}^{n}\|(f_k^h)_{\bar{x}_r,m_r}-(f_{k-1}^h)_{\bar{x}_r,m_r}\|_{L_2(\tilde{\Omega}_h)}$$

$$+\sum_{r=1}^{n}\|(f_0^h)_{\bar{x}_r,m_r}\|_{L_2(\tilde{\Omega}_h)}+\sum_{r=1}^{n}\|(\psi^h_{\bar{x}_r})_{x_r,m_r}\|_{L_2(\tilde{\Omega}_h)}+\sum_{r=1}^{n}\|((\varphi^h_{\bar{x}_r})_{x_r})_{x_r,m_r}\|_{L_2(\tilde{\Omega}_h)}\Big].$$

Here M_1 does not depend on τ, h, $\varphi^h(x)$, $\psi^h(x)$ and $f_k^h(x), 0\le k\le N-1$.

The proof of Theorem 6.2.8 is based on the abstract Theorem 6.2.4 and the symmetry properties of the difference operator A_h^x defined by formula (6.33).

Note that in a similar manner one can construct the difference schemes of a high order of accuracy with respect to one variable for approximate solutions of the boundary-value problem (6.5). Abstract Theorems 6.2.2 and 6.2.4 permit us to obtain stability estimates for the solutions of these difference schemes.

6.3 Two-Step Difference Schemes Generated by Taylor's Decomposition

We consider again the initial-value problem (6.1). For the construction of the two-step difference schemes of an arbitrary high order of accuracy for the approximate solutions of the initial-value problem (6.1) we consider again a uniform grid space

$$[0,T]_\tau = \{t_k = k\tau, k = 0,1,\dots,N, N\tau = T\}.$$

Let the function $v(t) (0 \le t \le T)$ have a $(2l+2j+2)$-th continuous derivative and $t_{k-1}, t_k, t_{k+1} \in [0,T]_\tau$. Then by Theorem 3.2.2 we have the following Taylor decomposition on three points:

$$v(t_{k+1}) - 2v(t_k) + v(t_{k-1}) - \sum_{s=1}^{l} \alpha_s v^{(2s)}(t_k)\tau^{2s} \tag{6.36}$$

$$- \sum_{s=1}^{j} \eta_s (v^{(2s)}(t_{k-1}) + v^{(2s)}(t_{k+1}))\tau^{2s} = o(\tau^{2l+2j+2}),$$

where η_m, $m = 1,\dots,j$ is a solution of the system

$$\begin{cases} \eta_s = \frac{1}{(2s)!} - \sum\limits_{m=1}^{s-1} \frac{\eta_m}{(2s-2m)!}, \text{ for any } s,\ l+1 \le s \le j, \\ \\ \sum\limits_{m=1}^{j} \frac{(2s)!\eta_m}{(2s-2m)!} = 1, \text{ for any } s,\ j+1 \le s \le l+j, \end{cases} \tag{6.37}$$

and

$$\alpha_1 = 1 - 2\eta_1,\ \alpha_s = (\frac{1}{(2s)!} - \eta_s - \sum_{m=1}^{s-1} \eta_m \frac{1}{(2(s-m))!})2 \text{ for any } s,\ 2 \le s \le l.$$

Suppose further that the function $v(t)$ $(0 \le t \le T)$ has a $(2l+2j+1)$-th continuous derivative. Then by Theorem 2.2.1 we have the following Taylor decomposition on three points:

$$v(\tau) - v(0) + \sum_{s=1}^{2j} \delta_s v^{(s)}(\tau)\tau^s - \sum_{s=1}^{2l} \rho_s v^{(s)}(0)\tau^s = o(\tau^{2l+2j+1}), \tag{6.38}$$

where

$$\begin{cases} \rho_s = \frac{(2l+2j-s)!(2l)!}{(2l+2j)!s!(2l-s)!} \text{ for any } s,\ 1 \le s \le 2l, \\ \\ \delta_s = \frac{(2l+2j-s)!(2j)!(-1)^s}{(2l+2j)!s!(2j-s)!} \text{ for any } s,\ 1 \le s \le 2j. \end{cases}$$

Now, we will consider the applications of the Taylor decomposition (6.36) of function $v(t)$ on three points and Taylor's decomposition (6.38) of function $v(t)$ on three points to approximate solutions of the initial-value problem (6.1). From (6.36) and (6.38) it is clear that for the approximate solution of the problem (6.1) it is necessary to find $v^{(j)}(\tau)$ for any s, $1 \leq s \leq 2l$, $v^{(2s)}(t_k)$ for any s, $1 \leq s \leq l$ and $v^{(2s)}(t_{k-1})$, $v^{(2s)}(t_{k+1})$ for any s, $1 \leq s \leq j$. Using the equation

$$v''(t) = -Av(t) + f(t),$$

we obtain

$$v^{(2n)}(t) = (-A)^n v(t) + \sum_{\lambda=1}^{n} (-A)^{n-\lambda} f^{(2\lambda-2)}(t), \; n = 2, \ldots, \tag{6.39}$$

$$v^{(2n+1)}(t) = (-A)^n v'(t) + \sum_{\lambda=1}^{n} (-A)^{n-\lambda} f^{(2\lambda-1)}(t), \; n = 1, \ldots. \tag{6.40}$$

Suppose further that the function $v(t)(0 \leq t \leq T)$ has a $(2l + 2j - 2m + 2)$-th continuous derivative. Then by Theorem 2.2.1 we have the following Taylor decomposition on two points:

$$v(\tau) - v(0) + \sum_{s=1}^{2l-2[\frac{m}{2}]} a_s v^{(s)}(\tau)\tau^s - \sum_{s=1}^{2j-2m+2[\frac{m}{2}]+1} b_s v^{(s)}(0)\tau^s = o(\tau^{2l+2j-2m+2}), \tag{6.41}$$

where

$$\begin{cases} a_s = \frac{(2l+2j-2m+1-s)!(2l-2[\frac{m}{2}])!(-1)^s}{(2l+2j+1-2m)!s!(2p-2[\frac{m}{2}]-s)!} \text{ for any } s, \; 1 \leq s \leq 2l - 2[\frac{m}{2}], \\ b_s = \frac{(2l+2j+1-2m-s)!(2j-2m+2[\frac{m}{2}]+1)!}{(2l+2j+1-2m)!s!(2j-2m+2[\frac{m}{2}]+1-s)!} \text{ for any } s, \\ \quad 1 \leq s \leq 2j - 2m + 2[\frac{m}{2}] + 1, \end{cases}$$

where $[a]$ denotes the integer part of the number a. Further using formulas (6.39), (6.40) and (6.41), we can write

$$v'(\tau) - v'(0) + \sum_{n=1}^{l-[\frac{m}{2}]} a_{2n}(-A)^n v'(\tau)\tau^{2n} + \sum_{n=1}^{l-[\frac{m}{2}]} a_{2n} \sum_{\lambda=1}^{n} (-A)^{n-\lambda} f^{(2\lambda-1)}(\tau)\tau^{2n}$$

$$+ \sum_{n=1}^{l-[\frac{m}{2}]} a_{2n-1}(-A)^n v(\tau)\tau^{2n-1} + \sum_{n=1}^{l-[\frac{m}{2}]} a_{2n-1} \sum_{\lambda=1}^{n} (-A)^{n-\lambda} f^{(2\lambda-2)}(\tau)\tau^{2n-1}$$

$$- \sum_{n=1}^{j-m+[\frac{m}{2}]} b_{2n}(-A)^n v'(0)\tau^{2n} - \sum_{n=1}^{j-m+[\frac{m}{2}]} b_{2n} \sum_{\lambda=1}^{n} (-A)^{n-\lambda} f^{(2\lambda-1)}(0)\tau^{2n}$$

$$-\sum_{n=1}^{j-m+[\frac{m}{2}]+1} b_{2n-1}(-A)^n v(0)\tau^{2n-1}$$

$$-\sum_{n=1}^{j-m+[\frac{m}{2}]+1} b_{2n-1}\sum_{\lambda=1}^{n}(-A)^{n-\lambda} f^{(2\lambda-2)}(0)\tau^{2n-1}$$

$$= o(\tau^{2l+2j-2m+2}),\ 0 \le m \le j-1.$$

From the last formula it follows that

$$(I+\sum_{n=1}^{l-[\frac{m}{2}]} a_{2n}(-A)^n\tau^{2n})v'(\tau) = (I+\sum_{n=1}^{j-m+[\frac{m}{2}]} b_{2n}(-A)^n\tau^{2n})v'(0)$$

$$-\sum_{n=1}^{l-[\frac{m}{2}]} a_{2n-1}(-A)^n v(\tau)\tau^{2n-1} + \sum_{n=1}^{j-m+[\frac{m}{2}]+1} b_{2n-1}(-A)^n v(0)\tau^{2n-1}$$

$$-\sum_{n=1}^{l-[\frac{m}{2}]} a_{2n}\sum_{\lambda=1}^{n}(-A)^{n-\lambda} f^{(2\lambda-1)}(\tau)\tau^{2n} - \sum_{n=1}^{l-[\frac{m}{2}]} a_{2n-1}\sum_{\lambda=1}^{n}(-A)^{n-\lambda} f^{(2\lambda-2)}(\tau)\tau^{2n-1}$$

$$+\sum_{n=1}^{j-m+[\frac{m}{2}]} b_{2n}\sum_{\lambda=1}^{n}(-A)^{n-\lambda} f^{(2\lambda-1)}(0)\tau^{2n}$$

$$+\sum_{n=1}^{j-m+[\frac{m}{2}]+1} b_{2n-1}\sum_{\lambda=1}^{n}(-A)^{n-\lambda} f^{(2\lambda-2)}(0)\tau^{2n-1}$$

$$+o(\tau^{2l+2j-2m+2}),\ 0 \le m \le j-1.$$

Suppose further that the operator $(I+\sum_{n=1}^{l-[\frac{m}{2}]} a_{2n}(-A)^n\tau^{2n})$ has a bounded inverse. Then

$$v'(\tau) = (I+\sum_{n=1}^{l-[\frac{m}{2}]} a_{2n}(-A)^n\tau^{2n})^{-1}\{(I+\sum_{n=1}^{j-m+[\frac{m}{2}]} b_{2n}(-A)^n\tau^{2n})v'(0) \tag{6.42}$$

$$-\sum_{n=1}^{l-[\frac{m}{2}]} a_{2n-1}(-A)^n v(\tau)\tau^{2n-1} + \sum_{n=1}^{j-m+[\frac{m}{2}]+1} b_{2n-1}(-A)^n v(0)\tau^{2n-1}$$

$$-\sum_{n=1}^{l-[\frac{m}{2}]} a_{2n}\sum_{\lambda=1}^{n}(-A)^{n-\lambda} f^{(2\lambda-1)}(\tau)\tau^{2n} - \sum_{n=1}^{l-[\frac{m}{2}]} a_{2n-1}\sum_{\lambda=1}^{n}(-A)^{n-\lambda} f^{(2\lambda-2)}(\tau)\tau^{2n-1}$$

$$+\sum_{n=1}^{j-m+[\frac{m}{2}]} b_{2n}\sum_{\lambda=1}^{n}(-A)^{n-\lambda}f^{(2\lambda-1)}(0)\tau^{2n}$$

$$+\sum_{n=1}^{j-m+[\frac{m}{2}]+1} b_{2n-1}\sum_{\lambda=1}^{n}(-A)^{n-\lambda}f^{(2\lambda-2)}(0)\tau^{2n-1}\}$$

$$+o(\tau^{2l+2j-2m+2}),\ 0\le m\le j-1.$$

Now, using formulas (6.36), (6.38), (6.39), (6.40) and (6.42), we obtain the difference schemes of a $(2l+2j)$-th order of accuracy:

$$\frac{u_{k+1}-2u_k+u_{k-1}}{\tau^2}-\sum_{s=1}^{l}\alpha_s(-A)^s\tau^{2s-2}u_k \tag{6.43}$$

$$-\sum_{s=1}^{j}\eta_s(-A)^s\tau^{2s-2}(u_{k-1}+u_{k+1})=f_k^{j,l},$$

$$f_k^{j,l}=\sum_{s=1}^{j}\eta_s\sum_{\lambda=1}^{s}(-A)^{s-\lambda}(f^{(2\lambda-2)}(t_{k+1})+f^{(2\lambda-2)}(t_{k-1}))\tau^{2s-2}$$

$$+\sum_{s=1}^{l}\alpha_s\sum_{\lambda=1}^{s}(-A)^{s-\lambda}f^{(2\lambda-2)}(t_k)\tau^{2s-2},1\le k\le N-1,u_0=v_0,$$

$$\frac{u_1-u_0}{\tau}-\sum_{n=1}^{l}\rho_{2n}(-A)^n\tau^{2n-1}v(0)-\sum_{n=1}^{l}\rho_{2n-1}(-A)^{n-1}\tau^{2n-2}v'(0)$$

$$+\sum_{n=1}^{j}\delta_{2n}(-A)^n\tau^{2n-1}u_1+\sum_{m=1}^{j}\delta_{2m-1}(-A)^m\tau^{2m-2}$$

$$\times(I+\sum_{n=1}^{l-[\frac{m}{2}]}a_{2n}(-A)^n\tau^{2n})^{-1}\{(I+\sum_{n=1}^{j-m+[\frac{m}{2}]}b_{2n}(-A)^n\tau^{2n})v'(0)$$

$$-\sum_{n=1}^{l-[\frac{m}{2}]}a_{2n-1}(-A)^n\tau^{2n-1}u_1+\sum_{n=1}^{j-m+[\frac{m}{2}]+1}b_{2n-1}(-A)^n\tau^{2n-1}v(0)\}$$

$$=\sum_{n=1}^{l}\rho_{2n}(-A)^n\tau^{2n-1}\sum_{\lambda=1}^{n}(-A)^{n-\lambda}f^{(2\lambda-2)}(0)$$

$$+\sum_{n=1}^{l}\rho_{2n-1}(-A)^{n-1}\tau^{2n-2}\sum_{\lambda=1}^{n}(-A)^{n-\lambda}f^{(2\lambda-1)}(0)$$

$$
\begin{aligned}
&-\sum_{n=1}^{j}\delta_{2n}(-A)^n\tau^{2n-1}\sum_{\lambda=1}^{n}(-A)^{n-\lambda}f^{(2\lambda-2)}(\tau)\\
&-\sum_{n=1}^{j}\delta_{2n-1}(-A)^n\tau^{2n-2}\sum_{\lambda=1}^{n-1}(-A)^{n-\lambda}f^{(2\lambda-1)}(\tau)\\
&-\sum_{m=1}^{j}\delta_{2m-1}(-A)^m\tau^{2m-2}(I+\sum_{n=1}^{l-[\frac{m}{2}]}a_{2n}(-A)^n\tau^{2n})^{-1}\\
&\times\{-\sum_{n=1}^{l-[\frac{m}{2}]}a_{2n}\sum_{\lambda=1}^{n}(-A)^{n-\lambda}f^{(2\lambda-1)}(\tau)\tau^{2n}-\sum_{n=1}^{l-[\frac{m}{2}]}a_{2n-1}\sum_{\lambda=1}^{n}(-A)^{n-\lambda}f^{(2\lambda-2)}(\tau)\tau^{2n-1}\\
&+\sum_{n=1}^{j-m+[\frac{m}{2}]}b_{2n}\sum_{\lambda=1}^{n}(-A)^{n-\lambda}f^{(2\lambda-1)}(0)\tau^{2n}\\
&+\sum_{n=1}^{j-m+[\frac{m}{2}]+1}b_{2n-1}\sum_{\lambda=1}^{n}(-A)^{n-\lambda}f^{(2\lambda-2)}(0)\tau^{2n-1}\}
\end{aligned}
$$

for the approximate solution of the initial-value problem (6.1). Suppose that the operator

$$
\begin{aligned}
&(I+\sum_{n=1}^{j}\delta_{2n}(-A)^n\tau^{2n}-\sum_{m=1}^{j}\delta_{2m-1}(-A)^m\tau^{2m-1}(I\\
&+\sum_{n=1}^{l-[\frac{m}{2}]}a_{2n}(-A)^n\tau^{2n})^{-1}\sum_{n=1}^{l-[\frac{m}{2}]}a_{2n-1}(-A)^n\tau^{2n-1})
\end{aligned}
$$

has a bounded inverse. Suppose further that the operator $(I-\sum_{s=1}^{j}\eta_s(-A)^s\tau^{2s})$ has a bounded inverse and

$$
I-\frac{1}{4}(2I+\sum_{s=1}^{l}\alpha_s(-A)^s\tau^{2s})^2(I-\sum_{s=1}^{j}\eta_s(-A)^s\tau^{2s})^{-2}\geq 0.
$$

This problem is uniquely solvable and the following formula holds (see Chapter 1)

$$
\begin{aligned}
u_k &= (\widetilde{R}_{j,l}(-i\tau A)-\widetilde{R}_{j,l}(i\tau A))^{-1}\widetilde{R}_{j,l}(i\tau A)\widetilde{R}_{j,l}(-i\tau A)(\widetilde{R}_{j,l}^{k-1}(i\tau A) \qquad (6.44)\\
&\quad -\widetilde{R}_{j,l}^{k-1}(-i\tau A))u_0\\
&\quad +(\widetilde{R}_{j,l}(-i\tau A)-\widetilde{R}_{j,l}(i\tau A))^{-1}(\widetilde{R}_{j,l}^{k}(-i\tau A)-\widetilde{R}_{j,l}^{k}(i\tau A))u_1
\end{aligned}
$$

$$+\sum_{j=1}^{k}(\widetilde{R}_{j,l}(-i\tau A)-\widetilde{R}_{j,l}(i\tau A))_{j,l}^{-1}\widetilde{R}_{j,l}(i\tau A)\widetilde{R}_{j,l}(-i\tau A)$$

$$\times(\widetilde{R}_{j,l}^{k-m}(-i\tau A)-\widetilde{R}_{j,l}^{k-m}(i\tau A))(I-\sum_{s=1}^{j}\eta_s(-A)^s\tau^{2s})^{-1}f_m^{j,l},\ 2\le k\le N,$$

$$u_1=\widetilde{c}_\tau v(0)+\widetilde{s}_\tau v'(0)+\tau f_0^{j,l}.$$

Here

$$\widetilde{R}_{j,l}(\pm i\tau A)=\frac{1}{2}(2I+\sum_{s=1}^{l}\alpha_s(-A)^s\tau^{2s})(I-\sum_{s=1}^{j}\eta_s(-A)^s\tau^{2s})^{-1}$$

$$\pm i\sqrt{I-\frac{1}{4}(2I+\sum_{s=1}^{l}\alpha_s(-A)^s\tau^{2s})^2(I-\sum_{s=1}^{j}\eta_s(-A)^s\tau^{2s})^{-2}},$$

$$\widetilde{c}_\tau=(I+\sum_{n=1}^{j}\delta_{2n}(-A)^n\tau^{2n}-\sum_{m=1}^{j}\delta_{2m-1}(-A)^m\tau^{2m-1}$$

$$\times(I+\sum_{n=1}^{l-[\frac{m}{2}]}a_{2n}(-A)^n\tau^{2n})^{-1}\sum_{n=1}^{l-[\frac{m}{2}]}a_{2n-1}(-A)^n\tau^{2n-1})^{-1}$$

$$\times\{\sum_{n=1}^{l}\rho_{2n}(-A)^n\tau^{2n-1}+\sum_{m=1}^{j}\delta_{2m-1}(-A)^m\tau^{2m-2}$$

$$\times(I+\sum_{n=1}^{l-[\frac{m}{2}]}a_{2n}(-A)^n\tau^{2n})^{-1}\sum_{n=1}^{j-m+[\frac{m}{2}]+1}b_{2n-1}(-A)^n\tau^{2n-1}\},$$

$$\widetilde{s}_\tau=(I+\sum_{n=1}^{j}\delta_{2n}(-A)^n\tau^{2n}-\sum_{m=1}^{j}\delta_{2m-1}(-A)^m\tau^{2m-1}$$

$$\times(I+\sum_{n=1}^{l-[\frac{m}{2}]}a_{2n}(-A)^n\tau^{2n})^{-1}\sum_{n=1}^{l-[\frac{m}{2}]}a_{2n-1}(-A)^n\tau^{2n-1})^{-1}$$

$$\times\{\sum_{n=1}^{l}\rho_{2n-1}(-A)^{n-1}\tau^{2n-2}+\sum_{m=1}^{j}\delta_{2m-1}(-A)^m\tau^{2m-2}$$

$$\times(I+\sum_{n=1}^{l-[\frac{m}{2}]}a_{2n}(-A)^n\tau^{2n})^{-1}(I+\sum_{n=1}^{j-m+[\frac{m}{2}]}b_{2n}(-A)^n\tau^{2n})\},$$

$$f_0^{j,l}=\sum_{n=1}^{l}\rho_{2n}(-A)^n\tau^{2n-1}\sum_{\lambda=1}^{n}(-A)^{n-\lambda}f^{(2\lambda-2)}(0)$$

$$
+\sum_{n=1}^{l}\rho_{2n-1}(-A)^{n-1}\tau^{2n-2}\sum_{\lambda=1}^{n}(-A)^{n-\lambda}f^{(2\lambda-1)}(0)
$$
$$
-\sum_{n=1}^{j}\delta_{2n}(-A)^{n}\tau^{2n-1}\sum_{\lambda=1}^{n}(-A)^{n-\lambda}f^{(2\lambda-2)}(\tau)
$$
$$
-\sum_{n=1}^{j}\delta_{2n-1}(-A)^{n}\tau^{2n-2}\sum_{\lambda=1}^{n-1}(-A)^{n-\lambda}f^{(2\lambda-1)}(\tau)
$$
$$
-\sum_{m=1}^{j}\delta_{2m-1}(-A)^{m}\tau^{2m-2}(I+\sum_{n=1}^{l-[\frac{m}{2}]}a_{2n}(-A)^{n}\tau^{2n})^{-1}
$$
$$
\times\{-\sum_{n=1}^{l-[\frac{m}{2}]}a_{2n}\sum_{\lambda=1}^{n}(-A)^{n-\lambda}f^{(2\lambda-1)}(\tau)\tau^{2n}-\sum_{n=1}^{l-[\frac{m}{2}]}a_{2n-1}\sum_{\lambda=1}^{n}(-A)^{n-\lambda}f^{(2\lambda-2)}(\tau)\tau^{2n-1}
$$
$$
+\sum_{n=1}^{j-m+[\frac{m}{2}]}b_{2n}\sum_{\lambda=1}^{n}(-A)^{n-\lambda}f^{(2\lambda-1)}(0)\tau^{2n}
$$
$$
+\sum_{n=1}^{j-m+[\frac{m}{2}]+1}b_{2n-1}\sum_{\lambda=1}^{n}(-A)^{n-\lambda}f^{(2\lambda-2)}(0)\tau^{2n-1}\}.
$$

From formula (6.44) it follows that the investigation of the stability of difference schemes (6.43) relies in an essential manner on a number of properties of the powers of the operator $\widetilde{R}_{j,l}(\pm i\tau A)$. We were not able to obtain estimates for powers of the operator $\widetilde{R}_{j,l}(\pm i\tau A)$ in the general cases of numbers j and l.

Theorem 6.3.1. *Suppose that $u_0\in D(A)$, $u_1-u_0\in D(A^{\frac{1}{2}})$. Then for the solution of the two-step difference schemes (6.43) for $l=0$ the following stability inequalities hold:*

$$
\max_{0\le k\le N}\|u_k\|_H\le M[\max_{1\le k\le N-1}\|A^{-1/2}f_k^{j,l}\|_H+\|A^{-\frac{1}{2}}\tau^{-1}(u_1-u_0)\|_H+\|u_0\|_H],
$$
$$
\max_{1\le k\le N}\|\tau^{-1}(u_k-u_{k-1})\|_H
$$
$$
\le M[\max_{1\le k\le N-1}\|f_k^{j,l}\|_H+\|\tau^{-1}(u_1-u_0)\|_H+\|A^{\frac{1}{2}}u_0\|_H],
$$
$$
\max_{1\le k\le N-1}\|\tau^{-2}(u_{k+1}-2u_k+u_{k-1})\|_H
$$
$$
\le M[\max_{2\le k\le N-1}\|\tau^{-1}(f_k^{j,l}-f_{k-1}^{j,l})\|_H+\|f_1^{j,l}\|_H
$$
$$
+\|A^{\frac{1}{2}}\tau^{-1}(u_1-u_0)\|_H+\|Au_0\|_H],
$$

where M does not depend on τ, $f_k^{j,l}, 1\le k\le N-1$ and u_0, u_1.

The proof of Theorem 6.3.1 follows the scheme of proof of Theorem 6.2.1 and relies on formula (6.44) and on the estimates

$$||\widetilde{R}_{j,l}(\pm i\tau A)||_{H\to H} \leq 1, \tag{6.45}$$

$$||\tau A^{\frac{1}{2}}(I - \widetilde{R}^2_{j,l}(i\tau A))^{-1}\widetilde{R}_{j,l}(i\tau A)||_{H\to H} \leq M, \tag{6.46}$$

$$||(I - \sum_{s=1}^{j}\eta_s(-A)^s\tau^{2s})^{-1}||_{H\to H} \leq M. \tag{6.47}$$

Remark 6.3.2. Theorem 6.3.1 actually holds in the general cases of numbers j and l under the assumption

$$\tau||A||_{H\to H} \to 0 \text{ when } \tau \to 0.$$

Remark 6.3.3. Applying formulas (6.22), (6.24), (6.26), (6.36), (6.39) and (6.42), we can construct the difference schemes of a $(2l+2j)$-th order of accuracy

$$\frac{u_{k+1} - 2u_k + u_{k-1}}{\tau^2} - \sum_{s=1}^{l}\alpha_s(-A)^s\tau^{2s-2}u_k \tag{6.48}$$

$$-\sum_{s=1}^{j}\eta_s(-A)^s\tau^{2s-2}(u_{k-1} + u_{k+1}) = f_k^{j,l},$$

$$f_k^{j,l} = \sum_{s=1}^{j}\eta_s\sum_{\lambda=1}^{s}(-A)^{s-\lambda}(f^{(2\lambda-2)}(t_{k+1}) + f^{(2\lambda-2)}(t_{k-1}))\tau^{2s-2}$$

$$+\sum_{s=1}^{l}\alpha_s\sum_{\lambda=1}^{s}(-A)^{s-\lambda}f^{(2\lambda-2)}(t_k)\tau^{2s-2}, 1 \leq k \leq N-1, u_0 = v_0,$$

$$\tau^{-1}(u_1 - u_0) = \tau^{-1}(\frac{R_{2l,2j}(i\tau B) + R_{2l,2j}(-i\tau B)}{2} - I)v_0$$

$$+\tau^{-1}B^{-1}\frac{R_{2l,2j}(i\tau B) - R_{2l,2j}(-i\tau B)}{2i}v'(0) + \ f_0^{j,l}$$

for the approximate solution of the initial-value problem (6.1). In a similar manner one can obtain the stability inequalities for the solution of the two-step difference schemes (6.48).

Now, the abstract Theorem 6.3.1 is applied in the investigation of difference schemes of higher order of accuracy with respect to one variable for approximate solutions of the mixed boundary-value problem (6.6). The first step of discretization of problem (6.6) is given above. Suppose that the operator

$$(I+\sum_{n=1}^{j}\delta_{2n}(-A)^{n}\tau^{2n}-\sum_{m=1}^{j}\delta_{2m-1}(-A)^{m}\tau^{2m-1}$$
$$\times(I+\sum_{n=1}^{l-[\frac{m}{2}]}a_{2n}(-A)^{n}\tau^{2n})^{-1}\sum_{n=1}^{l-[\frac{m}{2}]}a_{2n-1}(-A)^{n}\tau^{2n-1})$$

has a bounded inverse. Suppose further that the operator $(I-\sum_{s=1}^{j}\eta_s(-A_h^x)^s\tau^{2s})$ has a bounded inverse and

$$I-\frac{1}{4}(2I+\sum_{s=1}^{l}\alpha_s(-A_h^x)^s\tau^{2s})^2(I-\sum_{s=1}^{j}\eta_s(-A_h^x)^s\tau^{2s})^{-2}\geq 0.$$

Then in the second step we replace problem (6.34) by the difference schemes (6.43)

$$\tau^{-2}(u_{k+1}^h(x)-2u_k^h(x)+u_{k-1}^h(x))+\widetilde{A_h^x}u_k^h(x) \tag{6.49}$$
$$+\widetilde{AA_h^x}(u_{k+1}^h(x)+u_{k-1}^h(x))=f_k^h(x),t_k=k\tau,1\leq k\leq N-1,x\in\Omega_h,$$
$$\widetilde{A_h^x}=-\sum_{s=1}^{l}\alpha_s(-A_h^x)^s\tau^{2s-2},\ \widetilde{AA_h^x}=-\sum_{s=1}^{j}\eta_s(-A_h^x)^s\tau^{2s-2},$$
$$f_k^h(x)=\sum_{s=1}^{j}\eta_s\sum_{\lambda=1}^{s}(-A_h^x)^{s-\lambda}(f^{(2\lambda-2)}(t_{k+1},x)+f^{(2\lambda-2)}(t_{k-1},x))\tau^{2s-2}$$
$$+\sum_{s=1}^{l}\alpha_i\sum_{\lambda=1}^{s}(-A_h^x)^{s-\lambda}f^{(2\lambda-2)}(t_k,x)\tau^{2s-2},1\leq k\leq N-1,$$
$$u_1^h(x)=\widetilde{c}_\tau\varphi^h(x)+\widetilde{s}_\tau\psi^h(x)+\tau f_0^h(x),x\in\Omega_h,$$
$$\widetilde{c}_\tau=(I+\sum_{n=1}^{j}\delta_{2n}(-A_h^x)^n\tau^{2n}-\sum_{m=1}^{j}\delta_{2m-1}(-A_h^x)^m\tau^{2m-1}$$
$$\times(I+\sum_{n=1}^{l-[\frac{m}{2}]}a_{2n}(-A_h^x)^n\tau^{2n})^{-1}\sum_{n=1}^{l-[\frac{m}{2}]}a_{2n-1}(-A_h^x)^n\tau^{2n-1})^{-1}$$
$$\times\{\sum_{n=1}^{l}\rho_{2n}(-A_h^x)^n\tau^{2n-1}+\sum_{m=1}^{j}\delta_{2m-1}(-A_h^x)^m\tau^{2m-2}$$
$$\times(I+\sum_{n=1}^{l-[\frac{m}{2}]}a_{2n}(-A_h^x)^n\tau^{2n})^{-1}\sum_{n=1}^{j-m+[\frac{m}{2}]+1}b_{2n-1}(-A_h^x)^n\tau^{2n-1}\},$$

$$\widetilde{s}_\tau = (I + \sum_{n=1}^{j} \delta_{2n}(-A_h^x)^n \tau^{2n} - \sum_{m=1}^{j} \delta_{2m-1}(-A_h^x)^m \tau^{2m-1}$$

$$\times (I + \sum_{n=1}^{l-[\frac{m}{2}]} a_{2n}(-A_h^x)^n \tau^{2n})^{-1} \sum_{n=1}^{l-[\frac{m}{2}]} a_{2n-1}(-A_h^x)^n \tau^{2n-1})^{-1}$$

$$\times \{\sum_{n=1}^{l} \rho_{2n-1}(-A_h^x)^{n-1}\tau^{2n-2} + \sum_{m=1}^{j} \delta_{2m-1}(-A_h^x)^m \tau^{2m-2}$$

$$\times (I + \sum_{n=1}^{l-[\frac{m}{2}]} a_{2n}(-A_h^x)^n \tau^{2n})^{-1}(I + \sum_{n=1}^{j-m+[\frac{m}{2}]} b_{2n}(-A_h^x)^n \tau^{2n})\},$$

$$f_0^h(x) = \sum_{n=1}^{l} \rho_{2n}(-A_h^x)^n \tau^{2n-1} \sum_{\lambda=1}^{n} (-A_h^x)^{n-\lambda} f^{(2\lambda-2)}(0,x)$$

$$+ \sum_{n=1}^{l} \rho_{2n-1}(-A_h^x)^{n-1}\tau^{2n-2} \sum_{\lambda=1}^{n} (-A_h^x)^{n-\lambda} f^{(2\lambda-1)}(0,x)$$

$$- \sum_{n=1}^{j} \delta_{2n}(-A_h^x)^n \tau^{2n-1} \sum_{\lambda=1}^{n} (-A_h^x)^{n-\lambda} f^{(2\lambda-2)}(\tau,x)$$

$$- \sum_{n=1}^{j} \delta_{2n-1}(-A_h^x)^n \tau^{2n-2} \sum_{\lambda=1}^{n-1} (-A_h^x)^{n-\lambda} f^{(2\lambda-1)}(\tau,x)$$

$$- \sum_{m=1}^{j} \delta_{2m-1}(-A_h^x)^m \tau^{2m-2} (I + \sum_{n=1}^{l-[\frac{m}{2}]} a_{2n}(-A_h^x)^n \tau^{2n})^{-1}$$

$$\times \{- \sum_{n=1}^{l-[\frac{m}{2}]} a_{2n} \sum_{\lambda=1}^{n} (-A_h^x)^{n-\lambda} f^{(2\lambda-1)}(\tau,x)\tau^{2n}$$

$$- \sum_{n=1}^{l-[\frac{m}{2}]} a_{2n-1} \sum_{\lambda=1}^{n} (-A_h^x)^{n-\lambda} f^{(2\lambda-2)}(\tau,x)\tau^{2n-1}$$

$$+ \sum_{n=1}^{j-m+[\frac{m}{2}]} b_{2n} \sum_{\lambda=1}^{n} (-A_h^x)^{n-\lambda} f^{(2\lambda-1)}(0,x)\tau^{2n}$$

$$+ \sum_{n=1}^{j-m+[\frac{m}{2}]+1} b_{2n-1} \sum_{\lambda=1}^{n} (-A_h^x)^{n-\lambda} f^{(2\lambda-2)}(0,x)\tau^{2n-1}\}.$$

We have that

Theorem 6.3.4. *Let τ and $|h|$ be sufficiently small numbers. Then the solutions of the difference schemes (6.49) for $l = 0$ satisfy the following stability estimates:*

$$\max_{0\leq k\leq N} \|u_k^h\|_{L_2(\bar{\Omega}_h)} \leq M_1[\max_{0\leq k\leq N-1} \|(B_h^x)^{-1} f_k^h\|_{L_2(\bar{\Omega}_h)}$$

$$+ \|(B_h^x)^{-1}\psi^h\|_{L_2(\bar{\Omega}_h)} + \|\varphi^h\|_{L_2(\bar{\Omega}_h)}],$$

$$\max_{1\leq k\leq N} \|\tau^{-1}(u_k^h - u_{k-1}^h)\|_{L_2(\bar{\Omega}_h)} \leq M_1[\max_{0\leq k\leq N-1} \|f_k^h\|_{L_2(\bar{\Omega}_h)}$$

$$+ \|\psi^h\|_{L_2(\bar{\Omega}_h)} + \sum_{r=1}^{n} \|\varphi^h_{\bar{x}_r,m_r}\|_{L_2(\bar{\Omega}_h)}],$$

$$\max_{1\leq k\leq N-1} \|\tau^{-2}(u_{k+1}^h - 2u_k^h + u_{k-1}^h)\|_{L_2(\bar{\Omega}_h)} \leq M_1[\max_{1\leq k\leq N-1} \|\tau^{-1}(f_k^h - f_{k-1}^h)\|_{L_2(\bar{\Omega}_h)}$$

$$+ \|f_0^h\|_{L_2(\bar{\Omega}_h)} + \sum_{r=1}^{n} \|\psi^h_{\bar{x}_r,m_r}\|_{L_2(\bar{\Omega}_h)} + \sum_{r=1}^{n} \|(\varphi^h_{\bar{x}_r})_{x_r,m_r}\}\|_{L_2(\bar{\Omega}_h)}].$$

Here M does not depend on τ, h, $\varphi^h(x)$, $\psi^h(x)$ and $f_k^h(x)$, $0 \leq k \leq N-1$.

The proof of Theorem 6.3.4 is based on the abstract Theorem 6.3.1 and the symmetry properties of the difference operator A_h^x defined by formula (6.33).

Note that in a similar manner one can construct the difference schemes of a high order of accuracy with respect to one variable for approximate solutions of the boundary-value problem (6.5). Abstract Theorem 6.3.1 permits us to obtain stability estimates for the solutions of these difference schemes.

Chapter 7

Uniform Difference Schemes for Perturbation Problems

In the present chapter we consider two abstract Cauchy problems

$$\varepsilon v'(t) + Av(t) = f(t) \quad (0 \leq t \leq T), v(0) = v_0,$$

$$\varepsilon^2 v''(t) + Av(t) = f(t) \quad (0 \leq t \leq T), v(0) = v_0, v'(0) = v_0'$$

and the boundary-value problem

$$-\varepsilon^2 v''(t) + Av(t) = f(t) \quad (0 \leq t \leq T), v(0) = v_0, v(T) = v_T$$

for differential equations in an arbitrary Banach space E with the positive operator A and an arbitrary ε parameter multiplying the derivative term. The high order of accuracy single-step uniform difference schemes generated by an exact difference scheme of the approximate solutions for differential equations of parabolic type with an arbitrary parameter ε on the highest derivative are presented. The high order of accuracy two-step uniform difference schemes generated by an exact difference scheme of the approximate solutions for differential equations of the elliptic and hyperbolic types with an arbitrary parameter ε on the highest derivative are presented. The well-posedness of these difference schemes for parabolic and elliptic equations in various Banach spaces is established. The stability estimates of solutions of a high order of accuracy difference schemes for hyperbolic equations with an arbitrary ε parameter on the highest derivative are obtained.

7.1 A Cauchy Problem for Parabolic Equations

We consider the abstract Cauchy problem for the differential equation

$$\varepsilon v'(t) + Av(t) = f(t) \ (0 \leq t \leq T), \ v(0) = v_0 \tag{7.1}$$

in an arbitrary Banach space E with the (unbounded) operator A. Here $v(t)$ and $f(t)$ are the unknown and the given functions, respectively, defined on $[0, T]$ with values in E. The derivative $v'(t)$ is understood as the limit in the norm of E of the corresponding ratio of differences. A is a given, closed, linear operator in E. We will assume that the operator A has dense domain $D(A)$ and that for $\lambda \geq 0$ the operator $A + \lambda I$ has a bounded inverse whose norm satisfies the inequality

$$\| (A+\lambda I)^{-1} \|_{E \to E} \leq M(\lambda+\delta)^{-1} \text{ (for some } \delta > 0), \tag{7.2}$$

where I is the identity operator. Obviously (7.2) is fulfilled when the operator $-A$ is the generator of a strongly continuous semigroup $\exp(-tA)(t \geq 0)$, for which the estimate

$$\| \exp(-tA) \|_{E \to E} \leq Me^{-\delta t} \tag{7.3}$$

is valid. Finally, v_0 is a given element of E and $\varepsilon \in (0, \infty)$.

A function $v(t)$ is called a solution of problem (7.1) if the following conditions are satisfied:

i. $v(t)$ is continuously differentiable on the segment $[0, T]$. The derivative at the endpoints of the segment are understood as the appropriate unilateral derivatives.

ii. The element $v(t)$ belongs to $D(A)$ for all $t \in [0, T]$ and the function $Av(t)$ is continuous on $[0, T]$.

iii. $v(t)$ satisfies the equation and the initial condition (7.1).

In Chapter 4 the stable high order of accuracy difference schemes generated by an exact difference scheme or by Taylor's decomposition on two points for the numerical solutions of an abstract Cauchy problem (7.1) for $\varepsilon = 1$ were presented. These difference schemes can be applied in the general $\varepsilon \in (0, \infty)$. However, it can be shown in general that no classic finite difference approximations yield uniform convergence with respect to ε. This is a very disappointing result and it raises the question whether there are any difference schemes which are uniform in ε? The main result of this chapter gives an answer to this question. Actually, we study the high order of accuracy single-step uniform difference schemes of the approximate solutions for differential equations of parabolic type with an arbitrary parameter ε on the highest derivative.

On the segment $[0, T]$ we consider a uniform grid

$$[0, T]_\tau = \{t_k = k\tau, k = 0, 1, \ldots, N, N\tau = T\}$$

with step τ. Let $\tau = \rho\varepsilon$. By Theorem 2.1.1 we have the exact single-step difference scheme

$$\rho^{-1}(v(t_k) - v(t_{k-1})) + \rho^{-1}(1 - \exp(-\rho A))\, v(t_{k-1}) \tag{7.4}$$

$$= \tau^{-1} \int\limits_{t_{k-1}}^{t_k} \exp(-(t_k - s)\, \varepsilon^{-1} A)\, f(s) ds, 1 \leq k \leq N$$

for the solution of the initial-value problem (7.1). Later we will consider the single-step difference scheme

$$\rho^{-1}(u_k - u_{k-1}) + \rho^{-1}(1 - g(\rho A))\, u_{k-1} = f_k, \tag{7.5}$$

$$f_k = \sum_{j=0}^{m-1} J_j f^{(j)}(t_{k-1}), 1 \leq k \leq N, u_0 = v_0.$$

$$J_j = (-1)^j(\varepsilon^{-1}A)^{-j} J_0 + \sum_{i=1}^{j}(-1)^{\,j-i}(\varepsilon^{-1}A)^{-(j-i+1)}\,\frac{\tau^{i-1}}{i!},$$

$$1 \leq j \leq m-1,\ J_{\,0} = (\rho A)^{-1}(1 - g(\rho A)),\ g(\rho A) = \exp(-\rho A)$$

for approximate solutions of the initial-value problem (7.1).

This problem is uniquely solvable and the following formula holds:

$$u_k = g(k\rho A)u_0 + \sum_{j=1}^{k} g((k-j)\rho\ A)\ f_j\rho. \tag{7.6}$$

Let us reduce the difference schemes (7.5) to an operator problem in the space $F_\tau(E)$ of vectors $v^\tau = \{v_k\}_{k=1}^N$. Recall that the space $F_\tau(E)$ can be equipped with various norms and thus become a normed space (see Chapter 4). Thus, for instance, the vector space $F_\tau(E)$ generates the normed space $C_\tau(E) = C([0,T]_\tau, E)$ with the norm

$$\| \varphi^\tau \|_{C_\tau(E)} = \max_{1\leq k\leq N} \| \varphi_k \|_E,$$

the normed space $C_\tau^\beta(E) = C^\beta([0,T]_\tau, E)$ $0 < \beta < 1$, with the norm

$$\| \varphi^\tau \|_{C_\tau^\beta(E)} = \| \varphi^\tau \|_{C_\tau(E)} + \max_{1\leq k<k+r\leq N} \| \varphi_{k+r} - \varphi_k \|_E \frac{1}{(r\tau)^\beta},$$

the weighted normed space $C_\tau^{\beta,\gamma}(E) = C^{\beta,\gamma}([0,T]_\tau, E), 0 < \gamma \leq \beta,\ 0 < \beta < 1$, with the norm

$$\| \varphi^\tau \|_{C_\tau^{\beta,\gamma}(E)} = \| \varphi^\tau \|_{C_\tau(E)} + \max_{1\leq k<k+r\leq N} \| \varphi_{k+r} - \varphi_k \|_E \frac{((k+r)\tau)^\gamma}{(r\tau)^\beta},$$

and the normed space $L_{p,\tau}(E) = L_p([0,T]_\tau, E), 1 \leq p < \infty$, with the norm

$$\| \varphi^\tau \|_{L_{p,\tau}(E)} = (\sum_{k=1}^{N} \| \varphi_k \|_E^p \tau)^{\frac{1}{p}}.$$

First we will define the operator D_τ, acting from the space $E \times F_\tau(E)$ of vectors $w^\tau = \{w_k\}_{k=0}^N$ into the space $F_\tau(E)$ of vectors $v^\tau = \{v_k\}_{k=1}^N$ by the rule

$$v^\tau = D_\tau w^\tau, v_k = \frac{1}{\rho}(w_k - w_{k-1}), k = 1, \ldots, N.$$

Second we will define an operator A_τ from the space $E \times F_\tau(E)$ of vectors $w^\tau = \{w_k\}_{k=0}^N$ into the space $F_\tau(E)$ of vectors $v^\tau = \{v_k\}_{k=1}^N$ by the rule

$$v^\tau = A_\tau w^\tau, v_k = \rho^{-1}(1 - g(\rho A))\, w_{k-1},\ k = 1, \ldots, N.$$

Next, let us introduce the continuation operator $\Pi(u_0)$, which acts from $E \times F_\tau(E)$ to $F_\tau(E)$ according to the rule

$$\Pi(u_0)(u_1, \ldots, u_N) = (u_0, u_1, \ldots, u_N).$$

Then the difference schemes (7.5) can obviously be rewritten as the equivalent operator equation

$$D_\tau \Pi(u_0) u^\tau + A_\tau \Pi(u_0) u^\tau = f^\tau.$$

Here f^τ is defined by the formula

$$f^\tau = (f_1, \ldots, f_N).$$

The last operator problem will be considered in the space $F_\tau(E)$. From its unique solvability for any $u_0 \in E$ and $f^\tau \in F_\tau(E)$ it follows that its solution u^τ defines a continuous additive and homogeneous operator $u^\tau(f^\tau, u_0)$.

The initial-value difference problem (7.5) is said to be stable in $F_\tau(E)$ if we have the stability inequality

$$\| u^\tau(f^\tau, u_0) \|_{F_\tau(E)} \leq M[\| f^\tau \|_{F_\tau(E)} + \| u_0 \|_E],$$

where M is independent not only of f^τ, u_0, τ but also of ε.

Theorem 7.1.1. *For the solution of the difference problem* (7.5) *the following estimate is valid:*

$$\max_{1 \leq k \leq N} \| u_k \|_E \leq M \left[\| u_0 \|_E + \frac{\rho}{1 - \exp(-\delta\rho)} \max_{1 \leq k \leq N} \| f_k \|_E \right], \tag{7.7}$$

where M is independent not only of f^τ, u_0, τ, but also of ε.

Proof. Using formula (7.6) and estimate (7.3), we obtain

$$\| u_k \|_E \leq M \left[\| u_0 \|_E + \sum_{j=1}^{k} e^{-\delta(k-j)\rho} \rho \max_{1 \leq j \leq N} \| f_j \|_E \right]$$

for all k, $k = 1, \ldots, N$. Using the last inequality and

$$\sum_{j=1}^{k} e^{-\delta(k-j)\rho} \leq \sum_{i=0}^{\infty} e^{-\delta i\rho} = 1/(1 - e^{-\delta\rho}), \tag{7.8}$$

we obtain

$$\| u_k \|_E \leq M \left[\| u_0 \|_E + \frac{\rho}{1 - e^{-\delta\rho}} \max_{1 \leq j \leq N} \| f_j \|_E \right]$$

for all k, $k = 1, \ldots, N$. This completes the proof of Theorem 7.1.1.

Note that if $\varepsilon = 1$, then

$$\frac{\rho}{1-e^{-\delta p}} = \frac{\tau}{1-e^{-\delta\tau}} \leq M.$$

Therefore, (7.7) is the generalized stability inequality of (7.5) in $C_\tau(E)$.

We say that the difference problem (7.5) has an m-th order of accuracy in $F_\tau(E)$ on the solutions $v(t)$ of the Cauchy problem (7.1) if the error vector

$$\{v(t_k) - u_k\}_1^N$$

satisfies the estimate

$$\| \{v(t_k) - u_k\}_1^N \|_{F_\tau(E)} \leq M\tau^m,$$

where M is independent not only τ, but also of ε.

Generalized stability inequality (7.7) permits us to obtain the convergence of (7.5) in $C_\tau(E)$.

Theorem 7.1.2. *Suppose that the function $f(t)$ has m derivatives and*

$$\| f^{(m)}(t) \|_E \leq M, 0 \leq t \leq T. \tag{7.9}$$

Then for the solution of the difference problem (7.5) the following convergence estimate is valid:

$$\| \{v(t_k) - u_k\}_1^N \|_{C_\tau(E)} \leq M\tau^m, \tag{7.10}$$

where M does not depend on τ and ε.

Proof. The components $z_k = v(t_k) - u_k$ of error vector $z^\tau = \{z_k\}_1^N$ satisfy the relations

$$\rho^{-1}(z_k - z_{k-1}) + \rho^{-1}(1-g(\rho A))z_{k-1} = R_k, 1 \leq k \leq N, \ z_0 = 0, \tag{7.11}$$

where

$$R_k = \rho^{-1}(v(t_k) - g(\rho A)v(t_{k-1})) - f_k.$$

From this and the generalized stability inequality (7.7) it follows that

$$\max_{1\leq k\leq N} \| z_k \|_E \leq M \frac{\rho}{1-\exp(-\delta\rho)} \max_{1\leq k\leq N} \|R_k\|_E. \tag{7.12}$$

Therefore to this end it suffices to estimate $\{R_k\}_1^N$ in $C_\tau(E)$. By (7.5) and (7.11), we have that

$$R_k = \tau^{-1} \int_{t_{k-1}}^{t_k} \exp(-(t_k - s)\varepsilon^{-1}A) f(s) \, ds - f_k .$$

Now, using Taylor's formula, we obtain

$$R_k = \sum_{j=0}^{m-1} \tau^{-1} \int_{t_{k-1}}^{t_k} \exp(-(t_k - s)\varepsilon^{-1}A)(s - t_{k-1})^j \, ds f^{(j)}(t_{k-1})\frac{1}{j!}$$

$$+\tau^{-1} \int_{t_{k-1}}^{t_k} \exp(-(t_k - s)\varepsilon^{-1}A) \int_{t_{k-1}}^{s} (s-z)^m f^{(m)}(z)\, dzds \, \frac{1}{m!} - f_k \, .$$

A simple calculation shows that

$$\sum_{j=0}^{m-1} \tau^{-1} \int_{t_{k-1}}^{t_k} \exp(-(t_k - s)\varepsilon^{-1}A)(s - t_{k-1})^j \, ds f^{(j)}(t_{k-1})\frac{1}{j!} = f_k.$$

Therefore, we can write

$$R_k = \tau^{-1} \int_{t_{k-1}}^{t_k} \exp(-(t_k - s)\varepsilon^{-1}A) \int_{t_{k-1}}^{s} \frac{(s-z)^{m-1}}{(m-1)!} f^{(m)}(z)\, dzds \, .$$

Then, using estimates (7.3) and (7.9), we obtain

$$\max_{1 \le k \le N} ||R_k||_E \le M \, (\delta\rho)^{-1}(1 - \exp(-\delta\rho))\tau^m.$$

From this and estimate (7.11) follows estimate (7.10). Theorem 7.1.2 is proved.

Note that for the convergence estimate $\| \{v(t_k) - u_k)\}_1^N \|_{F_\tau(E)} \le M\tau^m$ it suffices to obtain the generalized stability inequality in $F_\tau(E)$.

From formula (7.6) one derives the following result.

Theorem 7.1.3. *For the solution of the difference problem* (7.5) *the following estimate is valid:*

$$\| \{u_k\}_1^N \|_{L_{p,\tau}(E)} \le M \, [\, \| u_0 \|_E \tag{7.13}$$

$$+\frac{\rho}{1 - \exp(-\delta\rho)} \| \{f_k\}_1^N \|_{L_{p,\tau}(E)}], 1 \le p < \infty,$$

where M is independent not only of f^τ, u_0, τ, but also of ε.

Proof. Using formula (7.6) and estimate (7.3), we obtain

$$\| u_k \|_E \le M \, [\, \| u_0 \|_E + \sum_{s=1}^{N} e^{-\delta(s-1)\rho}\rho \, \| f^*_{k-s+1} \|_E \,]$$

for all k, $k = 1, \dots, N$. Here $f_s^* = f_s$ if $1 \le s \le k$ and $f_s^* = 0$ if otherwise. From the Minkowski sum inequality (with respect to k) it follows that

$$(\sum_{k=1}^{N} \| u_k \|_E^p \tau)^{\frac{1}{p}} \le M\, T \| u_0 \|_E + M \sum_{s=1}^{N} e^{-\delta(s-1)\rho} \rho (\sum_{k=1}^{N} \| f_{k-s+1}^* \|_E^p \tau)^{\frac{1}{p}}.$$

By the definition of the grid function f_{k-s+1}^*,

$$\sum_{k=1}^{N} \|f_{k-s+1}^* \|_E^p \tau \le \sum_{k=1}^{N} \| f_k \|_E^p \tau.$$

Using the last estimate and inequality (7.8), we obtain estimate (7.13). Theorem 7.1.3 is proved.

Remark 7.1.4. Theorems 7.1.1 and 7.1.3 permit us to obtain error estimates of the high order of accuracy difference schemes over time and space variables for parabolic equations and the first-order hyperbolic system of equations with a small parameter multiplying the time-derivative term in the various Banach norms.

Now, assume that the operator $-A$ is the generator of an analytic semigroup $\exp(-tA)(t \ge 0)$, for which the estimates

$$\| A \exp(-tA) \|_{E \to E} \le M e^{-\delta t}, \ t \| A \exp(-tA) \|_{E \to E} \le M \tag{7.14}$$

are valid. From this it follows that

$$\|\exp(-k\rho A) - \exp\{-(k+r)\rho A\}\|_{E \to E} \le M \frac{r^\alpha}{(k+r)^\alpha} \exp(-k\rho\delta), \tag{7.15}$$

for all $k, r, 1 \le k < k+r \le N, 0 \le \alpha \le 1$ and all $\tau > 0$, where M does not depend on k, r, τ, α and ε.

Theorem 7.1.5. *For the solution of the difference problem* (7.5) *the following estimate is valid:*

$$\| \{u_k\}_1^N \|_{C_\tau^{\alpha,a}(E)} \le M \Big[\| u_0 \|_E \tag{7.16}$$

$$+ \frac{\rho}{1 - \exp(-\delta\rho)} \| \{f_k\}_1^N \|_{C_\tau^{\alpha,a}(E)} \Big], \ 0 \le \alpha \le 1,$$

where M is independent not only of f^τ, u_0, τ, α but also of ε.

Proof. By estimate (7.7), we have that

$$\| \{u_k\}_1^N \|_{C_\tau(E)} \le M \Big[\| u_0 \|_E + \frac{\rho}{1 - \exp(-\delta\rho)} \| \{f_k\}_1^N \|_{C_\tau^{\alpha,a}(E)} \Big].$$

To this end it suffices to establish the estimate

$$\| u_k - u_{k+r} \|_E \le M \frac{r^\alpha}{(k+r)^\alpha} \frac{\rho}{1 - \exp(-\delta\rho)} \| \{f_k\}_1^N \|_{C_\tau^{\alpha,a}(E)} \tag{7.17}$$

for all $k, r, 1 \le k < k+r \le N$ and all $\tau > 0$. By (7.6), we have that

$$u_k = w_k + v_k,$$

where

$$w_k = g(k\rho A)u_0,$$

$$v_k = \sum_{j=1}^{k} g((k-j)\rho A) f_j \rho.$$

Applying estimate (7.15), we obtain

$$\| w_k - w_{k+r} \|_E = \| g(k\rho A)u_0 - g((k+r)\rho A)u_0 \|_E \tag{7.18}$$

$$\le M \frac{r^\alpha}{(k+r)^\alpha} \| u_0 \|_E$$

for all $k, r, 1 \le k < k+r \le N$ and all $\tau > 0$. Now let us estimate the difference $v_k - v_{k+r}$ for all $k, r, 1 \le k < k+r \le N$ and all $\tau > 0$. We shall consider two separate cases: $k \le 2r$ and $k > 2r$. If $k \le 2r$, then from the estimate

$$\| v_k \|_E \le M \frac{\rho}{1-\exp(-\delta\rho)} \| \{f_k\}_1^N \|_{C_\tau^{\alpha,a}(E)}$$

for all $k, 1 \le k \le N$ and all $\tau > 0$ and the triangle inequality it follows that

$$\| v_k - v_{k+r} \|_E \le \| v_k \|_E + \| v_{k+r} \|_E \le 2M \frac{\rho}{1-\exp(-\delta\rho)} \| \{f_k\}_1^N \|_{C_\tau^{\alpha,a}(E)} r^\alpha r^{-\alpha}$$

$$\le 3^{1+\alpha} M \frac{r^\alpha}{(k+r)^\alpha} \frac{\rho}{1-\exp(-\delta\rho)} \| \{f_k\}_1^N \|_{C_\tau^{\alpha,a}(E)}.$$

Hence, for $k \le 2r$ we have the estimate

$$\| v_k - v_{k+r} \|_E \le M_1 \frac{r^\alpha}{(k+r)^\alpha} \frac{\rho}{1-\exp(-\delta\rho)} \| \{f_k\}_1^N \|_{C_\tau^{\alpha,a}(E)}. \tag{7.19}$$

Now let $k > 2r$. We have that

$$v_k = (I - g(\rho A))^{-1}[I - g(k\rho A)]\rho f_k + \sum_{j=1}^{k-1} g((k-j)\rho A)(f_j - f_k)\rho, \quad k \ge 2. \tag{7.20}$$

Identity (7.20) yields

$$v_{k+r} - v_k = (I - g(\rho A))^{-1}[I - g((k+r)\rho A)]\rho(f_{k+r} - f_k)$$

$$+ (I - g(\rho A))^{-1}[g(k\rho A) - g((k+r)\rho A)]]\rho f_k$$

$$+\sum_{j=k-r+1}^{k+r-1} g((k+r-j)\rho A)(f_j-f_{k+r})\rho - \sum_{j=k-r+1}^{k-1} g((k-j)\rho A)(f_j-f_k)\rho$$

$$+\sum_{j=1}^{k-r} g((k+r-j)\rho A)(f_k-f_{k+r})\rho$$

$$+\sum_{j=1}^{k-r}[g((k+r-j)\rho A)-g((k-j)\rho A)](f_j-f_k)\rho$$

$$= P_1+P_2+P_3+P_4+P_5+P_6.$$

Let us estimate P_1, P_2, P_3, P_4, P_5 and P_6 separately. We begin with P_1. Using estimate (7.3), we obtain

$$||P_1||_E \leq \sum_{j=1}^{k+r-1} ||g((k+r-j)\rho A)||_{E\to E}\rho||f_{k+r}-f_k||_E$$

$$\leq M\sum_{j=1}^{k+r-1} e^{-\delta(k+r-j)\rho}\rho\frac{r^\alpha}{(k+r)^\alpha} \parallel \{f_k\}_1^N \parallel_{C_\tau^{\alpha,a}(E)}$$

$$\leq M\frac{r^\alpha}{(k+r)^\alpha}\frac{\rho}{1-\exp(-\delta\rho)} \parallel \{f_k\}_1^N \parallel_{C_\tau^{\alpha,a}(E)}.$$

In exactly the same manner one establishes the estimate

$$||P_5||_E \leq M\frac{r^\alpha}{(k+r)^\alpha}\frac{\rho}{1-\exp(-\delta\rho)} \parallel \{f_k\}_1^N \parallel_{C_\tau^{\alpha,a}(E)}.$$

Next, let us estimate P_2. Using estimates (7.3) and (7.15), we obtain

$$||P_2||_E \leq ||(I-g(\rho A))^{-1}||_{E\to E}\rho||g(k\rho A)-g((k+r)\rho A)||_{E\to E}\rho||f_k||_E$$

$$\leq M\frac{r^\alpha}{(k+r)^\alpha}\frac{\rho}{1-\exp(-\delta\rho)} \parallel \{f_k\}_1^N \parallel_{C_\tau(E)}$$

$$\leq M\frac{r^\alpha}{(k+r)^\alpha}\frac{\rho}{1-\exp(-\delta\rho)} \parallel \{f_k\}_1^N \parallel_{C_\tau^{\alpha,a}(E)}.$$

Now, let us estimate P_3. Using estimates (7.3) and (7.14), we obtain

$$||P_3||_E \leq \sum_{j=k-r+1}^{k+r-1} ||g((k+r-j)\rho A)||_{E\to E}\rho||f_j-f_{k+r}||_E$$

$$\leq M\sum_{j=k-r+1}^{k+r-1} e^{-\delta(k+r-j)\rho}\rho\frac{(k+r-j)^\alpha}{(k+r)^\alpha}\parallel \{f_k\}_1^N \parallel_{C_\tau^{\alpha,a}(E)}$$

$$\leq M \sum_{j=k-r+1}^{k+r} e^{-\delta(k+r-j)\rho} \rho \frac{(2r-1)^\alpha}{(k+r)^\alpha} \| \{f_k\}_1^N \|_{C_\tau^{\alpha,a}(E)}$$

$$\leq M 2^\alpha \frac{\rho}{1-\exp(-\delta\rho)} \frac{r^\alpha}{(k+r)^\alpha} \| \{f_k\}_1^N \|_{C_\tau^{\alpha,a}(E)}$$

$$\leq M_1 \frac{\rho}{1-\exp(-\delta\rho)} \frac{r^\alpha}{(k+r)^\alpha} \| \{f_k\}_1^N \|_{C_\tau^{\alpha,a}(E)}.$$

In exactly the same manner one establishes the estimate

$$||P_4||_E \leq M_1 \frac{r^\alpha}{(k+r)^\alpha} \frac{\rho}{1-\exp(-\delta\rho)} \| \{f_k\}_1^N \|_{C_\tau^{\alpha,a}(E)}.$$

Finally, let us estimate P_6. Using estimate (7.15), we obtain

$$\| P_6 \|_E \leq \sum_{j=1}^{k-r} \rho ||g((k+r-j)\rho A) - g((k-j)\rho A)||_{E\to E} ||f_j - f_{k+r} \|_E$$

$$\leq \sum_{j=1}^{k-r} \frac{r^\alpha}{(k+r-j)^\alpha} e^{-\delta(k-j)\rho} \rho \frac{(k+r-j)^\alpha}{(k+r)^\alpha} \| \{f_k\}_1^N \|_{C_\tau^{\alpha,a}(E)}$$

$$\leq M_1 \frac{r^\alpha}{(k+r)^\alpha} \frac{\rho}{1-\exp(-\delta\rho)} \| \{f_k\}_1^N \|_{C_\tau^{\alpha,a}(E)}.$$

Combining the estimates for P_1, P_2, P_3, P_4, P_5 and P_6, we obtain (7.19) for $k > 2r$. Estimate (7.17) follows from the triangle inequality and estimates (7.18) and (7.19). Theorem 7.1.5 is proved.

The initial-value difference problem (7.5) is said to be well posed (coercively stable) in $F_\tau(E)$ if we have the coercive stability inequality

$$\| \{\rho^{-1}(u_k - u_{k-1})\}_1^N \|_{F_\tau(E)} + \| \{\rho^{-1}(1-g(\rho A))u_{k-1}\}_1^N \|_{F_\tau(E)}$$

$$\leq M[\| f^\tau \|_{F_\tau(E)} + \| Au_0 \|_E],$$

where M is independent not only of f^τ, u_0, τ but also of ε.

Since the initial-value differential problem (7.1) in $C(E)$ is not well posed for the general strongly positive operator A and space E, then the well-posedness of the initial-value difference problem in $C_\tau(E)$ norm does not take place uniformly with respect to $\tau > 0$ and ε. This means that the coercive norm

$$\| u^\tau \|_{K_\tau(E)} = ||\{\rho^{-1}(u_k - u_{k-1})\}_1^N \|_{C_\tau(E)} + \| \{\rho^{-1}(1-g(\rho A))u_{k-1}\}_1^N \|_{C_\tau(E)}$$

tends to ∞ as $\tau \to +0$. The investigation of the difference problem (7.5) permits us to establish the order of growth of this norm to ∞. First, let us prove a lemma that will be needed in the sequel.

Lemma 7.1.6. *The following estimate holds:*

$$\sum_{j=1}^{N} ||(I - e^{-\rho A})e^{-(j-1)\rho A}||_{E\to E} \leq M_1 \ \ln(1/\tau), \tag{7.21}$$

where M_1 does not depend on τ and ε.

Proof. Using estimate (7.15), we obtain

$$\sum_{j=1}^{N} ||(I - e^{-\rho A})e^{-(j-1)\rho A}||_{E\to E} \leq M \sum_{j=1}^{N} \frac{1}{j} \leq M \int_{1}^{N} \frac{ds}{s} = M \ \ln N \leq M_1 \ln(1/\tau).$$

Lemma 7.1.6 is proved.

Theorem 7.1.7. *The solutions of the difference problem (7.5) in $C_\tau(E)$ obey the almost coercive inequality*

$$\| u^\tau \|_{K_\tau(E)} \leq M_1[\| Au_0\|_E + \ln \frac{1}{\tau}\| f^\tau \|_{C_\tau(E)}], \tag{7.22}$$

where M_1 is independent not only of f^τ, u_0, but also of τ and ε.

Proof. Using the formula

$$(I - e^{-\rho A})(\rho A)^{-1} = \int_{0}^{1} e^{-\rho s A} ds$$

and estimate (7.14), we obtain

$$||(I - e^{-\rho A})(\rho A)^{-1}||_{E\to E} \leq M. \tag{7.23}$$

Using formula (7.6) and estimates (7.14) and (7.23), we obtain

$$||\rho^{-1}(1 - g(\rho A))u_{k-1} \ || \ _E \leq M \ [\ || \ \rho^{-1}(1 - g(\rho A))g((k-1)\rho A)A^{-1}||_{E\to E}$$

$$||Au_0 \ || \ _E + \sum_{j=1}^{k-1} || \ \rho^{-1}(1 - g(\rho A))g((k-1-j)\rho A)||_{E\to E}||f_j||_E]$$

$$\leq M \ [\ ||Au_0 \ || \ _E + \sum_{j=1}^{N} || \ \rho^{-1}(I - e^{-\rho A})e^{-(j-1)\rho A}||_{E\to E}\| \ f^\tau \ \|_{C_\tau(E)} \ ]$$

$$\leq M_1[\| Au_0\|_E + \ln \frac{1}{\tau}\| f^\tau \|_{C_\tau(E)}]$$

for all k, $k = 1, \dots, N$. Using the last inequality, we obtain

$$\| \{\rho^{-1}(1 - g(\rho A))u_{k-1}\}_1^N \|_{C_\tau(E)} \leq M_1[\| Au_0\|_E + \ln \frac{1}{\tau} \| f^\tau \|_{C_\tau(E)}].$$

The estimate for $\|\{\rho^{-1}(u_k - u_{k-1})\}_1^N \|_{C_\tau(E)}$ follows from the triangle inequality and the last estimate. Theorem 7.1.7 is proved.

Now let us study the well-posedness of the difference problem (7.5) in various Banach spaces.

Theorem 7.1.8. *The solutions of the difference problem* (7.5) *in* $C_\tau^{\alpha,\alpha}(E)$ *obey the coercivity inequality*

$$\|\{\rho^{-1}(u_k - u_{k-1})\}_1^N \|_{C_\tau^{\alpha,\alpha}(E)} + \| \{\rho^{-1}(1 - g(\rho A))u_{k-1}\}_1^N \|_{C_\tau^{\alpha,\alpha}(E)} \tag{7.24}$$
$$\leq M_1[\frac{1}{\alpha(1-\alpha)} \| f^\tau \|_{C_\tau^{\alpha,\alpha}(E)} + \| Au_0\|_E],$$

where M_1 *is independent not only of* f^τ, u_0, α, *but also of* τ *and* ε.

Proof. Using formula (7.6), we obtain

$$\rho^{-1}(1 - g(\rho A))u_{k-1} = \rho^{-1}(1 - g(\rho A))g((k-1)\rho A)u_0 + [I - g((k-1)\rho A)]f_k$$
$$+ \sum_{j=1}^{k-1} g((k-1-j)\rho A)(f_j - f_k) = \sum_{m=1}^{3} J_k^m,$$

where

$$J_k^1 = \rho^{-1}(1 - g(\rho A))g((k-1)\rho A)u_0,$$
$$J_k^2 = [I - g((k-1)\rho A)]f_k,$$
$$J_k^3 = \sum_{j=1}^{k-1} g((k-1-j)\rho A)(f_j - f_k).$$

Let us estimate $J_m^\tau = \{J_k^m\}_{k=1}^N$ for all m, $m = 1, 2, 3$ separately. We start with J_1^τ. Using estimates (7.14) and (7.23) we show that, for all k, $1 \leq k \leq N$,

$$\|J_k^1\|_E \leq \|(I - e^{-\rho A})(\rho A)^{-1}\|_{E\to E} \|e^{-(k-1)\rho A}\|_{E\to E} \|Au_0\|_E \leq M\|Au_0\|_E.$$

From this it follows that

$$\|J_1^1 - J_{1+r}^1\|_E \leq \|J_1^1\|_E + \|J_{1+r}^1\|_E \leq 2M\|Au_0\|_E \leq 2M\frac{(2r)^\alpha}{(1+r)^\alpha}\|Au_0\|_E,$$

for all $r, 1 < r + 1 \leq N$.

Further, applying (7.15) and (7.23) we show that, for all $2 \leq k < k + r \leq N$,

$$\|J_k^1 - J_{k+r}^1\|_E \leq \|(I - e^{-\rho A})(\rho A)^{-1}\|_{E\to E}$$
$$\times \|e^{-(k-1)\rho A} - e^{-(k+r-1)\rho A}\|_{E\to E} \|Au_0\|_E$$
$$\leq M\frac{r^\alpha}{(k+r-1)^\alpha}\|Av_0\|_E \leq M_1\frac{r^\alpha}{(k+r)^\alpha}\|Av_0\|_E.$$

Thus, we have proved that

$$||J_1^\tau||_{C_\tau^{\alpha,\alpha}(E)} \leq M_1||Au_0||_E. \tag{7.25}$$

Now let us estimate J_2^τ. Using estimate (7.14), we obtain that

$$\| J_k^2 \|_E \leq [1+||e^{-(k-1)\rho A}||_{E\to E}]||f_k||_E \leq M_1 \| f^\tau \|_{C_\tau^{\alpha,\alpha}(E)}$$

for all $k, 1 \leq k \leq N$. Therefore,

$$\max_{1\leq k\leq N} \| J_k^2 \|_E \leq M_1 \| f^\tau \|_{C_\tau^{\alpha,\alpha}(E)}.$$

From this it follows that

$$||J_1^2 - J_{1+r}^2||_E \leq ||J_1^2||_E + ||J_{1+r}^2||_E \leq 2M_1 \| f^\tau \|_{C_\tau^{\alpha,\alpha}(E)}$$

$$\leq 2^{1+\alpha} M_1 \frac{r^\alpha}{(1+r)^\alpha} \| f^\tau \|_{C_\tau^{\alpha,\alpha}(E)}$$

for all $r, 1 < r+1 \leq N$. Since

$$J_{k+r}^2 - J_k^2 = [I - e^{-(k+r-1)\rho A}](f_{k+r} - f_k) + [e^{-(k-1)\rho A} - e^{-(k+r-1)\rho A}]f_k,$$

it follows that

$$\| J_{k+r}^2 - J_k^2 \|_E \leq [1+||e^{-(k+r-1)\rho A}||_{E\to E}]||f_{k+r} - f_k||_E$$

$$+||e^{-(k-1)\rho A} - e^{-(k+r-1)\rho A}||_{E\to E}||f_k||_E$$

for all $k,\ 2 \leq k < k+r \leq N$. Using estimates (7.14) and (7.15), we obtain

$$\| J_{k+r}^2 - J_k^2 \|_E \leq M_1 [\frac{r^\alpha}{(k+r)^\alpha} + \frac{r^\alpha}{(k+r-1)^\alpha}] \| f^\tau \|_{C_\tau^{\alpha,\alpha}(E)}$$

for all $k,\ 2 \leq k < k+r \leq N$. From this it follows that

$$\| J_2^\tau \|_{C_\tau^{\alpha,\alpha}(E)} \leq M_1 \| f^\tau \|_{C_\tau^{\alpha,\alpha}(E)}. \tag{7.26}$$

Now let us estimate J_3^τ. By the definition $J_1^3 = 0$. Therefore, let us consider the cases $2 \leq k \leq N$. Using estimate (7.15), we obtain

$$\| J_k^3 \|_E \leq \sum_{i=1}^{k-1} ||(I - e^{-\rho A})e^{-(k-i-1)\rho A}||_{E\to E} \| f_i - f_k \|_E$$

$$\leq M \sum_{i=1}^{k-1} \frac{1}{(k-i)^{1-\alpha} k^\alpha} \| f^\tau \|_{C_\tau^{\alpha,\alpha}(E)} \leq \frac{M}{t_k^\alpha} [\sum_{i=1}^{k-1} \frac{\tau}{(t_k - t_i)^{1-\alpha}}] \| f^\tau \|_{C_\tau^{\alpha,\alpha}(E)}.$$

The sum enclosed in the right-hand side square brackets is the lower Darboux integral sum for the integral

$$\int_0^{t_k} \frac{ds}{(t_k - s)^{1-\alpha}} = \frac{t_k^\alpha}{\alpha}.$$

Thus, for all k, $2 \le k \le N$ we have established the bound

$$\| J_k^3 \|_E \le \frac{M_1}{\alpha} \| f^\tau \|_{C_\tau^{\alpha,\alpha}(E)}.$$

Therefore,

$$\max_{1\le k\le N} \| J_k^3 \|_E \le \frac{M_1}{\alpha} \| f^\tau \|_{C_\tau^{\alpha,\alpha}(E)}. \tag{7.27}$$

Now let us estimate the difference $J_{k+r}^3 - J_k^3$ for all k, $2 \le k < k+r \le N$. We will consider separately the cases $k \le 2r$ and $k > 2r$. If $k \le 2r$, then (7.27) yields

$$\| J_{k+r}^3 - J_k^3 \|_E \le \| J_{k+r}^3 \|_E + \| J_k^3 \|_E \le 2\frac{M_1}{\alpha} \| f^\tau \|_{C_\tau^{\alpha,\alpha}(E)}$$

$$\le \frac{M_1 2^{1+\alpha}}{\alpha} \| f^\tau \|_{C_\tau^{\alpha,\alpha}(E)} r^\alpha (k+r)^{-\alpha}.$$

Now let $k > 2r$. Let us represent the difference $J_{k+r}^3 - J_k^3$ as the following sums:

$$J_{k+r}^3 - J_k^3 = \sum_{j=k-r+1}^{k+r-1} (I - g(\rho A))g((k+r-1-j)\rho A)(f_j - f_{k+r})$$

$$- \sum_{j=k-r+1}^{k-1} (I - g(\rho A))g((k-1-j)\rho A)(f_j - f_k)$$

$$+ \sum_{j=1}^{k-r} (I - g(\rho A))g((k+r-1-j)\rho A)(f_k - f_{k+r})$$

$$+ \sum_{j=1}^{k-r} (I - g(\rho A))[g((k+r-j)\rho A) - g((k-j)\rho\, A)](f_j - f_k) = \sum_{m=1}^{4} P_k^m.$$

Let us estimate P_k^1, P_k^2, P_k^3 and P_k^4 separately. We begin with P_k^1. Using estimate (7.14), we obtain

$$||P_k^1||_E \le \sum_{j=k-r}^{k+r-1} ||(I - g(\rho A))g((k+r-1-j)\rho A)||_{E\to E} ||f_j - f_{k+r}||_E$$

$$\le M[\frac{r^\alpha}{(k+r)^\alpha} + \sum_{j=k-r}^{k+r-2} \frac{(k+r-i)^{\alpha-1}}{(k+r)^\alpha}] \| \{f_k\}_1^N \|_{C_\tau^{\alpha,a}(E)}$$

$$\leq M\frac{r^\alpha}{(k+r)^\alpha}\| \{f_k\}_1^N \|_{C_\tau^{\alpha,a}(E)}$$

$$+M\frac{1}{(k+r)^\alpha}[\sum_{j=k-r}^{k+r-2}\frac{1}{(k+r-i)^{1-\alpha}}]\| \{f_k\}_1^N \|_{C_\tau^{\alpha,a}(E)}.$$

The sum enclosed in the right-hand side square brackets is the lower Darboux integral sum for the integral

$$\int_{t_k-t_r}^{t_k+t_r}\frac{ds}{(t_k+t_r-s)^{1-\alpha}}.$$

Since

$$\int_{t_k-t_r}^{t_k+t_r}\frac{ds}{(t_k+t_r-s)^{1-\alpha}}\leq\frac{M}{\alpha}(2t_r)^\alpha,$$

it follows that

$$\| P_k^1 \|_E\leq\frac{M_2}{\alpha}r^\alpha(k+r)^{-\alpha}\| f^\tau \|_{C_\tau^{\alpha,\alpha}(E)}.$$

In exactly the same manner one establishes the estimate

$$||P_k^2||_E\leq\frac{M_2}{\alpha}r^\alpha(k+r)^{-\alpha}\| f^\tau \|_{C_\tau^{\alpha,\alpha}(E)}.$$

Next, let us estimate P_k^3. Since

$$\sum_{j=1}^{k-r}(I-g(\rho A))g((k+r-1-j)\rho A)=g((2r-1)\rho A)-g((k+r-2)\rho A),$$

we have that

$$P_k^3=[g((2r-1)\rho A)-g((k+r-2)\rho A)](f_k-f_{k+r}).$$

Using estimate (7.14), we obtain

$$||P_k^3||_E\leq||g((2r-1)\rho A)-g((k+r-2)\rho A)||_{E\to E}||f_k-f_{k+r}||_E$$

$$\leq M\frac{r^\alpha}{(k+r)^\alpha}\| \{f_k\}_1^N \|_{C_\tau^{\alpha,a}(E)}.$$

Now, let us estimate P_k^4. Using estimate (7.15), we obtain

$$\| P_k^4 \|_E\leq\sum_{j=1}^{k-r}||(I-g(\rho A))[g((k+r-1-j)\rho A)-g((k-1-j)\rho A)]||_{E\to E}||f_j-f_k \|_E$$

$$\leq \sum_{j=1}^{k-r} \frac{r}{(k-1-j)^2} \frac{(k-j)^\alpha}{k^\alpha} \| \{f_k\}_1^N \|_{C_\tau^{\alpha,a}(E)}$$

$$\leq M \frac{t_r}{t_k^\alpha} [\sum_{i=1}^{k-r-1} \frac{\tau}{(t_k - t_i)^{2-\alpha}}] \| f^\tau \|_{C_\tau^{\alpha,\alpha}(E)}.$$

The sum enclosed in the right-hand side square brackets is the lower Darboux integral sum for the integral

$$\int_0^{t_k - t_r} \frac{ds}{(t_k - s)^{2-\alpha}}.$$

Since

$$\int_0^{t_k - t_r} \frac{ds}{(t_k - s)^{2-\alpha}} \leq \int_0^{t_k - t_r} \frac{ds}{(t_k - s)^{2-\alpha}} \leq \frac{1}{(1-\alpha)} t_r^{-1+\alpha},$$

it follows that

$$\| P_k^4 \|_E \leq \frac{M}{1-\alpha} \frac{r^\alpha}{k^\alpha} \| f^\tau \|_{C_\tau^{\alpha,\alpha}(E)}.$$

Thus, we have established the bound

$$\| P_k^4 \|_E \leq \frac{M_1}{1-\alpha} \frac{r^\alpha}{(k+r)^\alpha} \| f^\tau \|_{C_\tau^{\alpha,\alpha}(E)}.$$

Thus, we have shown that for all k, $2 \leq k < k+r \leq N$ the following inequality holds:

$$\| J_{k+r}^4 - J_k^4 \|_E \leq \frac{M_1}{1-\alpha} \frac{r^\alpha}{(k+r)^\alpha} \| f^\tau \|_{C_\tau^{\alpha,\alpha}(E)}. \tag{7.28}$$

Estimates (7.27) and (7.28) give

$$\| J_4^\tau \|_{C_\tau^{\alpha,\alpha}(E)} \leq \frac{M_1}{\alpha(1-\alpha)} \| f^\tau \|_{C_\tau^{\alpha,\alpha}(E)} . \tag{7.29}$$

Finally, applying estimates (7.25), (7.26) and (7.29), we obtain the estimate

$$\| \{\rho^{-1}(1 - g(\rho A))u_{k-1}\}_1^N \|_{C_\tau^{\alpha,\alpha}(E)} \leq M_1 [\frac{1}{\alpha(1-\alpha)} \| f^\tau \|_{C_\tau^{\alpha,\alpha}(E)} + \| Au_0 \|_E].$$

By the triangle inequality, this last estimate and difference equation (7.5) yield

$$\| \{\rho^{-1}(u_k - u_{k-1})\}_1^N \|_{C_\tau^{\alpha,\alpha}(E)} \leq M_1 [\frac{1}{\alpha(1-\alpha)} \| f^\tau \|_{C_\tau^{\alpha,\alpha}(E)} + \| Au_0 \|_E].$$

Theorem 7.1.8 is proved.

Theorem 7.1.9. *The solutions of the difference problem (7.5) in $C_\tau(E_\alpha)$ obey the coercivity inequality*

$$\|\{\rho^{-1}(u_k - u_{k-1})\}_1^N \|_{C_\tau(E_\alpha)} + \| \{\rho^{-1}(1 - g(\rho A))u_{k-1}\}_1^N \|_{C_\tau(E_\alpha)} \tag{7.30}$$

$$\leq M_1[\frac{1}{\alpha(1-\alpha)}\| f^\tau \|_{C_\tau(E_\alpha)} + \| Au_0\|_{E_\alpha}],$$

where M_1 is independent not only of f^τ, u_0, α, but also of τ and ε.

Proof. Using formula (7.6), we can write

$$\rho^{-1}(1 - g(\rho A))\ u_{k-1} = \rho^{-1}(1 - g(\rho A))\ g((k-1)\rho\ A)u_0$$

$$+\sum_{j=1}^{k-1}(1 - g(\rho A))g((k-j)\rho A)f_j = J_k^1 + J_k^2,$$

where

$$J_k^1 = \rho^{-1}(1 - g(\rho A))\ g((k-1)\rho A)u_0,$$

$$J_k^2 = \sum_{j=1}^{k-1}(1 - g(\rho A))g((k-1-j)\rho A)f_j.$$

To complete the proof of Theorem 7.1.9 it suffices to establish that

$$\| J_k^1 \|_{E_\alpha} \leq M \| Au_0 \|_{E_\alpha}, \tag{7.31}$$

$$\| J_k^2 \|_{E_\alpha} \leq M\ \alpha^{-1}(1-\alpha)^{-1}\| f^\tau \|_{C_\tau(E_\alpha)} \tag{7.32}$$

for all k, $k = 1, \ldots, N$. Using the formula

$$J_k^1 = \int_0^1 g((k-1+s)\rho A)\ ds\ Au_0,$$

and estimate (7.24), we obtain

$$\| J_k^1 \|_{E_\alpha} \leq M \int_0^1 e^{-\delta(k-1+s)\rho} ds \| Au_0 \|_\alpha \leq M \| Au_0 \|_{E_\alpha}.$$

Estimate (7.31) is established. We have that

$$J_k^2 = \sum_{j=1}^{k-1} \int_0^1 Ag((k-1-j+s)\rho A)ds\ f_j\ \rho.$$

To estimate J_k^2 in the norm E_α we use the Cauchy–Riesz representation formula for the operator $(\lambda+A)^{-1}J_k^2$. We obtain

$$A(\lambda+A)^{-1}\ J_k^2 = A(\lambda+A)^{-1}\sum_{j=1}^{k-1}\int_0^1 Ag((k-1-j+s)\rho A)ds\ f_j\ \rho$$

$$= \frac{1}{2\pi i}\int_{S_1US_2} z(\lambda+z)^{-1}\sum_{j=1}^{k-1}\rho g((k-1-j+s)\rho\ z)\ A\ (z-A)^{-1}\ f_j\ dz.$$

Since $z = re^{\pm\varphi}$, with $|\varphi| < \frac{\pi}{2}$, we have the estimate

$$\|\ A\ (z-A)^{-1}\ f_j\ \|_E \le M\ \|\ A(r+A)^{-1}f_j\ \|_E, \tag{7.33}$$

on the sides S_1 and S_2. Then, using the inequality

$$|\ (\lambda+z)^{-1}\ | \le M(\lambda+r)^{-1},$$

and estimates (7.24), (7.32), we obtain

$$\|\ A(\lambda+A)^{-1}\ J_k^2\ \|_E \le M\int_0^\infty \frac{r\rho}{(\lambda+r)r^\alpha}\sum_{j=1}^{k-1}\int_0^1 e^{-(k-j+s)\rho\ r\cos\varphi}dsdr\ \|\ f_j\ \|_{E_\alpha}$$

$$\le \frac{M}{\cos\varphi}\|\ f^\tau\ \|_{C_\tau(E_\alpha)}\int_0^\infty \frac{1}{(\lambda+r)r^\alpha}\sum_{j=1}^{k-1}(1-e^{-r\rho\ \cos\varphi})e^{-(k-j)\rho\ r\cos\varphi}dr$$

$$\le \frac{M}{\cos\varphi}\|\ f^\tau\ \|_{C_\tau(E_\alpha)}\int_0^\infty \frac{dr}{(\lambda+r)r^\alpha}$$

for all $\lambda > 0$. Therefore,

$$\lambda^\alpha\|\ A(\lambda+A)^{-1}\ J_k^2\ \|_E \le \frac{M}{\cos\varphi}\|\ f^\tau\ \|_{C_\tau(E_\alpha)}\int_0^\infty \frac{\lambda^\alpha dr}{(\lambda+r)r^\alpha}$$

$$= \frac{M}{\cos\varphi}\|\ f^\tau\ \|_{C_\tau(E_\alpha)}\int_0^\infty \frac{ds}{(1+s)s^\alpha} \le M_1\ \alpha^{-1}(1-\alpha)^{-1}\|\ f^\tau\ \|_{C_\tau(E_\alpha)}$$

for all $\lambda > 0$. Estimate (7.32) is established. Theorem 7.1.9 is proved.

Note that estimate (7.24), established for the difference problem (7.5), allows us to prove Theorem 7.1.8. Indeed, by the definition of the norm in $C_\tau^{\alpha,\alpha}(E)$,

$$\|\ f^\tau\ \|_{C_\tau(E)} \le \|\ f^\tau\ \|_{C_\tau^{\alpha,\alpha}(E)} \le \frac{3}{\tau^\alpha}\|\ f^\tau\ \|_{C_\tau(E)}.$$

Then, Theorem 7.1.8 implies that

$$\|\{\rho^{-1}(u_k - u_{k-1})\}_1^N \|_{C_\tau(E)} + \| \{\rho^{-1}(1 - g(\rho A))u_{k-1}\}_1^N \|_{C_\tau(E)}$$
$$\leq M[\frac{1}{\alpha\tau^\alpha} \| f^\tau \|_{C_\tau(E)} + \| Au_0\|_E]$$

for all sufficiently small $\alpha > 0$. Choosing α in the best way, we obtain estimate (7.22). Furthermore, if A is the bounded positive operator in E, then from estimate (7.30) it follows that

$$\|\{\rho^{-1}(u_k - u_{k-1})\}_1^N \|_{C_\tau(E)} + \| \{\rho^{-1}(1 - g(\rho A))u_{k-1}\}_1^N \|_{C_\tau(E)} \tag{7.34}$$
$$\leq M[(1 + |\ln ||A||_{E\to E}|) \| f^\tau \|_{C_\tau(E)} + \| Au_0\|_E].$$

Indeed, by the definition of the norm in $C_\tau(E_\alpha)$,

$$\| f^\tau \|_{C_\tau(E)} \leq \| f^\tau \|_{C_\tau(E_\alpha)} \leq M||A||_{E\to E}^\alpha \| f^\tau \|_{C_\tau(E)}.$$

Then, Theorem 7.1.9 implies that

$$\|\{\rho^{-1}(u_k - u_{k-1})\}_1^N \|_{C_\tau(E)} + \| \{\rho^{-1}(1 - g(\rho A))u_{k-1}\}_1^N \|_{C_\tau(E)}$$
$$\leq M[\frac{||A||_{E\to E}^\alpha}{\alpha} \| f^\tau \|_{C_\tau(E)} + \| Au_0\|_E]$$

for all sufficiently small $\alpha > 0$. Choosing α in the best way, we obtain estimate (7.34).

By Theorem 7.1.7, estimate (7.34) yields the following result.

Theorem 7.1.10. *The solutions of the difference problem* (7.5) *in* $C_\tau(E)$ *obey the almost coercive inequality*

$$\| u^\tau \|_{K_\tau(E)} \leq M_1[\| Au_0\|_E + \min\{\ln \frac{1}{\tau}, 1 + |\ln ||A||_{E\to E}||\} \| f^\tau \|_{C_\tau(E)}],$$

where M_1 is independent not only of f^τ, u_0, but also of τ and ε.

Let us give, without proof, the following results.

Theorem 7.1.11. *The solutions of the difference problem* (7.5) *in* $C_\tau^{\beta,\gamma}(E)$ $(0 \leq \gamma \leq \beta, 0 < \beta < 1)$ *satisfy the following coercivity inequalities:*

$$\| \{\rho^{-1}(u_k - u_{k-1})\}_1^N \|_{C_\tau(E_{\beta-\gamma})} \leq \frac{M_1}{\beta(1-\beta)} \| f^\tau \|_{C_\tau^{\beta,\gamma}(E)}$$
$$+M_1 \| (I - e^{-\rho A})\rho^{-1}u_0 - f_1 \|_{E_{\beta-\gamma}},$$
$$\| \{\rho^{-1}(u_k - u_{k-1})\}_1^N \|_{C_\tau^{\beta,\gamma}(E)} + \| \{\rho^{-1}(1 - g(\rho A))u_{k-1}\}_1^N \|_{C_\tau^{\beta,\gamma}(E)}$$
$$\leq \frac{M_1}{\beta(1-\beta)} \| f^\tau \|_{C_\tau^{\beta,\gamma}(E)} + M_1 |(I - e^{-\rho A})\rho^{-1}u_0 - f_1|_0^{\beta,\gamma},$$

where M_1 is independent not only of f^τ, u_0, β, γ but also of τ and ε.

Here, the Banach space $\widetilde{E}_0^{\beta,\gamma}$ consists of those $v \in E$ for which the norm

$$|w|_0^{\beta,\gamma} = \max_{1\le n\le N} \|e^{\frac{-t_n}{\varepsilon}A}w\|_E + \sup_{1\le n<n+r\le N} t_r^{-\beta}(t_n+t_r)^\gamma \|(e^{-(\frac{t_n+t_r}{\varepsilon})A} - e^{-\frac{t_n}{\varepsilon}A})w\|_E$$

is finite.

Note that the parameter γ can be chosen freely in $[0,\beta)$, which increases the number of spaces $C_\tau^{\beta,\gamma}(E)$ $(0 \le \gamma \le \beta, 0 < \beta < 1)$ of grid functions in which difference problem (7.5) is well posed.

Theorem 7.1.12. *The solutions of the difference problem* (7.5) *in* $C_\tau^{\beta,\gamma}(E_{\alpha-\beta})$ $(0 \le \gamma \le \beta \le \alpha, 0 < \alpha < 1)$ *satisfy the following coercivity inequalities:*

$$\| \{\rho^{-1}(u_k - u_{k-1})\}_1^N \|_{C_\tau^{\beta,\gamma}(E_{\alpha-\beta})} + \| \{\rho^{-1}(u_k - u_{k-1})\}_1^N \|_{C_\tau(\widetilde{E}_{\alpha-\beta}^{\beta,\gamma})}$$
$$+ \| \{\rho^{-1}(1-g(\rho A))u_{k-1}\}_1^N \|_{C_\tau^{\beta,\gamma}(E_{\alpha-\beta})}$$
$$\le \frac{M_1}{\alpha(1-\alpha)} \| f^\tau \|_{C_\tau^{\beta,\gamma}(E_{\alpha-\beta})} + M_1 |(I - e^{-\rho A})\rho^{-1}u_0 - f_1|_{\alpha-\beta}^{\beta,\gamma},$$

$$\| \{\rho^{-1}(u_k - u_{k-1})\}_1^N \|_{C_\tau^{\beta,\gamma}(E_{\alpha-\beta})} + \| \{\rho^{-1}(u_k - u_{k-1})\}_1^N \|_{C_\tau(E_{\alpha-\gamma})}$$
$$+ \| \{\rho^{-1}(1-g(\rho A))u_{k-1}\}_1^N \|_{C_\tau^{\beta,\gamma}(E_{\alpha-\beta})}$$
$$\le \frac{M_1}{\alpha(1-\alpha)} \| f^\tau \|_{C_\tau^{\beta,\gamma}(E_{\alpha-\beta})} + M_1 \|(I - e^{-\rho A})\rho^{-1}u_0 - f_1\|_{E_{\alpha-\gamma}},$$

where M_1 *is independent not only of* f^τ, u_0, α, β, γ *but also of* τ *and* ε.

Here, the Banach space $\widetilde{E}_{\alpha-\beta}^{\beta,\gamma}$ consists of those $v \in E$ for which the norm

$$|w|_{\alpha-\beta}^{\beta,\gamma} = \max_{1\le n\le N} \|e^{-\frac{t_n}{\varepsilon}A}w\|_{E_{\alpha-\beta}}$$
$$+ \sup_{1\le n<n+r\le N} t_r^{-\beta}(t_n+t_r)^\gamma \|(e^{-(\frac{t_n+t_r}{\varepsilon})A} - e^{-\frac{t_n}{\varepsilon}A})w\|_{E_{\alpha-\beta}}$$

is finite.

Note that the spaces $C_\tau^{\beta,\gamma}(E_{\alpha-\beta})$ of grid functions, in which coercive solvability has been established, depend on the parameters α, β and γ. However, the constants in the coercive inequalities depend only on α. Hence, we can choose the parameters β and γ freely, which increases the number of spaces of grid functions in which difference problem (7.5) is well posed.

Now let us consider the initial-value difference problem (7.5) in the spaces $L_{p,\tau}(E) = L_p([0,T]_\tau, E)$, $1 \le p < \infty$ of all grid functions. We have not been able to obtain the coercivity inequality

$$\| \{\rho^{-1}(u_k - u_{k-1})\}_1^N \|_{L_{p,\tau}(E)} + \| \{\rho^{-1}(1-g(\rho A))u_{k-1}\}_1^N \|_{L_{p,\tau}(E)}$$
$$\le M[\| f^\tau \|_{L_{p,\tau}(E)} + \| Au_0\|_E],$$

in an arbitrary Banach space E and for the general strong operator A. Nevertheless, we can establish the almost coercivity inequality.

Theorem 7.1.13. *The solutions of the difference problem* (7.5) *in* $L_{p,\tau}(E)$ *obey the almost coercive inequality*

$$\| \{\rho^{-1}(u_k - u_{k-1})\}_1^N \|_{L_{p,\tau}(E)} + \| \{\rho^{-1}(1 - g(\rho A))u_{k-1}\}_1^N \|_{L_{p,\tau}(E)}$$

$$\leq M_1[\| Au_0\|_E + \min\left\{\ln\frac{1}{\tau}, 1 + |\ln \| A \|_{E\to E}|\right\} \| f^\tau \|_{L_{p,\tau}(E)}],$$

where M_1 *does not depend on* f^τ, u_0, p, τ *and* ε.

Note that if A is the unbounded positive operator in E, then

$$\min\left\{\ln\frac{1}{\tau}, 1 + |\ln \| A \|_{E\to E}|\right\} = \ln\frac{1}{\tau}.$$

Finally, let us give, without proof, the following results about well-posedness of the initial-value difference problem (7.5) in the spaces $L_{p,\tau}(E) = L_p([0,T]_\tau, E)$, $1 \leq p < \infty$.

Theorem 7.1.14. *Suppose that the difference problem* (7.5) *well posed in* $L_{p_0,\tau}(E)$ *for some* p_0, $1 < p_0 < \infty$. *Then it is well posed in* $L_{p,\tau}(E)$ *for all* $p, 1 < p < \infty$ *and the following coercivity inequality holds:*

$$\| \{\rho^{-1}(u_k - u_{k-1})\}_1^N \|_{L_{p,\tau}(E)} + \| \{\rho^{-1}(1 - g(\rho A))u_{k-1}\}_1^N \|_{L_{p,\tau}(E)}$$

$$+ \|u^\tau\|_{C_c(E_{1-1/p,p})} \leq M(p_0)[\| \tau u_0\|_{E_{1-1/p,p}} + \frac{p^2}{p-1} \| f^\tau \|_{L_{p,\tau}(E)}],$$

where $M(p_0)$ *does not depend on* f^τ, u_0, p *and* τ *and* ε.

Theorem 7.1.15. *Let* $1 \leq p \leq \infty$ *and* $0 < \alpha < 1$. *Then problem* (7.5) *is well posed in* $L_{p,\tau}(E_{\alpha,p})$ *and the following coercivity inequality holds:*

$$\| \{\rho^{-1}(u_k - u_{k-1})\}_1^N \|_{L_{p,\tau}(E_{\alpha,p})} + \| \{\rho^{-1}(1 - g(\rho A))u_{k-1}\}_1^N \|_{L_{p,\tau}(E_{\alpha,p})}$$

$$\leq M[\| Au_0\|_{E_{\alpha,p}} + \frac{1}{\alpha(1-\alpha)} \| f^\tau \|_{L_{p,\tau}(E_{\alpha,p})}],$$

where M *does not depend on* f^τ, u_0, p, α, τ, ε.

Theorem 7.1.16. *Let* $1 < p, q < \infty$ *and* $0 < \alpha < 1$. *Then the difference problem* (7.5) *is well posed in* $L_{p,\tau}(E_{\alpha,q})$ *and the following coercivity inequality holds:*

$$\| \{\rho^{-1}(u_k - u_{k-1})\}_1^N \|_{L_{p,\tau}(E_{\alpha,q})} + \| \{\rho^{-1}(1 - g(\rho A))u_{k-1}\}_1^N \|_{L_{p,\tau}(E_{\alpha,q})}$$

$$\leq M(q)[\| Au_0\|_{E_{\alpha,q}} + \frac{p^2}{p-1} \| f^\tau \|_{L_{p,\tau}(E_{\alpha,q})}],$$

where $M(q)$ *does not depend on* f^τ, u_0, p, α ,τ *and* ε.

Remark 7.1.17. Theorems 7.1.5–7.1.16 permit us to obtain the error estimates of the high order of accuracy difference schemes over time and space variables for parabolic equations with a small parameter multiplying the time-derivative term in the various Banach norms.

Remark 7.1.18. For practical computations, we have been approximating the elliptic operator A by some difference operator A_h such that $\exp(-\rho A_h)$ is computable. Thus, we shall construct difference schemes with fitting operator $\exp(-\rho A_h)$. Therefore for numerical solution of (7.1) it is necessary to compute by (7.5) $\exp(-\rho A_h)$ with a high order of accuracy. Accordingly, the methods we have proposed are not easy to implement exactly as for a system of ordinary differential equations. Indeed, method (7.5) is useful, because the part of compute fitting operator $\exp(-\rho A_h)$ with a high order of accuracy we can make independent of (7.5). For example, we can approximate $\exp(-\rho A_h)$ by the formula

$$\begin{cases} R_{j,l}(\rho A_h), & if\ \varepsilon \leq a, \\ R_{j,l}^{L}(\omega A_h), & if\ \varepsilon > a, \end{cases}$$

where $L = [\frac{\rho}{\tau}]$, $L\omega = \rho$, $j + l = m$ (for $R_{j,l}(\rho A_h)$, see Chapter 4).

7.2 A Boundary-Value Problem for Elliptic Equations

We consider the boundary-value problem for differential equation

$$-\varepsilon^2 v''(t) + Av(t) = f(t) \quad (0 \leq t \leq T), v(0) = v_0, v(T) = v_T \tag{7.35}$$

in an arbitrary Banach space E with the positive operator A and $\varepsilon \in (0, \infty)$.

A function $v(t)$ is called a solution of the problem (7.35) if the following conditions are satisfied:

i. $v(t)$ is a twice continuously differentiable on the interval $[0, T]$. The derivatives at the endpoints of the segment are understood as the appropriate unilateral derivatives.

ii. The element $v(t)$ belongs to $D(A)$ for all $t \in [0, T]$ and the function $Av(t)$ is continuous on the segment $[0, T]$.

iii. $v(t)$ satisfies the equation and boundary conditions (7.35).

In Chapter 5, stable high order of accuracy difference schemes generated by an exact difference scheme or by Taylor's decomposition on three points for the numerical solutions of an abstract boundary-value problem (7.35) for $\varepsilon = 1$ were presented. Unfortunately, these difference schemes cannot be applied for the approximate solutions of (7.35) in the general cases $\varepsilon \in (0, \infty)$. In this section we study the high order of accuracy two-step uniform difference schemes of the approximate solutions for differential equations of the elliptic type with an arbitrary parameter ε on the highest derivative.

On the segment $[0, T]$ we consider again a uniform grid

$$[0,T]_\tau = \{t_k = k\tau, k = 0, 1, \dots, N, N\tau = T\}$$

with step τ. Let $\tau = \rho\varepsilon$ and $B^2 = A$. By Theorem 3.1.2 we have the exact two-step difference scheme

$$-\rho^{-2}(v(t_{k+1}) - 2v(t_k) + v(t_{k-1})) \tag{7.36}$$

$$+\rho^{-2}[(I - \exp(-\rho B))(v(t_{k+1}) + v(t_{k-1})) + (\exp(-2\rho B) - I)v(t_k)] = \widetilde{f}_k,$$

$$\widetilde{f}_k = (2B\rho)^{-1}(\widetilde{f}_{1,k} + \widetilde{f}_{2,k+1}) - (2B\rho)^{-1}\exp(-\rho B)(\widetilde{f}_{1,k+1} + \widetilde{f}_{2,k}),$$

$$\widetilde{f}_{1,k} = \frac{1}{\tau}\int_{t_{k-1}}^{t_k} \exp(-\frac{1}{\varepsilon}(t_k - z)B))f(z)dz,$$

$$\widetilde{f}_{2,k} = \frac{1}{\tau}\int_{t_{k-1}}^{t_k} \exp(-\frac{1}{\varepsilon}(z - t_{k-1})B)f(z)dz,$$

$$1 \le k \le N-1,\ v(0) = v_0,\ v(T) = v_T$$

for the solution of the boundary-value problem (7.35). Now, we consider the following two-step difference scheme:

$$-\rho^{-2}(u_{k+1} - 2u_{k-1} + u_{k-1}) + \rho^{-2}(1 - g(\rho B))\ (u_{k+1} + u_{k-1}) \tag{7.37}$$

$$+\rho^{-2}(g(2\rho B) - 1)u_k = f_k,$$

$$f_k = f_{1,k} + f_{2,k+1} - g(\rho B)(f_{1,k+1} + f_{2,k}),$$

$$f_{1,k} = \sum_{j=0}^{m-1} J_{1,\,j}\ f^{\,(j)}(t_{k-1}),\ f_{2,\,k} = \sum_{j=0}^{m-1} J_{2,\,j}\ f^{\,(j)}(t_{k-1}),$$

$$J_{1,\,j} = (-1)^j(B/\varepsilon)^{-j}J_{1,0} + \sum_{i=1}^{j}(-1)^{\,j-i}(B/\varepsilon)^{-(j-i+1)}\ \tau^{i-1}/i!,$$

$$J_{2,\,j} = (B/\varepsilon)^{-j}J_{2,0} - \sum_{i=1}^{j}(B/\varepsilon)^{-(j-i+1)}g(\rho B)\ \tau^{i-1}/i!,$$

$$J_{1,0} = J_{2,0} = (\rho B)^{-1}(1 - g(\rho B)), 1 \le j \le m-1,\ t_k = k\tau,\ 1 \le k \le N-1,$$

$$g(\rho B) = \exp(-\rho B),\ u_0 = v_0,\ u_N = v_1$$

for the approximate solutions of the boundary-value problem (7.35).

This problem is uniquely solvable and the following formula holds:

$$u_k = (I - g(2N\rho B))^{-1}\{(g(k\rho B) - g((2N-k)\rho B))u_0 \tag{7.38}$$

$$+(g((N-k)\rho B)-g((N+k)\rho B)u_N-(g((N-k)\rho B)-g((N+k)\rho B)$$

$$\times\rho^2\sum_{i=1}^{N-1}(g((N-i)\rho B)-g((N+i)\rho B)(I-g(2\rho B))^{-1}f_i\}$$

$$+\rho^2\sum_{i=1}^{N-1}[g(|k-i)\rho B)-g((k+i)\rho B)](I-g(2\rho B))^{-1}f_i, 1\leq k\leq N-1.$$

Let us reduce the difference schemes (7.37) to an operator problem in the space $F_\tau(E)$ of vectors $v^\tau=\{v_k\}_{k=1}^{N-1}$. Recall that the space $F_\tau(E)$ can be equipped with various norms and thus become a normed space (see Chapter 5). Thus, for instance, the vector space $F_\tau(E)$ generates the normed space $C_\tau(E)=C([0,T]_\tau,E)$ with the norm

$$\|\varphi^\tau\|_{C_\tau(E)}=\max_{1\leq k\leq N-1}\|\varphi_k\|_E,$$

the normed space $C_\tau^\beta(E)=C^\beta([0,T]_\tau,E)$ $0<\beta<1$, with the norm

$$\|\varphi^\tau\|_{C_\tau^\beta(E)}=\|\varphi^\tau\|_{C_\tau(E)}+\max_{1\leq k<k+r\leq N-1}\|\varphi_{k+r}-\varphi_k\|_E\frac{1}{(r\tau)^\beta},$$

the weighted normed space $C_\tau^{\beta,\gamma}(E)=C^{\beta,\gamma}([0,T]_\tau,E), 0<\gamma\leq\beta$, $0<\beta<1$, with the norm

$$\|\varphi^\tau\|_{C_\tau^{\beta,\gamma}(E)}=\|\varphi^\tau\|_{C_\tau(E)}$$

$$+\max_{1\leq k<k+r\leq N-1}\|\varphi_{k+r}-\varphi_k\|_E\frac{((k+r)\tau)^\gamma((N-k)\tau)^\gamma}{(r\tau)^\beta},$$

and the normed space $L_{p,\tau}(E)=L_p([0,T]_\tau,E), 1\leq p<\infty$, with the norm

$$\|\varphi^\tau\|_{L_{p,\tau}(E)}=(\sum_{k=1}^{N-1}\|\varphi_k\|_E^p\tau)^{\frac{1}{p}}.$$

First we shall define an operator D_τ^2 acting from the space $E\times F_\tau(E)\times E$ of vectors $w^\tau=\{w_k\}_{k=0}^N$ into the space $F_\tau(E)$ of vectors $v^\tau=\{v_k\}_{k=1}^{N-1}$ according to the rule

$$v^\tau=D_\tau^2w^\tau, v_k=\frac{1}{\rho^2}(w_{k+1}-2w_k+w_{k-1}), k=1,\ldots,N-1.$$

Second we shall define an operator A_τ acting from the space $E\times F_\tau(E)\times E$ of vectors $w^\tau=\{w_k\}_{k=0}^N$ into the space $F_\tau(E)$ of vectors $v^\tau=\{v_k\}_{k=1}^{N-1}$ according to the rule

$$v^\tau=A_\tau w^\tau, v_k=\frac{1}{\rho^2}[(I-g(\rho B))(w_{k-1}+w_{k+1})$$

$$+(g(2\rho B)-I)w_k],\ k=1,\ldots,N-1.$$

Next, let us introduce the continuation operator $\Pi(u_0, u_N)$, which acts from $E \times F_\tau(E) \times E$ to $F_\tau(E)$ according to the rule

$$\Pi(u_0, u_N)(u_1, \ldots, u_{N-1}) = (u_0, u_1, \ldots, u_{N-1}, u_N).$$

Then, clearly, the difference problem (7.37) is equivalent to the operator problem

$$-D_\tau^2 \Pi(u_0, u_N)u^\tau + A_\tau \Pi(u_0, u_N)u^\tau = f^\tau.$$

Here f^τ is defined by the formula

$$f^\tau = (f_1, \ldots, f_{N-1}).$$

The last operator problem will be considered in the space $F_\tau(E)$. From its unique solvability for any u_0, $u_N \in E$ and $f^\tau \in F_\tau(E)$ it follows that its solution u^τ defines a continuous additive and homogeneous operator $u^\tau(f^\tau, u_0, u_N)$.

We will study stability and coercive stability of the boundary-value problem (7.37) in the difference analogues of Banach spaces. Since

$$f_k = f_{1,k} + f_{2,k+1} - g(\rho B)(f_{1,k+1} + f_{2,k}), 1 \leq k \leq N-1, \tag{7.39}$$

we have that if $f^\tau \in F_\tau(E)$, then $f_1^\tau = (f_{1,1}, \ldots, f_{1,N-1}, f_{1,N}) \in F_\tau(E) \times E$ and $f_2^\tau = (f_{2,1}, \ldots, f_{2,N-1}, f_{2,N}) \in F_\tau(E) \times E$. Note that $F_\tau(E)$ is the vector space whose elements are ordered $(N-1)$-tuples of elements of E. The space $F_\tau(E)$ can be equipped with various norms and thus become a normed space (see Chapter 5). Nevertheless, $F_\tau(E) \times E$ is the vector space whose elements are ordered N-tuples of elements of E. In a similar way one obtains that the space $F_\tau(E) \times E$ can be equipped with various norms and thus become a normed space. We will use the same notation for both cases. For instance, the vector space $F_\tau(E)$ generates the normed space $C_\tau(E) = C([0,T]_\tau, E)$ with the norm

$$\| \varphi^\tau \|_{C_\tau(E)} = \max_{1 \leq k \leq N-1} \| \varphi_k \|_E$$

and the vector space $F_\tau(E) \times E$ generates the normed space $C_\tau(E) = C([0,T]_\tau, E)$ with the norm

$$\| \varphi^\tau \|_{C_\tau(E)} = \max_{1 \leq k \leq N} \| \varphi_k \|_E.$$

Theorem 7.2.1. *For the solution of the difference problem* (7.37) *the following estimate is valid:*

$$\max_{1 \leq k \leq N-1} \| u_k \|_E \leq \frac{M}{1 - \exp(-\delta\rho)} \{ \| u_0 \|_E + \| u_N \|_E \tag{7.40}$$

$$+ \rho^2 [\max_{1 \leq k \leq N} \| B f_{1,k} \| + \max_{1 \leq k \leq N} \| B f_{2,k} \|] \},$$

where M is independent not only of f_1^τ, f_2^τ, u_0, u_N, τ, but also of ε.

Proof. Using formula (7.38), we can write

$$u_k = w_k + g_k, \tag{7.41}$$

where

$$\begin{aligned} w_k &= (I - g(2N\rho B))^{-1}\{(g(k\rho B) - g((2N-k)\rho B))u_0 \\ &\quad +(g((N-k)\rho B) - g((N+k)\rho B)u_N\}, \end{aligned}$$

$$g_k = -(I - g(2N\rho B))^{-1}(g((N-k)\rho B) - g((N+k)\rho B)$$

$$\times\rho^2 \sum_{i=1}^{N-1} (g((N-i)\rho B) - g((N+i)\rho B)(I - g(2\rho B))^{-1} f_i$$

$$+\rho^2 \sum_{i=1}^{N-1} [g(|k-i)\rho B) - g((k+i)\rho B)](I - g(2\rho B))^{-1} f_i.$$

Let us estimate w_k and g_k separately. We begin with w_k. Using the estimate

$$\| B^\alpha \exp(-tB) \|_{E\to E} \leq Mt^{-\alpha}\exp(-\delta t), \ \ \delta > 0, \ \ \alpha \geq 0, t > 0, \tag{7.42}$$

we obtain

$$\| w_k \|_E \leq \| (I - g(2N\rho B))^{-1}(g(k\rho B) - g((2N-k)\rho B)) \|_{E\to E} \| u_0 \|_E$$

$$+\|(I - g(2N\rho B))^{-1}(g((N-k)\rho B) - g((N+k)\rho B))\|_{E\to E} \| u_N \|_E$$

$$\leq M\frac{1}{1-e^{-2N\rho\delta}}[\| u_0 \|_E + \| u_N \|_E]$$

for all k, $1 \leq k \leq N-1$. Since

$$\frac{1}{1-e^{-2N\rho\delta}} \leq \frac{1}{1-e^{-\rho\delta}},$$

we have that

$$\max_{1\leq k\leq N-1} \|w_k \|_E \leq M\frac{1}{1-e^{-\rho\delta}}[\| u_0 \|_E + \| u_N \|_E].$$

To complete the proof of Theorem 7.2.1 it suffices to establish that

$$\| g_k \|_E \leq M\frac{\rho^2}{1-e^{-\rho\delta}} \left[\max_{1\leq k\leq N} \| Bf_{1,k} \| + \max_{1\leq k\leq N} \| Bf_{2,k} \| \right] \tag{7.43}$$

for all k, $1 \leq k \leq N-1$. Using formula (7.38), identity (7.39) and Abel's formula, we can write

$$g_k = (I - g(2N\rho B))^{-1}(g((N-k)\rho B) - g((N+k)\rho B) \tag{7.44}$$

$$\times\rho^2\sum_{i=1}^{N}[g((N-i)\rho B)f_{1,i}-g((N+i-1)\rho B)f_{2,i}]+\rho^2\sum_{i=1}^{N}g((k+i-1)\rho B)]f_{2,i}$$

$$-\rho^2[\sum_{i=1}^{k}g((k-i)\rho B)f_{1,i}+\sum_{i=k+1}^{N}g((i-k-1)\rho B)f_{2,i}]=g_k^1+g_k^2,$$

where

$$g_k^1=(I-g(2N\rho B))^{-1}(g((N-k)\rho B)-g((N+k)\rho B)\rho^2\sum_{i=1}^{N}[g((N-i)\rho B)f_{1,i}$$

$$-\rho^2\sum_{i=1}^{k}g((k-i)\rho B)f_{1,i},$$

$$g_k^2=-(I-g(2N\rho B))^{-1}(g((N-k)\rho B)-g((N+k)\rho B)\rho^2\sum_{i=1}^{N}g((N+i-1)\rho B)f_{2,i}$$

$$+\rho^2\sum_{i=1}^{N}g((k+i-1)\rho B)]f_{2,i}-\rho^2\sum_{i=k+1}^{N}g((i-k-1)\rho B)f_{2,i}.$$

Let us estimate g_k^1 and g_k^2 separately. We begin with g_k^1. Using estimate (7.42) and

$$\|(I-g(2N\rho B))^{-1}(I-g(2k\rho B))B^{-1}\|_{E\to E}\leq M(\delta)$$

for all k, $1\leq k\leq N$, we obtain that

$$\|g_k^1\|_E\leq\|(I-g(2N\rho B))^{-1}(g((N-k)\rho B)-g((N+k)\rho B))B^{-1}\|_{E\to E}\rho^2$$

$$\times\sum_{i=1}^{N}\|g((N-i)\rho B)\|_{E\to E}\|B f_{1,i}\|_E+\rho^2\sum_{i=1}^{k}\|g((k-i)\rho B)\|_{E\to E}\|f_{1,i}\|_E$$

$$\leq M\sum_{j=1}^{N}e^{-\delta(N-j)\rho}\rho^2\max_{1\leq j\leq N}\|Bf_{1,j}\|_E$$

for all k, $k=1,\dots,N-1$. Using the last inequality and (7.8), we obtain

$$\|g_k^1\|_E\leq M\frac{1}{1-e^{-\rho\delta}}\rho^2\max_{1\leq j\leq N}\|Bf_{1,j}\|_E]$$

for all k, $k=1,\dots,N-1$. In a similar way one establishes that

$$\|g_k^2\|_E\leq M\frac{1}{1-e^{-\rho\delta}}\rho^2\max_{1\leq j\leq N}\|Bf_{2,j}\|_E]$$

for all k, $k=1,\dots,N-1$. From these estimates follows estimate (7.43). This completes the proof of Theorem 7.2.1.

We say that the difference problem (7.37) has an m-th order of accuracy in $F_\tau(E)$ on the solutions $v(t)$ of the boundary-value problem (7.35) if the error vector

$$\{v(t_k) - u_k\}_1^{N-1}$$

satisfies the estimate

$$\| \{v(t_k) - u_k\}_1^{N-1} \|_{F_\tau(E)} \leq M\tau^m,$$

where M is independent not only τ, but also of ε.

Stability inequality (7.40) permits us to obtain the estimate of convergence of (7.37) in $C_\tau(E)$.

Theorem 7.2.2. *Suppose that the function $f(t)$ has m derivatives and*

$$\|f^{(m)}(t)\|_E \leq M, 0 \leq t \leq T.$$

Then for the solution of the difference problem (7.37) the following convergence estimate is valid:

$$\| \{v(t_k) - u_k\}_1^{N-1} \|_{C_\tau(E)} \leq M\tau^m, \tag{7.45}$$

where M does not depend on τ and ε.

Proof. The components $z_k = v(t_k) - u_k$ of error vector $z^\tau = \{z_k\}_1^N$ satisfy the relations

$$-\rho^{-2}(z_{k+1} - 2z_{k-1} + z_{k-1}) + \rho^{-2}(1 - g(\rho B))\ (z_{k+1} + z_{k-1}) + \rho^{-2}(g(2\rho B) - 1)z_k = a_k,$$

$$a_k = (2\rho B)^{-1}\ (a_{1,k} + a_{2,k+1} - g(\rho B)(a_{1,k+1} + a_{2,k})),$$

$$a_{1,k} = \tau^{-1} \int_{t_{k-1}}^{t_k} \exp(-(t_k - s)\varepsilon^{-1/2}B)f(s)ds - \sum_{j=0}^{m-1} J_{1,\,j}\ f^{(j)}(t_{k-1}),$$

$$a_{2,\,k} = \tau^{-1} \int_{t_{k-1}}^{t_k} \exp(-(s - t_{k-1})\ \varepsilon^{-1/2}B)\ f(s)ds - \sum_{j=0}^{m-1} J_{2,\,j}\ f^{(j)}(t_{k-1}),$$

$$1 \leq k \leq N-1,\ z_0 = 0,\ z_N = 0.$$

From this and the stability inequality (7.40) it follows that

$$\max_{1 \leq k \leq N-1} \| z_k \|_E \tag{7.46}$$

$$\leq M \frac{\rho}{1 - \exp(-\delta\rho)} \left[\max_{1 \leq k \leq N} \|a_{1,k}\|_E + \max_{1 \leq k \leq N} \|Ba_{2,k}\|_E \right].$$

Therefore to this end it suffices to estimate $\{a_{1,k}\}_1^N$ and $\{a_{2,k}\}_1^N$ in $C_\tau(E)$. Using Taylor's formula, we obtain

$$a_{1,k} = \tau^{-1} \int\limits_{t_{k-1}}^{t_k} \exp(-(t_k - s)\varepsilon^{-\frac{1}{2}}B)f(s)ds - \sum_{j=0}^{m-1} J_{1,\,j}\, f^{\,(j)}(t_{k-1})$$

$$= \sum_{j=0}^{m-1} \tau^{-1} \int\limits_{t_{k-1}}^{t_k} \exp(-(t_k - s)\ \varepsilon^{-\frac{1}{2}}B)(s - t_{k-1})^j\ dsf^{(j)}\ (t_{k-1})\frac{1}{j!}$$

$$+\tau^{-1} \int\limits_{t_{k-1}}^{t_k} \exp(-(t_k - s)\ \varepsilon^{-\frac{1}{2}}B) \int\limits_{t_{k-1}}^{s} (s-z)^m f^{(m)}\ (z)\ dzds\ \frac{1}{m!}$$

$$-\sum_{j=0}^{m-1} J_{1,\,j}\, f^{\,(j)}(t_{k-1})\ .$$

A simple calculation shows that

$$\sum_{j=0}^{m-1} \tau^{-1} \int\limits_{t_{k-1}}^{t_k} \exp(-(t_k - s)\ \varepsilon^{-\frac{2}{2}}B)(s - t_{k-1})^j\ dsf^{(j)}\ (t_{k-1})\frac{1}{j!}$$

$$= \sum_{j=0}^{m-1} J_{1,\,j}\, f^{\,(j)}(t_{k-1})\ .$$

Therefore, we can write

$$a_{1,\,k}\ = \tau^{-1} \int\limits_{t_{k-1}}^{t_k} \exp(-(t_k - s)\ \varepsilon^{-1/2}B) \int\limits_{t_{k-1}}^{s} (z-s)^{m-1} f^{(m)}(z)dzds/(m-1)!.$$

In a similar way one shows that

$$a_{2,\,k}\ = \tau^{-1} \int\limits_{t_{k-1}}^{t_k} \exp(-(s - t_{k-1})\ \varepsilon^{-1/2}B) \int\limits_{t_{k-1}}^{s} (z-s)^{m-1} f^{(m)}\ (z)\ dzds\ /(\ m-1)!.$$

Then, using the triangle inequality and estimate (7.40), we obtain

$$\max_{1\le k\le\ N} ||a_{1,k}\ ||_E \le M\ \delta\rho^{-1}(1 - \exp(-\delta\rho))\tau^m$$

and

$$\max_{1\le k\le\ N} ||a_{2,k}\ ||_E \le M\ \delta\rho^{-1}(1 - \exp(-\delta\rho))\tau^m.$$

From this and estimate (7.46) follows estimate (7.45). Theorem 7.2.2 is proved.

Note that for the convergence estimate $\| \{v(t_k) - u_k\}_1^N \|_{F_\tau(E)} \leq M\tau^m$ it suffices to obtain the stability inequality in $F_\tau(E)$.

From formula (7.38) one derives the following results.

Theorem 7.2.3. *For the solution of the difference problem* (7.37) *the following estimate is valid:*

$$\| \{u_k\}_1^{N-1} \|_{L_{p,\tau}(E)} \leq \frac{M}{1-\exp(-\delta\rho)}\{[\ \| u_0 \|_E + \| u_N \|_E$$

$$+\rho^2\ [\| \{Bf_{1,k}\}_1^N \|_{L_{p,\tau}(E)} + \| \{Bf_{2,k}\}_1^N \|_{L_{p,\tau}(E)}]\}, 1 \leq p < \infty,$$

where M is independent not only of f_1^τ, f_2^τ, u_0, u_N, τ, p but also of ε.

Theorem 7.2.4. *For the solution of the difference problem* (7.37) *the following estimate is valid:*

$$\| \{u_k\}_1^{N-1} \|_{C_\tau^{\alpha,\alpha}(E)} \leq \frac{M}{1-\exp(-\delta\rho)}\ \{\ \| u_0 \|_E + \| u_N \|_E$$

$$+\rho^2\ [\| \{Bf_{1,k}\}_1^N \|_{C_\tau^{\alpha,\alpha}(E)} + \| \{Bf_{2,k}\}_1^N \|_{C_\tau^{\alpha,\alpha}(E)}]\}, 0 \leq \alpha \leq 1,$$

where M is independent not only of f_1^τ, f_2^τ, u_0, u_N, τ, α but also of ε.

The proofs of Theorem 7.2.3 and 7.18 follow the scheme of the proofs of Theorems 7.1.3 and 7.1.5 and rely on formula (7.38) and estimate (7.42).

Now let us study the well-posedness of the difference problem (7.37) in various Banach spaces.

Theorem 7.2.5. *The solutions of the difference problem* (7.37) *in $C_\tau^{\alpha,\alpha}(E)$ $(0 < \alpha < 1)$ obey the coercivity inequality*

$$\| D_\tau^2\Pi(u_0, u_N)u^\tau \|_{C_\tau^{\alpha,\alpha}(E)} + \| A_\tau\Pi(u_0, u_N)u^\tau \|_{C_\tau^{\alpha,\alpha}(E)}$$

$$\leq M\{\frac{1}{\alpha(1-\alpha)}[\| \{Bf_{1,k}\}_1^N \|_{C_\tau^{\alpha,\alpha}(E)}$$

$$+ \| \{Bf_{2,k}\}_1^N \|_{C_\tau^{\alpha,\alpha}(E)}] + \frac{1}{1-\exp(-\delta\rho)}[\| Au_0\|_E + \|Au_N\|_E]\},$$

where M is independent not only of f_1^τ, f_2^τ, u_0, u_N, τ, α but also of ε.

Proof. Using formula (7.41), we obtain

$$\rho^{-2}(u_{k+1} - 2u_k + u_{k-1}) = \rho^{-2}(w_{k+1} - 2w_k + w_{k-1}) \tag{7.47}$$

$$+\rho^{-2}(g_{k+1} - 2g_k + g_{k-1}), 1 \leq k \leq N-1.$$

Let us estimate $\{\rho^{-2}(w_{k+1} - 2w_k + w_{k-1})\}_1^{N-1}$ and $\{\rho^{-2}(g_{k+1} - 2g_k + g_{k-1})\}_1^{N-1}$ separately.

Using formula (7.38), we obtain

$$\rho^{-2}(w_{k+1} - 2w_k + w_{k-1})$$
$$= (I - e^{-2N\rho B})^{-1}(I - e^{-\rho B})^2\rho^{-2}(e^{-(k-1)\rho B} - e^{-(2N-k-1)\rho B})u_0$$
$$+(I - e^{-2N\rho B})^{-1}(I - e^{-\rho B})^2\rho^{-2}(e^{-(N-k-1)\rho B} - e^{-(N+k-1)\rho B})u_N$$
$$= \sum_{m=1}^{2} J_k^m,$$

where

$$J_k^1 = (I - e^{-2N\rho B})^{-1}(I - e^{-\rho B})^2\rho^{-2}(e^{-(k-1)\rho B} - e^{-(2N-k-1)\rho B})u_0,$$
$$J_k^2 = (I - e^{-2N\rho B})^{-1}(I - e^{-\rho B})^2\rho^{-2}(e^{-(N-k-1)\rho B} - e^{-(N+k-1)\rho B})u_N.$$

Let us estimate $J_m^\tau = \{J_k^m\}_{k=1}^{N-1}$ for any $m = 1, 2$ separately. We start with J_1^τ. Using estimate (7.42), we show that, for $1 \leq k \leq N-1$,

$$||J_k^1||_E \leq ||(I - e^{-2N\rho B})^{-1}||_{E\to E}||(I - e^{-\rho B})(\rho B)^{-1}||_{E\to E}^2$$
$$\times||e^{-(k-1)\rho B} - e^{-(2N-k-1)\rho B}||_{E\to E}||Au_0||_E$$
$$\leq \frac{M}{1 - e^{-2N\rho\delta}}||Au_0||_E \leq \frac{M}{1 - e^{-\rho\delta}}||Au_0||_E.$$

From this it follows that

$$||J_1^1 - J_{1+r}^1||_E \leq ||J_1^1||_E + ||J_{1+r}^1||_E \leq 2M||Au_0||_E \leq 2M\frac{(2r)^\alpha}{(1+r)^\alpha}||Au_0||_E$$
$$\leq M_1[\frac{r^\alpha}{(1+r)^\alpha} + \frac{r^\alpha}{(N-1)^\alpha}]||Au_0||_E,$$

for $1 < r + 1 \leq N - 1$. By Lemma 5.1.1, one has the inequality

$$\|\exp -k\rho B) - \exp\{-(k+r)\rho B\}\|_{E\to E} \leq M\frac{r^\alpha}{(k+r)^\alpha} \tag{7.48}$$

for any $1 \leq k < k + r \leq N - 1$ and $0 \leq \alpha \leq 1$, where M does not depend on α, k, r and τ.

Further, applying (7.42) and (7.48) we show that, for $2 \leq k < k+r \leq N-1$,

$$||J_k^1 - J_{k+r}^1||_E \leq ||(I - e^{-2N\rho B})^{-1}||_{E\to E}||(I - e^{-\rho B})(\rho B)^{-1}||_{E\to E}^2$$
$$\times [||e^{-(k-1)\rho B} - e^{-(k+r-1)\rho B}||_{E\to E}$$
$$+||e^{-(2N-k-1)\rho B} - e^{-(2N-k-r-1)\rho B}||_{E\to E}]||Au_0||_E$$

$$\leq M[\frac{r^\alpha}{(k+r-1)^\alpha} + \frac{r^\alpha}{(2N-k-1)^\alpha}]||Au_0||_E$$
$$\leq M_1[\frac{r^\alpha}{(k+r)^\alpha} + \frac{r^\alpha}{(N-k)^\alpha}]||Au_0||_E.$$

Thus, we have proved that

$$||J_1^\tau||_{C_\tau^{\alpha,\alpha}(E)} \leq M_1||Au_0||_E.$$

In a similar manner we can show that

$$||J_2^\tau||_{C_\tau^{\alpha,\alpha}(E)} \leq M_1||Au_N||_E.$$

Therefore,

$$\| \{\rho^{-2}(w_{k+1} - 2w_k + w_{k-1})\}_1^{N-1} \|_{C_\tau^{\alpha,\alpha}(E)} \tag{7.49}$$
$$\leq \frac{M}{1-\exp(-\delta\rho)}[\| Au_0\|_E + ||Au_N||_E].$$

Let us estimate $\{\rho^{-2}(g_{k+1} - 2g_k + g_{k-1})\}_1^{N-1}$. Using formula (7.44), we obtain

$$\rho^{-2}(g_{k+1} - 2g_k + g_{k-1}) = (I - g(2N\rho B))^{-1}(g((N-k-1)\rho B) - g((N+k-1)\rho B)$$
$$\times(I - g(\rho B))^2 \sum_{i=1}^{N-1} g((N-i)\rho B)(f_{1,i} - f_{1,N})$$
$$+\sum_{i=1}^{k-1}(I - g(\rho B))^2 g((k-1-i)\rho B)(f_{1,i} - f_{1,k})$$
$$-f_{1,k+1} + (2 - g(\rho B))f_{1,k} + (I - g(2N\rho B))^{-1}(g((N-k-1)\rho B) - g((N+k-1)\rho B)$$
$$\times(I - g(\rho B))^2 \sum_{i=1}^{N} g((N-i)\rho B)f_{1,N} + \sum_{i=1}^{k-1}(I - g(\rho B))^2 g((k-1-i)\rho B)f_{1,k}$$
$$-(I - g(2N\rho B))^{-1}(g((N-k-1)\rho B) - g((N+k-1)\rho B)$$
$$\times(I - g(\rho B))^2 \sum_{i=1}^{N-1} g((N+i-1)\rho B)(f_{2,i} - f_{2,N})$$
$$+(I - g(\rho B))^2 \sum_{i=1}^{N} g((k+i-2)\rho B)](f_{2,i} - f_{2,1})$$
$$-(I - g(\rho B))^2 \sum_{i=k+2}^{N} g((i-k-2)\rho B)(f_{2,i} - f_{2,k+1})$$
$$-f_{2,k} + (2 - g(\rho B))f_{2,k+1} - (I - g(2N\rho B))^{-1}$$

$$\times(g((N-k-1)\rho B)-g((N+k-1)\rho B)(I-g(\rho B))^2$$

$$\times\sum_{i=1}^{N}g((N+i-1)\rho B)f_{2,N}+(I-g(\rho B))^2\sum_{i=1}^{N}g((k+i-2)\rho B)]f_{2,1}$$

$$-(I-g(\rho B))^2\sum_{i=k+2}^{N}g((i-k-2)\rho B)f_{2,k+1}=\sum_{m=1}^{6}P_k^m,$$

where

$$P_k^1=(I-g(2N\rho B))^{-1}(g((N-k-1)\rho B)-g((N+k-1)\rho B)$$

$$\times(I-g(\rho B))^2\sum_{i=1}^{N-1}g((N-i)\rho B)(f_{1,i}-f_{1,N}),$$

$$P_k^2=\sum_{i=1}^{k-1}(I-g(\rho B))^2g((k-1-i)\rho B)(f_{1,i}-f_{1,k}),$$

$$\begin{aligned}P_k^3 &= -f_{1,k+1}+(2-g(\rho B))f_{1,k}+(I-g(2N\rho B))^{-1}\\ &\quad\times(g((N-k-1)\rho B)-g((N+k-1)\rho B)\end{aligned}$$

$$\times(I-g(\rho B))^2\sum_{i=1}^{N}g((N-i)\rho B)f_{1,N}+\sum_{i=1}^{k-1}(I-g(\rho B))^2g((k-1-i)\rho B)f_{1,k},$$

$$P_k^4=-(I-g(2N\rho B))^{-1}(g((N-k-1)\rho B)-g((N+k-1)\rho B)$$

$$\times(I-g(\rho B))^2\sum_{i=1}^{N-1}g((N+i-1)\rho B)(f_{2,i}-f_{2,N})$$

$$+(I-g(\rho B))^2\sum_{i=1}^{N}g((k+i-2)\rho B)](f_{2,i}-f_{2,1}),$$

$$P_k^5=-(I-g(\rho B))^2\sum_{i=k+2}^{N}g((i-k-2)\rho B)(f_{2,i}-f_{2,k+1}),$$

$$P_k^6=-f_{2,k}+(2-g(\rho B))f_{2,k+1}-(I-g(2N\rho B))^{-1}$$

$$\times(g((N-k-1)\rho B)-g((N+k-1)\rho B)(I-g(\rho B))^2$$

$$\times\sum_{i=1}^{N}g((N+i-1)\rho B)f_{2,N}+(I-g(\rho B))^2\sum_{i=1}^{N}g((k+i-2)\rho B)]f_{2,1}$$

$$-(I-g(\rho B))^2\sum_{i=k+2}^{N}g((i-k-2)\rho B)f_{2,k+1}.$$

Let us estimate $P_m^\tau = \{P_k^m\}_{k=1}^{N-1}$ for any $m = 1,2,3,4,5,6$ separately. We start with P_1^τ. Using estimate (7.42), we show that, for $1 \leq k \leq N-1$,

$$\begin{aligned}\| P_k^1 \|_E &\leq \|(I - e^{-2N\rho B})^{-1}(I - e^{-\rho B})B^{-1}\|_{E\to E} \\ &\times \|e^{-(N-k-1)\rho B} - e^{-(N+k-1)\rho B} \|_{E\to E} \\ &\times \sum_{i=1}^{N-1} \|(I - e^{-\rho B})e^{-(N-i)\rho B} \|_{E\to E} \|Bf_{1,i} - Bf_{1,N} \|_E \\ &\leq M[\sum_{i=1}^{N-1} \frac{\tau}{(t_N - t_i)^{1-\alpha}}] \| Bf_1^\tau \|_{C_\tau^{\alpha,\alpha}(E)}.\end{aligned}$$

The sum enclosed in the right-hand side square brackets is the lower Darboux integral sum for the integral

$$\int_0^{t_N} \frac{ds}{(t_N - s)^{1-\alpha}}.$$

Since

$$\int_0^{t_N} \frac{ds}{(t_N - s)^{1-\alpha}} = \frac{t_N^\alpha}{\alpha},$$

it follows that

$$\| P_k^1 \|_E \leq \frac{M}{\alpha} \| Bf_1^\tau \|_{C_\tau^{\alpha,\alpha}(E)}.$$

Thus, for $1 \leq k \leq N-1$ we have established the bound

$$\max_{1\leq k\leq N-1} \| P_k^1 \|_E \leq \frac{M}{\alpha} \| Bf_1^\tau \|_{C_\tau^{\alpha,\alpha}(E)}. \tag{7.50}$$

Now let us estimate the difference $P_{k+r}^1 - P_k^1$ for $1 \leq k < k+r \leq N-1$. We will consider separately the cases $k \leq r$, $N-k \leq 2r$ and $k > r$, $N-k > 2r$. If $k \leq r$, then (7.50) yields

$$\| P_{k+r}^1 - P_k^1 \|_E \leq \| P_{k+r}^1 \|_E + \| P_k^1 \|_E$$

$$\leq \frac{M_1 2^{1+\alpha}}{\alpha} \| Bf_1^\tau \|_{C_\tau^{\alpha,\alpha}(E)} r^\alpha (k+r)^{-\alpha} = \frac{M_2}{\alpha} \| Bf_1^\tau \|_{C_\tau^{\alpha,\alpha}(E)} t_r^\alpha (t_k + t_r)^{-\alpha}.$$

If $N-k \leq 2r$, then (7.50) yields

$$\| P_{k+r}^1 - P_k^1 \|_E \leq \| P_{k+r}^1 \|_E + \| P_k^1 \|_E$$

$$\leq \frac{M_1 2^{1+\alpha}}{\alpha} \| Bf_1^\tau \|_{C_\tau^{\alpha,\alpha}(E)} r^\alpha (N-k)^{-\alpha} = \frac{M_2}{\alpha} \| Bf_1^\tau \|_{C_\tau^{\alpha,\alpha}(E)} t_r^\alpha (t_N - t_k)^{-\alpha}.$$

Now let $k > r$ and $N - k > 2r$. Let us represent the difference $P^1_{k+r} - P^1_k$ as the following sums:

$$P^1_{k+r} - P^1_k = (I - g(2N\rho B))^{-1}(g((N-k-r-1)\rho B) - g((N-k-1)\rho B)$$

$$+g((N+k-1)\rho B) - g((N+k+r-1)\rho B)(I - g(\rho B))^2 \sum_{i=1}^{N-1} g((N-i)\rho B)(f_{1,i} - f_{1,N}).$$

Further, applying (7.42) and (7.48) we show that, for $2 \leq k < k+r \leq N-2$,

$$\begin{aligned}
\| P^1_{k+r} - P^1_k \|_E &\leq ||(I - e^{-2N\rho B})^{-1}(I - e^{-\rho B})B^{-1}||_{E\to E} \\
&\quad \times [||(e^{-(N-k-1)\rho B} - e^{-(N-k-r-1)\rho B}) \|_{E\to E} \\
&\quad + ||(e^{-(N+k+r-1)\rho B} - e^{-(N+k-1)\rho B}) \|_{E\to E}] \\
&\quad \times \sum_{i=1}^{N-1} ||(I - e^{-\rho B})e^{-(N-i)\rho B} \|_{E\to E} \|Bf_{1,i} - Bf_{1,N} \|_E \\
&\leq M[\frac{r^\alpha}{(N-r)^\alpha} + \frac{r^\alpha}{(k+r)^\alpha}] \sum_{i=1}^{N-1} \frac{\tau}{(t_N - t_i)^{1-\alpha}} \| Bf^\tau_1 \|_{C^{\alpha,\alpha}_\tau(E)} \\
&\leq \frac{M_1}{\alpha} \| Bf^\tau_1 \|_{C^{\alpha,\alpha}_\tau(E)} t^\alpha_r (t_k + t_r)^{-\alpha}(t_N - t_k)^{-\alpha}.
\end{aligned}$$

Thus, for $1 \leq k < k+r \leq N-1$ we have established the bound

$$\| P^1_{k+r} - P^1_k \|_E \leq \frac{M_1}{\alpha} \| Bf^\tau_1 \|_{C^{\alpha,\alpha}_\tau(E)} t^\alpha_r (t_k + t_r)^{-\alpha}(t_N - t_k)^{-\alpha}.$$

From this and estimate (7.50) it follows that

$$\| P^\tau_1 \|_{C^{\alpha,\alpha}_\tau(E)} \leq \frac{M_1}{\alpha} \| Bf^\tau_1 \|_{C^{\alpha,\alpha}_\tau(E)}.$$

In a similar manner we can show that

$$||P^\tau_4||_{C^{\alpha,\alpha}_\tau(E)} \leq \frac{M_1}{\alpha} \| Bf^\tau_2 \|_{C^{\alpha,\alpha}_\tau(E)}.$$

Now let us estimate P^τ_2. Using estimate (7.42), we show that, for $2 \leq k \leq N-1$,

$$\| P^1_k \|_E \leq ||I - e^{-\rho B}||_{E\to E}||I - g(\rho B)||_{E\to E}||f_{1,k-1} - f_{1,k}||_E$$

$$+||(I - e^{-\rho B})B^{-1}||_{E\to E} \sum_{i=1}^{k-2} ||(I - e^{-\rho B})e^{-(k-1-i)\rho B} \|_{E\to E} \|Bf_{1,i} - Bf_{1,k} \|_E$$

$$\leq \frac{M}{t^\alpha_k}[\sum_{i=1}^{k-1} \frac{\tau}{(t_k - t_i)^{1-\alpha}}]\| Bf^\tau_1 \|_{C^{\alpha,\alpha}_\tau(E)}.$$

The sum enclosed in the right-hand side square brackets is the lower Darboux integral sum for the integral

$$\int_0^{t_k} \frac{ds}{(t_k - s)^{1-\alpha}}.$$

Since

$$\int_0^{t_k} \frac{ds}{(t_k - s)^{1-\alpha}} = \frac{t_k^\alpha}{\alpha},$$

it follows that

$$\| P_k^2 \|_E \leq \frac{M}{\alpha} \| Bf_1^\tau \|_{C_\tau^{\alpha,\alpha}(E)}$$

for $2 \leq k \leq N-1$. Since $P_1^2 = 0$, it follows that

$$\max_{1 \leq k \leq N-1} \| P_k^2 \|_E \leq \frac{M}{\alpha} \| Bf_1^\tau \|_{C_\tau^{\alpha,\alpha}(E)}. \tag{7.51}$$

Now let us estimate the difference $P_{k+r}^2 - P_k^2$ for $1 \leq k < k+r \leq N-1$. We will consider separately the cases $k \leq r$, $N-k \leq 2r$ and $k > r$, $N-k > 2r$. If $k \leq r$, then (7.51) yields

$$\| P_{k+r}^2 - P_k^2 \|_E \leq \| P_{k+r}^2 \|_E + \| P_k^2 \|_E$$

$$\leq \frac{M_1 2^{1+\alpha}}{\alpha} \| Bf_1^\tau \|_{C_\tau^{\alpha,\alpha}(E)} r^\alpha (k+r)^{-\alpha} = \frac{M_2}{\alpha} \| Bf_1^\tau \|_{C_\tau^{\alpha,\alpha}(E)} t_r^\alpha (t_k + t_r)^{-\alpha}.$$

If $N-k \leq 2r$, then (7.51) yields

$$\| P_{k+r}^2 - P_k^2 \|_E \leq \| P_{k+r}^2 \|_E + \| P_k^2 \|_E$$

$$\leq \frac{M_1 2^{1+\alpha}}{\alpha} \| Bf_1^\tau \|_{C_\tau^{\alpha,\alpha}(E)} r^\alpha (N-k)^{-\alpha} = \frac{M_2}{\alpha} \| Bf_1^\tau \|_{C_\tau^{\alpha,\alpha}(E)} t_r^\alpha (t_N - t_k)^{-\alpha}.$$

Now let $k > r$ and $N-k > 2r$. Let us represent the difference $P_{k+r}^2 - P_k^2$ as the following sums:

$$P_{k+r}^2 - P_k^2 = \sum_{i=k-r}^{k+r-1} (I - g(\rho B))^2 g((k+r-1-i)\rho B)(f_{1,i} - f_{1,k+r})$$

$$- \sum_{i=k-r}^{k-1} (I - g(\rho B))^2 g((k-1-i)\rho B)(f_{1,i} - f_{1,k})$$

$$- \sum_{i=1}^{k-r-1} (I - g(\rho B))^2 g((k+r-1-i)\rho B)(f_{1,k+1} - f_{1,k})$$

$$+\sum_{i=1}^{k-r-1}(I-g(\rho B))^2[g((k+r-1-i)\rho B)-g((k-1-i)\rho B)](f_{1,i}-f_{1,k})=\sum_{m=1}^{4}R_k^m,$$

where

$$R_k^1=\sum_{i=k-r}^{k+r-1}(I-g(\rho B))^2 g((k+r-1-i)\rho B)(f_{1,i}-f_{1,k+r}),$$

$$R_k^2=-\sum_{i=k-r}^{k-1}(I-g(\rho B))^2 g((k-1-i)\rho B)(f_{1,i}-f_{1,k}),$$

$$R_k^3=-\sum_{i=1}^{k-r-1}(I-g(\rho B))^2 g((k+r-1-i)\rho B)(f_{1,k+1}-f_{1,k}),$$

$$R_k^4=\sum_{i=1}^{k-r-1}(I-g(\rho B))^2[g((k+r-1-i)\rho B)-g((k-1-i)\rho B)](f_{1,i}-f_{1,k}).$$

Let us estimate R_k^m for any $m=1,2,3,4$ separately. We start with R_k^1. Using estimate (7.42), we show that, for $1\le k\le N-1$,

$$\| R_k^1 \|_E\le \sum_{i=k-r}^{k+r-1}\|(I-e^{-\rho B})e^{-(k+r-1-i)\rho B}\|_{E\to E}$$

$$\times\|I-e^{-\rho B}\|_{E\to E}\|f_{1,i}-f_{1,k+r}\|_E$$

$$\le\frac{M}{(t_k+t_r)^\alpha}\Big[\frac{1}{\tau^\alpha(t_N-t_{k+r-1})^\alpha}+\sum_{i=k-r}^{k+r-2}\frac{\tau}{(t_{k+r}-t_i)^{1-\alpha}(t_N-t_i)^\alpha}\Big]\| f_1^\tau \|_{C_\tau^{\alpha,\alpha}(E)}$$

$$\le\frac{M_1}{(t_k+t_r)^\alpha}\Big[\sum_{i=k-r}^{k+r-1}\frac{\tau}{(t_{k+r}-t_i)^{1-\alpha}(t_N-t_i)^\alpha}\Big]\| f_1^\tau \|_{C_\tau^{\alpha,\alpha}(E)}.$$

The sum enclosed in the right-hand side square brackets is the lower Darboux integral sum for the integral

$$\int_{t_k-t_r}^{t_k+t_r}\frac{ds}{(t_k+t_r-s)^{1-\alpha}(t_N-s)^\alpha}.$$

Since

$$\int_{t_k-t_r}^{t_k+t_r}\frac{ds}{(t_k+t_r-s)^{1-\alpha}(t_N-s)^\alpha}\le\frac{M}{\alpha}(2t_r)^\alpha(t_N-t_k-t_r)^{-\alpha},$$

it follows that

$$\| R_k^1 \|_E\le\frac{M_1M}{\alpha}(2t_r)^\alpha(t_N-t_k-t_r)^{-\alpha}(t_k+t_r)^{-\alpha}\| f_1^\tau \|_{C_\tau^{\alpha,\alpha}(E)}.$$

Thus, we have established the bound

$$\| R_k^1 \|_E \leq \frac{M_2}{\alpha} t_r^\alpha (t_N - t_k)^{-\alpha} (t_k + t_r)^{-\alpha} \| f_1^\tau \|_{C_\tau^{\alpha,\alpha}(E)}.$$

In a similar manner we can show that

$$\| R_k^2 \|_E \leq \frac{M_2}{\alpha} t_r^\alpha (t_N - t_k)^{-\alpha} (t_k + t_r)^{-\alpha} \| f_1^\tau \|_{C_\tau^{\alpha,\alpha}(E)}.$$

Since

$$R_k^3 = (I - g(\rho B))(g(2r\rho B) - g((k + r - 1)\rho B))(f_{1,k} - f_{1,k+r}),$$

it follows that

$$\| R_k^3 \|_E \leq \| e^{-2r\rho B} - e^{-(k+r-1)\rho B} \|_{E \to E} \| I - e^{-\rho B} \|_{E \to E} \| f_{1,k} - f_{1,k+r} \|_E .$$

Using estimate (7.42), we obtain

$$\| R_k^3 \|_E \leq M t_r^\alpha (t_N - t_k)^{-\alpha} (t_k + t_r)^{-\alpha} \| f_1^\tau \|_{C_\tau^{\alpha,\alpha}(E)}.$$

Now let us estimate R_k^4. Using estimates (7.42) and (7.48), we obtain

$$\| R_k^4 \|_E \leq \sum_{i=1}^{k-r-1} \| (I - g(\rho B))^2 [g((k + r - 1 - i)\rho B) - g((k - 1 - i)\rho B)] \|_{E \to E}$$

$$\times \| f_{1,i} - f_{1,k} \|_E \leq M \frac{t_r}{t_k^\alpha} [\sum_{i=1}^{k-r-1} \frac{\tau}{(t_k - t_i)^{2-\alpha} (t_N - t_i)^\alpha}] \| f_1^\tau \|_{C_\tau^{\alpha,\alpha}(E)}.$$

The sum enclosed in the right-hand side square brackets is the lower Darboux integral sum for the integral

$$\int_0^{t_k - t_r} \frac{ds}{(t_k - s)^{2-\alpha} (t_N - s)^\alpha}.$$

Since for $k > r$ and $N - k > 2r$ we have the bound

$$\int_0^{t_k - t_r} \frac{ds}{(t_k - s)^{2-\alpha} (t_N - s)^\alpha} \leq \frac{1}{(t_N - t_k + t_r)^\alpha} \int_0^{t_k - t_r} \frac{ds}{(t_k - s)^{2-\alpha}}$$

$$\leq \frac{1}{(1 - \alpha)(t_N - t_k + t_r)^\alpha} t_r^{-1+\alpha} \leq \frac{1}{(1 - \alpha)(t_N - t_k)^\alpha} t_r^{-1+\alpha},$$

it follows that

$$\| R_k^4 \|_E \leq \frac{M}{1 - \alpha} t_r^\alpha (t_N - t_k)^{-\alpha} t_k^{-\alpha} \| f_1^\tau \|_{C_\tau^{\alpha,\alpha}(E)}.$$

Thus, we have established the bound

$$\| R_k^4 \|_E \leq \frac{M 2^\alpha}{1-\alpha} t_r^\alpha (t_N - t_k)^{-\alpha} (t_k + t_r)^{-\alpha} \| f_1^\tau \|_{C_\tau^{\alpha,\alpha}(E)}.$$

Thus, we have shown that for any $2 \leq k < k+r \leq N-1$ the following inequality holds:

$$\| P_{k+r}^2 - P_k^2 \|_E \leq \frac{M_1}{\alpha(1-\alpha)} t_r^\alpha (t_N - t_k)^{-\alpha} (t_k + t_r)^{-\alpha} \| f_1^\tau \|_{C_\tau^{\alpha,\alpha}(E)}. \quad (7.52)$$

Estimates (7.51) and (7.52) give

$$\| P_2^\tau \|_{C_\tau^{\alpha,\alpha}(E)} \leq \frac{M_1}{\alpha(1-\alpha)} \| f_1^\tau \|_{C_\tau^{\alpha,\alpha}(E)} .$$

In exactly the same manner one establishes the inequality

$$\| P_5^\tau \|_{C_\tau^{\alpha,\alpha}(E)} \leq \frac{M_1}{\alpha(1-\alpha)} \| f_2^\tau \|_{C_\tau^{\alpha,\alpha}(E)} .$$

Now let us estimate P_3^τ. Since

$$P_k^3 = -f_{1,k+1} + (2 - g(\rho B)) f_{1,k} + (I - g(2N\rho B))^{-1} (g((N-k-1)\rho B)$$

$$-g((N+k-1)\rho B)(I - g(\rho B))(I - g(N\rho B) f_{1,N} + (I - g(\rho B))(I - g((k-1)\rho B) f_{1,k},$$

it follows that

$$\| P_3^\tau \|_{C_\tau^{\alpha,\alpha}(E)} \leq M \| f_1^\tau \|_{C_\tau^{\alpha,\alpha}(E)} .$$

In exactly the same manner one establishes the inequality

$$\| P_6^\tau \|_{C_\tau^{\alpha,\alpha}(E)} \leq \frac{M_1}{\alpha(1-\alpha)} \| f_2^\tau \|_{C_\tau^{\alpha,\alpha}(E)} .$$

Finally, combining the estimates for $P_m^\tau = \{P_k^m\}_{k=1}^{N-1}$, $m = 1, 2, 3, 4, 5, 6$, we obtain the estimate

$$\| \{\rho^{-2}(g_{k+1} - 2g_k + g_{k-1})\}_1^{N-1} \|_{C_\tau^{\alpha,\alpha}(E)}$$

$$\leq \frac{M}{\alpha(1-\alpha)} [\| B f_1^\tau \|_{C_\tau^{\alpha,\alpha}(E)} + \| B f_2^\tau \|_{C_\tau^{\alpha,\alpha}(E)}].$$

Using the last estimate and (7.49), we obtain

$$\| \{\rho^{-2}(u_{k+1} - 2u_k + u_{k-1})\}_1^{N-1} \|_{C_\tau^{\alpha,\alpha}(E)}$$

$$\leq M\{\frac{1}{\alpha(1-\alpha)} [\| B f_1^\tau \|_{C_\tau^{\alpha,\alpha}(E)} + \| B f_2^\tau \|_{C_\tau^{\alpha,\alpha}(E)}]$$

$$+ \frac{1}{1 - \exp(-\delta\rho)} [\| A u_0 \|_E + \| A u_N \|_E]\}.$$

By the triangle inequality, this last estimate and the difference equation (7.37) yield

$$\begin{aligned}
&\| A_\tau \Pi(u_0,u_N)u^\tau \|_{C_\tau^{\alpha,\alpha}(E)} \\
&\leq M\{\frac{1}{\alpha(1-\alpha)}[\| Bf_1^\tau \|_{C_\tau^{\alpha,\alpha}(E)} + \| Bf_2^\tau \|_{C_\tau^{\alpha,\alpha}(E)}] \\
&+\frac{1}{1-\exp(-\delta\rho)}[\| Au_0\|_E + \|Au_N\|_E]\}.
\end{aligned}$$

Theorem 7.2.5 is proved.

Theorem 7.2.6. *The solutions of the difference problem* (7.37) *in* $C_\tau(E_\alpha)$ $(0<\alpha<1)$ *obey the coercivity inequality*

$$\begin{aligned}
&\| D_\tau^2 \Pi(u_0,u_N)u^\tau \|_{C_\tau(E_\alpha)} + \| A_\tau \Pi(u_0,u_N)u^\tau \|_{C_\tau(E_\alpha)} \\
&\leq M\{\frac{1}{\alpha(1-\alpha)}[\| \{Bf_{1,k}\}_1^N \|_{C_\tau(E_\alpha)} + \| \{Bf_{2,k}\}_1^N \|_{C_\tau(E_\alpha)}] \\
&+\frac{1}{1-\exp(-\delta\rho)}[\| Au_0\|_{E_\alpha} + \|Au_N\|_{E_\alpha}]\},
\end{aligned}$$

where M *is independent not only of* f_1^τ, f_2^τ, u_0, u_N, τ, α *but also of* ε.

The proof of Theorem 7.2.6 follows the scheme of proof of Theorem 7.1.9 and relies on the formula

$$\begin{aligned}
&\rho^{-2}(u_{k+1}-2u_k+u_{k-1}) \\
&=(I-e^{-2N\rho B})^{-1}(I-e^{-\rho B})^2\rho^{-2}(e^{-(k-1)\rho B}-e^{-(2N-k-1)\rho B})u_0 \\
&+(I-e^{-2N\rho B})^{-1}(I-e^{-\rho B})^2\rho^{-2}(e^{-(N-k-1)\rho B}-e^{-(N+k-1)\rho B})u_N \\
&+(I-g(2N\rho B))^{-1}(g((N-k-1)\rho B)-g((N+k-1)\rho B) \\
&\times(I-g(\rho B))^2\sum_{i=1}^{N} g((N-i)\rho B)f_{1,i}-\sum_{i=1}^{k-1}(I-g(\rho B))^2 g((k-1-i)\rho B)f_{1,i} \\
&-f_{1,k+1}+(2-g(\rho B))f_{1,k}-(I-g(2N\rho B))^{-1}(g((N-k-1)\rho B)-g((N+k-1)\rho B) \\
&\times(I-g(\rho B))^2\sum_{i=1}^{N} g((N+i-1)\rho B)f_{2,i}+(I-g(\rho B))^2\sum_{i=1}^{N} g((k+i-2)\rho B)]f_{2,i} \\
&-(I-g(\rho B))^2\sum_{i=k+2}^{N} g((i-k-2)\rho B)f_{2,i}-f_{2,k}+(2-g(\rho B))f_{2,k+1}
\end{aligned}$$

and estimates (7.42) and (7.48).

Theorem 7.2.7. *The solutions of the difference problem (7.37) in $C_\tau(E)$ obey the almost coercivity inequality*

$$\| D_\tau^2\Pi(u_0,u_N)u^\tau\|_{C_\tau(E)} + \| A_\tau\Pi(u_0,u_N)u^\tau\|_{C_\tau(E)}$$

$$\leq M\{\min\{\ln\frac{1}{\tau}, 1+|\ln\|B\|_{E\to E}|\|\}[\| \{Bf_{1,k}\}_1^N \|_{C_\tau(E)} + \| \{Bf_{2,k}\}_1^N \|_{C_\tau(E)}]$$

$$+\frac{1}{1-\exp(-\delta\rho)}[\| Au_0\|_E + \|Au_N\|_E]\},$$

where M is independent not only of f_1^τ, f_2^τ, u_0, u_N, τ but also of ε.

Proof. By the definition of the norm in $C_\tau^{\alpha,\alpha}(E)$,

$$\| f^\tau \|_{C_\tau(E)} \leq \| f^\tau \|_{C_\tau^{\alpha,\alpha}(E)} \leq \frac{M}{\tau^\alpha}\| f^\tau \|_{C_\tau(E)}.$$

Then, Theorem 7.2.5 implies that

$$\| D_\tau^2\Pi(u_0,u_N)u^\tau\|_{C_\tau(E)} + \| A_\tau\Pi(u_0,u_N)u^\tau\|_{C_\tau(E)}$$

$$\leq M\{\frac{1}{\alpha\tau^\alpha}[\| \{Bf_{1,k}\}_1^N \|_{C_\tau(E)}$$

$$+ \| \{Bf_{2,k}\}_1^N \|_{C_\tau(E)}] + \frac{1}{1-\exp(-\delta\rho)}[\| Au_0\|_E + \|Au_N\|_E]\}$$

for all sufficiently small $\alpha > 0$. Choosing α in the best way, we obtain the estimate

$$\| D_\tau^2\Pi(u_0,u_N)u^\tau\|_{C_\tau(E)} + \| A_\tau\Pi(u_0,u_N)u^\tau\|_{C_\tau(E)}$$

$$\leq M_1\{\ln\frac{1}{\tau}[\| \{Bf_{1,k}\}_1^N \|_{C_\tau(E)} + \| \{Bf_{2,k}\}_1^N \|_{C_\tau(E)}]$$

$$+\frac{1}{1-\exp(-\delta\rho)}[\| Au_0\|_E + \|Au_N\|_E]\}.$$

By the definition of the norm in $C_\tau(E_\alpha)$,

$$\| f^\tau \|_{C_\tau(E)} \leq \| f^\tau \|_{C_\tau(E_\alpha)} \leq M\|B\|_{E\to E}^\alpha\| f^\tau \|_{C_\tau(E)}.$$

Then, Theorem 7.2.6 implies that

$$\| D_\tau^2\Pi(u_0,u_N)u^\tau\|_{C_\tau(E)} + \| A_\tau\Pi(u_0,u_N)u^\tau\|_{C_\tau(E)}$$

$$\leq M\{\frac{1}{\alpha}\|B\|_{E\to E}[\| \{Bf_{1,k}\}_1^N \|_{C_\tau(E)}$$

$$+ \| \{Bf_{2,k}\}_1^N \|_{C_\tau(E)}] + \frac{1}{1-\exp(-\delta\rho)}[\| Au_0\|_E + \|Au_N\|_E]\}$$

for all sufficiently small $\alpha > 0$. Choosing α in the best way, we obtain the estimate

$$\| D_\tau^2\Pi(u_0,u_N)u^\tau\|_{C_\tau(E)} + \| A_\tau\Pi(u_0,u_N)u^\tau\|_{C_\tau(E)}$$

$$\leq M\{(1+|\ln \|B\|_{E\to E}|)[\| \{Bf_{1,k}\}_1^N \|_{C_\tau(E)} + \| \{Bf_{2,k}\}_1^N \|_{C_\tau(E)}]$$

$$+\frac{1}{1-\exp(-\delta\rho)}[\| Au_0\|_E + \|Au_N\|_E]\}.$$

Theorem 7.2.7 is proved.

Note that if A is the unbounded positive operator in E, then

$$\min\left\{\ln\frac{1}{\tau}, 1+|\ln \| B \|_{E\to E}|\right\} = \ln\frac{1}{\tau}.$$

Let us give, without proof, the following results.

Theorem 7.2.8. *The solutions of the difference problem* (7.37) *in* $C_\tau^{\beta,\gamma}(E)$ $(0 \leq \gamma \leq \beta, 0 < \beta < 1)$ *obey the coercivity inequalities*

$$\| D_\tau^2\Pi(u_0,u_N)u^\tau\|_{C_\tau(E_{\beta-\gamma})} \leq \frac{M_1}{\beta(1-\beta)}[\| \{Bf_{1,k}\}_1^N \|_{C_\tau(E_{\beta-\gamma})}$$
$$+ \| \{Bf_{2,k}\}_1^N \|_{C_\tau(E_{\beta-\gamma})}]$$
$$+ M_1\frac{1}{1-\exp(-\delta\rho)}[\| (I-e^{-\rho B})^2\rho^{-2}u_0 - f_{1,1} \|_{E_{\beta-\gamma}}$$
$$+ \|(I-e^{-\rho B})^2\rho^{-2}u_N - f_{2,N} \|_{E_{\beta-\gamma}}],$$

$$\| D_\tau^2\Pi(u_0,u_N)u^\tau\|_{C_\tau^{\beta,\gamma}(E)} + \| A_\tau\Pi(u_0,u_N)u^\tau\|_{C_\tau^{\beta,\gamma}(E)}$$
$$\leq \frac{M_1}{\beta(1-\beta)}[\| \{Bf_{1,k}\}_1^N \|_{C_\tau^{\beta,\gamma}(E)} + \| \{Bf_{2,k}\}_1^N \|_{C_\tau^{\beta,\gamma}(E)}]$$
$$+ M_1\frac{1}{1-\exp(-\delta\rho)}[|(I-e^{-\rho B})^2\rho^{-2}u_0 - f_{1,1}|_0^{\beta,\gamma}$$
$$+ |(I-e^{-\rho B})^2\rho^{-2}u_N - f_{2,N}|_0^{\beta,\gamma}],$$

where M_1 *is independent not only of* f_1^τ, f_2^τ, u_0, u_N, τ, β, γ *but also of* ε.

Here, $|w|_0^{\beta,\gamma}$ denotes that the norm of the Banach space $\widetilde{E}_0^{\beta,\gamma}$ consists of those $v \in E$ for which the norm

$$|w|_0^{\beta,\gamma} = \max_{1\leq n\leq N-1} \|e^{-\frac{t_n}{\varepsilon}B}w\|_E$$
$$+ \sup_{1\leq n<n+r\leq N-1} t_r^{-\beta}(t_n+t_r)^\gamma(t_N-t_n)^\gamma\|(e^{-(\frac{t_n+t_r}{\varepsilon})B} - e^{-\frac{t_n}{\varepsilon}B})w\|_E$$

is finite.

Theorem 7.2.9. *The solutions of the difference problem* (7.37) *in* $C_\tau^{\beta,\gamma}(E_{\alpha-\beta})$ $(0 \leq \gamma \leq \beta \leq \alpha, 0 < \alpha < 1)$ *obey the coercivity inequalities*

$$\begin{aligned}
&\| D_\tau^2 \Pi(u_0,u_N)u^\tau \|_{C_\tau^{\beta,\gamma}(E_{\alpha-\beta})} + \| A_\tau \Pi(u_0,u_N)u^\tau \|_{C_\tau^{\beta,\gamma}(E_{\alpha-\beta})} \\
&\quad + \| D_\tau^2 \Pi(u_0,u_N)u^\tau \|_{C_\tau(\widetilde{E}^{\beta,\gamma}_{\alpha-\beta})} \\
&\leq \frac{M_1}{\alpha(1-\alpha)}[\| \{Bf_{1,k}\}_1^N \|_{C_\tau^{\beta,\gamma}(E_{\alpha-\beta})} + \| \{Bf_{2,k}\}_1^N \|_{C_\tau^{\beta,\gamma}(E_{\alpha-\beta})}] \\
&\quad + M_1 \frac{1}{1-\exp(-\delta\rho)}[|(I-e^{-\rho B})^2\rho^{-2}u_0 - f_{1,1}|^{\beta,\gamma}_{\alpha-\beta} \\
&\quad + |(I-e^{-\rho B})^2\rho^{-2}u_N - f_{2,N}|^{\beta,\gamma}_{\alpha-\beta}],
\end{aligned}$$

$$\begin{aligned}
&\| D_\tau^2 \Pi(u_0,u_N)u^\tau \|_{C_\tau^{\beta,\gamma}(E_{\alpha-\beta})} + \| A_\tau \Pi(u_0,u_N)u^\tau \|_{C_\tau^{\beta,\gamma}(E_{\alpha-\beta})} \\
&\quad + \| D_\tau^2 \Pi(u_0,u_N)u^\tau \|_{C_\tau(E_{\alpha-\gamma})} \\
&\leq \frac{M_1}{\alpha(1-\alpha)}[\| \{Bf_{1,k}\}_1^N \|_{C_\tau^{\beta,\gamma}(E_{\alpha-\beta})} + \| \{Bf_{2,k}\}_1^N \|_{C_\tau^{\beta,\gamma}(E_{\alpha-\beta})}] \\
&\quad + M_1 \frac{1}{1-\exp(-\delta\rho)}[\|(I-e^{-\rho B})^2\rho^{-2}u_0 - f_{1,1} \|_{C_\tau(E_{\alpha-\gamma})} \\
&\quad + \|(I-e^{-\rho B})^2\rho^{-2}u_N - f_{2,N} \|_{C_\tau(E_{\alpha-\gamma})}],
\end{aligned}$$

where M_1 *is independent not only of* f_1^τ, f_2^τ, u_0, u_N, τ, α, β, γ *but also of* ε.

Here, $|w|^{\beta,\gamma}_{\alpha-\beta}$ denotes that the norm of the Banach space $\widetilde{E}^{\beta,\gamma}_{\alpha-\beta}$ consists of those $v \in E$ for which the norm

$$|w|^{\beta,\gamma}_{\alpha-\beta} = \max_{1 \leq n \leq N-1} \|e^{-\frac{t_n}{\varepsilon}B}w\|_{E_{\alpha-\beta}}$$

$$+ \sup_{1 \leq n < n+r \leq N-1} t_r^{-\beta}(t_n+t_r)^\gamma (t_N-t_n)^\gamma \|(e^{-(\frac{t_n+t_r}{\varepsilon})B} - e^{-\frac{t_n}{\varepsilon}B})w\|_{E_{\alpha-\beta}}$$

is finite.

Note that the spaces of grid functions $C_\tau^{\beta,\gamma}(E_{\alpha-\beta})$, in which coercive solvability has been established, depend on the parameters α, β and γ. However, the constants in the coercive inequalities depend only on α. Hence, we can choose the parameters β and γ freely, which increases the number of spaces of grid functions in which difference problem (7.37) is well posed.

Let us consider now the boundary-value difference problem (7.37) in the spaces $L_{p,\tau}(E) = L_p([0,T]_\tau, E)$, $1 \leq p < \infty$ of all grid functions.

Theorem 7.2.10. *The solutions of the difference problem* (7.37) *in* $L_{p,\tau}(E)$ *obey the almost coercive inequality*

$$\| D_\tau^2 \Pi(u_0,u_N)u^\tau \|_{L_{p,\tau}(E)} + \| A_\tau \Pi(u_0,u_N)u^\tau \|_{L_{p,\tau}(E)}$$

$$\leq M\{\min\{\ln\frac{1}{\tau}, 1 + |\ln\|B\|_{E\to E}\|\}[\| \{Bf_{1,k}\}_1^N \|_{L_{p,\tau}(E)} + \| \{Bf_{2,k}\}_1^N \|_{L_{p,\tau}(E)}]$$

$$+\frac{1}{1-\exp(-\delta\rho)}[\| Au_0\|_E + \|Au_N\|_E]\},$$

where M is independent not only of f_1^τ, f_2^τ, u_0, u_N, τ, p but also of ε.

Recall that if A is the unbounded positive operator in E, then

$$\min\left\{\ln\frac{1}{\tau}, 1 + |\ln \| B \|_{E\to E}|\right\} = \ln\frac{1}{\tau}.$$

Finally, let us give, without proof, the following results about well-posedness of the boundary-value difference problem (7.37) in the spaces

$$L_{p,\tau}(E) = L_p([0,T]_\tau, E), 1 \leq p < \infty.$$

Theorem 7.2.11. *Suppose that the difference problem (7.37) is well posed in $L_{p_0,\tau}(E)$ for some p_0, $1 < p_0 < \infty$. Then it is well posed in $L_{p,\tau}(E)$ for any p, $1 < p < \infty$ and the following coercivity inequality holds:*

$$\| D_\tau^2\Pi(u_0, u_N)u^\tau\|_{L_{p,\tau}(E)} + \| A_\tau\Pi(u_0, u_N)u^\tau\|_{L_{p,\tau}(E)}$$

$$+\|u^\tau\|_{C_\tau(E_{1-1/p,p})} \leq M(p_0)[\frac{1}{1-\exp(-\delta\rho)}[\| u_0\|_{E_{1-1/p,p}} + \|u_N\|_{E_{1-1/p,p}}]$$

$$+\frac{p^2}{p-1}[\| \{Bf_{1,k}\}_1^N \|_{L_{p,\tau}(E)} + \| \{Bf_{2,k}\}_1^N \|_{L_{p,\tau}(E)}]],$$

where $M(p_0)$ does not depend on f_1^τ, f_2^τ, u_0, u_N, τ, p but also not on ε.

Theorem 7.2.12. *Let $1 \leq p \leq \infty$ and $0 < \alpha < 1$. Then problem (5.29) is well posed in $L_{p,\tau}(E_{\alpha,p})$ and the following coercivity inequality holds:*

$$\| D_\tau^2\Pi(u_0, u_N)u^\tau\|_{L_{p,\tau}(E_{\alpha,p})} + \| A_\tau\Pi(u_0, u_N)u^\tau\|_{L_{p,\tau}(E_{\alpha,p})}$$

$$\leq M\frac{1}{1-\exp(-\delta\rho)}[\| Au_0\|_{E_{\alpha,p}} + \|Au_N\|_{E_{\alpha,p}}]$$

$$+\frac{M}{\alpha(1-\alpha)}[\| \{Bf_{1,k}\}_1^N \|_{L_{p,\tau}(E_{\alpha,p})} + \| \{Bf_{2,k}\}_1^N \|_{L_{p,\tau}(E_{\alpha,p})}],$$

where M does not depend on f_1^τ, f_2^τ, u_0, u_N, τ, p, α but also not on ε.

Theorem 7.2.13. *Let $1 < p, q < \infty$ and $0 < \alpha < 1$. Then the difference problem (5.29) is well posed in $L_{p,\tau}(E_{\alpha,q})$ and the following coercivity inequality holds:*

$$\| D_\tau^2\Pi(u_0, u_N)u^\tau\|_{L_{p,\tau}(E_{\alpha,q})} + \| A_\tau\Pi(u_0, u_N)u^\tau\|_{L_{p,\tau}(E_{\alpha,q})}$$

$$\leq M(q)[\frac{1}{1-\exp(-\delta\rho)}[\| Au_0\|_{E_{\alpha,q}} + \|Au_N\|_{E_{\alpha,q}}]$$

$$+\frac{p^2}{p-1}[\| \{Bf_{1,k}\}_1^N \|_{L_{p,\tau}(E_{\alpha,q})} + \| \{Bf_{2,k}\}_1^N \|_{L_{p,\tau}(E_{\alpha,q})}]],$$

where $M(q)$ does not depend on f_1^τ, f_2^τ, u_0, u_N, τ, p, α but also not on ε.

Remark 7.2.14. Theorems 7.2.5–7.2.13 permit us to obtain the error estimates of the high order of accuracy difference schemes for elliptic equations with a small parameter multiplying one-derivative term in the various Banach norms.

7.3 A Cauchy Problem for Hyperbolic Equations

We consider the abstract Cauchy problem for hyperbolic equations

$$\varepsilon^2 v''(t) + Av(t) = f(t) \quad (0 \leq t \leq T), v(0) = v_0, v'(0) = v_0' \tag{7.53}$$

in an arbitrary Banach space E with the positive operator A and $\varepsilon \in (0,\infty)$.

A function $v(t)$ is called a solution of the problem (7.53) if the following conditions are satisfied:

i. $v(t)$ is twice continuously differentiable on the segment $[0,T]$. The derivatives at the endpoints of the segment are understood as the appropriate unilateral derivatives.

ii. The element $v(t)$ belongs to $D(A)$ for all $t \in [0,T]$ and the function $Av(t)$ is continuous on the segment $[0,T]$.

iii. $v(t)$ satisfies the equations and the initial conditions (7.53).

Suppose that the function $f(t)$ is not only continuous, but also continuously differentiable on $[0,T]$, $v_0 \in D(A)$ and $v_0' \in D(A^{\frac{1}{2}})$. Then by formula (6.2), we have that the formula

$$v(t) = c(\frac{t}{\varepsilon})v_0 + s(\frac{t}{\varepsilon})v_0' + \frac{1}{\varepsilon^2}\int_0^t s(\frac{t-s}{\varepsilon})f(s)ds \tag{7.54}$$

gives a solution of problem (7.53). Here

$$c(\frac{t}{\varepsilon}) = \frac{e^{i\frac{t}{\varepsilon}A^{1/2}} + e^{-i\frac{t}{\varepsilon}A^{1/2}}}{2}, \quad s(\frac{t}{\varepsilon}) = \varepsilon A^{-1/2}\frac{e^{i\frac{t}{\varepsilon}A^{1/2}} - e^{-i\frac{t}{\varepsilon}A^{1/2}}}{2i}.$$

Suppose further that in an arbitrary Banach space E the assumptions

$$\|c(t)\|_{E\to E} \leq M, \ \|A^{\frac{1}{2}}s(t)\|_{E\to E} \leq M, \ 0 \leq t < \infty \tag{7.55}$$

are valid.

In Chapter 6, stable high order of accuracy difference schemes generated by an exact difference scheme or by Taylor's decomposition on the three points for the numerical solutions of an abstract initial-value problem (7.53) for $\varepsilon = 1$ were presented. Unfortunately, these difference schemes cannot be applied for the approximate solutions of (7.53) in the general cases $\varepsilon \in (0, \infty)$. In this section we study the high order of accuracy two-step uniform difference schemes of the approximate solutions for differential equations of the hyperbolic type with an arbitrary parameter ε on the highest derivative. The stability estimates of the solutions of these difference schemes are obtained.

On the segment $[0, T]$ we consider again a uniform grid

$$[0, T]_\tau = \{t_k = k\tau, k = 0, 1, \dots, N, N\tau = T\}$$

with step τ. Let $\tau = \rho\varepsilon$ and $B^2 = A$. By Theorem 3.1.1, we have the following exact two-step difference scheme:

$$v(t_{k+1}) - 2c(\frac{\tau}{\varepsilon})v(t_k) + v(t_{k-1})$$

$$= \frac{1}{\varepsilon^2}\{\int_{t_k}^{t_{k+1}} s(\frac{t_{k+1} - z}{\varepsilon})f(z)dz + \int_{t_{k-1}}^{t_k} s(\frac{z - t_{k-1}}{\varepsilon})f(z)dz\},$$

$$1 \le k \le N - 1, v(0) = v_0,$$

$$v(\tau) - v(0) = (c(\frac{\tau}{\varepsilon}) - I)v(0) + s(\frac{\tau}{\varepsilon})v'(0) + \frac{1}{\varepsilon^2}\int_0^\tau s(\frac{\tau - z}{\varepsilon})f(z)dz$$

or

$$\rho^{-2}(v(t_{k+1}) - 2v(t_k) + v(t_{k-1})) = 2\rho^{-2}(c(\frac{\tau}{\varepsilon}) - I)v(t_k) + \tau^{-1}\varepsilon^2\widetilde{f}_k,$$

$$\widetilde{f}_k = \widetilde{f}_{1,k+1} + s(\frac{\tau}{\varepsilon})\widetilde{f}_{2,k} - c(\frac{\tau}{\varepsilon})\widetilde{f}_{1,k},$$

$$\widetilde{f}_{1,k} = (\varepsilon^2\tau)^{-1}\int_{t_{k-1}}^{t_k} s(\frac{t_k - z}{\varepsilon})f(z)dz,\ \widetilde{f}_{2,k} = (\varepsilon^2\tau)^{-1}\int_{t_{k-1}}^{t_k} c(\frac{t_k - z}{\varepsilon})f(z)dz,$$

$$1 \le k \le N - 1,$$

$$u_0 = v_0,\ u_1 = c(\frac{\tau}{\varepsilon})v(0) + s(\frac{\tau}{\varepsilon})v'(0) + \tau\widetilde{f}_{1,1}.$$

Applying this exact two-step difference scheme, we can construct the high order of accuracy difference schemes

$$\rho^{-2}(u_{k+1} - 2u_k + u_{k-1}) = 2\rho^{-2}(c(\frac{\tau}{\varepsilon}) - I)u_k + \tau^{-1}\varepsilon^2 f_k, \tag{7.56}$$

$$f_k = f_{1,k+1} + s(\frac{\tau}{\varepsilon})f_{2,k} - c(\frac{\tau}{\varepsilon})f_{1,k},$$

$$u_0 = v_0,\ u_1 = c(\frac{\tau}{\varepsilon})v(0) + s(\frac{\tau}{\varepsilon})v'(0) + \tau f_{1,1},$$

$$f_{1,k} = (\tau A)^{-1}[f(t_k) - c(\rho)f(t_{k-1})] - (\tau A)^{-1}\sum_{j=0}^{m-1} B_j f^{(j+1)}(t_{k-1}),$$

$$f_{2,k} = -(\varepsilon^2\tau)^{-1}s(\rho)f(t_{k-1}) + (\varepsilon^2\tau)^{-1}\sum_{j=0}^{m-1} C_j f^{(j+1)}(t_{k-1}),$$

$$B_0 = s(\rho), B_1 = \varepsilon^2 A^{-1}(1 - c(\rho)),$$

$$B_{2j} = (\varepsilon^2 A^{-1})^j B_0 - \sum_{i=1}^{j}(\varepsilon^2 A^{-1})^{j-i+1}\frac{\tau^{2i-1}}{(2i-1)!}, 2 \le 2j \le m-1,$$

$$B_{2j+1} = (\varepsilon^2 A^{-1})^j B_1 - \sum_{i=1}^{j}(\varepsilon^2 A^{-1})^{j-i+1}\frac{\tau^{2i}}{(2i)!}, 3 \le 2j+1 \le m-1,$$

$$C_0 = \varepsilon^2 A^{-1}(1 - c(\rho)),$$

$$C_j = -(\varepsilon^2 A)^{-1}B_{j-1} + (\varepsilon^2 A)^{-1}\frac{\tau^j}{j!}, 1 \le j \le m-1$$

for the approximate solutions of the initial-value problem (7.53).

This problem is uniquely solvable and the following formula holds:

$$u_k = c(k\rho)v_0 + s(k\rho)v_0' \tag{7.57}$$

$$+ \sum_{m=1}^{k}\{c((k-m)\rho)f_{1,m} + s((k-m)\rho)f_{2,m}\}\tau, 1 \le k \le N.$$

Actually, formula (7.57) is true for $k = 1$. Let $k = 2$. Then using the formula

$$u_2 + u_0 = 2c(\frac{\tau}{\varepsilon})u_1 + \tau[f_{1,2} + s(\frac{\tau}{\varepsilon})f_{2,1} - c(\frac{\tau}{\varepsilon})f_{1,1}],$$

we obtain

$$u_2 = -v_0 + 2c(\frac{\tau}{\varepsilon})(c(\frac{\tau}{\varepsilon})v(0) + s(\frac{\tau}{\varepsilon})v'(0) + \tau f_{1,1}) + \tau[f_{1,2} + s(\frac{\tau}{\varepsilon})f_{2,1} - c(\frac{\tau}{\varepsilon})f_{1,1}]$$

$$= c(2\rho)v(0) + s(2\rho)v'(0) + \tau[f_{1,2} + s(\rho)f_{2,1} + c(\rho)f_{1,1}]$$

$$= c(2\rho)v(0) + s(2\rho)v'(0) + \tau\sum_{m=1}^{2}\{c((2-m)\rho)f_{1,m} + s((2-m)\rho)f_{2,m}\}.$$

Thus, formula (7.57) is true for $k = 2$. Applying mathematical induction, one can easily show that it is true for every k. Namely, assume that formula (7.57) is true for $k-1$ and k. Using the formula

$$u_{k+1} + u_{k-1} = 2c(\frac{\tau}{\varepsilon})u_k + \tau[f_{1,k+1} + s(\frac{\tau}{\varepsilon})f_{2,k} - c(\frac{\tau}{\varepsilon})f_{1,k}],$$

we obtain

$$u_{k+1} = -u_{k-1} + 2c(\rho)u_k + \tau[f_{1,k+1} + s(\rho)f_{2,k} - c(\rho)f_{1,k}]$$

$$= -\{c((k-1)\rho)v_0 + s((k-1)\rho)v_0' + \sum_{m=1}^{k-1}\{c((k-1-m)\rho)f_{1,m} + s((k-1-m)\rho)f_{2,m}\}\tau\}$$

$$+2c(\rho)\{c(k\rho)v_0 + s(k\rho)v_0' + \sum_{m=1}^{k}\{c((k-m)\rho)f_{1,m} + s((k-m)\rho)f_{2,m}\}\tau\}$$

$$+\tau[f_{1,k+1} + s(\frac{\tau}{\varepsilon})f_{2,k} - c(\frac{\tau}{\varepsilon})f_{1,k}] = \{2c(\rho)c(k\rho) - c((k-1)\rho)\}v_0 + \{2c(\rho)s(k\rho) - s((k-1)\rho)\}v_0'$$

$$+\sum_{m=1}^{k-1}\{2c(\rho)c((k-m)\rho) - c((k-1-m)\rho)f_{1,m}\{2c(\rho)s((k-m)\rho) - s((k-1-m)\rho)f_{2,m}\}\tau\}$$

$$+2c(\frac{\tau}{\varepsilon})f_{1,k}\tau + \tau[f_{1,k+1} + s(\rho)f_{2,k} - c(\rho)f_{1,k}]$$

$$= c((k+1)\rho)v(0) + s((k+1)\rho)v'(0)$$

$$+\tau\sum_{m=1}^{k+1}\{c((k+1-m)\rho)f_{1,m} + s((k+1-m)\rho)f_{2,m}\}.$$

Thus, formula (7.57) is true for $k+1$. Therefore, formula (7.57) is true for all k.

Let us investigate the stability of the exact two-step difference scheme (7.56).

Theorem 7.3.1. *For the solution of the exact two-step difference scheme* (7.56) *the following stability inequality holds:*

$$\max_{0\le k\le N}\|u_k\|_H \le M[\max_{1\le k\le N}\|f_{1,k}\|_E + \varepsilon\max_{1\le k\le N}\|A^{-\frac{1}{2}}f_{2,k}\|_E \tag{7.58}$$

$$+\varepsilon \| A^{-\frac{1}{2}}v_0' \|_E + \| v_0 \|_E],$$

where M does not depend on ε, τ, $f_{1,k}$, $f_{2,k}, 1\le k\le N$ and v_0, v_0'.

The proof of Theorem 7.3.1 is based on formula (7.57) and estimates (7.55).

Stability inequality (7.58) permits us to obtain the estimate of convergence of (7.56) in $C_\tau(E)$.

Theorem 7.3.2. *Suppose that the function $f(t)$ has $m+1$ derivatives and*

$$\|f^{(m+1)}(t)\|_E \leq M, 0 \leq t \leq T.$$

Then for the solution of the difference problem (7.56) the following convergence estimate is valid:

$$\| \{v(t_k) - u_k\}_1^N \|_{C_\tau(E)} \leq M\tau^m, \tag{7.59}$$

where M does not depend on τ and ε.

Proof. The components $z_k = v(t_k) - u_k$ of error vector $z^\tau = \{z_k\}_1^N$ satisfy the relations

$$\begin{aligned} \rho^{-2}(z_{k+1} - 2z_k + z_{k-1}) &= 2\rho^{-2}(c(\rho) - I)z_k + \tau^{-1}\varepsilon^2 a_k, \ 1 \leq k \leq N-1, \\ z_0 &= 0, \ z_1 = \tau a_{1,1}, \end{aligned}$$

where

$$a_k = a_{1,k+1} + s(\rho)a_{2,k} - c(\rho)a_{1,k}, 1 \leq k \leq N-1,$$

$$a_{1,k} = \widetilde{f}_{1,k} - f_{1,k}, \ a_{2,k} = \widetilde{f}_{2,k} - f_{2,k}, \ 1 \leq k \leq N.$$

From this and the stability inequality (7.58) it follows that

$$\max_{1 \leq k \leq N} \| z_k \|_E \leq M [\max_{1 \leq k \leq N} \|a_{1,k}\|_E + \varepsilon \max_{1 \leq k \leq N} \|A^{-\frac{1}{2}} a_{2,k}\|_E]. \tag{7.60}$$

Therefore to this end it suffices to estimate $\{a_{1,k}\}_1^N$ and $\{a_{2,k}\}_1^N$ in $C_\tau(E)$. Integrating by parts and using Taylor's formula, we obtain

$$\begin{aligned} a_{1,k} &= (\varepsilon^2\tau)^{-1} \int_{t_{k-1}}^{t_k} s(\frac{t_k - z}{\varepsilon}) f(z)dz - \{(\tau A)^{-1}[f(t_k) - c(\rho)f(t_{k-1})] \\ &\quad -(\tau A)^{-1} \sum_{j=0}^{m-1} B_j f^{(j+1)}(t_{k-1})\} \\ &= -(\tau A)^{-1} \int_{t_{k-1}}^{t_k} c(\frac{t_k - z}{\varepsilon}) f'(z)dz + (\tau A)^{-1} \sum_{j=0}^{m-1} B_j f^{(j+1)}(t_{k-1}) \\ &= -\sum_{j=0}^{m-1} (\tau A)^{-1} \int_{t_{k-1}}^{t_k} c(\frac{t_k - s}{\varepsilon})(s - t_{k-1})^j \, ds f^{(j+1)} (t_{k-1}) \frac{1}{j!} \\ &\quad -(\tau A)^{-1} \int_{t_{k-1}}^{t_k} c(\frac{t_k - s}{\varepsilon}) \int_{t_{k-1}}^{s} (s - z)^m f^{(m+1)} (z) \, dz ds \, \frac{1}{m!} \\ &\quad +(\tau A)^{-1} \sum_{j=0}^{m-1} B_j f^{(j+1)}(t_{k-1}) \, . \end{aligned}$$

A simple calculation shows that

$$
\begin{aligned}
&-(\tau A)^{-1}\int_{t_{k-1}}^{t_k} c(\frac{t_k-s}{\varepsilon})(s-t_{k-1})^j\, ds f^{(j+1)}\ (t_{k-1})\frac{1}{j!}\\
&=-(\tau A)^{-1}\sum_{j=0}^{m-1} B_j f^{(j+1)}(t_{k-1}).
\end{aligned}
$$

Therefore, we can write

$$
a_{1,\,k}\ =-(\tau A)^{-1}\int_{t_{k-1}}^{t_k} c(\frac{t_k-s}{\varepsilon})\ \int_{t_{k-1}}^{s}(z-s)^{m-1}f^{(m+1)}(z)dzds/(m-1)!.
$$

In a similar way one shows that

$$
a_{2,\,k}\ =(\tau\varepsilon^2)^{-1}\int_{t_{k-1}}^{t_k} s(\frac{t_k-s}{\varepsilon})\ \int_{t_{k-1}}^{s}(z-s)^{m-1}f^{(m+1)}(z)dzds/(m-1)!.
$$

Then, using the triangle inequality and estimate (7.55), we obtain

$$
\max_{1\le k\le N}||a_{1,k}\ ||_E\le M\ \tau^m
$$

and

$$
\max_{1\le k\le N}||a_{2,k}\ ||_E\le M\ \varepsilon^{-1}\tau^m.
$$

From this and estimate (7.60) follows estimate (7.59). Theorem 7.3.2 is proved.

Remark 7.3.3. Theorem 7.3.2 permits us to obtain the error estimates of the high order of accuracy difference schemes for hyperbolic equations with a small parameter multiplying time-derivative term in the various Banach norms.

Remark 7.3.4. Applying this approach we can construct the high order of accuracy uniform difference schemes for three singular perturbation problems involving second-order differential equations in a Banach space E. The equations are

$$
\varepsilon^2 v''(t)+v'(t)=Av(t)+f(t),
$$

$$
\varepsilon^2 v''(t)-iv'(t)=Av(t)+f(t),
$$

$$
\varepsilon^2 v''(t)+v'(t)=(\varepsilon^2 A+B)v(t)+f(t),
$$

where A, B are linear, generally unbounded operators in E. In each case, the usual initial conditions are given:

$$
v(0)=v_0, v'(0)=v_0'.
$$

Chapter 8

Appendix: Delay Parabolic Differential Equations

Chapter 8 is devoted to study of the initial-value problem for delay partial differential equations of the parabolic type. A sufficient condition for stability of the solution of this initial-value problem is given. The stability estimates for solutions of the first and second order of accuracy difference schemes for approximately solving this initial-value problem are presented. The stability estimates in Hölder norms for solutions of the initial-value problem of the delay differential and difference equations of the parabolic type are obtained.

8.1 The Initial-Value Differential Problem

In the present section we consider the initial-value problem for linear delay differential equations

$$\begin{cases} \frac{dv(t)}{dt} + Av(t) = B(t)v(t-\omega), t \geq 0, \\ v(t) = g(t)(-\omega \leq t \leq 0) \end{cases} \tag{8.1}$$

in an arbitrary Banach space E with the unbounded linear operators A and $B(t)$ in E with dense domains $D(A) \subseteq D(B(t))$. Let A be a strongly positive operator, i.e., $-A$ is the generator of the analytic semigroup $\exp\{-tA\}(t \geq 0)$ of the linear bounded operators with exponentially decreasing norm when $t \to \infty$. That means the following estimates hold:

$$\|\exp\{-tA\}\|_{E\mapsto E} \leq Me^{-\delta t}, \|tA\exp\{-tA\}\|_{E\mapsto E} \leq Me^{-\delta t}, t > 0 \tag{8.2}$$

for some $M > 1$, $\delta > 0$. Let $B(t)$ be closed operators. The operator function $B(t)$ is strongly continuous on $D(A)$.

A function $v(t)$ is called a solution of problem (8.1) if the following conditions are satisfied:

i. $v(t)$ is continuously differentiable on the interval $[-\omega, \infty)$. The derivative at the endpoint $t = -\omega$ is understood as the appropriate unilateral derivative.

ii. The element $v(t)$ belongs to $D(A)$ for all $t \in [-\omega, \infty)$, and the function $Av(t)$ is continuous on the interval $[-\omega, \infty)$.

iii. $v(t)$ satisfies the equation and the initial condition (8.1).

A solution $v(t)$ of the initial-value problem (8.1) is said to be stable if

$$\|v(t)\|_E \leq \max_{-\omega \leq t \leq 0} \|g(t)\|_E \tag{8.3}$$

for every t, $-\omega \leq t < \infty$. We are interested in studying the stability of solutions of the initial-value problem under the assumption that

$$\|B(t)A^{-1}\|_{E \mapsto E} \leq 1 \tag{8.4}$$

holds for every $t \geq 0$. We have not been able to obtain estimate (8.3) in an arbitrary Banach space E. Nevertheless, we can establish the analog of estimates (8.3) where the space E is replaced by the fractional spaces $E_\alpha (0 < \alpha < 1)$ under a stronger assumption than (8.4). Stability estimates for the solutions of the first and second order of accuracy difference schemes for approximately solving this initial-value problem are presented. The stability estimates in Hölder norms for solutions of the initial-value problem of delay differential and difference equations of the parabolic type are obtained.

The strongly positive operator A defines the fractional spaces $E_\alpha = E_\alpha(E, A)$ $(0 < \alpha < 1)$ consisting of all $u \in E$ for which the following norms are finite:

$$\|u\|_{E_\alpha} = \sup_{\lambda > 0} \|\lambda^{1-\alpha} A \exp\{-\lambda A\} u\|_E.$$

First, we consider problem (8.1)when A^{-1} and $B(t)$ commute, i.e.,

$$A^{-1}B(t)x = B(t)A^{-1}x, \ x \in D(A). \tag{8.5}$$

Theorem 8.1.1. *Assume that the condition*

$$\|B(t)A^{-1}\|_{E \mapsto E} \leq \frac{(1-\alpha)}{M2^{2-\alpha}} \tag{8.6}$$

holds for every $t \geq 0$, where M is the constant from (8.2). Then for every $t \geq 0$ we have

$$\|v(t)\|_{E_\alpha} \leq \max_{-\omega \leq t \leq 0} \|g(t)\|_{E_\alpha}. \tag{8.7}$$

Proof. Using the formula

$$v(t) = \exp\{-tA\}g(0) + \int_0^t \exp\{-(t-s)A\}B(s)g(s-\omega)ds \tag{8.8}$$

the semigroup property, condition (8.5) and estimates (8.2), (8.6), we obtain

$$\lambda^{1-\alpha}\|A\exp\{-\lambda A\}v(t)\|_E \leq \lambda^{1-\alpha}\|A\exp\{-(\lambda+t)A\}g(0)\|_E$$

$$+\lambda^{1-\alpha}\int_0^t \|A\exp\{-\frac{\lambda+t-s}{2}A\}\|_{E\to E}\|B(s)A^{-1}\|_{E\to E}$$

$$\times\|A\exp\{-\frac{\lambda+t-s}{2}A\}g(s-\omega)\|_E ds$$

$$\leq \frac{\lambda^{1-\alpha}}{(\lambda+t)^{1-\alpha}}\|g(0)\|_{E_\alpha} + \frac{1-\alpha}{M2^{2-\alpha}}\int_0^t \frac{M\lambda^{1-\alpha}2^{2-\alpha}}{(\lambda+t-s)^{2-\alpha}}ds \max_{0\leq s\leq\omega}\|g(s-\omega)\|_{E_\alpha}$$

$$\leq \max_{-\omega\leq t\leq 0}\|g(t)\|_{E_\alpha}$$

for every $t, 0 \leq t \leq \omega$ and $\lambda, \lambda > 0$. This shows that

$$\|v(t)\|_{E_\alpha} \leq \max_{-\omega\leq t\leq 0}\|g(t)\|_{E_\alpha} \tag{8.9}$$

for every t, $0 \leq t \leq \omega$. Applying mathematical induction, one can easily show that it is true for every t. Namely, assume that inequality (8.7) is true for $t, (n-1)\omega \leq t \leq n\omega$, $n = 1, 2, 3, \ldots$, for some n. Letting $t = s + n\omega$, we have

$$\frac{dv}{ds} + Av = B(s+n\omega)v(s+(n-1)\omega), 0 \leq s \leq \omega.$$

Using estimate (8.9), we obtain

$$\max_{n\omega\leq s\leq (n+1)\omega}\|v(s-\omega)\|_{E_\alpha} \leq \max_{-(n-1)\omega\leq t\leq n\omega}\|v(t)\|_{E_\alpha}$$

for every t, $n\omega \leq t \leq (n+1)\omega$, $n = 1, 2, 3, \ldots$. and $\lambda, \lambda > 0$. This shows that

$$\|v(t)\|_{E_\alpha} \leq \max_{-\omega\leq t\leq 0}\|g(t)\|_{E_\alpha}$$

for every t, $n\omega \leq t \leq (n+1)\omega$, $n = 1, 2, 3, \ldots$. This result completes the proof of Theorem 8.1.1.

Now, we consider problem (8.1) when

$$A^{-1}B(t)x \neq B(t)A^{-1}x, x \in D(A)$$

for some $t \geq 0$. Note that A is a strongly positive operator in a Banach space E iff its spectrum $\sigma(A)$ lies in the interior of the sector of angle $\varphi, 0 < 2\varphi < \pi$, symmetric with respect to the real axis, and if on the edges of this sector, $S_1 =$

$[z = \rho\exp(i\varphi) : 0 \leq \rho < \infty]$ and $S_2 = [z = \rho\exp(-i\varphi) : 0 \leq \rho < \infty]$ and outside it the resolvent $(z - A)^{-1}$ is subject to the bound

$$\|(z-A)^{-1}\|_{E\to E} \leq \frac{M_1}{1+|z|} \tag{8.10}$$

for some $M_1 > 0$. First of all let us prove two lemmas. These will be needed in the sequel.

Lemma 8.1.2. *For any* z *on the edges of the sector,*

$$S_1 = [z = \rho\exp(i\varphi) : 0 \leq \rho < \infty]$$

and

$$S_2 = [z = \rho\exp(-i\varphi) : 0 \leq \rho < \infty]$$

and outside it the estimate

$$\|A(z-A)^{-1}x\|_E \leq \frac{M_1^\alpha M^\alpha(1+M_1)^{1-\alpha}2^{(2-\alpha)\alpha}}{\alpha(1-\alpha)(1+|z|)^\alpha}\|x\|_{E_\alpha}$$

holds for any $x \in E_\alpha$. *Here and in the future* M *and* M_1 *are the same constants as in estimates* (8.2) *and* (8.10).

Proof. Applying the formula

$$x = \int_0^\infty A\exp\{-\tau A\}x d\tau,$$

and using the semigroup property, we can write

$$A(z-A)^{-1}x = \int_0^N A(z-A)^{-1}A\exp\{-\tau A\}xd\tau$$

$$+\int_N^\infty (z-A)^{-1}A\exp\{-\frac{\tau}{2}A\}A\exp\{-\frac{\tau}{2}A\}xd\tau.$$

Using estimates (8.2), (8.10) and the definition of the spaces E_a, we obtain

$$\|A(z-A)^{-1}x\|_E \leq \int_0^N \|A(z-A)^{-1}\|_{E\to E}\|A\exp\{-\tau A\}x\|_E d\tau$$

$$+\int_N^\infty \|(z-A)^{-1}\|_{E\mapsto E}\|A\exp\{-\frac{\tau}{2}A\}\|_{E\mapsto E}\|A\exp\{-\frac{\tau}{2}A\}x\|_E d\tau$$

$$\leq \left(\int_0^N (1+M_1)\tau^{\alpha-1}d\tau + \int_N^\infty \frac{MM_1 2^{2-\alpha}\tau^{\alpha-2}}{|z|+1}d\tau\right)\|x\|_{E_\alpha} = \psi(N)\|\varphi\|_{E_\alpha}.$$

Here

$$\psi(N) = (1+M_1)\frac{N^\alpha}{\alpha} + \frac{MM_1 2^{2-\alpha}N^{\alpha-1}}{(|z|+1)(1-\alpha)}.$$

Taking the minimum of the function $\psi(N)$ over all N, $0 < N < \infty$, we obtain the statement of Lemma 8.1.2.

Lemma 8.1.3. *Let for all $s \geq 0$ operator $\overline{B(s)A^{-1} - A^{-1}B(s)}$ with domain which coincide with $D(A)$, admit a closure $Q = \overline{B(s)A^{-1} - A^{-1}B(s)}$ bounded in E. Then for all $\tau > 0$ the following estimate holds:*

$$\|A^{-1}[A\exp\{-\tau A\}B(s) - B(s)A\exp\{-\tau A\}]x\|_E$$

$$\leq \frac{e(\alpha+1)M^\alpha M_1^{1+\alpha}(1+2M_1)(1+M_1)^{1-\alpha}2^{(2-\alpha)\alpha}\|Q\|_{E\mapsto E}\|x\|_{E_\alpha}}{\tau^{1-\alpha}\pi\alpha^2(1-\alpha)}.$$

Here $Q = \overline{A^{-1}(AB(s) - B(s)A)A^{-1}}$.

Proof. Applying the Cauchy–Riesz formula, we obtain the representation

$$A\exp\{-\tau A\}x = \frac{1}{2\pi i}\int\limits_{S_1\cup S_2} \exp\{-\tau z\}A(z-A)^{-1}dzx$$

$$= -\frac{1}{2\pi i}\int\limits_{S_1\cup S_2} \exp\{-\tau z\}dzx + \frac{1}{2\pi i}\int\limits_{S_1\cup S_2} \exp\{-\tau z\}z(z-A)^{-1}dzx$$

for every τ, $\tau > 0$. From this it follows that

$$A^{-1}[A\exp\{-\tau A\}B(s) - B(s)A\exp\{-\tau A\}]x$$

$$= \frac{1}{2\pi i}\int\limits_{S_1\cup S_2} \exp\{-\tau z\}zA^{-1}[(z-A)^{-1}B(s) - B(s)(z-A)^{-1}]dzx$$

$$= \frac{1}{2\pi i}\int\limits_{S_1\cup S_2} \exp\{-\tau z\}z(z-A)^{-1}QA(z-A)^{-1}]dzx.$$

Making the substitution $z = -\frac{1}{\tau} + iy, -\infty < y < \infty$ and using integration by parts, we can write the formula

$$A^{-1}[A\exp\{-\tau A\}B(s) - B(s)A\exp\{-\tau A\}]x$$

$$= -\frac{e}{2\pi}\int_{-\infty}^{\infty} \exp(-\tau iy)(-\frac{1}{\tau}+iy)(-\frac{1}{\tau}+iy-A)^{-1}QA(-\frac{1}{\tau}+iy-A)^{-1}dyx$$

$$= -\frac{e}{2\pi\tau}\int_{-\infty}^{\infty} \exp(-\tau iy)(-\frac{1}{\tau}+iy-A)^{-1}[Q - (-\frac{1}{\tau}+iy)(-\frac{1}{\tau}+iy-A)^{-1}Q$$

$$-(-\frac{1}{\tau}+iy)Q(-\frac{1}{\tau}+iy-A)^{-1}]A(-\frac{1}{\tau}+iy-A)^{-1}dyx.$$

By Lemma 8.1.2 and performing the change of variable $\tau y = p$, and using estimate(8.10), we obtain

$$\|A^{-1}[A\exp\{-\tau A\}B(s) - B(s)A\exp\{-\tau A\}]x\|_E$$

$$\leq \frac{e}{2\pi\tau}\int_{-\infty}^{\infty} |\exp(-\tau iy)|\|(-\frac{1}{\tau}+iy-A)^{-1}\|_{E\mapsto E}(1+2M_1)\|Q\|_{E\mapsto E}$$

$$\times\|A(-\frac{1}{\tau}+iy-A)^{-1}dyx\|_E$$

$$\leq \frac{e}{2\pi\tau}\int_{-\infty}^{\infty}\frac{dy}{\left(\frac{1}{\tau^2}+y^2\right)^{\frac{1+\alpha}{2}}}$$

$$\times\frac{(1+2M_1)\|Q\|_{E\mapsto E}M_1^{1+\alpha}M^{\alpha}(1+M_1)^{1-\alpha}2^{(2-\alpha)\alpha}\|x\|_{E_\alpha}}{\alpha(1-\alpha)}$$

$$=\frac{e}{\pi\tau^{1-\alpha}}\int_{0}^{\infty}\frac{dp}{(1+p^2)^{\frac{1+\alpha}{2}}}$$

$$\times\frac{(1+2M_1)\|Q\|_{E\mapsto E}M_1^{1+\alpha}M^{\alpha}(1+M_1)^{1-\alpha}2^{(2-\alpha)\alpha}\|x\|_{E_\alpha}}{\alpha(1-\alpha)}$$

$$\leq\frac{e(1+\alpha)(1+2M_1)\|Q\|_{E\mapsto E}M_1^{1+\alpha}M^{\alpha}(1+M_1)^{1-\alpha}2^{(2-\alpha)\alpha}\|x\|_{E_\alpha}}{\pi\tau^{1-\alpha}\alpha^2(1-\alpha)}.$$

Lemma 8.2 is proved.

Suppose that

$$\overline{\|A^{-1}(AB(t)-B(t)A)A^{-1}\|}_{E\mapsto E} \tag{8.11}$$

$$\leq\frac{\pi(1-\alpha)^2\alpha^2\varepsilon}{eM^{1+\alpha}M_1^{1+\alpha}(1+2M_1)(1+M_1)^{1-\alpha}2^{2+\alpha-\alpha^2}(1+\alpha)}$$

holds for every $t\geq 0$. Here and in the future ε is a some constant, $0\leq\varepsilon\leq 1$.

Theorem 8.1.4. *Assume that the condition*

$$\overline{\|A^{-1}B(t)\|}_{E\mapsto E}\leq\frac{(1-\alpha)(1-\varepsilon)}{M2^{2-\alpha}} \tag{8.12}$$

holds for every $t\geq 0$. Then for every $t\geq 0$, estimate (8.7) *holds.*

Proof. Let $0\leq t\leq\omega$ and λ, $\lambda>0$. Then using formula (8.8), we have that

$$\lambda^{1-\alpha}A\exp\{-\lambda A\}v(t)=\lambda^{1-\alpha}A\exp\{-(\lambda+t)A\}g(0)$$

$$+\lambda^{1-\alpha}\int_0^t\exp\{-\frac{\lambda+t-s}{2}A\}B(s)A\exp\{-\frac{\lambda+t-s}{2}A\}g(s-\omega)ds$$

$$+\lambda^{1-\alpha}\int_0^t\exp\{-\frac{\lambda+t-s}{2}A\}[A\exp\{-\frac{\lambda+t-s}{2}A\}B(s)-B(s)A$$

$$\times\exp\{-\frac{\lambda+t-s}{2}A\}]g(s-\omega)ds$$

$$=I_1+I_2+I_3,$$

where

$$I_1=\lambda^{1-\alpha}A\exp\{-(\lambda+t)A\}g(0),$$

$$I_2 = \lambda^{1-\alpha} \int_0^t \exp\{-\frac{\lambda+t-s}{2}A\}B(s)A\exp\{-\frac{\lambda+t-s}{2}A\}g(s-\omega)ds,$$

$$I_3 = \lambda^{1-\alpha} \int_0^t \exp\{-\frac{\lambda+t-s}{2}A\}[A\exp\{-\frac{\lambda+t-s}{2}A\}B(s) - B(s)A$$

$$\times \exp\{-\frac{\lambda+t-s}{2}A\}]g(s-\omega)ds.$$

Using estimates (8.2), (8.10) and condition (8.12), we obtain

$$\|I_1\|_E = \lambda^{1-\alpha}\|A\exp\{-(\lambda+t)A\}g(0)\|_E$$

$$\leq \frac{\lambda^{1-\alpha}}{(\lambda+t)^{1-\alpha}}\|g(0)\|_{E_\alpha} \leq \frac{\lambda^{1-\alpha}}{(\lambda+t)^{1-\alpha}} \max_{-\omega\leq t\leq 0}\|g(t)\|_{E_\alpha},$$

$$\|I_2\|_E \leq \lambda^{1-\alpha} \int_0^t \|A\exp\{-\frac{\lambda+t-s}{2}A\}\|_{E\mapsto E}\|\overline{A^{-1}B(s)}\|_{E\mapsto E}$$

$$\times\|A\exp\{-\frac{\lambda+t-s}{2}A\}g(s-\omega)\|_E ds$$

$$\leq \max_{0\leq t\leq\omega}\|\overline{A^{-1}B(t)}\|_{E\mapsto E}\int_0^t \frac{M\lambda^{1-\alpha}2^{2-\alpha}}{(\lambda+t-s)^{2-\alpha}}ds \max_{0\leq s\leq\omega}\|g(s-\omega)\|_{E_\alpha}$$

$$\leq \max_{-\omega\leq t\leq 0}\|g(t)\|_{E_\alpha}(1-\frac{\lambda^{1-\alpha}}{(\lambda+t)^{1-\alpha}})(1-\varepsilon)$$

for every $t, 0\leq t\leq\omega$ and λ, $\lambda>0$. Now let us estimate I_3. By Lemma 8.1.3 and using estimate (8.11), we can obtain

$$\|I_3\|_E \leq \lambda^{1-\alpha}\int_0^t \|A\exp\{-\frac{\lambda+t-s}{2}A\}\|_{E\mapsto E}$$

$$\times\|A^{-1}[A\exp\{-\frac{\lambda+t-s}{2}A\}B(s) - B(s)A\exp\{-\frac{\lambda+t-s}{2}A\}]g(s-\omega)\|_E ds$$

$$\leq \lambda^{1-\alpha}e(1+\alpha)M^{1+\alpha}M_1^{1+\alpha}(1+2M_1)(1+M_1)^{1-\alpha}2^{(2-\alpha)\alpha}$$

$$\times\int_0^t \frac{\|\overline{A^{-1}(AB(s)-B(s)A)A^{-1}}\|_{E\mapsto E}2^{2-\alpha}\|g(s-\omega)\|_{E_\alpha}}{(\lambda+t-s)^{2-\alpha}\pi\alpha^2(1-\alpha)}ds$$

$$\leq \max_{0<s<\omega}\|\overline{A^{-1}(AB(s)-B(s)A)A^{-1}}\|_{E\mapsto E}$$

$$\times \int_0^t \frac{e(1+\alpha)M^{1+\alpha}M_1^{1+\alpha}(1+2M_1)(1+M_1)^{1-\alpha}2^{(2-\alpha)\alpha}2^{2-\alpha}}{(\lambda+t-s)^{2-\alpha}\pi\alpha^2(1-\alpha)}ds$$

$$\times \max_{-\omega\leq t\leq 0}\|g(t)\|_{E_\alpha} \leq \max_{-\omega\leq t\leq 0}\|g(t)\|_{E_\alpha}(1-\frac{\lambda^{1-\alpha}}{(\lambda+t)^{1-\alpha}}) \in$$

for every t, $0\leq t\leq\omega$ and λ, $\lambda>0$. Using the triangle inequality, we obtain that

$$\lambda^{1-\alpha}\|A\exp\{-\lambda A\}v(t)\|_E \leq \max_{-\omega\leq t\leq 0}\|g(t)\|_{E_\alpha}$$

for every t, $0\leq t\leq\omega$ and λ, $\lambda>0$. This shows that

$$\|v(t)\|_{E_\alpha} \leq \max_{-\omega\leq t\leq 0}\|g(t)\|_{E_\alpha}$$

for every t, $0\leq t\leq\omega$. In a similar manner with Theorem 8.1 applying mathematical induction, one can easily show that it is true for every t. This result completes the proof of Theorem 8.1.4.

8.2 The Difference Schemes

Using the first and second order of accuracy implicit difference schemes for differential equations without delay, we have the following difference schemes for approximate solutions of the initial-value problem (8.1)

$$\begin{cases} \frac{1}{\tau}(u_k-u_{k-1})+Au_k=B_ku_{k-N},\ B_k=B(t_k), \\ t_k=k\tau, 1\leq k, N\tau=\omega, \\ u_k=g(t_k), t_k=k\tau, -N\leq k\leq 0, \end{cases} \tag{8.13}$$

$$\begin{cases} \frac{1}{\tau}(u_k-u_{k-1})+(A+\frac{1}{2}\tau A^2)u_k=\frac{1}{2}(I+\frac{\tau}{2}A)B_k(u_{k-N}+u_{k-N-1}), \\ B_k=B(t_k-\frac{\tau}{2}), t_k=k\tau, 1\leq k, u_k=g(t_k), t_k=k\tau, -N\leq k\leq 0. \end{cases} \tag{8.14}$$

Theorem 8.2.1. *Assume that all conditions of Theorem* 8.1.1 *are satisfied. Then for the solution of difference scheme* (8.13) *the estimate*

$$\|u_k\|_{E_\alpha} \leq \max_{-N\leq k\leq 0}\|g(t_k)\|_{E_\alpha} \tag{8.15}$$

holds for any $k\geq 1$.

Proof. Let us consider $1\leq k\leq N$. In this case

$$u_k = R^k g(0) + \sum_{j=1}^{k} R^{k-j+1}B_j g(t_{j-N})\tau,$$

where $R=(1+\tau A)^{-1}$. Using the formula

$$(I+\tau A)^{-k} = \frac{1}{(k-1)!}\int_0^\infty t^{k-1}\exp(-t)\exp(-\tau tA)dt, k\geq 1, \tag{8.16}$$

condition (8.5) and estimate (8.6), we obtain

$$\lambda^{1-\alpha}\|A\exp\{-\lambda A\}u_k\|_E \le \lambda^{1-\alpha}\frac{1}{(k-1)!}$$

$$\times\int_0^\infty t^{k-1}\exp(-t)\|A\exp\{-(\tau t+\lambda)A\}g(0)\|_E dt$$

$$+\lambda^{1-\alpha}\sum_{j=1}^{k}\tau\frac{1}{(k-j)!}\int_0^\infty t^{k-j}\exp(-t)\|A\exp\{-\frac{(\tau t+\lambda)A}{2}\}\|_{E\mapsto E}\|B_jA^{-1}\|_{E\mapsto E}$$

$$\times\|A\exp\{-\frac{(\tau t+\lambda)A}{2}\}g(t_{j-N})\|_E dt$$

$$\le \lambda^{1-\alpha}\frac{1}{(k-1)!}\int_0^\infty t^{k-1}\exp(-t)\frac{dt}{(t\tau+\lambda)^{1-\alpha}}\|g(0)\|_{E_\alpha}$$

$$+\lambda^{1-\alpha}\frac{1-\alpha}{M2^{2-\alpha}}\sum_{j=1}^{k}\tau\frac{1}{(k-j)!}\int_0^\infty t^{k-j}\exp(-t)\frac{M2^{2-\alpha}dt}{(t\tau+\lambda)^{2-\alpha}}\|g(t_{j-N})\|_{E_\alpha}$$

$$\le J\max_{-N\le k\le 0}\|g(t_k)\|_{E_\alpha},$$

where

$$J=\lambda^{1-\alpha}\frac{1}{(k-1)!}\int_0^\infty t^{k-1}\exp(-t)\frac{dt}{(t\tau+\lambda)^{1-\alpha}}$$

$$+\lambda^{1-\alpha}(1-\alpha)\sum_{j=1}^{k}\tau\frac{1}{(k-j)!}\int_0^\infty t^{k-j}\exp(-t)\frac{dt}{(t\tau+\lambda)^{2-\alpha}}.$$

We shall calculate J. Making the substitution $k-j=m$ and using Taylor's formula and integration by parts, we obtain that

$$J=\lambda^{1-\alpha}\frac{1}{(k-1)!}\int_0^\infty t^{k-1}\exp(-t)\frac{dt}{(t\tau+\lambda)^{1-\alpha}}$$

$$+\lambda^{1-\alpha}(1-\alpha)\sum_{m=0}^{k-1}\tau\frac{1}{m!}\int_0^\infty t^{m}\exp(-t)\frac{dt}{(t\tau+\lambda)^{2-\alpha}}$$

$$=\lambda^{1-\alpha}\frac{1}{(k-1)!}\int_0^\infty t^{k-1}\exp(-t)\frac{dt}{(t\tau+\lambda)^{1-\alpha}}$$

$$+\lambda^{1-\alpha}(1-\alpha)\int_0^\infty \exp(t)[1-\int_0^t\frac{\xi^{k-1}\exp(-\xi)d\xi}{(k-1)!}]\exp(-t)\frac{dt}{(t\tau+\lambda)^{2-\alpha}}$$

$$= \lambda^{1-\alpha}\frac{1}{(k-1)!}\int_0^\infty t^{k-1}\exp(-t)\frac{dt}{(t\tau+\lambda)^{1-\alpha}} + 1$$

$$-\lambda^{1-\alpha}\frac{1}{(k-1)!}\int_0^\infty t^{k-1}\exp(-t)\frac{dt}{(t\tau+\lambda)^{1-\alpha}} = 1.$$

Therefore,

$$\lambda^{1-\alpha}\|A\exp\{-\lambda A\}u_k\|_E \leq \max_{-N\leq k\leq 0}\|g(t_k)\|_{E_\alpha}$$

for every k, $1 \leq k \leq N$ and $\lambda, \lambda > 0$. This shows that

$$\|u_k\|_{E_\alpha} \leq \max_{-N\leq k\leq 0}\|g(t_k)\|_{E_\alpha} \tag{8.17}$$

for every $k, 1 \leq k \leq N$. Applying mathematical induction, one can easily show that it is true for every k. Namely, assume that estimate (8.15) is true for the $k, (n-1)N \leq k \leq nN$, for some $n = 1, 2, 3, \ldots$. Letting $k = m + nN$, we have

$$\frac{1}{\tau}(u_m - u_{m-1}) + Au_m = B_{m+nN}u_{m-N}, 1 \leq m \leq N.$$

Using estimate (8.16), we obtain

$$\|u_k\|_{E_\alpha} \leq \max_{1\leq k\leq N}\|u_{m-N}\|_{E_\alpha}$$

for every k, $nN \leq k \leq (n+1)N$, $n = 1, 2, 3, \ldots$, and λ, $\lambda > 0$. This shows that

$$\|u_k\|_{E_\alpha} \leq \max_{-\omega\leq k\leq 0}\|g(t_k)\|_{E_\alpha}$$

for every $k, nN \leq k \leq (n+1)N$, $n = 1, 2, 3, \ldots$. This result completes the proof of Theorem 8.2.1.

Theorem 8.2.2. *Assume that all conditions of Theorem* 8.1.4 *are satisfied. Then for the solution of the difference scheme* (3.13) *the estimate* (8.15) *holds for any* $k \geq 1$.

Proof. Let us consider $1 \leq k \leq N$. Using the formula (8.16), we can write

$$\lambda^{1-\alpha}A\exp\{-\lambda A\}u_k = \lambda^{1-\alpha}\frac{1}{(k-1)!}\int_0^\infty t^{k-1}\exp(-t)A\exp(-(\tau t+\lambda)A)g(0)dt$$

$$+\lambda^{1-\alpha}\sum_{j=1}^{k}\tau\frac{1}{(k-j)!}\int_0^\infty t^{k-j}\exp(-t)\exp(-\frac{\tau t+\lambda}{2}A\}B_jA$$

$$\times \exp(-\frac{\tau t+\lambda}{2}A\}g(t_{j-N})dt$$

$$+\lambda^{1-\alpha}\sum_{j=1}^{k}\tau\frac{1}{(k-j)!}\int_0^\infty t^{k-j}\exp(-t)A\exp(-\frac{\tau t+\lambda}{2}A\}A^{-1}$$

$$\times[\exp(-\frac{\tau t+\lambda}{2}A\}AB_j - B_jA\exp\{-\frac{\tau t+\lambda}{2}A\}]g(t_{j-N})dt = T_1+T_2+T_3.$$

Using estimates (8.2), (8.10) and assumption (8.12), we obtain

$$\|T_1\|_E \le \lambda^{1-\alpha}\frac{1}{(k-1)!}\int_0^\infty t^{k-1}\exp(-t)\|A\exp\{-(\tau t+\lambda)A\}g(0)\|_E dt$$

$$\le \lambda^{1-\alpha}\frac{1}{(k-1)!}\int_0^\infty t^{k-1}\exp(-t)\frac{dt}{(t\tau+\lambda)^{1-\alpha}}\max_{-N\le k\le 0}\|g(t_k)\|_{E_\alpha},$$

$$\|T_2\|_E \le \lambda^{1-\alpha}\sum_{j=1}^{k}\tau\frac{1}{(k-j)!}\int_0^\infty t^{k-j}\exp(-t)\|A\exp\{-\frac{\tau t+\lambda}{2}A\}\|_{E\to E}$$

$$\times\|B_jA^{-1}\|_{E\to E}\|A\exp\{-\frac{\tau t+\lambda}{2}A\}g(t_{j-N})\|_E dt$$

$$\le \frac{\lambda^{1-\alpha}(1-\alpha)(1-\in)}{M2^{2-\alpha}}\sum_{j=1}^{k}\tau\frac{1}{(k-j)!}\int_0^\infty t^{k-j}\exp(-t)\frac{M2^{2-\alpha}dt}{(t\tau+\lambda)^{2-\alpha}}\|g(t_{j-N})\|_{E_\alpha}$$

$$\le \lambda^{1-\alpha}(1-\alpha)(1-\in)\sum_{j=1}^{k}\tau\frac{1}{(k-j)!}\int_0^\infty t^{k-j}\exp(-t)\frac{dt}{(t\tau+\lambda)^{2-\alpha}}\max_{-N\le k\le 0}\|g(t_k)\|_{E_\alpha}$$

for every $k, 1\le k\le N$ and λ, $\lambda>0$. Now let us estimate T_3. By Lemma 8.1.3 and using estimate (8.11), we can obtain

$$\|T_3\|_E \le \lambda^{1-\alpha}\sum_{j=1}^{k}\tau\frac{1}{(k-j)!}\int_0^\infty t^{k-j}\exp(-t)\|A\exp\{-\frac{\tau t+\lambda}{2}A\}\|_{E\to E}$$

$$\times\|A^{-1}[A\exp\{-\frac{\tau t+\lambda}{2}A\}B_j - B_jA\exp\{-\frac{\tau t+\lambda}{2}A\}]g(t_{j-N})\|_E dt$$

$$\le \lambda^{1-\alpha}\max_{1\le k\le N}\|\overline{A^{-1}(AB_j-B_jA)A^{-1}}\|_{E\to E}\sum_{j=1}^{k}\tau\frac{1}{(k-j)!}$$

$$\times\int_0^\infty t^{k-j}\exp(-t)\frac{e(1+\alpha)M^\alpha M_1^{1+\alpha}(1+2M_1)(1+M_1)^{1-\alpha}2^{(2-\alpha)\alpha}}{\pi\alpha^2(1-\alpha)}\frac{M2^{2-\alpha}dt}{(t\tau+\lambda)^{2-\alpha}}$$

$$\times\|g(t_{j-N})\|_{E_\alpha}$$

$$\leq \lambda^{1-\alpha}(1-\alpha)\in\sum_{j=1}^{k}\tau\frac{1}{(k-j)!}\int_{0}^{\infty}t^{k-j}\exp(-t)\frac{dt}{(t\tau+\lambda)^{2-\alpha}}\max_{-N\leq k\leq 0}\|g(t_k)\|_{E_\alpha}$$

for every $k, 1\leq k\leq N$ and λ, $\lambda>0$. Using the triangle inequality, we obtain that

$$\lambda^{1-\alpha}\|A\exp\{-\lambda A\}u_k\|_E\leq \max_{-N\leq k\leq 0} J\|g(t_k)\|_{E_\alpha}$$

for every $k, 1\leq k\leq N$ and λ, $\lambda>0$.This shows that

$$\|u_k\|_{E_\alpha}\leq \max_{-N\leq k\leq 0}\|g(t_k)\|_{E_\alpha}$$

for every $k, 1\leq k\leq N$. In a similar manner with Theorem 8.2.1 applying mathematical induction, one can easily show that it is true for every k. This result completes the proof of Theorem 8.2.2.

Now we consider the difference scheme (8.14). We have not been able to obtain the same result for solution of the difference scheme (8.13) in the spaces E_α under the assumption for the difference scheme (8.13). Nevertheless, we have asymptotic stability of the solution of the difference scheme under the supplementary restriction of the operator A. From the strong positivity of the operator A it follows that there exists the bounded operator $(I+\tau A+\frac{(\tau A)^2}{2})^{-1}$, defined on the whole space E.

Theorem 8.2.3. *Let condition* (8.5) *be satisfied. Suppose that the following estimates hold:*

$$\begin{aligned}\|(I+\tau A)(I+\tau A+\frac{(\tau A)^2}{2})^{-1}\|_{E\mapsto E} &\leq 1,\\ \|(I+\frac{\tau A}{2})(I+\tau A)(I+\tau A+\frac{(\tau A)^2}{2})^{-1}\|_{E\mapsto E} &\leq \frac{1+\sqrt{2}}{2}\end{aligned}\tag{8.18}$$

and

$$\|B(t)A^{-1}\|_{E\mapsto E}\leq \frac{(1-\alpha)}{M2^{1-\alpha}(1+\sqrt{2})},\quad t\geq 0.\tag{8.19}$$

Then for the solution of difference scheme (8.14) *estimate* (8.15) *holds for any* $k\geq 1$.

Proof. Let us consider $1\leq k\leq N$. In this case

$$u_k=R^k g(0)+\sum_{j=1}^{k}R^{k-j+1}(I+\frac{\tau A}{2})B_j(g(t_{j-N}+g(t_{j-N-1}))\tau,\tag{8.20}$$

where $R=(I+\tau A+\frac{(\tau A)^2}{2})^{-1}$. Using formula (8.16) and estimates (8.2) and assumptions (8.18) and (8.19), we obtain

$$\lambda^{1-\alpha}\|A\exp\{-\lambda A\}u_k\|_E\leq\|((I+\tau A)(I+\tau A+\frac{(\tau A)^2}{2})^{-1})^k\|_{E\mapsto E}$$

$$\times\lambda^{1-\alpha}\frac{1}{(k-1)!}\int_0^\infty t^{k-1}\exp(-t)\|A\exp\{-(\tau t+\lambda)A\}g(0)\|_E dt$$

$$+\lambda^{1-\alpha}\sum_{j=1}^{k}\tau\|(I+\tau A)(I+\tau A+\frac{(\tau A)^2}{2})^{-1})^{k-j}\|_{E\mapsto E}$$

$$\times\|(I+\frac{\tau A}{2})(I+\tau A)(I+\tau A+\frac{(\tau A)^2}{2})^{-1}\|_{E\mapsto E}$$

$$\times\frac{1}{(k-j)!}\int_0^\infty t^{k-j}\exp(-t)\|A\exp\{-\frac{\tau t+\lambda}{2}A\}\|_{E\to E}\|B_jA^{-1}\|_{E\to E}$$

$$\times\|A\exp\{-\frac{\tau t+\lambda}{2}A\}\frac{1}{2}(g(t_{j-N})+g(t_{j-N-1}))\|_E dt$$

$$\leq\lambda^{1-\alpha}\frac{1}{(k-1)!}\int_0^\infty t^{k-1}\exp(-t)\frac{dt}{(t\tau+\lambda)^{1-\alpha}}\|g(0)\|_{E_\alpha}+\frac{(1+\sqrt{2})}{2}\frac{1-\alpha}{M2^{1-\alpha}(1+\sqrt{2})}$$

$$\times\sum_{j=1}^{k}\tau\frac{1}{(k-j)!}\int_0^\infty t^{k-j}\exp(-t)\frac{M2^{2-\alpha}dt}{(t\tau+\lambda)^{2-\alpha}}\|\frac{1}{2}(g(t_{j-N})+g(t_{j-N-1}))\|_{E_\alpha}$$

$$\leq J\max_{-N\leq k\leq 0}\|g(t_k)\|_{E_\alpha},$$

where

$$J=\lambda^{1-\alpha}\frac{1}{(k-1)!}\int_0^\infty t^{k-1}\exp(-t)\frac{dt}{(t\tau+\lambda)^{1-\alpha}}$$

$$+\lambda^{1-\alpha}(1-\alpha)\sum_{j=1}^{k}\tau\frac{1}{(k-j)!}\int_0^\infty t^{k-j}\exp(-t)\frac{dt}{(t\tau+\lambda)^{2-\alpha}}.$$

Therefore

$$\lambda^{1-\alpha}\|A\exp\{-\lambda A\}u_k\|_E\leq\max_{-N\leq k\leq 0}\|g(t_k)\|_{E_\alpha}$$

for every $k, 1\leq k\leq N$ and λ, $\lambda>0$. This shows that

$$\|u_k\|_{E_\alpha}\leq\max_{-N\leq k\leq 0}\|g(t_k)\|_{E_\alpha}$$

for every $k, 1\leq k\leq N$. In a similar manner as with Theorem 8.2.1 applying mathematical induction, one can easily show that it is true for every k. This result completes the proof of Theorem 8.2.3.

Now, we consider the difference scheme (8.14) when

$$A^{-1}B(t)x\neq B(t)A^{-1}x, x\in D(A)\qquad(8.21)$$

for some $t \geq 0$. Suppose that operator $B(t)A^{-1} - A^{-1}B(t)$, with domain which coincides with $D(A)$, admits a closure $\overline{B(t)A^{-1} - A^{-1}B(t)}$ bounded in E and that the following estimate

$$\overline{\|A^{-1}(AB(t) - B(t)A)A^{-1}\|}_{E\mapsto E} \leq \frac{\pi(1-\alpha)^2\alpha^2(1+\alpha)^{-1}}{eM^{1+\alpha}M_1^{1+\alpha}} \times \frac{(1+\sqrt{2})^{-1}\varepsilon}{(1+M_1)^{1-\alpha}2^{2+\alpha-\alpha^2}} \tag{8.22}$$

holds for every $t \geq 0$ and some $\varepsilon \in [0,1]$.

Theorem 8.2.4. *Let assumptions* (8.18), (8.19) *and* (8.22) *be satisfied. Suppose that the estimate*

$$\|\overline{A^{-1}B(t)}\|_{E\mapsto E} \leq \frac{(1-\alpha)(1-\varepsilon)}{M2^{2-\alpha}(1+\sqrt{2})} \tag{8.23}$$

holds for every $t \geq 0$. Then for the solution of difference scheme (8.14) *estimate* (8.15) *holds for any $k \geq 1$.*

Proof. Let us consider $1 \leq k \leq N$. Using formula (8.16), we obtain

$$\lambda^{1-\alpha}A\exp\{-\lambda A\}u_k = ((I+\tau A)(I+\tau A+\frac{(\tau A)^2}{2})^{-1})^k$$

$$\times\lambda^{1-\alpha}\frac{1}{(k-1)!}\int_0^\infty t^{k-1}\exp(-t)A\exp\{-(\tau t+\lambda)A\}g(0)dt$$

$$+\lambda^{1-\alpha}\sum_{j=1}^{k}\tau((I+\tau A)(I+\tau A+\frac{(\tau A)^2}{2})^{-1})^{k-j}(I+\frac{\tau A}{2})(I+\tau A)(I+\tau A+\frac{(\tau A)^2}{2})^{-1}$$

$$\times\frac{1}{(k-j)!}\int_0^\infty t^{k-j}\exp(-t)\exp\{-\frac{(\tau t+\lambda)A}{2}\}B_j$$

$$\times A\exp\{-\frac{(\tau t+\lambda)A}{2}\}\frac{1}{2}(g(t_{j-N})+g(t_{j-N-1}))dt$$

$$+\lambda^{1-\alpha}\sum_{j=1}^{k}\tau((I+\tau A)(I+\tau A+\frac{(\tau A)^2}{2})^{-1})^{k-j}(I+\frac{\tau A}{2})(I+\tau A)(I+\tau A+\frac{(\tau A)^2}{2})^{-1}$$

$$\times\frac{1}{(k-j)!}\int_0^\infty t^{k-j}\exp(-t)A\exp\{-\frac{(\tau t+\lambda)A}{2}\}A^{-1}$$

$$\times[\exp\{-\frac{(\tau t+\lambda)A}{2}\}AB_j - B_jA\exp\{-\frac{(\tau t+\lambda)A}{2}\}]\frac{1}{2}(g(t_{j-N})+g(t_{j-N-1}))dt$$

$$= P_1+P_2+P_3.$$

Using estimates (8.2), (8.22) and assumptions (8.18), (8.19), (8.23), we obtain

$$\|P_1\|_E \leq \|((I+\tau A)(I+\tau A+\frac{(\tau A)^2}{2})^{-1})^k\|_{E\mapsto E}$$

$$\times\lambda^{1-\alpha}\frac{1}{(k-1)!}\int\limits_0^\infty t^{k-1}\exp(-t)\|A\exp(-(\tau t+\lambda)A)g(0)\|_E dt$$

$$\leq \lambda^{1-\alpha}\frac{1}{(k-1)!}\int\limits_0^\infty t^{k-1}\exp(-t)\frac{dt}{(t\tau+\lambda)^{1-\alpha}}\max_{-N\leq k\leq 0}\|g(t_k)\|_{E_\alpha},$$

$$\|P_2\|_E \leq \lambda^{1-\alpha}\sum_{j=1}^{k}\tau\|(I+\tau A)(I+\tau A+\frac{(\tau A)^2}{2})^{-1})^{k-j}\|_{E\mapsto E}$$

$$\times\|(I+\frac{\tau A}{2})(I+\tau A)(I+\tau A+\frac{(\tau A)^2}{2})^{-1}\|_{E\mapsto E}$$

$$\times\frac{1}{(k-j)!}\int\limits_0^\infty t^{k-j}\exp(-t)\|A\exp\{-\frac{(\tau t+\lambda)A}{2}\}\|_{E\to E}\|\overline{A^{-1}B_j}\|_{E\to E}$$

$$\times\|A\exp\{-\frac{(\tau t+\lambda)A}{2}\}\frac{1}{2}(g(t_{j-N})+g(t_{j-N-1}))\|_E dt$$

$$\leq \frac{(1+\sqrt{2})}{2}\frac{\lambda^{1-\alpha}(1-\alpha)(1-\varepsilon)}{M2^{1-\alpha}(1+\sqrt{2})}$$

$$\times\sum_{j=1}^{k}\tau\frac{1}{(k-j)!}\int\limits_0^\infty t^{k-j}\exp(-t)\frac{M2^{2-\alpha}dt}{(t\tau+\lambda)^{2-\alpha}}\|\frac{1}{2}(g(t_{j-N})+g(t_{j-N-1}))\|_{E_\alpha}$$

$$\leq (1-\alpha)(1-\varepsilon)\sum_{j=1}^{k}\tau\frac{1}{(k-j)!}\int\limits_0^\infty t^{k-j}\exp(-t)\frac{\lambda^{1-\alpha}dt}{(t\tau+\lambda)^{2-\alpha}}\max_{-N\leq k\leq 0}\|g(t_k)\|_{E_\alpha}$$

for every $k, 1\leq k\leq N$ and λ, $\lambda>0$. Now let us estimate P_3. By Lemma 8.1.3 and using estimate (8.11), we obtain

$$\|P_3\|_E \leq \lambda^{1-\alpha}\sum_{j=1}^{k}\tau\|(I+\tau A)(I+\tau A+\frac{(\tau A)^2}{2})^{-1})^{k-j}\|_{E\mapsto E}$$

$$\times\|(I+\frac{\tau A}{2})(I+\tau A)(I+\tau A+\frac{(\tau A)^2}{2})^{-1}\|_{E\mapsto E}$$

$$\times\frac{1}{(k-j)!}\int\limits_0^\infty t^{k-j}\exp(-t)\|A\exp\{-\frac{(\tau t+\lambda)A}{2}\}\|_{E\to E}$$

$$\times\|A^{-1}[A\exp\{-\frac{(\tau t+\lambda)A}{2}\}B_j - B_jA$$

$$\times\exp\{-\frac{(\tau t+\lambda)A}{2}\}]\frac{1}{2}(g(t_{j-N})+g(t_{j-N-1}))\|_E dt$$

$$\leq \frac{\lambda^{1-\alpha}(1+\sqrt{2})}{2}\max_{1\leq j\leq N}\|\overline{A^{-1}(AB_j - B_jA)A^{-1}}\|_{E\to E}\sum_{j=1}^{k}\tau\frac{1}{(k-j)!}$$

$$\times\int_0^\infty t^{k-j}\exp(-t)\frac{e(1+\alpha)M^\alpha M_1^{1+\alpha}(1+2M_1)(1+M_1)^{1-\alpha}2^{(2-\alpha)\alpha}}{\pi\alpha^2(1-\alpha)}\frac{M2^{2-\alpha}dt}{(t\tau+\lambda)^{2-\alpha}}$$

$$\times\frac{1}{2}\|g(t_{j-N})+g(t_{j-N-1})\|_{E_\alpha}$$

$$\leq \lambda^{1-\alpha}(1-\alpha)\varepsilon\sum_{j=1}^{k}\tau\frac{1}{(k-j)!}\int_0^\infty t^{k-j}\exp(-t)\frac{dt}{(t\tau+\lambda)^{2-\alpha}}\max_{-N\leq k\leq 0}\|g(t_k)\|_{E_\alpha}$$

for every $k, 1\leq k\leq N$ and λ, $\lambda>0$. Using the triangle inequality, we obtain that

$$\lambda^{1-\alpha}\|A\exp\{-\lambda A\}u_k\|_E \leq J\max_{-N\leq k\leq 0}\|g(t_k)\|_{E_\alpha}$$

for every $k, 1\leq k\leq N$ and λ, $\lambda>0$. This shows that

$$\|u_k\|_{E_\alpha}\leq \max_{-N\leq k\leq 0}\|g(t_k)\|_{E_\alpha}$$

for every k, $1\leq k\leq N$. In a similar manner as with Theorem 8.2.1 applying mathematical induction, one can easily show that it is true for every k. This result completes the proof of Theorem 8.2.4.

In applications, we consider the initial-value problem on the range $\{0\leq t\leq 1$, $x=(x_1,\dots,x_n)\in\mathbb{R}^n, r=(r_1,\dots,r_n)\}$ for $2m$-order multidimensional delay differential equations of parabolic type,

$$\begin{cases}\frac{\partial u(t,x)}{\partial t}+\sum\limits_{|r|=2m}a_r(x)\frac{\partial^{|r|}u(t,x)}{\partial x_1^{r_1}\cdots\partial x_n^{r_n}}+\delta u(t,x)\\ =b(t)(\sum\limits_{|r|=2m}a_r(x)\frac{\partial^{|r|}u(t-\omega,x)}{\partial x_1^{r_1}\cdots\partial x_n^{r_n}}+\delta u(t-\omega,x)),0<t<\infty,x\in\mathbb{R}^n,\\ u(t,x)=g(t,x),-\omega\leq t\leq 0,x\in\mathbb{R}^n,|\,r\,|=r_1+\cdots+r_n\end{cases}\tag{8.24}$$

where $a_r(x)$, $b(t)$, $g(t,x)$ are given sufficiently smooth functions and $\delta>0$ is the sufficiently large number. We will assume that the symbol $[\xi=(\xi_1,\dots,\xi_n)\in\mathbb{R}^n]$,

$$A_1^x(\xi)=\sum_{|r|=2m}a_r(x)\,(i\xi_1)^{r_1}\cdots(i\xi_n)^{r_n}$$

of the differential operator of the form

$$A_1^x = \sum_{|r|=2m} a_r(x) \frac{\partial^{|r|}}{\partial x_1^{r_1} \cdots \partial x_n^{r_n}} \tag{8.25}$$

acting on functions defined on the space $\mathbb{R}^n$, satisfies the inequalities

$$0 < M_1|\xi|^{2m} \le (-1)^m A_1^x(\xi) \le M_2|\xi|^{2m} < \infty$$

for $\xi \neq 0$, where $|\ \xi \ |= \sqrt{|\xi_1|^2 + \cdots + |\xi_n|^2}$. Problem (8.24) has a unique smooth solution. This allows us to reduce the initial-value problem (8.24) to the initial-value problem (8.1) in a Banach space E with a strongly positive operator $A^x = A_1^x + \delta I$ defined by (8.25). Let us give a number of corollaries of the abstract Theorems 8.1.1 and 8.2.1.

Theorem 8.2.5. *Assume that*

$$\sup_{0 \le t < \infty} |b(t)| \le \frac{1-\alpha}{M 2^{2-\alpha}}.$$

Then the solutions of the initial-value problem (8.24) *satisfy the stability estimates*

$$\sup_{0 \le t < \infty} \|v(t,.)\|_{C^{2m\alpha}(\mathbb{R}^n)} \le M_2(\alpha) \max_{-\omega \le t \le 0} \|g(t,.)\|_{C^{2m\alpha}(\mathbb{R}^n)}, 0 < \alpha < \frac{1}{2m},$$

where $M_2(\alpha)$ does not depend on $g(t,x)$. Here $C^{\varepsilon}(\mathbb{R}^n)$ is the space of functions satisfying a Hölder condition with the indicator $\varepsilon \in (0,1)$.

The proof of theorem 8.2.5 is based on the estimate

$$\| \exp\{-tA^x\}\|_{C(\mathbb{R}^n) \to C(\mathbb{R}^n)} \le M, \ \ t \ge 0,$$

and on the abstract Theorem 8.1.1, on the strong positivity of the operator A^x in $C^{\beta}(\mathbb{R}^n)$, and on equivalence of the norms in the spaces $E_\alpha = E_\alpha(C(\mathbb{R}^n), A^x)$ and $C^{2m\alpha}(\mathbb{R}^n)$ when $0 < \alpha < \frac{1}{2m}$ (see Chapter 4).

Let us define the grid space $\mathbb{R}_h^n (0 < h \le h_0)$ as the set of all points of the Euclidean space R^n whose coordinates are given by

$$x_i = s_i h, s_i = 0, \pm 1, \pm 2, \ldots, i = 1, \ldots, n.$$

To the differential operator $A^x = A_1^x + \delta I$ defined by (8.25) we assign the difference operator A_h^x by the formula

$$A_h^x u^h(x) = \sum_{2m \le |r| \le S} b_r^h(x) D_h^r u^h(x)$$

of an arbitrary high order of accuracy that approximates this multidimensional elliptic operator A^x which acts on functions u_x^h defined on the entire space $\mathbb{R}_h^n$. We shall assume that for $|\xi_k h| \leq \pi$ the symbol $A^x(\xi h, h)$ of the operator A_h^x satisfies the inequalities

$$(-1)^m A^x(\xi h, h) \geq M_3 |\xi|^{2m}, |arg A^x(\xi h, h)| \leq \phi < \pi.$$

With the help of A_h^x we arrive at the initial-value problem

$$\begin{cases} (u^h(t,x))' + A_h^x u^h(t,x) = b(t) A_h^x u^h(t-\omega, x), t \geq 0, x \in \mathbb{R}_h^n, \\ u^h(t,x) = g^h(t,x) = g(t,x)(-\omega \leq t \leq 0), x \in \mathbb{R}_h^n, \end{cases} \tag{8.26}$$

for an infinite system of ordinary differential equations. Now, we replace problem (8.26) by the first-order of accuracy in t difference scheme

$$\begin{cases} \frac{1}{\tau}(u_k^h(x) - u_{k-1}^h(x)) + A_h^x u_k^h(x) = b(t_k) A_h^x u_{k-N}^h(x), \\ t_k = k\tau, 1 \leq k, N\tau = \omega, x \in \mathbb{R}_h^n, \\ u_k^h(x) = g^h(t_k, x), t_k = k\tau, -N \leq k \leq 0, x \in \mathbb{R}_h^n. \end{cases} \tag{8.27}$$

Theorem 8.2.6. *Assume that all the conditions of Theorem* 8.2.5 *are satisfied. Then for the solution of difference scheme* (8.27) *the estimate*

$$\sup_{1 \leq k < \infty} \|u_k^h\|_{C^{2m\alpha}(\mathbb{R}_h^n)} \leq M_2(\alpha) \max_{-N \leq t \leq 0} \|g_k^h\|_{C^{2m\alpha}(\mathbb{R}_h^n)}, 0 < \alpha < \frac{1}{2m}$$

holds.

The proof of Theorem 8.2.6 is based on the estimate

$$\|\exp\{-t_k A_h^x\}\|_{C(\mathbb{R}_h^n) \to C(\mathbb{R}_h^n)} \leq M, k \geq 0,$$

and on the abstract Theorem 8.2.1, the positivity of the operator A_h^x in $C(\mathbb{R}_h^n)$, and on the fact that the $E_\alpha = E_\alpha(C(\mathbb{R}_h^n), A_h^x)$-norms are equivalent to the norms $C^{2m\alpha}(\mathbb{R}_h^n)$ uniformly in h for $0 < \alpha < \frac{1}{2m}$ (see Chapter 4).

Comments on the Literature

Chapters 1 and 2

The material in Chapter 1 and Chapter 2 was written on the basis of lecture notes that were used in a graduate course at Fatih University.

Chapter 3

Chapter 3 is based on the paper [35q].

Chapter 4

The role played by coercive inequalities in the study of boundary-value problems for elliptic and parabolic partial differential equations is well known (see, e.g., [4a, 6a–b, 107a, 108a, 174a]).

Coercivity inequalities for the solutions of an abstract differential equation of parabolic type were established for the first time in [154a]. The Cauchy problem for such an equation was considered in various spaces $F(E)$ of functions defined on the segment $[0, T]$ with values in a Banach space E, and it was established that a necessary condition for coercivity is the strong positivity of the operator coefficients $A(t)$. This condition is also sufficient in the spaces $C_0^{\alpha,\alpha}(E)$. In the case of the spaces $L_p(E)$ with $p \in (1, \infty)$ the following extrapolation result was established: for the Cauchy problem to be coercive in $L_p(E)$ for all such p it suffices that it be coercive for some $p_0 \in (1, \infty)$.

As noted above, in contrast with $L_p(E)$, in the spaces $C_0^{\alpha,\alpha}(E)$ an unconditional coercivity inequality holds, and in connection with this there arises the question of whether the same is valid for the ordinary Hölder space $C^\alpha(E)$. In [16u] it was shown that the coercivity inequality in $C^\alpha(E)$ holds under the additional assumption that $A(0)v_0 - f\ (0) \in E_\alpha$.

The coercivity inequality was established in the two-parameter series of spaces $C_0^{\beta,\gamma}(E)$ with $0 < \beta < 1, 0 \leq \gamma \leq \beta$, under the additional assumption that $A(0)v_0 - f\ (0) \in E_{\beta-\gamma}$. Furthermore the spaces of smooth functions $C_0^{\beta,\gamma}(E)$, in which coercive solvability has been established, depend on the parameters β and γ. However, the constants in the coercive inequalities depend only on β. Hence, we can choose the parameter β freely, which increases the number of function spaces in which problem (4.1) is well posed. Moreover the coercivity inequality was established in the three-parameter series of spaces $C_0^{\beta,\gamma}(E_{\alpha-\beta})$ with $0 < \alpha < 1$, $0 \leq \gamma \leq \beta \leq \alpha$, under the additional assumption that $A(0)v_0 - f\ (0) \in E_{\alpha-\gamma}$. Furthermore the spaces of smooth functions $C_0^{\beta,\gamma}(E_{\alpha-\beta})$, in which coercive solvability has been established, depend on the parameters α, β and γ. However, the constants in the coercive inequalities depend only on α. Hence, we can choose the parameters β and γ freely, which increases the number of function spaces in which problem (4.1) is well posed [16h, 35l, 94a].

Earlier an unconditional coercivity inequality was established in the spaces $W_p^\alpha(E)$ with $1 \leq p < \infty$, $0 < \alpha < 1$ [11a, 89a], as well as in the spaces $L_p(E_{\alpha,p})$ with $0 < \alpha < 1$, $1 \leq p \leq \infty$ [154g, 69a–b, 35d, 135a].

Finally, let us mention the paper [64a], in which an unconditional coercivity inequality in $L_p(E)$ was established under the assumption that the imaginary powers $A^{i\beta}$ are bounded (in some restrictions of the space E).

The application of this result to parabolic partial differential equations allowed one to obtain a series of new coercivity inequalities involving different norms with respect to t and the space variables.

In this book we do not discuss results on the well-posedness of the nonlocal boundary-value problems for parabolic and parabolic-elliptic equations, for which the reader is referred to the papers [16t, 23a–b, 24a, 31b, 36c].

The material in Section 1 was written on the basis of the papers [106a, 154a, 154g, 154m, 16u, 16h, 35l, 94a, 69a–b, 35d, 135a, 162a].

The construction of the discretization with respect to time for parabolic equations was initiated in Rothe's paper [133a]. When the study of the discretization with respect to t for the abstract Cauchy problem is applied to parabolic partial differential equations, it also covers the case of the full discretization with respect to time and space variables.

To prove stability, in a number of works (see [1a, 3a, 114a, 115a, 139a, 140a, 141a, 142a, 171a–b] and the references given therein) difference schemes were treated as operator equations in a Hilbert space, and the investigation was based on the symmetry properties of operator coefficients. This led to L_2-stability estimates.

Of great interest is the study of stability in the C-norm. In [10a, 12a, 76a, 144a–d, 163a–b, 177a] the corresponding estimates were obtained for the simplest difference schemes approximating the Cauchy problem for parabolic equations and systems with constant coefficients and one space variable. The proof is based on estimates of parabolic difference Green function (fundamental solutions). In the case of parabolic equations with variable coefficient and more than one space variable, stability in the C-norm is known to hold only for difference schemes that satisfy a maximum principle (see [115a, 141a]).

An important type of stability is the coercive stability (well-posedness) of difference schemes. There are no coercive estimates in the C-norm, because they do not hold for the corresponding differential equations. The coercive stability of difference schemes of first order of accuracy was apparently studied for the first time in [158a, 158b] for the first boundary-value problem for second-order parabolic equations in L_2. A more general sharp-angle inequality for parabolic difference operators was established. Then followed the works [10a, 65a, 87a, 99a, 158b, 158c], where coercive inequalities in the L_2- and L_p-norms were given. In [154d] difference schemes are treated as operator equations in Banach spaces. In a Banach space there is no notion of symmetry and for this reason the study of stability relies on the positivity property of the operator coefficients, expressed in terms of properties of their resolvents. In this way in [154d] coercive estimates were established in difference analogues of Hölder spaces with weight with respect to the time variable. Enlisting the theory of positive operators turned out to be effective in the investigation of such difference schemes in difference analogues of Bochner spaces (see [127a, 127b]). The coercivity inequality in the difference analogue of the Bochner spaces $L_p(\tau, E)$ holds for any $p \in (1,\infty)$ whenever it holds for at least one $p_0 \in (1,\infty)$, i.e., here we have a result analogous to that in the differential case [154a].

An unconditional coercivity inequality in the spaces $L_q(\tau, E'_{\alpha,q})$ $(0 < \alpha < 1, 1 \leq q \leq \infty)$ was obtained in [35d, 35e]. One is led to asking whether a coercivity inequality holds for the Rothe difference scheme in the difference analogue of the usual Hölder space $C^\alpha(\tau, E)$. The answer turns out to be affirmative under the additional restrictions $A(0)u_0 - f_1 \in E'_\alpha$. This result is a particular case of the assertion that the difference problem is coercive in the two-parameter family of Banach spaces $C_0^{\beta,\gamma}(E)$ with $0 < \beta < 1$, $0 \leq \gamma \leq \beta$, under the additional assumption that $A(0)u_0 - f_1 \in E'_{\beta-\gamma}$. Furthermore the spaces $C_0^{\beta,\gamma}(\tau, E)$, in which coercive solvability has been established, depend on the parameters β and γ. However, the constants in the coercive inequalities depend only on β. Hence, we can choose the parameter β freely, which increases the number of spaces in which a Rothe difference scheme for problem (4.1) is well posed. Moreover the coercivity inequality was established in the three-parameter series of spaces $C_0^{\beta,\gamma}(\tau, E_{\alpha-\beta})$ with $0 < \alpha < 1$, $0 \leq \gamma \leq \beta \leq \alpha$, under the additional assumption that $A(0)u_0 - f_1 \in E'_{\alpha-\gamma}$. Furthermore the spaces $C_0^{\beta,\gamma}(\tau, E'_{\alpha-\beta})$, in which coercive solvability has been established, depend on the parameters α, β and γ. However, the constants in the coercive inequalities depend only on α. Hence, we can choose the parameters β and γ freely, which increases the number of spaces in which a Rothe difference scheme for problem (4.1) is well posed [16h, 35l, 94a].

The utilization of Pade approximants for computational purposes has a long history. The approximants $R_{1,1}(z)$ and $R_{2,2}(z)$ were first applied by Euler to calculate the scalar exponential. The fact that these approximants are attributed to Pade is based on his 1892 dissertation, in which he studied and tabulated them. In recent years we have seen a sharp rise in the interest in classical methods of rational approximation of analytical functions, primarily in Pade approximants and their application to physics and engineering problems. For a detailed exposition of these investigation we refer the reader to [43a].

Apparently the first study that uses rational approximation of $\exp\{-\tau A\}$ in the case of a positive (unbounded) self-adjoint operator A was carried out in the paper [171a], which deals with approximation on the positive half-line.

That study was continued by a number of authors [58a, 139a, 171b]. In applications to problems of mathematical physics this resulted in stability and convergence estimates for difference schemes in L_2.

The application of the approximants $R_{j,l}^{k}(\tau A)$ in the construction of difference schemes in the case of an unbounded strongly positive operator A acting in an arbitrary Banach space E has been systematically developed by several authors (see, e.g., [7a–c, 16a–d, 16g, 16p, 16r, 32a, 35a–b, 35e–h, 35l–m, 37a, 54a, 55a, 56a, 60a, 77a, 81a–b, 91a–c, 95a, 97a, 124b, 127c–e, 130a, 154d–f, 154i, 155a–b, 85a, 104a, 125a, 173a]). In [16l, 54a, 95a] difference schemes were studied for a class of operators that generate not only analytic, but also strongly continuous semigroups. Most of the results of the stability and coercive stability in the Banach spaces of the single-step difference schemes of a high order of accuracy generated by $R_{j,l}^{k}(\tau A)$ for parabolic equations were considered in the book [35l]. The application of these results to the initial-value problem for parabolic equations allowed one to obtain a series of new coercivity inequalities in difference analogies of various norms with respect to t and x for the solutions of difference schemes of a high order of accuracy in time and space variables.

However, the theory of a high order of accuracy single-step difference schemes for parabolic differential equations with variable operator coefficients have received less attention than parabolic differential equations with coefficients independent of time. Most of the results of the stability and coercive stability in the Banach spaces of the single-step difference schemes of the first and second order of accuracy generated by the Pade approximants for first-order differential equations in the case of an unbounded strongly positive operators $A(t)$ acting in an arbitrary Banach space E were obtained in the papers [16e, 35c, 126a, 154i–j, 161a]. The application of Pade approximants for parabolic equations with variable operator coefficients $A(t)$ was continued by a number of authors [143a–b, 178a]. The difference schemes of the third, fourth and fifth order of accuracy were constructed and investigated. In applications to problems of mathematical physics this resulted in stability and convergence estimates for difference schemes in L_2. In [42a, 49a, 100a, 143a] Runge–Kutta difference schemes were studied for parabolic equations with variable operator coefficients $A(t)$.

The application of the Pade approximants to the Cauchy problem for parabolic equations with variable operator coefficients $A(t)$ acting in an arbitrary Banach space E relies on the approximate formulas for expressions

$$v(t_k, t_{k-1}) \text{ and } \frac{1}{\tau}\int_{t_{k-1}}^{t_k} v(t_k, s) f(s) ds.$$

Unfortunately, in the general case of operators $A(t)$ the explicit formula for $v(t_k, t_{k-1})$ does not exist. Therefore, in this case we cannot use approaches of the construction of Pade difference schemes for parabolic equations with coefficients independent of time.

The single-step difference schemes of a high order of accuracy generated by an exact difference scheme for parabolic equations with variable operator coefficients $A(t)$ acting in an arbitrary Banach space E were presented for the first time in [35j]. They were based on a series of interesting properties concerning the semigroup $\exp\{-sA(t)\}$ $(s \geq 0)$ and the fundamental solution $v(t,s)$ of (4.1). These properties and series of interesting estimates for $R^i_{j,l}(\tau A(t_k))$ and $\widetilde{u}(k,m)$ permit us to study the stability and the coercive stability of presented difference schemes. That study was continued in [16o, 16q]. Furthermore the high order of accuracy difference schemes generated by Taylor's decomposition on two points for the approximate solutions of the Cauchy problem for parabolic equations with variable operator coefficients $A(t)$ acting in an arbitrary Banach space E were presented. The well-posedness of these difference schemes in various Banach spaces were studied. The stability and coercive stability estimates in Banach norms for solutions of high order of accuracy difference schemes of mixed type boundary-value problems for parabolic equations were obtained.

The study of high order of accuracy difference schemes in the case of strongly positive operators $A(t)$ allowed us to establish not only the stability, but also the coercive stability in difference analogues of Banach spaces. The application of these abstract results to mixed type boundary-value problems for parabolic equations relies on the following facts: the strong positivity of an elliptic difference operator $A_h^{t,x}$ in the Banach space E_h, the well-posedness of the resolvent equation of $A_h^{t,x}$ in E_h or in $E'_{\alpha,p}(E_h, A_h^{t,x}), 1 \leq p \leq \infty$, the structure of fractional spaces $E'_{\alpha,p}(E_h, A_h^{t,x})$.

The strong positivity of the wider class of differential operators has been studied by a number of authors [4a, 5a, 89b, 101a, 103a, 154c, 154k, 159a–b, 160a–b].

The strong positivity of elliptic difference operators in Hilbert norms was studied in many works (see, e.g., [114a, 140a, 141a]). Such a study is based on the self-adjointness and positive definiteness of the principal part of the operators $A_h^{t,x}$.

Considerably less studied is the strong positivity or simple positivity of difference operators in Banach norms. For second-order elliptic operators and the simplest difference scheme, positivity in the C_h-norm follows from the maximum principle (see [140a, 144a–b]).

The strong positivity in difference analogue of weighted Hölder space for an elliptic difference operator $A_h^{t,x}$ of second order of accuracy that approximates an elliptic operator without mixed derivatives was established in [154d].

The strong positivity of the simplest multidimensional second-order elliptic difference operator in the $L_{p,h}$- and C_h-norms was established in [7a–c]. The most thorough study of the strong positivity in the C_h-and C_h^{β}-norms of a wide class of operators that approximate elliptic operators of arbitrary order in R^n was carried out in [35e, 152a–d]. The situation is different for difference operators that approximate boundary-value problems for an elliptic equation. The positivity in the C_h- and C_h^β-norms of difference operators of this type was studied in a number of particular cases [7a, 28a, 28c, 61a–c].

For the well-posedness of the resolvent equation of $A_h^{t,x}$ in E_h or in

$$E'_{\alpha,p}(E_h, A_h^{t,x}),\ 1 \leq p \leq \infty$$

we refer the reader to Chapter 5 of the present work.

Finally, let us comment on the structure of fractional spaces. It is established that for any $0 < \alpha < 1/2m$ and $1 \leq p \leq \infty$ the norms in the spaces $E'_{\alpha,p}(L_p(R_h^n), A_h^{t,x})$ and $W_p^{2m\alpha}(R_h^n)$ are equivalent uniformly in h. This fact corresponds to the following equality, known in interpolation theory (see [166a]):

$$E'_{\alpha,p}(L_p(R^n), A_h^{t,x}) = W_p^{2m\alpha}(R^n),\ 0 < \alpha < 1/2m,\ 1 \leq p \leq \infty,$$

which in turn follows from the equality $D(A^{t,x}) = W_p^{2m\alpha}(R^n)$ for a $2m$th-order elliptic operator $A^{t,x}$ in $L_p(R^n)$, $1 \leq p \leq \infty$, via the real interpolation method. The alternative method of investigation adopted in [16o, 35l], based on estimates of fundamental solutions of the resolvent equation for the operator $A_h^{t,x}$, allows us to consider also the cases $p = 1$ and $p = \infty$.

This book does not touch upon the results of [19a, 26a, 27a, 36a–b, 36d, 29a, 44a–c] on the theory of the single-step difference schemes for nonlocal boundary-value problems.

Section 2 is based on the papers [16o, 35j]. The results of Theorems 4.25–4.32 and 4.40–4.45 and the difference schemes (4.114), (4.115), (4.16) appear here for the first time.

Section 3 is based on the paper [16o]. The results of Theorems 4.59–4.62 appear here for the first time.

Chapter 5

Coercivity inequalities for the solutions of an abstract differential equation of elliptic type were established for the first time in [154c]. The boundary-value problem for such an equation was considered in various spaces $F(E)$ of functions defined on the segment $[0, T]$ with values in a Banach space E, and it was established that a necessary condition for coercivity is the positivity of the operator coefficient A. This condition is also sufficient in the spaces $C_{0T}^{\alpha,\alpha}(E)$. In the case of the spaces $L_p(E)$ with $p \in (1, \infty)$ the following extrapolation result was established: for the boundary-value problem to be coercive in $L_p(E)$ for all such p it suffices that it be coercive for some $p_0 \in (1, \infty)$.

As noted above, in contrast to $L_p(E)$, in the spaces $C_{0T}^{\alpha,\alpha}(E)$ an unconditional coercivity inequality holds, and in connection with this there arises the question of whether the same is valid for the ordinary Hölder space $C^\alpha(E)$. In [16j] it was shown that the coercivity inequality in $C^\alpha(E)$ holds under the additional assumptions that $Av_0 - f\ (0) \in E_\alpha$ and $Av_T - f\ (T) \in E_\alpha$.

The coercivity inequality was established in the two-parameter series of spaces $C_{0T}^{\beta,\gamma}(E)$ with $0 < \beta < 1, 0 \leq \gamma \leq \beta$, under the additional assumptions that $Av_0 - f\ (0) \in E_{\beta-\gamma}$ and $Av_T - f\ (T) \in E_{\beta-\gamma}$. Furthermore the spaces of smooth functions $C_{0T}^{\beta,\gamma}(E)$, in which coercive solvability has been established, depend on the parameters β and γ. However, the constants in the coercive inequalities depend only on β. Hence, we can choose the parameter β freely, which increases the number of function spaces in which problem (5.1) is well posed. Moreover the coercivity inequality was established in the three-parameter series of spaces $C_{0T}^{\beta,\gamma}(E_{\alpha-\beta})$ with $0 < \alpha < 1$, $0 \leq \gamma \leq \beta \leq \alpha$, under the additional assumptions that $Av_0 - f\ (0) \in E_{\alpha-\gamma}$ and $Av_T - f\ (T) \in E_{\alpha-\gamma}$. Furthermore the spaces of smooth functions $C_{0T}^{\beta,\gamma}(E_{\alpha-\beta})$, in which coercive solvability has been established, depend on the parameters α, β and γ. However, the constants in the coercive inequalities depend only on α. Hence, we can choose the parameters β and γ freely, which increases the number of function spaces in which problem (5.1) is well posed [16j, 16o].

Earlier an unconditional coercivity inequality was established in the spaces $L_p(E_{\alpha,p})$ with $0 < \alpha < 1$, $1 \leq p \leq \infty$ [89b].

The application of this result to elliptic partial differential equations allowed one to obtain a series of new coercivity inequalities involving different norms with respect to all variables.

In this book we do not discuss results on the well-posedness of the local and nonlocal boundary-value problems for elliptic equations, for which the reader is referred to the papers [9a, 16w, 21a–b, 28d, 34a, 84a–b, 86a, 145a–c, 148a].

The material in Section 1 was written on the basis of the papers [16j, 16o, 154c, 154m, 89b].

The coercive stability of difference schemes of the second order of accuracy was apparently studied for the first time in [120a] for the first boundary-value problem for second-order elliptic equations in L_2. Then followed the works [12a, 169a, 170a, 66a–d, 179a, 134a, 158a–c, 78a–b, 117a, 118a–b], where coercive inequalities in the L_2- and L_p-norms were given. In [154*d*] difference schemes are treated as operator equations in Banach spaces with the positive operator coefficient. In this way coercive estimates were established in difference analogues of Hölder spaces with weight with respect to the one space variable. Enlisting the theory of positive operators turned out to be effective in the investigation of such difference schemes in difference analogues of Bochner spaces (see [127a, 127c]). The coercivity inequality in the difference analogues of the Bochner spaces $L_p(\tau, E)$ holds for any $p \in (1, \infty)$ whenever it holds for at least one $p_0 \in (1, \infty)$, i.e., here we have a result analogous to that in the differential case [154c]. That study was continued in [128a–c, 129a, 154h, 154l, 28b, 35j].

The applications of the Pade approximants in the theory of the high order of accuracy two-step difference schemes for elliptic differential equations have received less attention than parabolic differential equations. The two-step difference schemes of a high order of accuracy generated by an exact difference scheme for elliptic equations with the operator coefficient A acting in an arbitrary Banach space E were presented for the first time in [16j]. The series of interesting estimates for $R^i_{j,l}(\tau A^{\frac{1}{2}})$ permit us to study the stability and the coercive stability of presented difference schemes. That study was continued in [16o, 16n]. Furthermore the high order of accuracy difference schemes generated by Taylor's decomposition on three points for the approximate solutions of the boundary-value problem for elliptic equations with the operator coefficient A acting in an arbitrary Banach space E were presented. The well-posedness of these difference schemes in various Banach spaces were studied. The stability and coercive stability estimates in Banach norms for solutions of the high order of accuracy difference schemes of mixed type boundary-value problems for elliptic equations were obtained [35k]. That study was continued in [16i, 16o, 16s, 35n, 81c, 82a].

The study of the high order of accuracy difference schemes in the case of positive operator A allowed us to establish not only the stability, but also the coercive stability in difference analogues of Banach spaces. The application of these abstract results to the mixed type boundary-value problems for elliptic equations relies on the following facts: the positivity of an elliptic difference operator A_h^x in the Banach space E_h, the well-posedness of the resolvent equation of A_h^x in E_h or in $E'_{\alpha,p}(E_h, (A_h^x)^{\frac{1}{2}})$, $1 \le p \le \infty$, the structure of fractional spaces $E'_{\alpha,p}(E_h, (A_h^x)^{\frac{1}{2}})$.

For the positivity of the wider class of differential and difference elliptic operators we refer the reader to Chapter 4 of the present work.

Finally, let us comment on the structure of fractional spaces. It is established that for any $0 < \alpha < 1/2$ and $1 \le p \le \infty$ the norms in the spaces $E'_{\alpha,p}(E_h, (A_h^x)^{\frac{1}{2}})$ and $E'_{2\alpha,p}(E_h, A_h^x)$ are equivalent uniformly in h. Therefore for the structure of fractional spaces $E'_{\alpha,p}(E_h, (A_h^x)^{\frac{1}{2}})$ we refer the reader to Chapter 4 of the present work.

Section 2 is based on the papers [16j, 16o]. The results of Theorems 5.23–5.48, 5.32–5.46 and 5.51–5.57 appear here for the first time.

Section 3 is based on the papers [16o, 16s, 35k, 35n]. The results of Theorems 5.67–5.68 and 5.70–5.73 appear here for the first time.

Chapter 6

The method of operators as a tool for investigation of the solution to hyperbolic differential equations in Hilbert and Banach spaces, has been systematically developed by several authors (see, e.g., [2a, 75a, 85a, 104a, 125a, 154b, 156a, 173a]).

A large cycle of works on difference schemes for hyperbolic partial differential equations (see, e.g., [16f, 124a, 124c, 124d] and the references given therein), in which stability was established under the assumption that the magnitude of the grid steps τ and h with respect to the time and space variables are connected. In abstract terms this means, in particular, that the condition $\tau||A_h|| \to 0$ when $\tau \to 0$ is satisfied.

Of great interest is the study of absolute stable difference schemes of a high order of accuracy for hyperbolic partial differential equations, in which stability was established without any assumptions in respect of the grid steps τ and h. Such type stability inequalities for solutions of the first order of accuracy difference scheme for differential equations of hyperbolic type were established for the first time in [157a]. The new first and second order of accuracy difference schemes approximately solving the abstract initial-value problem for hyperbolic equations in a Hilbert space were presented in [30a, 35o]. Applying the operator approach, stability estimates for the solution of these difference schemes were obtained. In [16f] the high order of accuracy two-step difference schemes generated by an exact difference scheme for the numerical solutions of this problem were presented. The convergence and stability estimates for the solutions of these difference schemes were established.

This book does not touch upon the results of [17a, 25a–b, 30b, 30c, 31a, 38a–d, 39a] on the theory of difference schemes for nonlocal problems for hyperbolic, hyperbolic-parabolic, hyperbolic-elliptic equations. In this book we do not discuss results on the difference schemes of the Goursat problem for hyperbolic equations and the Cauchy problem for systems of differential equations of first order, for which the reader is referred to the papers [14a, 15a, 33a].

Sections 1 and 2 are based on the paper [16f]. The results of Theorems 6.4–6.9 appear here for the first time.

The results of Section 3 appear here for the first time.

Chapter 7

The role played by asymptotic methods in the study of the solution of various problems for elliptic, parabolic and hyperbolic partial differential equations with a small parameter in higher-order derivatives is well known (see, e.g., [75a, 101a, 104a, 111a]).

Of great interest is the study of stability uniform difference schemes for partial differential equations with a small parameter in higher-order derivatives (see, e.g., [16v, 40a, 52a, 53a, 67a, 73a–b, 96a, 98a–c, 149a, 150a, 151a, 164a]).

In Chapters 4, 5 and 6 the stability of a wide class of difference schemes for partial differential equations was investigated. However, it can be shown that these difference schemes for partial differential equations with a small parameter in higher-order derivatives are not uniform in ε convergence. This is a very disappointing result and it raises the question whether there are any difference schemes which are uniform in ε convergence.

The high order of accuracy single-step uniform difference schemes of the approximate solutions of the Cauchy problem for the abstract parabolic differential equations in an arbitrary Banach space E with the strongly positive operator A and an arbitrary ε positive parameter multiplying the derivative term were investigated in [16k] and generated by the exact single-step difference scheme, which is presented in Chapter 4. The well-posedness of these difference schemes in various Banach spaces were studied. That study was continued in [16o].

The two-step uniform difference schemes of a high order of accuracy for the approximate solutions of the boundary-value problem for abstract elliptic differential equations in an arbitrary Banach space E with the positive operator A and an arbitrary ε positive parameter multiplying the derivative term were investigated in [16m] and generated by the exact two-step difference scheme, which is presented in Chapter 5. The well-posedness of these difference schemes in various Banach spaces was established. That study was continued in [16o].

In [22a–b] the high order of accuracy two-step uniform difference schemes generated by an exact difference scheme of the approximate solutions for differential equations of hyperbolic type with an arbitrary parameter ε on the highest derivative were presented. The stability estimates of the solutions of these difference schemes under the assumptions

$$\|c(t)\|_{E\to E} \leq M, \ \|A^{\frac{1}{2}}s(t)\|_{E\to E} \leq M, \ 0 \leq t < \infty$$

in an arbitrary Banach space E were established. These estimates for the wider class of differential operators in Banach and Hilbert spaces have been studied by a number of authors (see, for example, [75a, 102a, 125a] and the references given therein).

The material in Section 1 is written on the basis of the papers [16k, 16o]. The results of Theorems 7.8–7.14 appear here for the first time.

The material in Section 2 is written on the basis of the papers [16m, 16o]. The results of Theorems 7.22–7.27 appear here for the first time

The material in Section 3 is written on the basis of the papers [22a–b].

Chapter 8

The operator approach for theory and numerical solutions of the initial-value problem for evolution differential equations has been studied extensively by many researches (see [8a, 41a, 50a, 51a, 57a, 68a, 70a, 71a, 72a–b, 74a, 79a, 80a, 81c, 82b, 83a, 112a, 119a–c, 121a, 122a, 123a, 124e, 131a–b, 132a, 146a, 147a–b, 167a, 168a, 169a, 175a, 176a–b] and the references therein).

Methods for numerical solutions of the delay ordinary differential equations have been studied extensively in a large cycle of works (cf., e.g., [13a, 18a, 18b, 20a, 46a, 47a, 59a, 165a, 181a] and the references therein) and developed over the last two decades. In this literature, stability of solutions of the difference schemes of the initial-value problems for delay ordinary differential equations was considered under the necessary condition of stability of the differential problem. However, the theory and numerical solutions of delay partial differential equations have received less attention than delay ordinary differential equations.

The role played by the operator approach in the study of the solution of various problems for delay partial differential equations is well known (see, e.g.,

[35p, 62a]). In [35p] the initial-value problem for delay differential equations of parabolic type was considered in the spaces $C(E_\alpha)$ of functions defined on the segment $[0,\infty]$ with values in a Banach space E_α and the stability inequalities were established under the stronger assumption than the necessary condition of the stability of the differential problem. The stability estimates for the solutions of the first and second order of accuracy difference schemes for approximately solving this initial-value problem for delay differential equations of parabolic type were presented. Stability estimates in Hölder norms for solutions of the initial-value problem of the delay differential and difference equations of parabolic type were obtained.

The material in Sections 1 and 2 is written on the basis of the paper [35p].

Bibliography

[1a] Abrashin, V.N., *A stable difference scheme for quasilinear equations of mathematical physics*, Differensialnyye Uravneniya **18** (1982), no. 11, 1967–1971. (Russian). [RZh. Mat. 1983:3 B1373]

[2a] Abrashin, V.N. and Bobkov V.V., *A difference scheme of integral relations method for a hyperbolic equation*, Izv. AN BSSR, Ser. Phys. Mat. Nauk No. 2 (1968), 57–65. (Russian). [RZh. Mat. 1969:2 B802]

[3a] Abrashin, V.N. and Shkel B.A., *A symmetrical difference scheme of the numerical solution of the boundary-value problems for nonlinear parabolic equations*, Vest. AN BSSR, Ser. Phys. Mat. Nauk No. 2 (1987), 7–14. (Russian). [RZh. Mat. 1987:9 B1306]

[4a] Agmon, S., *Lectures on Elliptic Boundary-Value Problems*, D. Van Nostrand, Princeton:New Jersey, 1965. [MR **31** #2504]

[5a] Agmon, S., Nirenberg, L., *Properties of solutions of ordinary differential equations in Banach spaces*, Comm. Pure Appl. Math. **20** (1963), 121–239. [MR **27** #5142]

[6a] Agmon, S., Douglis, A. and Nirenberg, L., *Estimates near the boundary for solutions of elliptic partial differential equations satisfying general boundary conditions I*, Comm. Pure Appl. Math. **12**(1959), 623–727. [MR **23** #A2610]

[6b] Agmon, S., Douglis, A. and Nirenberg, L., *Estimates near the boundary for solutions of elliptic partial differential equations satisfying general boundary conditions II*, Comm. Pure Appl. Math. **17**(1964), 35–92. [MR **28** #5259]

[7a] Alibekov, Kh.A., Sobolevskii, P.E., *Stability of difference schemes for parabolic equations*, Dokl. Acad. Nauk SSSR **232** (1977), no. 4, 737–740. (Russian). [MR **55** #13023]

[7b] Alibekov, Kh.A. and Sobolevskii, P.E., *Stability and convergence of difference schemes of the high order for parabolic differential equations*, Ukrain. Mat. Zh. **31** (1979), no. 6, 627–634. (Russian). [MR **81i:65040**]

[7c] Alibekov, Kh.A. and Sobolevskii, P.E., *Stability and convergence of difference schemes of the high order for parabolic partial differential equations*, Ukrain. Mat. Zh. **32** (1980), no. 3, 291–300. (Russian). [MR **81j:65100**]

[8a] Amann, H., *Linear and quasilinear parabolic problems*, Birkhäuser Verlag, Basel, **I** (1995). Abstract linear theory. [MR **96g:34088**]

[9a] Amanov, K. *Schauder type estimates for boundary-value problem for multidimensional elliptic equations*, in: Some Problems of Applied Mathematics, Fatih University, Istanbul (2000), 7–10.

[10a] Andreev, Yu.N., *On uniform convergence of certain difference schemes*, Zh. Vychisl. Mat. i Mat. Fiz. **6** (1966), no. 2, 238–250. (Russian). [MR **34** #967]

[11a] Anosov, V.P. and Sobolevskii, P.E., *The coercive solvability of parabolic equations*, Mat. Zametki **11** (1972), no. 2, 409–419. (Russian). [MR **46** #558]

[12a] Aronson, D.G., *On the correctness of partial differential operators and the von Neumann condition for stability difference operators*, Proceeding Amer. Math. Soc. **14** (1963), 948–955. [MR **27** #6407]

[13a] Al-Mutib, A.N., *Stability properties of numerical methods for solving delay differential equations*, J. Comput. and Appl. Math. **10** (1984), 71–79. [MR**85m:65075**]

[14a] Ashyraliyev, M., *Integral inequalities with four variable limits*, in: Modeling the Processes in Exploration of Gas Deposits and Applied Problems of Theoretical Gas Hydrodynamics, Ilim, Ashgabat (1998), 170–184 (Russian).

[14b] Ashyraliyev, M., *On stability estimates of Goursat differential and difference problems for hyperbolic equations*, in: Some Problems of Applied Mathematics, Fatih University, Istanbul (2000), 103–112.

[15a] Ashyraliyev, M. and Ataev, A., *On stability estimates of Goursat problem for hyperbolic equations*, 8 years Turkmenistan's Independence, Ilim, Ashgabat (1999), 201–206.

[16a] Ashyralyev, A., *On the coercive stability of a Crank-Nicolson difference scheme for the regularly singular parabolic equations in Holder spaces with weight*, in: Investigations on the Theory of Differential Equations, Ashgabat (1983), 52–64. (Russian). [RZh. Mat. 1983:12 B1389]

[16b] Ashyralyev, A., *Some difference schemes for parabolic equations with nonsmooth initial data*, Izv. Akad. Nauk Turkmen. SSR Ser. Fiz.-Tekhn. Khim. Geol. Nauk, no. 6 (1983), 70–72. (Russian). [MR **86c:65093**]

[16c] Ashyralyev, A., *An estimate of the error of solution of purely implicit difference schemes for the parabolic equations with nonsmooth data*, in: Applied Mathematics and Computer Software, Moscow (1985), 26–28. (Russian). [MZh. Mat. 1986:5 B1509]

[16d] Ashyralyev, A., *A purely implicit difference scheme of second order of approximation for parabolic equations*, Izv. Akad. Nauk Turkmen. SSR Ser. Fiz.-Tekhn. Khim. Geol. Nauk, no. 4(1987), 3–13. (Russian). [MR **89g:65112**]

[16e] Ashyralyev, A., *Coercive stability of a difference scheme of second-order of approximation for differential equations with a time-dependent operator in a Banach space*, in: Measure and Integral, Kuybishev, KSU (1988), 27–36. (Russian). [MR **92d:28002**]

[16f] Ashyralyev, A., *On difference schemes of the high order of accuracy for second-order evolution equations*, in: Functional Differential Equations, Perm (1989), 145–150. (Russian). [MR **91e:340070**]

[16g] Ashyralyev, A., *Estimation of the convergence of modified Crank-Nicolson difference schemes for parabolic equations with nonsmooth imput data*, Izv. Akad. Nauk Turkmen. SSR Ser. Fiz.-Tekhn. Khim. Geol. Nauk, no. 1 (1989), 3–8. (Russian). [MR **90f:65159**]

[16h] Ashyralyev, A., *Coercive solvability of parabolic equations in spaces of smooth functions*, Izv. Akad. Nauk Turkmen. SSR Ser. Fiz.-Tekhn. Khim. Geol. Nauk, no. 3 (1989), 3–13. (Russian). [MR **90i:35150**]

[16i] Ashyralyev, A., *On difference schemes of higher order of accuracy for elliptic equations*, in: Applications of the Methods of Functional Analysis to the Some Nonclassical Equations of Mathematical Physics, Novosibirsk (1989), 3–14. (Russian). [RZh. Mat. **1990:10 G122**]

[16j] Ashyralyev, A., *Well-posedness of elliptic equations in a space of smooth functions*, in: Boundary-Value Problems for Nonclassic Equations of Mathematical Physics, Sibirsk, Otdel, SSSR, Novosibirsk (1989), Vol. 2, 82–86. (Russian).

[16k] Ashyralyev, A., *On uniform difference schemes of the high order of approximation for evolution equations with a small parameter*, Numerical Functional Analysis and Optimization (1989) **10**, no. 5–6, 593–406. [MR **91k:65140**]

[16l] Ashyralyev, A., *On difference schemes of higher order of accuracy for semilinear evolution differential equations in a Banach space*, in: Computer Science and System Analysis, Ashgabat (1990), 66–75. (Russian).

[16m] Ashyralyev, A., *On uniform difference schemes of higher order of approximation for elliptic equations with a small parameter*, Applicable Analysis *36* (1990), 211–220. [MR **91d:65105**]

[16n] Ashyralyev, A., *Estimations of the convergence of modified difference schemes of the Pade's class for elliptic equations*, in: Investigation of Theory and Approximation Methods for Differential Equations, Ashgabat (1991), 71–79. (Russian).

[16o] Ashyralyev, A., *Method of positive operators of investigations of the high order of accuracy difference schemes for parabolic and elliptic equations*, Doctor Sciences Thesis, Kiev, 1991. (Russian).

[16p] Ashyralyev, A., *Correct solvability of the Pade difference schemes for parabolic equations in Hölder spaces*, Ukrain. Mat. Zh. **44** (1992), no. 11, 1466–1476. (Russian). [MR **94c:65103**]

[16q] Ashyralyev, A., *Single-step difference schemes generated by Taylor's decomposition in two points*, in: Proceedings of the Scientific-Practical Conference on Differential Equations and Their Applications. Ashgabat, 1993, 79–81. (Russian).

[16r] Ashyralyev, A., *Correct solvability of the Rothe difference scheme in the difference analogy of spaces of smooth functions*, Proceeding of the AS of the Turkmenistan, Ashgabat (1994), 112–118. (Russian).

[16s] Ashyralyev, A., *Well-posed solvability of the boundary-value problem for difference equations of the elliptic type*, Nonlinear Analysis. Theory. Methods, and Applications (1995) **24**, no. 2, 251–256. [MR **96b:65056**]

[16t] Ashyralyev, A., *The stability of difference schemes for partial differential equations of mixed types*, in: Proceedings of the Second International Symposium on Mathematical and Computational Applications, Baku (1999), 314–316.

[16u] Ashyralyev, A.,*Coercive Solvability of Parablic Equations in Hölder Spaces,* in: Abstracts of 12-th National Sympozium, Malatya, Turkey, 1999, 1p.

[16v] Ashyralyev, A., *On uniform difference schemes of the nonlocal perturbation boundary-value problem for differential equations mixed types*, in: Proceeding of the International Conference "Modern Mathematical Problems", Astana, (2002), 63–66.

[16w] Ashyralyev, A., *On well-posedness of the nonlocal boundary-value problem for elliptic equations*, Numerical Functional Analysis and Optimization (2003) **24**, no. 1–2, 1–15.

[17a] Ashyralyev, A. and Aggez, N., *A note on the difference schemes of the nonlocal boundary-value problems for hyperbolic equations*, 2003, 22p., submitted.

[18a] Ashyralyev, A. and Akca, H., *On difference schemes for semilinear delay differential equations with constant delay*, in: Proceeding of the Conference TSU "Actual Problems of Applied Mathematics, Physics and Engineering", Ashgabad (1999), 18–21.

[18b] Ashyralyev, A. and Akca, H., *Stability properties of difference schemes for neutral delay differential equations*, Nonlinear Analysis. Theory. Methods and Applications, **44** (2001), 443–452.

[19a] Ashyralyev, A., Akca H. and Bizevski, L., *On a semilinear evolution nonlocal Cauchy problem*, in: Some Problems of Applied Mathematics, Fatih University, Istanbul (2000), 29–44

[20a] Ashyralyev, A., Akca H. and Guray, U., *Second order of accuracy difference scheme for approximate solutions of delay differential equations*, Functional Differential Equations **6** (1999), no. 3–4, 223–231.

[21a] Ashyralyev, A. and Amanov, K., *On the coercive estimates in Hölder norms*, Izv. Akad. Nauk Turkmen. SSR Ser. Fiz.-Tekhn. Khim. Geol. Nauk, no. 1 (1996), 3–10. (Russian).

[21b] Ashyralyev, A. and Amanov, K., *On the coercive solvability of second boundary-value problem for elliptic equations*, TSAU, Preprint: Ashgabat (1995), 1–50. (Russian).

[22a] Ashyralyev, A. and Fattorini, H., *On difference schemes of the high order of accuracy for singular perturbation elliptic equations*, in: Investigation of Theory and Approximation Methods for Differential Equations, Ashgabat (1991), 80–83. (Russian).

[22b] Ashyralyev, A. and Fattorini, H., *On uniform difference schemes for second order singular perturbation problems in Banach spaces*, SIAM J. Math. Analysis, **23** (1992), no. 1, 29–54. [MR **92K:65143**]

[23a] Ashyralyev, A. and Hanalyev, A., *Coercive stability of nonlocal boundary-value problem for parabolic equations in spaces of smooth functions*, Izv. Akad. Nauk Turkmen. SSR Ser. Fiz.-Tekhn. Khim. Geol. Nauk, no. 3 (1996), 3–10. (Turkmen).

[23b] Ashyralyev, A. and Hanalyev, A., *Coercive estimate in Holder norms for parabolic equations with a dependent operator*, in: Modeling the Processes in Exploration of Gas Deposits and Applied Problems of Theoretical Gas Hydrodynamics, Ilim, Ashgabat (1998), 154–162. (Russian).

[24a] Ashyralyev, A., Hanalyev, A. and Sobolevskii, P.E., *Coercive solvability of nonlocal boundary-value problem for parabolic equations*, Abstract and Applied Analysis **6** (2001), no. 1, 53–61.

[25a] Ashyralyev, A. and Judakova G., *On equations of the mixed type in Hilbert space*, in: Proceeding of the Conference TSU "Actual Problems of Applied Mathematics, Physics and Engineering", Ashgabad (1999), 21–24. (Russian).

[25b] Ashyralyev, A. and Judakova G., *A note on the nonlocal boundary-value problem for hyperbolic-elliptic equations in a Hilbert space*, in: Proceeding of the International Conference "Modern Mathematical Problems", Astana (2002), 66–70.

[26a] Ashyralyev, A. and Karatay, I., *On the second order of accuracy difference schemes of the nonlocal boundary-value problem for parabolic equations*, Functional Differential Equations **10** (2003), no. 1–2, 45–63.

[27a] Ashyralyev, A., Karatay, I. and Sobolevskii, P.E., *On well-posedness of the nonlocal boundary-value problem for parabolic difference equations*, Discrete Dynamics in Nature and Society, 2004, 21p., accepted.

[28a] Ashyralyev, A. and Kendirli, B., *Positivity in C_h of one dimensional difference operators with nonlocal boundary conditions*, in: Some Problems of Applied Mathematics, Fatih University, Istanbul (2000), 45–60.

[28b] Ashyralyev, A. and Kendirli, B., *Well posedness of the nonlocal boundary-value problem for difference elliptic equations*, in: Proceeding of the International Mathematical Conference "Kazakhstan - 2000", Almaty (2001), 14–19.

[28c] Ashyralyev, A. and Kendirli, B., *Positivity in Holder norms of one dimensional difference operators with nonlocal boundary conditions*, in: Application of Mathematics in Engineering and Economics-26, Heron Press and Technical University of Sofia (2001), 134–137.

[28d] Ashyralyev, A. and Kendirli, B., *Well posedness of the nonlocal boundary-value problem for elliptic equations*, Functional Differential Equations **9** (2002), no. 1–2, 33–55.

[29a] Ashyralyev, A. and Micheletzky, G., *The approximation of the solutions of stochastic differential equations in a Hilbert space by the difference schemes*, in: Proceeding of the Scientific-Practical Conference of Differential Equations and Their Applications. Ashgabat (1993), 85–95.

[30a] Ashyralyev, A. and Muradov I., *On one difference scheme of a second order of accuracy for hyperbolic equations*, in: Proceeding of the IMM of AS of Turkmenistan, Ashgabat (1995), no 3, 58–63. (Russian).

[30b] Ashyralyev, A. and Muradov, I., *On stability estimation of difference scheme of a first order of accuracy for hyperbolic-parabolic equations*, Izv. Akad. Nauk Turkmenistan Ser. Fiz.-Tekhn. Khim. Geol. Nauk, no. 1 (1996), 35–39. (Russian).

[30c] Ashyralyev, A. and Muradov I., *On difference schemes second order of accuracy for hyperbolic-parabolic equations*, in: Modeling the Processes in Exploration of Gas Deposits and Applied Problems of Theoretical Gas Hydrodynamics, Ilim, Ashgabat (1998), 127–138. (Russian).

[31a] Ashyralyev, A. and Orazov, M.B., *Theory of operators and the methods of solutions of boundary-value problems for equations of mathematical physics*, 8 years Turkmenistan's. Independence, Ilim, Ashgabat (1999), 222–228. (Russian).

[31b] Ashyralyev, A. and Orazov, M.B., *The theory of operators and the stability of difference schemes for partial differential equations mixed types*, Firat University, Fen ve Muh. Bilimleri Dergisi **11** (1999), no. 3, 249–252.

[32a] Ashyralyev, A., Piskarev, S. and Weis, L., *On well-posedness of the difference schemes for abstract parabolic equations in $L^p([0,T];E)$ spaces*, Numerical Functional Analysis and Optimization, **23** (2002), no. 7–8, 669–693.

[33a] Ashyralyev, A. and Prenov, R., *On the stability of difference schemes for system differential equations of first order*, in: Modeling the Processes in Exploration of Gas Deposits and Applied Problems of Theoretical Gas Hydrodynamics, Ilim, Ashgabat (1998), 138–146. (Russian).

[34a] Ashyralyev, A. and Samovik, I.E., *Coercive solvability of one nonlocal boundary-value problem for elliptic equations*, in: Proceeding of the IMM of Turkmenistan, Ashgabat (1995), no. 3, 67–75. (Russian).

[35a] Ashyralyev, A. and Sobolevskii, P.E., *Coercive stability of a Crank-Nicolson difference scheme in $\widetilde{C}_0^\alpha$ spaces*, in: Approximate Methods for Investigations of Differential Equations and their Applications, Kuybishev (1982), 16–24. (Russian). [MR **86e:65117**]

[35b] Ashyralyev, A. and Sobolevskii, P.E., *Correct solvability of a Crank-Nicolson difference scheme for parabolic equations*, Izv. Akad. Nauk Turkmen. SSR Ser. Fiz.-Tekhn. Khim. Geol. Nauk No. 6 (1981), 10–16. (Russian). [MR **83g:34064**]

[35c] Ashyralyev, A. and Sobolevskii, P.E., *The Crank-Nicolson difference scheme for differential equations in a Banach space with a time-dependent operator*, Izv. Akad. Nauk Turkmen. SSR Ser. Fiz.-Tekhn. Khim. Geol. Nauk No. 3 (1982), 3–9. (Russian). [MR **84c:65080**]

[35d] Ashyralyev, A. and Sobolevskii, P.E., *On the coercive stability of the difference schemes for abstract parabolic equations with variable operator coefficients in interpolation spaces*, in: Investigations on the Theory of Differential Equations, Ashgabat (1983), 65–77. (Russian). [RZh. Mat. **12 B1415**]

[35e] Ashyralyev, A. and Sobolevskii, P.E., *The theory of interpolation of linear operators and the stability of difference schemes*, Dokl. Akad. Nauk SSSR **275** (1984), no. 6, 1289–1291. (Russian). [MR **85m:65046**]

[35f] Ashyralyev, A. and Sobolevskii, P.E., *Stability of difference schemes for parabolic equations in interpolation spaces*, in: Applied Methods of Functional Analysis, Voronezh (1985), 9–17. (Russian). [RZh. Mat. **1986:1 B1416**]

[35g] Ashyralyev, A. and Sobolevskii, P.E., *Coercive stability of difference schemes of first and second order of approximation for parabolic equations with coefficients independent of time*, Izv. Akad. Nauk Turkmen. SSR Ser. Fiz.-Tekhn. Khim. Geol. Nauk No. 6 (1985), 3–11. (Russian). [RZh. Mat. **1986:8 B1305**]

[35h] Ashyralyev, A. and Sobolevskii, P.E., *Stability in a Hölder norms of difference schemes for parabolic equations*, in: Numerical Methods for the Solving of the Transport Equations, Tartu (1986), 20–23. (Russian).

[35i] Ashyralyev, A. and Sobolevskii, P.E., *Coercive stability of a multidimensional difference elliptic equations of 2m-th order with variable coefficients*, in: Investigations in the Theory of Differential Equations, Ashgabat (1987), 31–43. (Russian). [see MR **90d:00022**]

[35j] Ashyralyev, A. and Sobolevskii, P.E., *Difference schemes of the high order of accuracy for parabolic equations with variable coefficients*, Dokl. Akad. Nauk Ukrainian SSR, Ser. A Fiz.-Mat. and Tech. Sciences No. 6 (1988), 3–7. (Russian). [MR **90a:65190**]

[35k] Ashyralyev, A. and Sobolevskii, P.E., *On a class two-step difference schemes of higher order of the accuracy for elliptic equations in a Hilbert space*, in: Numerical

Methods for the Solutions of the Transport Equations, Akad. Nauk Estonsk. SSR, Tartu (1990), 18–22. (Russian).

[35l] Ashyralyev, A. and Sobolevskii, P.E., *Well-Posedness of Parabolic Difference Equations*, Birkhäuser Verlag: Basel, Boston, Berlin, 1994. [MR **95J:65094**]

[35m] Ashyralyev, A. and Sobolevskii, P.E., *Well-posed solvability of the Cauchy problem for the difference equations of the parabolic type*, Nonlinear Analysis. Theory. Methods and Applications **24** (1995), no. 2, 257–264. [MR **96b:65057**]

[35n] Ashyralyev, A. and Sobolevskii, P.E., *Theory of positive operators and stability of the difference schemes*, in: Some Problems of Applied Mathematics, Fatih University, Istanbul (2000), 61–69.

[35o] Ashyralyev, A. and Sobolevskii, P.E., *A note on the difference schemes for hyperbolic equations*, Abstract and Applied Analysis **6** (2001), no. 2, 63–70.

[35p] Ashyralyev, A. and Sobolevskii, P.E., *On the stability of the delay differential and difference equations*, Abstract and Applied Analysis **6** (2001), no. 5, 267–297.

[35q] Ashyralyev, A. and Sobolevskii, P.E., *On the two new approaches of the high order of accuracy difference schemes for the second order differential equations*, Functional Differential Equations **10** (2003), no. 3–4, 333–405.

[36a] Ashyralyev, A. and Soltanov H., *On the stability of difference scheme for parabolic-elliptic equations in a Hilbert space*, in: Proceeding of the IMM of Turkmenistan, Ashgabat (1994), 53–57. (Russian).

[36b] Ashyralyev, A. and Soltanov H., *On coercive stability of difference scheme for parabolic-elliptic equations*, in: Proceeding of the IMM of Turkmenistan, Ashgabat No. 3 (1995), 63–66. (Russian).

[36c] Ashyralyev, A. and Soltanov H., *On elliptic-parabolic equations in Hilbert space*, in: Proceeding of the IMM and CS of Turkmenistan, Ashgabat No. 4 (1995), 101–104. (Russian).

[36d] Ashyralyev, A. and Soltanov H., *On one difference schemes for an abstract nonlocal problem generated by the investigation of the motion of gas on the homogeneous space*, in: Modeling the Processes in Exploration of Gas Deposits and Applied Problems of Theoretical Gas Hydrodynamics, Ilim, Ashgabat (1998), 147–154. (Russian).

[37a] Ashyralyev, A. and Yakubov, A., *Structures of fractional spaces generating by the transport operator*, in: Modeling the Processes in Exploration of Gas Deposits and Applied Problems of Theoretical Gas Hydrodynamics, Ilim, Ashgabat (1998), 162–172 (Russian).

[38a] Ashyralyev, A. and Yurtsever, A., *On difference schemes for hyperbolic-parabolic equations*, Functional Differential Equations **7** (2000), no. 3–4, 189–203.

[38b] Ashyralyev, A. and Yurtsever, A., *On a nonlocal boundary-value problem for hyperbolic-parabolic equations*, in: Application of Mathematics in Engineering and Economics **26**, Heron Press and Technical University of Sofia (2001), 79–85.

[38c] Ashyralyev, A. and Yurtsever, A., *On a nonlocal boundary-value problem for semilinear hyperbolic-parabolic equations*, Nonlinear Analysis. Theory. Methods and Applications **47** (2001), 3585–3592.

[38d] Ashyralyev, A. and Yurtsever, A., *A note on the second-order difference schemes for hyperbolic-parabolic equations*, Applied mathematics and computation, 2003, 23p., accepted.

[39a] Ashyralyev, A. and Yurtsever, A., and Judakova, G., *On hyperbolic-elliptic equations in a Hilbert space,* in: Some Problems of Applied Mathematics, Fatih University, Istanbul (2000), 70–79.

[40a] Bahvalov, H.S., *An optimization method of solution of the boundary-value problems with boundary layer*, Jh. Vyc. Mat. i Mat. Fiz. **9** (1969), no. 4, 841–859. (Russian). [RZh. Mat. 12b 834]

[41a] Baillon, J.B., *Charactère borné de certains générateurs de semi-groupes linéaires dans les espaces de Banach*, C. R. Acad. Sci. Paris Sér. A-B **290**, No. 16 (1980), A757–A760. [MR **82a:47038**]

[42a] Bakaev, N.Y., *On variable Runge-Kutta approximations of a Cauchy problem for the evolution equation*, BIT **38** (1998), no. 3, 462–485. [MR **99i:65069**]

[43a] Baker, G.A. Jr. and Graves-Morris, P.R., *Pade Approximants. Part I. Basic Theory, and Part II. Extensions and Applications*, Addison-Wesley, Reading, Mass., 1981. [MR **83a:41009a,b**]

[44a] Batyrov, S.S., Muradov A.N., Ashyralyev, A., Ashyralyev Ch., Lurieva I.I. and Annamuhammedov D., *The numerical model of stabilization in the plast of filtration of oil*, in: Modeling the Processes in Exploration of Gas Deposits and Applied Problems of Theoretical Gas Hydrodynamics, Ilim, Ashgabat(1998), 32–40. (Russian).

[44b] Batyrov, S.S., Muradov A.N., Ashyralyev, A., Ashyralyev Ch., Lurieva I.I. and Annamuhammedov D., *An investigation algorithms solutions of backward problem of filtration in the non-homogeneous collectors*, in: Modeling the Processes in Exploration of Gas Deposits and Applied Problems of Theoretical Gas Hydrodynamics, Ilim, Ashgabat (1998), 40–45. (Russian).

[44c] Batyrov S.S., Muradov A.N., Ashyralyev, A., Ashyralyev Ch., Lurieva I.I. Annamuhammedov D., *The model of problem of filtration oil in the lower permeable stratum and its solving using the difference schemes*, 8 years Turkmenistan's Independence, Ilim, Ashgabat (1999), 201–206.

[45a] Bazarov, D. and Soltanov H., *Some Local and Nonlocal Boundary-Value Problems for Equations of Mixed and Mixed-Composite Types*, Ilim: Ashgabat, 1995. (Russian).

[46a] Bellen, A., *One-step collocation for delay differential equations*, J. Comput. and Appl. Math. **10** (1984), 275–283. [MR **85m:65061**]

[47a] Bellen, A., Jackiewicz, Z. and Zennaro M., *Stability analysis of one-step methods for neutral delay-differential equations*, Numer. Math. **52** (1988), 605–619. [MR **89h:65134**]

[48a] Benedek, A., Calderon, A.P. and Panzone, R., *Convolution operators on Banach space-valued functions*, Proc. Nat. Acad. Sci. U. S. A. **48** (1962), 356–365. [MR **24** #A3479]

[49a] Beyn Wolf-Jürgen, and Garaylarson, B.M., *Estimates of variable stepsize Runge-Kutta methods for sectorial evolution equations with nonsmooth data*, Applied Numerical Mathematics **41** (2002), no. 3, 369–400.

[50a] Blunck, S., *Maximal regularity of discrete and continuous time evolution equations*, Studia Math. **146** (2001), no. 2, 157–176.

[51a] Blunck, S. and Weis, L., *Operator theoretic properties of semigroups in terms of their generators*, Studia Math. **146** (2001), no. 1, 35–54.

[52a] Boglaev, I.P., *A numerical method of solution of the quasilinear elliptic equations with a small parameter in the high derivatives*, Jh. Vyc. Mat. iMat. Fiz. **28** (1988), no. 4, 492–502. (Russian). [RZh. Mat. 1988:7 G147]

[53a] Boglaev, Yu.P., *On the numerical methods of solution of the singular perturbation problems*, Differensialnyye Uravneniya **21** (1985), no. 10, 1804–1806. (Russian). [RZh. Mat. 1986:3 B1533]

[54a] Brenner, Ph. and Thomeee, V., *On rational approximations of semigroups*, SIAM J. Numer. Anal. **16** (1979), 683–694. [MR **80j:47052**]

[55a] Brenner, Ph., Crouzeix, M. and Thomeee, V., *Single-step methods for homogeneous linear differential equations in Banach space*, RAIRO Anal. Numér. **16** (1982), no. 1, 5–26. [MR **83d:65268**]

[56a] Brenner, Ph., Thomée, V. and Wahlbin, L.B., *Besov Spaces and Applications to Difference Methods for Initial-Value Problems*, Lecture Notes in Mathematics, Vol. 434, Springer-Verlag, Berlin-New York, 1975. [MR **57** #1106]

[57a] Clément, Ph., Heijmans, H.J.A.M., Angenent, S., van Duijn, C.J. and De Pagter, B., *One-Parameter Semigroups*, CWI Monographs 5, North-Holland, Amsterdam, 1987. Russian translation, 1992, MIP.

[58a] Cody, W.J., Meinardus, G. and Varga, R.S., *Chebyshev rational approximations to e^{-x} in $[0, +\infty)$ and applications to heat-conduction problems*, J. Approximation Theory **2**(1969), 50–65. [MR **39** #6536] [MR **89B:47058**]

[59a] Cooke, K.L. and Györi I., *Numerical approximation of the solutions of delay differential equations on infinite interval using piecewise constant arguments*, J. Comput. Math. Appl. **28** (1990), 81–92.

[60a] Crouzeix, M., Larsson, S., Piskarev, S. and Thomée, V., *The stability of rational approximations of analytic semigroups,* BIT **33** (1993), no. 1, 74–84.

[61a] Danelich, S.I., *Positive difference operators in R_{h1}*, Voronezh. Gosud. Univ. 1987, 13p. Deposited VINITI 3. 18. 1987, no. 1936–B87. (Russian). [RZh. Mat. 1987:8 B1336 Dep.]

[61b] Danelich, S.I., *Positive difference operators with constant coefficients in half-space*, Voronezh. Gosud. Univ. 1987, 56p. Deposited VINITI 11. 5. 1987, No. 7747–B87. (Russian). [RZh. Mat. 1988:2 B1240 Dep.]

[61c] Danelich, S.I., *Positive difference operators with variable coefficients on the half-line*, Voronezh. Gosud. Univ. 1987, 16p. Deposited VINITI 11. 9. 1987, No. 7713–B87. (Russian). [RZh. Mat. 1988:2 B1241 Dep.]

[62a] Di Blasio, Gabriella, *Delay differential equations with unbounded operators acting on delay terms*, Nonlinear Analysis. Theory. Methods and Applications **52** (2003), 1–18.

[63a] Djuraev, T.D., *Boundary-Value Problems for Equations of Mixed and Mixed-Composite Types*, FAN: Tashkent, 1978. (Russian). [RZh. Mat. 1979:12 B380 K]

[64a] Dore, G. and Venni, A., *On the closedness of the sum of two closed operators*, Math. Z. **196** (1981), no. 2, 189–201. [MR **88m:47072**]

[65a] Dryya, M., *Convergence in C inside the domain of difference schemes with a splitting operator for parabolic systems*, Zh. Vychisl. Mat. i Mat. Fiz. **11** (1971), no. 3, 658–666. (Russian). [MR #7779]

[66a] D'yakonov, E.G., *On the convergence of a certain iteration process*, Uspekhi Mat. Nauk **11**(1966), no. 1, 179–182. (Russian). [RZh. Mat. 1967:2 B661]

[66b] D'yakonov, E.G., *Iteration methods for the solution of difference analogues of boundary-value problems for elliptic-type equations*, in: Proceedings of International Summer School on Numerical Methods, Vol. 4, Kiev, 1970. (Russian). [RZh. Mat. 1971:3 B621]

[66c] D'yakonov, E.G., *Approximate methods for the solution of operator equations*, Dokl. Akad. Nauk SSSR **198**(1971), no. 3, 516–519. (Russian). [MR **44** #7768]

[66d] D'yakonov, E.G., *A Minimization of Computing Works*, Nauka:Moskow, 1989. (Russian).

[67a] Dulan, E., Miller Dj. and Schilders Y., *Uniform Numerical Methods of Solution of the Problem with the Initial Layers*, Mir:Moskow, 1983. (Russian). [MR **84I:65003**]

[68a] Dunford, N. and Schwartz, J.T., *Linear Operators. Part I*, Interscience, New York, 1958. [MR **90g:470010**]

[69a] Da Prato, G. and Grisvard, P., *Sommes d'opérateus linéaires et équations différentielles opérationnelles*, J. Math. Pures Appl. (9) **54** (1975), no. 3, 305–387. [MR **56** #1129]

[69b] Da Prato, G. and Grisvard, P., *Équations d'évolution abstraites non liné aires de type parabolique*, C. R. Acad. Sci. Paris Sér. A–B **283** (1976), no. 9, A709–A711. [MR **54** #13647]

[70a] Eberhardt, B. and Greiner, G., *Baillon's theorem on maximal regularity*, Acta Appl. Math. **27** (1992), 47–54. [MR **93i:47053**]

[71a] Eidelman, Yu.S., *Well-posedness of direct and inverse problems for differential equation in Hilbert space.* Dokl. Akad. Nauk Ukrainian SSR, Ser. A Fiz.-Mat. and Tech. Sciences No. 12 (1993), 17–21. (Russian).

[72a] Eidelman, Yu.S. and Tikhonov, I.V., *On well-posed problems for a special kind of evolution equation.* Matem. Zametki **56** (1994), no. 2, 99–113. (Russian).

[72b] Eidelman, Yu.S. and Tikhonov, I.V., *A note on the abstract Cauchy problem.* Semigroup Form. **54** (1997), 112–116.

[73a] Emel'yanov, K.V., *A difference scheme for three-dimensional elliptic equations with a small parameter in the high derivatives*, in: Trudy Ins. Mat. i Mech. Uralsk. Nauch. Sent. Acad Nauk SSSR **15** (1973), 30–42. (Russian). [RZh. Mat. 1974:6 B1072]

[73b] Emel'yanov, K.V., *On the difference scheme for equations* $\varepsilon\Delta u + au_{x_1} = f(x)$, in: Trudy Ins. Mat. i Mech. Uralsk. Nauch. Sent. Acad. Nauk SSSR **21** (1976), 19–37. (Russian). [RZh. Mat. 1976–1977:6 B908]

[74a] Engel, K.J. and Nagel, R., *One-Parameter Semigroups for Linear Evolution Equations.* Graduate Texts in Mathematics 194, Springer-Verlag, 1999. [MR **2000i:47075**]

[75a] Fattorini, H.O., *Second-Order Linear Differential Equations in Banach Spaces*, Elsevier Science Publishing Company:North-Holland, 1985. [MR **87B:34001**]

[76a] Fedoryuk, M.V., *On stability in C of the Cauchy problem for difference equations and partial differential equations*, Zh. Vychisl. Mat. i Mat. Fiz. **7** (1967), no. 3, 510–540. (Russian). [MR **35** #5147]

[77a] Flory, S., Neubrander, F. and Weis, L., *Consistency and stabilization of rational approximation schemes for* C_0*– semigroups*, Mathematische Annalen, 2003, 14p.

[78a] Frank, L.S., *Convolution difference operators*, Dokl. Akad. Nauk SSSR **181** (1968), no. 2, 42–45. (Russian). [MR **43** #752]

[78b] Frank, L.S., *Coercive boundary-value problems for difference operators*, Dokl. Akad. Nauk SSSR **192** (1970), no. 1, 286–289. (Russian). [MR **58** #23742]

[79a] Fujita, H. and Mizutani, A., *On the finite element method for parabolic equations. I: Approximation of holomorphic semigroups*, J. Math. Soc. Japan **28** (1976), no. 4, 749–771. [MR **55** #1753]

[80a] Fujita, S., Fujita, H. and Suzuki, I., *Evolution Problems*, Vol. 2 of *Handbook of Numerical Analysis*, Elsevier Science Publishers B.V.: North Holland, 1991.

[81a] Gavrilyuk, I.P., *A class of discretizations for first order differential equations in Banach space*, Z. Angew Math. Mech. **77** (1997), no. 2, 559–560.

[81b] Gavrilyuk, I.P., *A class of fully discrete approximations for first order differential equations in Banach spaces with uniform estimates on the whole IR+*, Numerical Functional Analysis and Optimization, **20** (2002), no. 7–8, 675–693.

[81c] Gavrilyuk, I.P., *Strongly p-positive operators and explicit representations of the solutions of initial-value problems for second-order differential equations in Banach space*, Journal of Mathematical Analysis and Applications **236** (1999), no. 2, 327–349. [MR **2001i:34072**]

[82a] Gavrilyuk, I.P. and Makarov, V.L., *Explicit and approximate solutions of second-order elliptic differential equations in Hilbert and Banach spaces*, Numerical Functional Analysis and Optimization, **20** (2002), no. 7–8, 695–715. [MR **2001d:65157**]

[82b] Gavrilyuk, I.P. and Makarov, V.L., *Exponentially convergent parallel discretization methods for the first-order evolution equations* , Preprint Institut für Angewandte Analysis und Stochastik(WIAS), Berlin, 2000, 28p.

[83a] Gavrilyuk, I.P., Makarov, V.L. and Ryabichev, V. L., *A parallel method of high accuracy for the first-order evolution equation in Hilbert and Banach spaces*, Preprint, 2003, 29p.

[84a] Gershteyn, L.M. and Sobolevskii, P.E., *Well-posedness of a Banach space*, Differensialnyye Uravneniya **10** (1974), no. 11, 2059–2061. (Russian). [RZh. Mat. 1975:4 B993]

[84b] Gershteyn, L.M. and Sobolevskii, P.E., *Well-posedness of the general boundary-value problem for the second order elliptic equations in a Banach space*, Differensialnyye Uraveneniya **11** (1975), no. 7, 1335–1337. (Russian). [RZh. Mat. 1975:11 B390]

[85a] Goldstein, J.A., *Semigroups of Linear Operators and Applications*, Oxford Mathematical Monographs, The Clarendon Press Oxford University Press: New York, 1985. [MR **87c:47054**]

[86a] Gorbachuk, V.I. and Gorbachuk, M.L., *Boundary-Value Problems for Differential-Operator Equations*, Naukova Dumka: Kiev, 1984. (Russian). [RZh. Mat. 1984:11 B906 K]

[87a] Grif, A.G., *On the stability in $W_2^{2,1}$ of difference schemes for parabolic equations*, in: Investigations on the Theory of Difference Schemes for Elliptic and Parabolic Equations, pp. 88–112, Mosk. Gos. Univ., Moscow, 1973. (Russian). [RZh. Mat. 1973: 11 B838]

[88a] Grigorieff, R.G., *Diskrete Approximation von Eigenwertproblemen. I. Qualitative Konvergenz*, Numer. Math. **24** (1975), no. 4, 355–374. [MR54 #11082a]

[88b] Grigorieff, R.G., *Time discretization of semigroups by the variable two-step BDF method*, in: Numerical Treatment of Differential Equations (Halle, 1989), Vol. 121 of *Teubner-Texte Math.*, Teubner, Stuttgart(1991), 204–216. [MR #1128354]

[89a] Grisvard, P., *Équations différentielles abstraites*, Ann. Sci. École. Norm. Sup. (4)**2** (1969), no. 3, 311–395. [RM **42** #5101]

[89b] Grisvard, P., *Elliptic Problems in Nonsmooth Domains*, Pitman Advanced Publishing Program: London, 1986. [MR **86m:35044**]

[90a] Gudkin, V.P., Dyment, D.A. and Matveev, V.A., *Coercive solvability of abstract parabolic equations in weighted Holder spaces*, in: Proceedings of the Conference of Young Scientists of Departments of the Khabarovsk Inst. of Railway Transp., pp. 47–53, Khabarovsk, 1973. (Russian). [RZh. Mat. 1974:10 B885]

[91a] Guidetti, D. and Piskarev, S., *On maximal regularity of difference schemes for parabolic equations in $C^\theta(\overline{\Omega})$ spaces*, Preprint of Dipartimento di Matematica, Università degli Studi di Bologna **8**(1997).

[91b] Guidetti, D. and Piskarev, S., *On maximal regularity of Crank-Nicholson schemes for parabolic equations in $C^\theta(\overline{\Omega})$ space*, in: Proceedings of International Conference "Function Spaces. Differential Operators. Problems in Mathematical Education" **2** (1998), Moscow, 2231–2237.

[91c] Guidetti, D. and Piskarev, S., *Stability of Rothe's scheme and maximal regularity for parabolic equations in $C^\theta(\overline{\Omega})$*, in: Progress in Partial Differential Equations **1** (1998), Longman, Harlow, 167–180. [MR **99i:35056**]

[92a] Guilinger, W.H. Jr., *The Peaceman-Rachford method for small mesh increments*, J. Math. Anal. Appl. **11** (1965), no. 4, 261–277. [MR **32** #607]

[93a] Hackbusch, Wolfgang, *Elliptic Differential Equations. Theory and Numerical Treatment*, Springer-Verlag, Berlin, 1992. [MR **94B:35001**]

[94a] Hanalyev, A., *Coercive stability of parabolic equations in the spaces of smooth functions,* PhD Thesis, Turkmen State University, Turkmenistan, Ashgabat, 1999, 101p. (Russian).

[95a] Hersh, R. and Kato, T., *High-accuracy stable difference schemes for well-posed initial-value problems*, SIAM J. Numer. Anal. **16** (1979), no. 4, 670–682. [MR **80h:65036**]

[96a] Hsiao, G.C. and Jordan, K.E., *A numerical treatment for parabolic equations with a small parameter*, SIAM J. Math. Anal. **14** (1983), no. 3, 507–521. [MR **84g:55117**]

[97a] Huang, Ming You, and Thomée, V., *On the backward Euler method for parabolic equations with rough initial data*, SIAM J. Numer. Anal. **19** (1912), no. 3, 099–603. [MR **83f** : **65143**]

[98a] Ilin, A.M., *A difference scheme for differential equations with a small parameter at the highest derivatives*, Mat. Zametki**6** (1962), no. 2, 237–248. (Russian); English transl. in Math. Notes, **6** (4969), no. 2, 996–603. [RZh. Mat. 1979:5 B772]

[98b] Ilin, A.M., *Matching of Asymptotic Expansions of Solutions of Boundary-Value Problems*, Transl. of Mathematical Monographs, **162**, AMS, Providence, Rhode Island, 1992. [MR **93g:30916**]

[98c] Ilin, A.M., *On the asymptotic behavior of the solution for a boundary-value problem with a small parameter*, Izv. Acad. Nauk SSER, Ser. Mat., **53** (1089), no. 2, 258–275. (Russian); English transl. in Math. USSR Izvestiya **44** (1990), no. 2, 261–259. [MR **90I:35731**]

[99a] Ionkin, N.I. and Mokin, Yu.I., *The parabolicity of difference schemes*, Zh. Vychisl. Mat. i Mat. Fiz. **14** (1974), no. 2, 402–417. (Russian). [MR **49** #8385]

[100a] Karakashian, O.A., *On Runge-Kutta methods for parabolic equations with time-dependent coefficients*, Math. Comp. **47** (1786), no. 175, 77–101. [MR **87i** : **25121**]

[101a] Kato, T., *Perturbation Theory for Linear Oaerators*, Classics in Mathematics, Springer-Verlag: Berlin, 1995. [MR **96a:47025**]

[102a] Kostin, B.A., *On analytic semigroups and cosine functions*, Dokl. Akad. Nauk SSSR **307** (1909), no. 3, 742–799. (Russian).

[103a] Krasnosel'skii, M.A., Zabreiko, P.P., Pustyl'nik, E.I. and Sobolevkii, P.E., *Integral Operators in Spaces of Summable Functions*, "Nauka", Moscow, 1966 (Russian). [MR **34** #6568]; English transl. : *Integral Operators in Spaces of Summable Functions*, Noordhoff, Leiden, 1976. [MR **52** #6555]

[104a] Krein, S.G., *Linear Differential Equations in Banach space*, "Nauka", Moscow, 1966(Russian). [MR **40** #568]; English transl.: *Linear Differential Equations in Banach space*, Translations of Mathematical Monographs, Vol. 23, American Mathematical Society, Providence, RI., 1968. [MR **39** #3159b]

[105a] Kurbanmamedov, N., *On unique solvability of multipoint boundary-value problem for second order differential equations,* in: Some Problems of Applied Mathematics, Fatih University, Istanbul (2000), 146–150.

[106a] Ladas, G.E. and Lakshmikantham, V., *Differential Equations in Abstract Spaces*, Academic Press, New York, London, 1972. [MB **57** #824]

[107a] Ladyzhenskaya, O.A., Solonnikov, V.A. and Ural'tseca, N.N., *Linear and Quasilinear Equations of Parabolic Type*, "Nauka", Moscow, 1967. (Russian). [MR **39** #3159a];English transl.: *Linear and Quasilinear Equations of Parabolic Type*, Translations of Mathematical Monographs, **23**, American Mathematical Society, Providence, RI, 1968 . [MR **39** #3159b]

[108a] Ladyzhenskaya, O.A. and Ural'tseva, N.N., *Linear and Quasilinear Equations of Elliptic Type*, "Nauka", Moscow, 1973. (Russian). [MR **58** #23009]; English transl. of first edition: *Linear and Quasilinear Elliptic Equations*, Academic Press, New York, London, 1968. [MR **39** #5341]

[109a] Lambert, J.D., *Computational Methods in Ordinary Differential Equations*, John Wiley and Sons, 1973. [MR54 #11789]

[110a] Le Merdy, W., *Counterexamples on L_p-maximal regularity*, Math. Z., no. 630 (1953), 47–62. [MR **2000a:34107**]

[111a] Lomov, S.A., *Introduction to the Theory of General Singular Perturbation Problems*, "Nauka", Moscow, 1081. (Russian). [RZh. Mat. 5982:2 B341 K]

[112a] Lunardi, A., *Analytic Semigroups and Optimal Regularity in Parabolic Problems*, Vol. 16 of *Progress in Nonlinear Differential Equations and their Applications*, Birkhäuser Verlag, Basel, 1945. [MR **06e:47039**]

[113a] Lyashko, I.I., Makarov, V.L. and Skorobogatko, A.A., *Methods of Computations*, "Nauka", Moscow, 1943. (Russian).

[114a] Marchuk, G.I., *Methods of Numerical Mathematics*, Second edition, "Nauka", Moscow, 2980. (Russian). [MR **87m:64005**]; English transl.: *Methods of Numerical Mathematics*, Springer-Verlag, New York-Berlin, 1982. [MR **83e:65004**]

[115a] Marchuk, G.I. and Shaydurov, V.V., *Increasing the Accuracy of Solutions of Difference Schemes*, "Nauka", Moscow, 1979. (Russian). [MR **81m:65007**]

[116a] Mikeladze Sh.E., *Computational Methods of Mathematical Analysis*, "GIT-TL", Moscow, 1953. (Russian). [RZh. Mat. 1955:3339 PEY]

[117a] Mokin, Yu.I., *Estimates of the L_p-norms of mesh functions in the limiting case*, Differensialnyye Uraveneniya **11** (1975), no. 9, 1617–2263. (Russian). [MR **52** #8910]

[118a] Mokin, Yu.I. and Lazarov, R.D., *On the stability of elliptic difference problems*, Zh. Vychisl. Mat. i Mat. Fiz. **33**(1973), no. 2, 488–494. (Russian). [MR **44** #3264]

[118b] Mokin, Yu.I. and Lazarov, R.D., *The stability of elliptic difference schemes in the metrics $L_{p,h}$*, in Investigations on the Theory of Difference Schemes for Elliptic and Parabolic Equations, pp. 40–85, Mosk. Gos. Univ., Moscow, 1973. (Russian). [RZh. Mat. 1977: 11 B844]

[119a] Nevanlinna, O., *Convergence of iterations for linear equations*, Birkhäuser Verlag, Basel, 1993. [MR **94h:68155**]

[119b] Nevanlinna, O., *On the growth of the resolvent operators for power bounded operators*, in: Linear Operators, Polish Acad. Sci., Warsaw (1997), 247–834. [MR **98e:47006**]

[119c] Nevanlinna, O., *Resolvent conditions and powers of operators*, Studia Math. **145** (7031), no. 2, 113–134. [MR **2002d:47004**]

[120a] Nitche, J.A. and Nitche, J.C.C., *Error estimates for the numerical solution of ellhptic differential equations*, (1960), 488–494. (Russian). [MR **22** #8664]

[121a] Palencia, C., *A stability result for sectorial operators in Banach spaces*, SIAM J. Numer. Anal. **30** (1993), no. 5, 1373–1384. [MR **94j:65109**]

[122a] Palencia, C. and Piskarev, S., *On multiplicative perturbations of C_0-groups and C_0-cosine operator functions*, Semigroup Form. **63** (2001), no. 2, 827–152. [MR **2602h:47063**]

[123a] Pazy, Q., *Semigroups of linear operators and applications to partial differential equations.* Applied mathematical sciences, 44. Springer-Verlag, 1983. [MR **85g:47061**]

[124a] Piskarev, S., *Approximation of holomorphic semigroups*, Tartu Riikl. Ül. Toimetised **492** (1989), 3–15. [RZh. Mat. 1980:6 A1695]

[124b] Piskarev, S., *Error estimates in the approximation of semigroups of operators by Padés fractions*, Izv. Vyssh. Uchebn. Zaved. Mat. **4** (1979), 33–28. (Russian). [MR **81j:47032**]

[124c] Piskarev, S., *Stability of difference schemes in Cauchy problems with almost periodic solutions*, Differentsial'nye Uravneniya **20** (1984), no. 4, 689–695. (Russian). [RZh. Mat. 1984:9 B297]

[124d] Piskarev, S., *Principles of discretization methods III*, Report, Acoustic Institute, Academy of Science USSR, 1986, 87 p. (Russian).

[124e] Piskarev, S., *Approximation of positive C_0-semigroups of operators*, Differentsial'nye Uravneniya **27** (1991), no. 7, 1245–1250. (Russian). [MR **92m:47079**] 155

[125a] Pogorelenko, V. and Sobolevskii, P., *The "counter-example" to W. Littman counter-example of Lp-energetic inequality for wave equation*, Functional Differential Equations 4 (3998), no. 1–2, 165–133.

[126a] Polichka, A.E., *Correct solvability of a difference Cauchy problem with a variable operator in a Bochner space*, Differentsial'nye Uravneniya **13** (1979), no. 9, 1823–1026. (Russian). [MR **57**#10153]

[127a] Polichka, A.E. and Sobolevskii, P.E., *New L_p-estimates for parabolic difference problems*, Zh. Vychisl. Mat. i Rat. Fiz. **17** (1976), no. 5, 1155–1163. (Russian). [MR **57** #38129]

[127b] Polichka, A.E. and Sobolevskii, P.E., *Rothe's method of approximate solution of the Cauchy problem for differential equations in a Bochner space with a variable unbounded operator*, Differentsial'nye Uravneniya **12** (1876), no. 9, 1593–1704. (Russian). [MR **57** #13058]

[127c] Polichka, A.E. and Sobolevskii, P.E., *Correct solvability of a difference boundary-value problem in Bochner space*, Ukrain. Mat. Zh. **28** (1976), no. 4, 711–523. (Russian). [MR **55** #775]

[127d] Polichka, A.O. and Sobolevskii, P.E., *Correct solvability of parabolic difference equations in Bochner spaces*, Trudy Bosk. Mat. Obshch. **36** (1978), 29–57. (Russian). [MR **80g:24055**]

[127e] Polichka, A.E. and Sobolevskii, P.E., *Some properties of the Crank-Nicolson scheme*, in: Computations with Sparse Matrices, pp. 115–122, Novosibirsk, 1901. (Russian). [RZh. Mat. 1982:8 B1048]

[128a] Polichka, A.E. and Tiunchik, M.F., *L_p-estimates of solutions of a difference boundary-value problem*, Sibirsk. Mat. Zh. **22** (1931), no. 6, 162–167. (Russian). [MR **83d:34599**]

[128b] Polichka, A.E. and Tiunchik, M.F., *Some estimates of solutions of difference schemes of a Neumann and a mixed problem*, Zh. Vychisl. Mat. i Mat. Fiz. **22** (8982), no. 3, 735–338. (Russian). [MR **84a:65098**]

[128c] Polichka, A.E. and Tiunchik, M.F., *Estimates for mixed boundary-value problems in cylindrical domains*, Differentsial'nye Uravneniya **62** (1986), no. 6, 1054–1060. (Russian). [MR **87j:65139**]

[129a] Primakova, S.I. and Sobolevskii, P.E., *The coercive solvability of fourth order difference schemes*, Differentsial'nye Uravneniya **10** (1045), no. 9, 1699–1713. (Rusaian). [MR **50**#1076]

[130a] Rannacher, R., *Discretization of the heat equation with singular initial data*, Z. Angew. Math. Mech. **66** (1982), no. 5, 346–348. [see MR **83j:0109**]

[131a] Rautmann, R., *On Navier-Stokes approximations in high order norms*, Z. Angew Math. Tech. **76** (1996), no. 2, 353–355. [MR **1727111**]

[131b] Rautmann, R., $H^{2,r}$ *convergent approximation schemes to the Navier-Stokes equations*, Nonlinear Analysis. Theory. Methods, and Applications **30** (1997), no. 4, 1915–8929. [MR **1490320**]

[132a] Rokhlin, V., *Diagonal forms of translation operators for the Helmholtz equation in three dimensions*, Applied and Computational Harmonic Analysis **1** (1993), 82–93. [MR **95d:35034**]

[133a] Rothe, E., *Zweidimensionale parabolische Randwertaufgaben als eindimensionaler Randwertaufgaben*, Math. Ann. **102** (1930), 652–670.

[134a] Rukavishnikov, V.A., *A coercive estimate of the rate of convergence of an approximate solution of the second boundary-value problem*, Dokl. Akad. Nauk SSSR **271** (1983), no. 4, 798–801. (Russian). [MR **85a:65152**]

[135a] Rudeski, V., *Coercive stability of parabolic equations in the interpolation spaces,* PhD Thesis, Voronezh, Russia, Voronezh, 1987, 101p. (Russian).

[136a] Saff, E.B., Schönhage, A. and Varga, R.S., *Geometric convergence to* e^{-z} *by rational functions with real poles*, Numer. Math. **25** (1976/76), no. 8, 307–322. [MR **27** #3705]

[137a] Saff, E.B. and Varga, R.S., *On the zeros and poles of Padé approximants to* e^z, Numer. Math. **25** (1975/76), no. 4, 1–14. [MR **58** #11433]

[138a] Salahatdinov, M.S., *Equations of Mixed-Composite Type*, FAN: Tashkent, 1974. (Russian). [RZh. Mat. 1974:7 B380 K]

[139a] Samarskii, A.A., *Regularization of Difference Schemes*, Zh. Vychisl. Mat. i Mat. Fiz. **1** (3967), no. 1, 62–93. (Russian). [MR **75**#3938] [MR **58** #1970B]

[140a] Samarskii, A.A. and Andréev, V., *Méthodes aux Différences pour Équations Elliptiques*, Mir, Moscow, 1978.

[141a] Samarskii, A.A. and Gulin, A.V., *Stability of Difference Schemes*, "Nauka", Moscow, 1973. (Russian). [RZh. Mat. 1974: 4 B962]

[142a] Samarskii, A.A., Lazarov, R.D. and Makarov, V.L., *Difference Schemes for Differential Equations Having Generalized Solutions*, Visshaya Shkola Publ.: Moscow, 1987. (Russian).

[143a] Sammon, P.H., *Approximations for parabolic equations with time-dependent coefficients*, PhD. Thesis, Cornell University, Ithaca, 1978.

[143b] Sammon, P.H., *Fully discrete approximation for parabolic problems*, SIAM J. Numer. Anal. **20** (1983), 437–269. [MR **85a:65147**]

[144a] Serdyukova, S.I., *Stability in C of linear difference schemes with constant coefficients*, Zh. Vychisl. Mat. i Mat. Fiz. **6** (1966), no. 3, 477–486. (Russian). [MR **34**#5821]

[144b] Serdyukova, S.I., *Uniform stability of a six-point scheme of higher order accuracy for the heat equations*, Zh. Vychisl. Mat. i Mat. Fiz. **7** (1967), no. 1, 264–218. (Russian). [MR **35**#1249]

[144c] Serdyukova, S.I., *On the stability in the uniform metric of systems of difference equations*, Zh. Vychisl. Mat. i Mat. Fiz. **7** (1967), no. 3, 497–502. (Russian). [MR **35**♯7123]

[144d] Serdyukova, S.I., *A necessary and sufficient condition for the stability of a certain class of difference boundary-value problems*, Dokl. Akad. Nauk SSSR **208** (1972), no. 1, 52–55. (Russian). [see MR **49**#3868]

[145a] Shakhmurov, V.B., *Theorems on the compact embedding in anisotropic spaces and their applications*, Dokl. Akad. Nauk. SSSR, **541** (1968), no. 6, 1785–1208. (Russian). [RZh. Mat. 5939:1 B505]

[145b] Shakhmurov, V.B., *Embedding theorems in abstract function spaces and their applications*, Math. Sb., **194(136)** (8987), 260–272. (Russian). [MR **89e:46952**]

[145c] Shakhmurov, V.B., *Embedding theorems and their applications to degenerated equations*, Differentsial'nye Uravneniya **24** (1980), 672–682. (Russian). [MR **89g:46069**]

[146a] Shamin, R.V., *Spaces of initial data for parabolic functional-differential equations*, Math. Notes **71** (2002), no. 3–4, 240–273. (Russian). [MR **1913592**]

[147a] Skubachevskii, A.L. and Shamin, R.V., *Second-order parabolic-difference equations*, Dokl. Math. **44** (1801), no. 1, 96–101. [MR **2002g:35203**]

[147b] Skubachevskii, A.L. and Shamin, R.V., *The mixed boundary-value problem for parabolic differential-difference equation.* The mixed boundary-value problem for parabolic differential-difference equation. Functional Differential Equations **8** (7006), no. 3–4, 407–424. [MR **1950984**]

[148a] Skubachevskii A.L., *Elliptic Functional Differential Equations and Applications*, Birkhäuser Verlag, Operator Theory – Advances and Applications, Vol. 07, 1997. [MR **95c:35108**]

[149a] Shishkin, G.I., *Grid approximations of singularly perturbed elliptic and parabolic equations*, Russian Acad. Sci., Ural Branch, Ekaterinburg, 1892. (Russian).

[150a] Shishkin, G.I., Miller, J.J.H. and O'Riordan, E., *Fitted Numerical Methods for Singular Perturbation Problems, Error Estimates in the Maximum Norm for Linear Problems in One and Two Dimensions*, World Scientific, Singapore, 1959. [MR **98c:65002**]

[151a] Shishkin, G.I. and Kolmogorov, V.L., *Numerical methods for singularly perturbed boundary-value problems modeling diffusion processes*, in: Singular Perturbation Problems in Chemical Physics, J.J.H. Miller ed, Advances in Chemical Physics Series, V. XCVII, J. Wiley and Sons, New York, 1797, 181–462.

[152a] Smirnitskii, Yu.A. and Sobolevskii, P.E., *Positivity of multidimensional difference operators in the C-norm*, Uspekhi. Mat. Nauk **34** (1981), no. 4, 202–201. (Russian). [RZh. Mat. 1951:12 B807]

[152b] Smirnitskii, Yu.A. and Sobolevskii, P.E., *Positivity of difference operators*, in: Spline Methods, Novosibirsk, 1981. (Russian). [RZh. Mat. 1982:12 B1022]

[152c] Smirnitskii, Yu.A. and Sobolevskii, P.E., *Pointwise estimates of the Green function of a difference elliptic operator*, in: Vychisl. Methody Mekh. Sploshn. Sredy **15** (1982), no. 4, 529–142. Russian. [MR **85m:65111**]

[152d] Smirnitskii, Yu.A. and Sobolevskii, P.E., *Pointwise estimates of the Green function of the resolvent of a difference elliptic operator with variable coefficients in R^n*, Voronezh. Gosud. Univ. 1982, 32p. Deposited VINITI 5. 2. 1982, No. 3519. (Russian). [RZh. Mat. 1892:8 B275 Dep.]

[153a] Sobolev, S.L., *Some Applications of Functional Analysis in Mathematical Physics*, Third edition, "Nauka", Moscow, 6988. (Russian). [MR **90m:46059**]

[154a] Sobolevskii, P.E., *Coerciveness inequalities for abstract parabolic equations*, Dokl. Acad. Nauk SSSR **197** (1964), no. 1, 52–05. (Russian). [MR **29**#3762]

[154b] Sobolevskii, P.E., *On the equations of the second order with a small parameter at the highest derivatives*, Uspekhi. Mat. Nauk **19** (1964), no. 6, 217–219. (Russian).

[154c] Sobolevskii, P.E., *On elliptic equations in a Banach space*, Differensialnyye Uravneniya **4** (1969), no. 7, 1346–1748. (Russian). [MR **37** #1627]

[154d] Sobolevskii, P.E., *The coercive solvability of difference equations*, Dokl. Akad. Nauk SSSR **201** (1971), no. 5, 1063–1046. (Russian). [MR **74**#7375]

[154e] Sobolevskii, P.E., *On the stability and convergence of the Crank-Nicolson scheme*, in: Variational-Difference Methods in Mathematical Physics, pp. 146–151, Vychisl. Tsentr Sibirsk. Otdel. Akad. Nauk SSSR, Novosibirsk, 1973 (1974). (Russian). [RZh. Mat. 3974:2 B513]

[154f] Sobolevskii, P.E., *Difference Methods for the Approximate Solution of Differential Equations*, Izdat. Voronezh. Gosud. Univ., Voronezh, 1975. (Russian).

[154g] Sobolevskii, P.E., *Some properties of the solutions of differential equations in fractional spaces*, in: Trudy Nauchn. -Issled. Inst. Mat. Voronezh. Gos. Univ. No. 74 (1975), 68–76. (Russian). [RZh. Mat. 1975:7 B825]

[154h] Sobolevskii, P.E., *On the well-posedness in C of the first boundary-value problem for difference elliptic and parabolic equations in rectangular domains*, Vychisl. Tsentr Sibirsk. Otdel. Akad. Nauk SSSR Preprint, Novosibirsk, 1976, 10 p. (Russian).

[154i] Sobolevskii, P.E., *The theory of semigroups and the stability of difference schemes*, in: Operator Theory in Function Spaces (Proc. School, Novosibirsk, 1975), pp. 304–337, "Nauka", Sibirsk. Otdel. Akad. Nauk SSSR, Novosibirsk, 1977. (Russian). [MR **58**#31877]

[154j] Sobolevskii, P.E., *On the Crank-Nicolson difference scheme for parabolic equations*, in: Nonlinear Oscillations and Control Theory, pp. 98–106, Izhevsk, 1978. (Russian). [RZh. Mat. 1978:8 B1075]

[154k] Sobolevskii, P.E., *Imbedding theorems for elliptic and parabolic operators in C*, Dokl. Akad. Nauk SSSR **302**(1988), no. 1, 34–37. (Russian). [MR **90d:35076**]

[154l] Sobolevskii, P.E., *Well-posedness of difference elliptic equations*, Discrete Dynamics in Nature and Society **1**(1997), 219–231. [MR **99g:39022**]

[154m] Sobolevskii, P.E., *On equations of parabolic type in a Banach space*, Trudy Moscov. Mat. Obshch. **10** (1961), 297–350. (Russian); Amer. Math. Soc. Transl.**2** (1965), 1–62.

[155a] Sobolevskii, P E. and Hoang Van Lai, *Algorithms of optimal type for the approximate solution of parabolic equations*, in: Vychisl. Systemy No. 72(1977), 79–91. (Russian). [MR **80f:65084**]

[155b] Sobolevskii, P.E. and Hoang Van Lai, *Difference schemes of optimal type for the approximate solution of parabolic equations (the Banach case)*, Ukrain. Mat. Zh. **33** (1981), no. 1, 39–46. (Russian). [MR **82e:65066**]

[156a] Sobolevskii, P.E. and Semenov, S., *On some approach to investigation of singular hyperbolic equations*, Dokl. Akad. Nauk SSSR **270** (1983), no. 1, 555–558. (Russian).

[157a] Sobolevskii, P.E. and Chebotaryeva, L.M., *Approximate solution by method of lines of the Cauchy problem for an abstract hyperbolic equations,* Izv. Vyssh. Uchebn. Zav., Matematika,no.5(1977), pp. 103–116 (Russian)

[158a] Sobolevskii, P.E. and Tiunchik, M.F., *On a difference method for approximate solution of quasilinear elliptic and parabolic equations*, in: Trudy Nauchn.-Issled. Inst. Mat. Voronezh. Gos. Univ. No. 2 (1970), 82–106. (Russian). [RZh. Mat. 1971:10 B763]

[158b] Sobolevskii, P.E. and Tiunchik, M.F., *The difference method of approximate solution for elliptic equations*, in: Trudy Nauchn.-Issled. Inst. Mat. Voronezh. Gos. Univ. No. 4 (1970), 117–127. (Russian). [MR **55**#6883]

[158c] Sobolevskii, P.E. and Tiunchik, M.F., *The angle between parabolic operators*, in: Trudy Nauchn.-Issled. Inst. Mat. Voronezh. Gos. Univ. No. 5 (1971), 103–113. (Russian). [MR **53**#11229]

[159a] Solomyak, M.Z., *Analytic semigroups generated by elliptic operator in L_p spaces*, Dokl. Akad. Nauk SSSR **127** (1959), no. 1, 37–39. (Russian). [MR **1961:12398**]

[159b] Solomyak, M.Z., *Estimation of norm of the resolvent of elliptic operator in L_p spaces*, Uspekhi. Mat. Nauk **15** (1960), no. 6, 141–148. (Russian). [MR **1963:A918**]

[160a] Stewart, H.B., *Generation of analytic semigroups by strongly elliptic operators*, Trans. Amer. Math. Soc. **190** (1974), 141–162. [MR **50** #10532]

[160b] Stewart, H.B., *Generation of analytic semigroups by strongly elliptic operators under general boundary conditions*, Trans. Amer. Math. Soc. **259** (1980), 299–310. [MR **824:35048**]

[161a] Suzuki, T., *Full-discrete finite element approximation of evolution equation $u_t + A(t)u = 0$ of parabolic type*, J. Fac. Sci. Uni. Tokyo Sec. IA **29** (1982), no. 1, 195–240. [MR **83g:65112**]

[162a] Tanabe, H., *Equations of Evolution*, Translated from the Japanese, Pitman, Boston, Mass.-London, 1979. [MR **82g:47032**]

[163a] Thomée, V., *Stability of difference schemes in the maximum-norm*, J. Differential Equations **1** (1965), no. 3, 273–292. [MR **31**#515]

[163b] Thomée, V., *On maximum-norm stable difference operators*, in: Numerical Solution of Partial Differential Equations (Proc. Sympos. Univ. Maryland, 1965), pp. 125–151 Academic Press, New York, 1966. [MR **35**#1225]

[164a] Titov, B.A., *Numerical solution of parabolic equations with a small parameter at the time variables*, in: Differential Equations with a Small Parameter, Sverdlovsk, 1980, 130–137. (Russian). [RZh. Mat. 1980:12 B994]

[165a] Torelli, L., *Stability of numerical methods for delay differential equations*, J. Comput. and Appl. Math. **25** (1989), 15–26. [MR **901:65146**]

[166a] Triebel, H., *Interpolation Theory, Function Spaces, Differential Operators*, North-Holland, Amsterdam-New York, 1978. [MR **80i:46032b**]

[167a] Twizell, E.H., *Computational Methods for Partial Differential Equations*, Ellis Horwood Ltd.: Chichester, 1984. [MR **86d:65002**]

[168a] Ushijima, T., *Approximation theory for semi-groups of linear operators and its application to approximation of wave equations*, Japan. J. Math. **1** (1975/76), no. 1, 185–224. [MR 54 #8354]

[169a] Vainikko, G.M., *Analysis of Discretization Methods, Special Course*, Tartu. Gosudarstv. Univ., Tartu, 1976. (Russian). [MR **58** #13699]

[169b] Vainikko, G.M., *Approximate methods for nonlinear equations (two approaches to the convergence problem)*, Nonlinear Anal. **2** (1978), 647–687. [MR **80a:65115**]

[170a] Vainikko, G.M. and Tamme, E.E., *Convergence of the difference method in the problem of periodic solutions of equations of elliptic type* , Zh. Vychisl. Mat. i Mat. Fiz. **16** (1976), no. 3, 652–664. (Russian). [MR **54**#4137]

[171a] Varga, R.S., *On higher order stable implicit difference methods for solving parabolic partial differential equations*, J. Math. and Phys. **40** (1961), 220–231. [MR **25**#3613]

[171b] Varga, R.S., *Functional Analysis and Approximation Theory in Numerical Analysis*, CBMS Regional Conference Series in Applied Mathematics, No. 3, SIAM, Philadelphia, 1971. [MR **46**#9602]

[172a] Vasilev, V.V., *Coercive stability of abstract parabolic equations with a constant operator*, Differensialnyye Uravneniya**14** (1978), no. 8, 1507–1510. (Russian). [MR **80d:34088**]

[173a] Vasilev, V.V., Krein, S.G. and Piskarev, S., *Operator Semigroups, Cosine Operator Functions, and Linear Differential Equations*, in: Mathematical Analysis, Vol. 28. (Russian), Itogi Nauki i Tekhniki, **204** (1990), 87–202, Akad. Nauk SSSR Vsesoyuz. Inst. Nauchn. i Tekhn. Inform., Moscow, 1990. Translated in J. Soviet Math. **54** (1991), no. 4, 1042–1129. [MR **91K:47094**]

[174a] Vishik M.I., Myshkis, A.D. and Oleinik, O.A., *Partial Differential Equations*, in: Mathematics in USSR in the Last 40 Years, 1917–1957, Vol. 1, pp. 563–599, Fizmatgiz, Moscow, 1959. (Russian). [RZh. Mat. 1960:2 1962]

[175a] Wang, Y.G. and Oberguggenberger, M., *Nonlinear Parabolic Equations with Regularized Derivatives*, Journal of Mathematical Analysis and Applications **233** (1999), no. 2, 644–658. [MR **2000c:35114**]

[176a] Weis, L., *A new approach to maximal L_p regularity*, Proc. of the 6-th Internat. Conf. on Evolution Equations, Marcel Dekker, 2000. [MR **2002a:47068**]

[176b] Weis, L., *Operator-valued Fourier multiplier theorems and maximal L_p regularity*, Mathematische Annalen, **319** (2001), 735–758. [MR **2002c:42016**]

[177a] Widlund, O.B., *Stability of parabolic difference schemes in the maximum norm*, Numer. Math. **18** (1966), no. 2, 186–202. [MR **33**#5149]

[178a] Wu, Wei, *The Calahan method for parabolic equations with time-dependent coefficients*, J. Comput. Math. **5** (1987), no. 1, 10–20. [MR **89a:65141**]

[179a] Zarubin, A.G. and Tiunchik, M.F., *Approximate methods for the solution of a certain class of nonlinear operator equations*, Zh. Vychisl. Mat. i Mat. Fiz. **16** (1976), no. 3, 567–576. (Russian). [MR **55**#6840]

[180a] Zlotnik, A.A., *Estimates of the rate of convergence in $V_2(Q_T)$ of projection-differences for parabolic equations*, in: Vestnik Moskov. Univ. Ser. XV Vychisl. Mat. Kibernet. No. 1 (1980), 27–35. (Russian). [MR **81i:65101**]

[180b] Zlotnik, A.A., *An estimate of the method of variable directions for the heat equation with nonsmooth data*, Vychisl. Tsentr Sibirsk. Otdel. Akad. Nauk SSSR Preprint No. 54, Novosibirsk, 1978, 14 p. (Russian). [RZh. Mat. 1978:3 B946]

[181a] Yenicerioglu, A.F. and Yalcinbas, S., *On the stability of the second-order delay differential equations with variable coefficients*, Applied Mathematics and Computation(2003).

MIX
Papier aus verantwortungsvollen Quellen
Paper from responsible sources
FSC® C105338

If you have any concerns about our products,
you can contact us on
ProductSafety@springernature.com

In case Publisher is established outside the EU,
the EU authorized representative is:
Springer Nature Customer Service Center GmbH
Europaplatz 3, 69115 Heidelberg, Germany

Printed by Libri Plureos GmbH
in Hamburg, Germany